Student Solutions Manual

Kenny Fister
DeVry University

Charles Trinkel
Murray State University

to accompany

Technical Mathematics & Technical Mathematics with Calculus

Sixth Edition

Paul A. Calter
Vermont Technical College

Michael A. Calter
Wesleyan University

WILEY
John Wiley & Sons, Inc.

Cover Image: Tetra Images/Getty Images, Inc.

ISBN-13 978-0470-53494-6

SKY10021090_090920

CONTENTS

Chapter 1: Numerical Computation

CHAPTER 1

Exercise 1 ◊ The Real Numbers

Equality and Inequality

1. $7 \boxed{<} 10$

3. $-3 \boxed{<} 4$

5. $\frac{3}{4} \boxed{=} 0.75 \left[\text{divide 3 by } 4 \Rightarrow 4\overline{)3}\right]$

Absolute Value

7. $|4| = 4$

9. $-|-6| = -6$

11. $|12-5+8| - |-6| + |15| = |15| - |6| + |15|$
$= 15 - 6 + 15$
$= 24$

Significant Digits and Decimal Places

13. $78.3 \Rightarrow 3$

15. $4.008 \Rightarrow 4$ (the zeros are not used to locate the decimal point; thus, they are counted)

17. $20{,}000 \Rightarrow 1$ (How are the zeros used, to locate the decimal point or not locate the decimal point?)

19. $0.9972 \Rightarrow 4$

21. $39.5 \Rightarrow 1$

23. $5.882 \Rightarrow 3$

Rounding

25. $38.468 \Rightarrow 38.47$ (the discarded digit (8) is greater than 5)

27. $96.835001 \Rightarrow 96.84$ (since the discarded digit is 5, round to the nearest even number)

29. $398.372 \Rightarrow 398.37$ (the discarded digit (2) is less than 5)

31. $13.98 \Rightarrow 14.0$ (the discarded digit is greater than 5, the 9 increases to 10; 10 is greater than 5 thus 3 increases to 4)

33. $5.6501 \Rightarrow 5.7$ (5 is followed by a nonzero digit in any of the decimal places to the right)

35. $398.36 \Rightarrow 398.4$

37. $28{,}583 \Rightarrow 28{,}600$

39. $3{,}845{,}240 \Rightarrow 3{,}845{,}200$

41. $9.284 \Rightarrow 9.28$ (the procedure is the same as rounding to a given number of decimal places)

43. $0.04825 \Rightarrow 0.0482$ (the discarded digit is 5, round to the nearest even number)

45. $0.08375 \Rightarrow 0.0838$

47. $34.9274 \Rightarrow 34.927$

49. $4.03726 \Rightarrow 4.0373$

51. $5.937254 \Rightarrow 5.9373$

53. Evaluate the expressions in problems 7 through 12 by calculator.

7. $|4|$

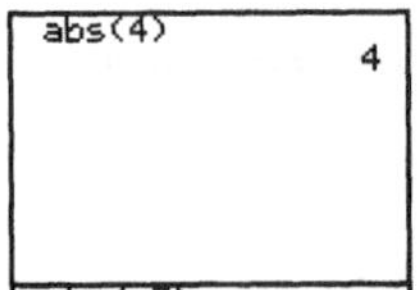

8. $|-3| = 3$

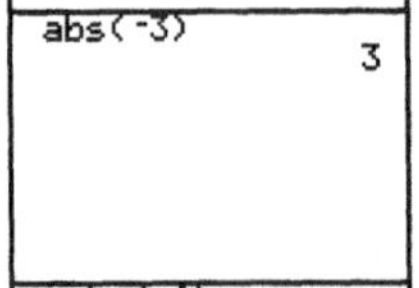

9. $-|-6| = -6$

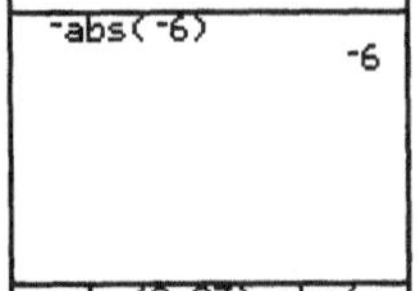

10. $-|9-23|-|-7+3|$

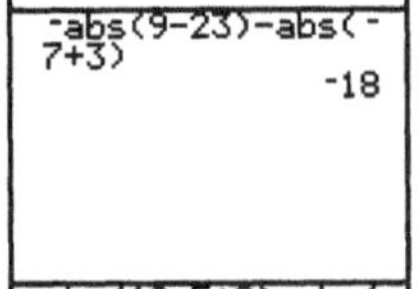

11. $|12-5+8|-|6|+|15|$

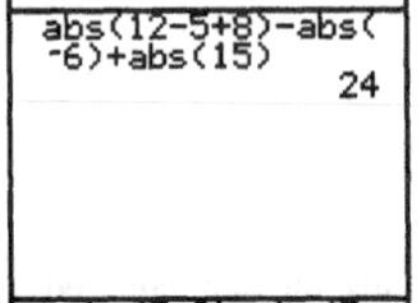

12. $-|3-9|-|5-11|+|21+4|$

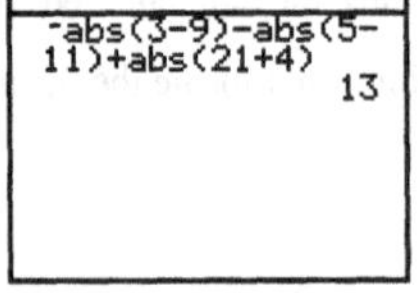

Exercise 2 ◊ Addition and Subtraction

Adding and Subtracting Signed Numbers

1. 926 + 863 = 1789

3. –576 + (–553) = –1129

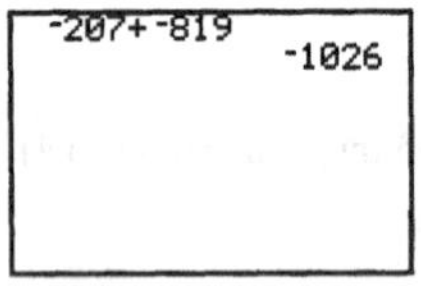

5. –575 – 275 = –850

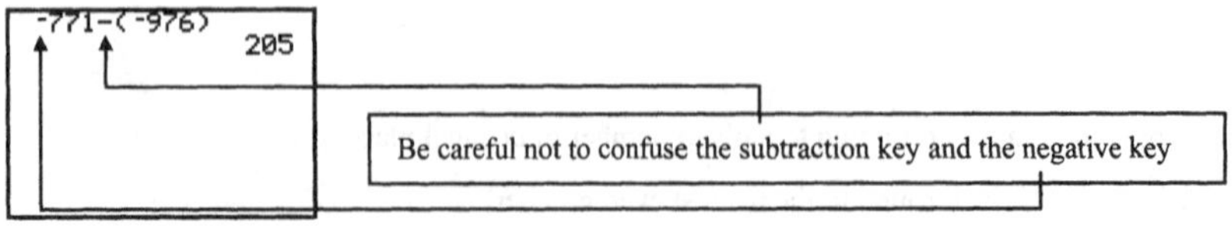

7. 1123 – (–704) = 1827

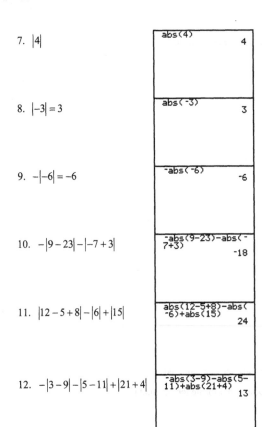

Adding and Subtracting Approximate Numbers

9. 4857 + 73.8 = 4930.8 ⇒ 4931 (round to the nearest integer as it has the least number of decimal places)

11. $296.44 + 296.997 = 593.437 \Rightarrow 593.44$

13. $0.000583 + 0.0008372 - 0.00173 = -0.0003098 \Rightarrow -0.00031$

Applications

15. $237{,}321 - 158{,}933 = 78{,}388 \Rightarrow 78{,}388$ mi^2 (rounded to 6 significant digits)

17. Radius to the outside of the pipe equals $10.6 + 2.125 + 4.8 = 17.525$ cm. The diameter is twice the radius; therefore $2(17.525) = 35.05 \Rightarrow 35.0$ cm (following the rules of rounding)

19. $27.3 + 4.0155 + 9.75 = 41.0655 \Rightarrow 41.1\ \Omega$

Exercise 3 ◊ Multiplication

Multiplying Signed Numbers

1. $4 \times (-2) = -8$

3. $(-24) \times (-5) = 120$

Multiplying Approximate Numbers

5. $3.967 \times 2.84 = 11.26628 \Rightarrow 11.3$ (keep the same number of significant digits as the factor that has the fewest number of significant digits (Rule 8); 2.84 has 3 significant digits)

7. $93.9 \times 0.0055908 = 0.52497612 \Rightarrow 0.525$

9. $69.0 \times (-258) = -17{,}802 \Rightarrow -17{,}800$

11. $2.86 \times (4.88 \times 2.97) \times 0.533 = 22.92278789 \Rightarrow 22.9$

Multiplying Exact and Approximate Numbers

13. $4 \times 2.55 = 10.2$

15. $-4.273 \times (-5) = 21.36$

Applications

17. Cost = 52.5 tons × \$63.25/ton = \$3320.625 ⇒ \$3320

19. Total tonnage = $3 \times 26 + 35 = 113$ tons

    Total value @ \$12.75/ton = 113 × \$12.75 = \$1440.75 ⇒ \$1441

21. Power = 4.7 A × 115.45 V = 542.615 W ⇒ 540 W

23. Weight = 1000×2.375 g = 2375 g

25. Number of degrees = 360 degrees/rev × 4.863 rev = 1750.68° ⇒ 1751°

Exercise 4 ◊ Division

Dividing Signed Numbers

1. $14 \div (-2) = -7$

3. $(-24) \div (-4) = 6$

Dividing Approximate Numbers

5. $947 \div 5.82 = 162.71477 \Rightarrow 163$ (round to 3 significant digits)

7. $-99.4 \div 286.5 = -0.3469458 \Rightarrow -0.347$

9. $5836 \div 8264 = 0.7061955 \Rightarrow 0.7062$

11. $94{,}840 \div 1.33876 = 70{,}841.67439 \Rightarrow 70{,}840$

Reciprocals

13. $693 \Rightarrow 693^{-1} \Rightarrow \frac{1}{693} = 0.0014430014 \Rightarrow 0.00144$

15. $-396 \Rightarrow -\frac{1}{396} = -0.00252525 \Rightarrow -0.00253$

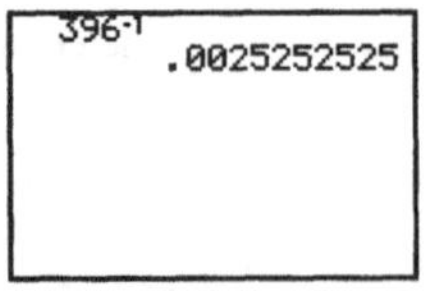

17. $-0.00573 = -\frac{573}{100{,}000} \Rightarrow -\frac{100{,}000}{573} = -174.520069 \Rightarrow -175$ (round to 3 significant digits)

19. $4.992 \Rightarrow 0.2003205 \Rightarrow 0.2003$

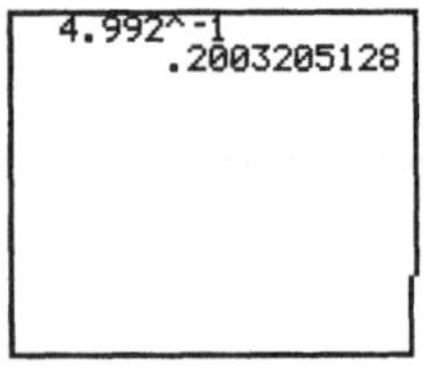

you may also use the [^] key and raise to the −1 power

Applications Involving Division

21. length = 1858.54 m ÷ 5 = 371.708 m

23. 245 ft ÷ 4.5 ft/d = 54.4444 ft/d ⇒ 54.4 ft/d

 54.4 ft/d ÷ 3 = 18.1333 ⇒ 18.1 ft

25. $\frac{1}{R} = \frac{1}{475} + \frac{1}{928}$

$$\frac{1}{R} = \frac{1403}{440{,}800}$$
$$1(440{,}800) = 1403R$$
$$\frac{440{,}800}{1403} = R$$
$$314.18389 = R$$
$$R = 314\Omega$$

27. $\sin\theta = \frac{1}{3.58}$

$$\sin\theta = 0.2793296$$
$$\sin\theta = 0.279$$

Exercise 5 ◊ Powers and Roots

Powers

1. $2^3 = 2 \times 2 \times 2 = 8$
3. $9^2 = 9 \times 9 = 81$
5. $10^3 = 1000$
7. $10^2 = 100$

Powers by Calculator

9. $(8.55)^3 = 8.55 \times 8.55 \times 8.55 = 625.026375 \Rightarrow 625$ (round to 3 significant digits since it is the fewest number of significant digits)

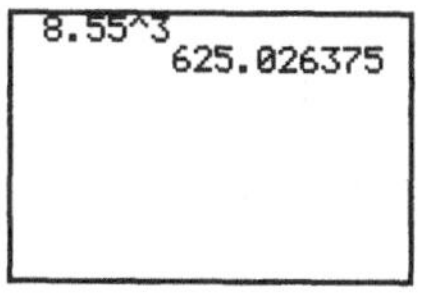

11. $(9.55)^3 = 870.98387 \Rightarrow 871$

13. $(3.95)^3 = 61.6298 \Rightarrow 61.6$

15. $(1.65)^4 = 7.412006 \Rightarrow 7.41$

17. $(12.5)^2 = 156.25 \Rightarrow 156$

19. $(2.26)^6 = 133.2449 \Rightarrow 133$

Negative Base

21. $(-3)^3 = (-3) \times (-3) \times (-3) = -27$ (a negative power will yield a negative output)

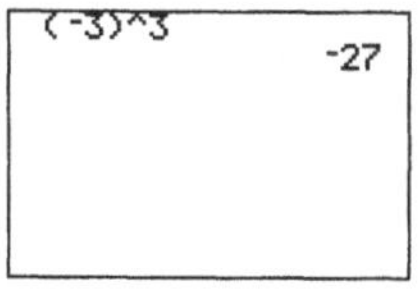

If you leave out the parentheses, you will still get the correct output.

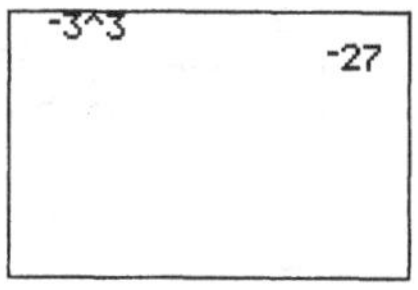

23. $(-4)^3 = (-4) \times (-4) \times (-4) = -64$

25. $(-8.01)^3 = (-8.01) \times (-8.01) \times (-8.01) = -513.9224 \Rightarrow -514$

27. $(-5.33)^3 = -151.419437 \Rightarrow -151$

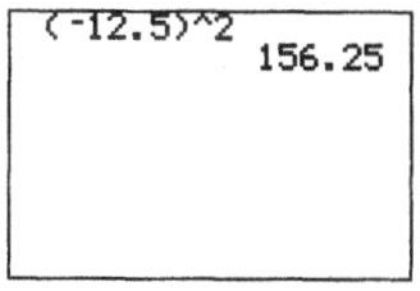

If you leave out the parentheses, you will get an incorrect output. Why?

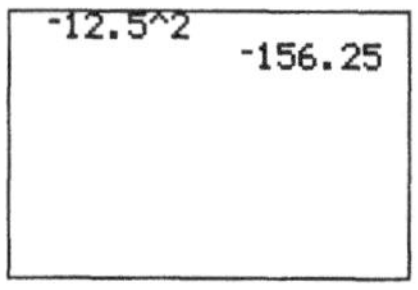

29. $(-1.33)^3 = -2.35263 \Rightarrow -2.35$

31. $(-2.84)^3 = -22.90630 \Rightarrow -22.9$

Negative Exponent

33. $1^{-3} = \dfrac{1}{1^3} = \dfrac{1}{1} = 1$

35. $10^{-2} = \dfrac{1}{10^2} = \dfrac{1}{100} = 0.01$

37. $(3.85)^{-2} = 0.067465 \Rightarrow 0.0675$

39. $(3.84)^{-3} = 0.17660 \Rightarrow 0.0177$

41. $(-5.37)^{-3} = -\left(\dfrac{1}{(5.37)^3}\right) = -\dfrac{1}{5.37 \times 5.37 \times 5.37} = -\dfrac{1}{154.85415} = -0.0064576 \Rightarrow -0.00646$

43. $(-1.85)^{-3} = -0.15793 \Rightarrow -0.158$

Fractional and Decimal Exponents

45. $(8.55)^{1/3} = 2.04482 \Rightarrow 2.04$ $\left[(8.55)^{1/3} \text{ is equivalent to } \sqrt[3]{(8.55)^1}\right]$

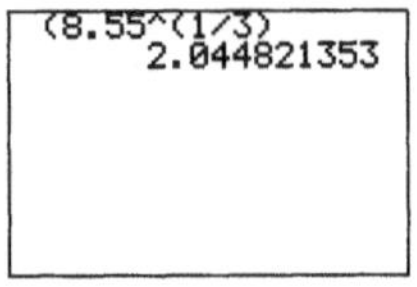

Caution!! If you omit the parentheses in the exponent, the results are not the same. Why?

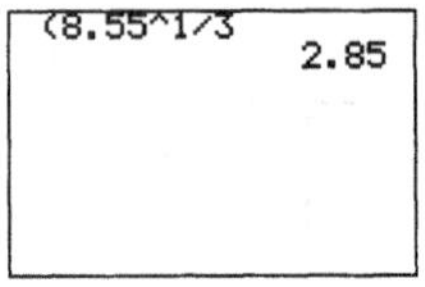

47. $(9.55)^{1/3} = 2.12162 \Rightarrow 2.12$

$(84.2)^{1/2}$ is equivalent to $\sqrt[2]{(84.2)^1} \Rightarrow \sqrt{84.2}$; which by calculator may be computed by:

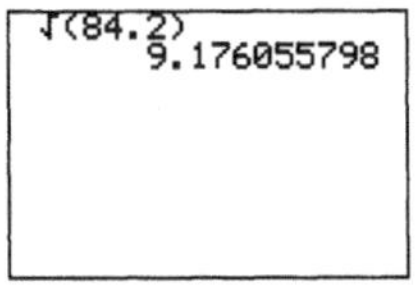

49. $(2.85)^{2/3} = 2.01015 \Rightarrow 2.01$

$(2.85)^{2/3}$ is equivalent to $\sqrt[3]{(2.85)^2} \Rightarrow \sqrt[3]{8.1225} \Rightarrow (8.1225)^{1/3} = 2.01015$

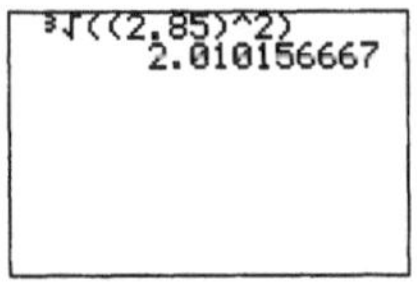

or

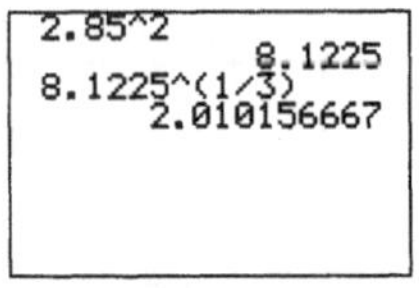

51. $(1.84)^{2/3} = 1.50156 \Rightarrow 1.50$

53. $(8.88)^{2.13} = 104.74169 \Rightarrow 105$

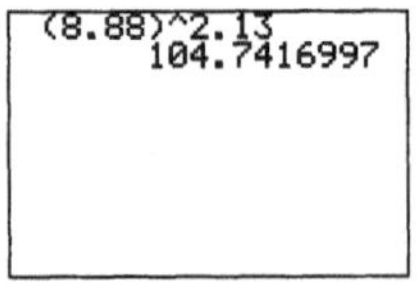

55. $(4.38)^{2.63} = 48.64933 \Rightarrow 48.6$

Applications Involving Powers

57. $S = 16(5.448)^2 = 16(29.6807) = 474.89126 \Rightarrow 470$ ft (since 16 has 2 significant digits, the fewest in the problem)

59. $V = (35.8)^3 = 45{,}882.712 \Rightarrow 45{,}900\text{ cm}^3$

61. $A = \$2000(1.0625)^{7.5} = \$2000(1.57567) = \$3151.3528 \Rightarrow \$3151.35$

Roots

63. $\sqrt[3]{27} = 3$

65. $\sqrt[3]{-27} = -3$

67. $\sqrt[5]{-32} = -2$

69. $\sqrt{1.863} = 1.36491 \Rightarrow 1.365$

By calculator:

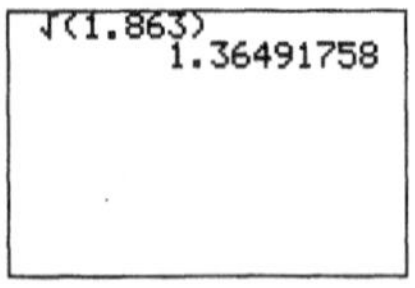

71. $\sqrt{772} = 27.78488 \Rightarrow 27.8$

73. $\sqrt[3]{7295} = 19.39434 \Rightarrow 19.39$

75. $\sqrt[5]{-18.4} = -1.79045 \Rightarrow -1.79$

By calculator, it may be done using the x root key (under MATH) or by raising -18.4 to the $\frac{1}{5}$ power

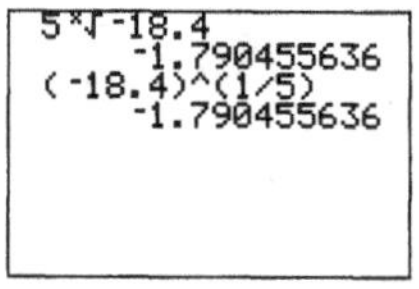

Applications of Roots

77. $T = 2\pi\sqrt{\frac{2.55}{32.0}}$ seconds $= 2\pi\sqrt{0.07968} \Rightarrow 2\pi(0.28228) \Rightarrow 1.77367 \Rightarrow 1.77\,s$

79. $B = \sqrt{(3.75)(9.83)} = \sqrt{36.8625} \Rightarrow 6.071449 \Rightarrow 6.07$

Exercise 6 ◊ Combined Operations
Combined Operations with Exact Numbers

1. $(37)(28) + (36)(64) = 1036 + 2304 \Rightarrow 3340$

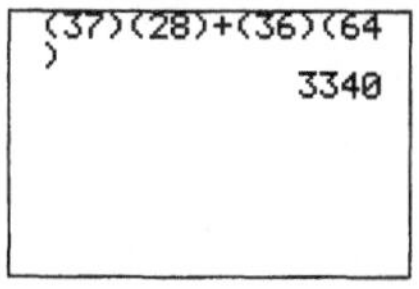

3. $(63 + 36)(37 - 97) = (99)(-60) \Rightarrow -5940$

5. $\frac{219}{73} + \frac{194}{97} = 3 + 2 \Rightarrow 5$

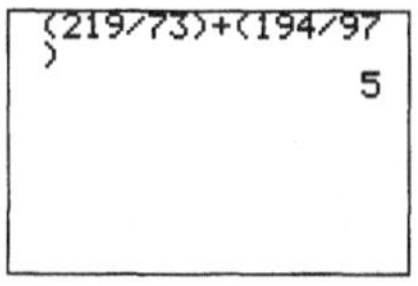

7. $\frac{647 + 688}{337 + 108} = \frac{1335}{445} \Rightarrow 3$

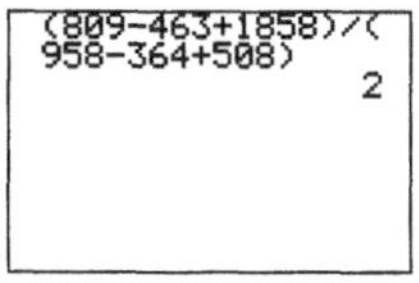

9. $(5 + 6)^2 = (11)^2 \Rightarrow 121$

11. $(423 - 420)^3 = 3^3 \Rightarrow 27$

13. $\left(\frac{141}{47}\right)^3 = (3)^3 \Rightarrow 27$

15. $\sqrt{(8)(72)} = \sqrt{576} \Rightarrow 24$

17. $\sqrt[4]{(27)(768)} = \sqrt[4]{20{,}736} \Rightarrow 12$

19. $\sqrt[4]{\dfrac{1136}{71}} = \sqrt[4]{16} \Rightarrow 2$

```
4×√(1136/71)
                 2
```

21. $\sqrt[4]{625} + \sqrt{961} - \sqrt[3]{216} = 5 + 31 - 6 \Rightarrow 30$

```
4×√(625)+√(961)-
3×√(216)
                30
(625)^(1/4)+(961
)^(1/2)-(216)^(1
/3)
                30
```

Combined Operations with Approximate Numbers

23. $(7.37)(3.28) + (8.36)(2.64) = 24.1736 + 22.0704 \Rightarrow 46.244 \Rightarrow 46.2$

25. $(63.5 + 83.6)(8.37 - 1.72) = (147.1)(6.65) \Rightarrow 978.215 \Rightarrow 978$

27. $\dfrac{583}{473} + \dfrac{946}{907} = 1.23255 + 1.04299 = 2.27554 \Rightarrow 2.28$

29. $\dfrac{6.47 + 8.604}{3.37 + 90.8} = \dfrac{15.074}{94.17} \Rightarrow 0.16007 \Rightarrow 0.160$

31. $(5.37 + 2.36)^2 = (7.73)^2 \Rightarrow 59.7529 \Rightarrow 59.8$

33. $(6.423 + 1.05)^2 = (7.473)^2 \Rightarrow 55.84572 \Rightarrow 55.8$

35. $\left(\dfrac{8.90}{4.75}\right)^2 = (1.87368)^2 \Rightarrow 3.51069 \Rightarrow 3.51$

37. $\sqrt[3]{657 + 553 - 842} = \sqrt[3]{368} \Rightarrow 7.16609 \Rightarrow 7.17$

39. $\sqrt[5]{(9.06)(4.86)(7.93)} = \sqrt[5]{349.17058} \Rightarrow 3.22557 \Rightarrow 3.23$

41. $\sqrt[4]{\dfrac{4.50}{7.81}} = \sqrt[4]{0.576184} \Rightarrow 0.87124 \Rightarrow 0.871$

43. $\sqrt[4]{528} + \sqrt{94.2} - \sqrt[3]{284} = 4.79356 + 9.70566 - 6.57313 \Rightarrow 7.92609 \Rightarrow 7.93$

Exercise 7 ◊ Scientific and Engineering Notation

Powers of 10

1. $10^5 = 100{,}000$

3. $10^{-5} = \dfrac{1}{10^5} = 0.00001$

5. $10^4 = 10{,}000$

7. $1{,}000{,}000 = 10^6$

9. $0.001 = 10^{-3}$

Converting Numbers to Scientific Notation

11. $186{,}000 \Rightarrow 1.86 \times 10^5$

13. $25{,}742 \Rightarrow 2.5742 \times 10^4$

15. $98.3 \times 10^3 \Rightarrow 9.83 \times 10^4$

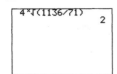

Converting Numbers from Scientific Notation

17. $2.85 \times 10^3 \Rightarrow 2850$

19. $9 \times 10^4 \Rightarrow 90{,}000$

21. $3.667 \times 10^{-3} \Rightarrow 0.003667$

Converting Numbers to Engineering Notation

23. $3.58 \times 10^2 \Rightarrow 358$

25. $0.134 \Rightarrow 134 \times 10^{-3}$

27. $0.00374 \Rightarrow 3.74 \times 10^{-3}$

Converting Numbers from Engineering Notation

29. $18{,}640 \times 10^{-3} \Rightarrow 18.64$

31. $7739 \times 10^{-3} \Rightarrow 7.739$

33. $2.66 \times 10^6 \Rightarrow 2{,}660{,}000$

Addition and Subtraction

35. $(75.0 \times 10^2) + 3210 = (75.0 \times 10^2) + (32.10 \times 10^2) \Rightarrow (75.0 + 32.1) \times 10^2 \Rightarrow 107.1 \times 10^2 \Rightarrow 1.070 \times 10^4$

37. $0.037 - (6.0 \times 10^{-3}) = (37 \times 10^{-3}) - (6.0 \times 10^{-3}) \Rightarrow (37 - 6.0) \times 10^{-3} \Rightarrow 31 \times 10^{-3} \Rightarrow 3.1 \times 10^{-2}$

Multiplication

39. $10^5 \cdot 10^2 = 10^{5+2} \Rightarrow 10^7$

41. $10^{-5} \cdot 10^{-4} = 10^{-5+(-4)} \Rightarrow 10^{-9}$

43. $10^{-1} \cdot 10^{-4} = 10^{-1+(-4)} \Rightarrow 10^{-5}$

45. $(5 \times 10^4)(8 \times 10^{-3}) = (5 \times 8)(10^4 \times 10^{-3}) \Rightarrow 40 \times 10^{4+(-3)} \Rightarrow 40 \times 10^1 \Rightarrow 4.0 \times 10^2$

47. $(2 \times 10^4)(30{,}000) = (2 \times 10^4)(3.0 \times 10^4) \Rightarrow (2 \times 3)(10^{4+4}) \Rightarrow 6 \times 10^8$

Division

49. $10^4 \div 10^6 = 10^{4-6} \Rightarrow 10^{-2}$

51. $10^{-3} \div 10^5 = 10^{-3-(5)} \Rightarrow 10^{-8}$

53. $(8 \times 10^4) \div (2 \times 10^2) = (8 \div 2)(10^{4-2}) \Rightarrow 4 \times 10^2$

55. $(3 \times 10^3) \div (6 \times 10^5) = (3 \div 6)(10^{3-5}) \Rightarrow 0.5 \times 10^{-2} \Rightarrow 5 \times 10^{-3}$

57. $(9 \times 10^4) \div (3 \times 10^{-2}) = (9 \div 3)(10^{4-(-2)}) \Rightarrow 3 \times 10^6$

Scientific and Engineering Notation on the Calculator

59. $(1.58 \times 10^2)(9.82 \times 10^3) = 1.55 \times 10^6$

By calculator: first, you need to change to scientific mode pressing MODE and then choosing Sci

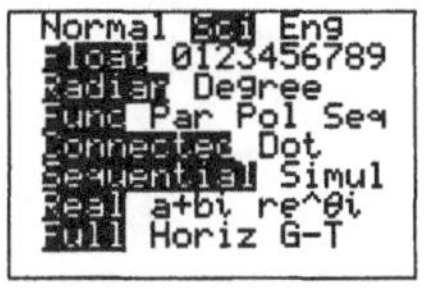

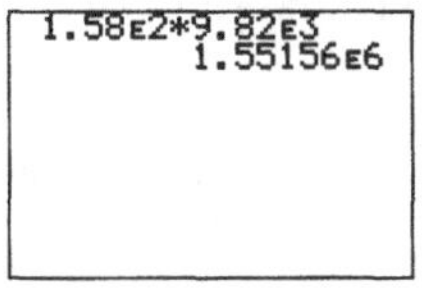

61. $(3.87 \times 10^{-2})(5.44 \times 10^5) = 2.11 \times 10^4$

63. $(5.6 \times 10^2)(3.1 \times 10^{-1}) = 1.7 \times 10^2$

Applications

65. $R = (4.98 \times 10^5) + (2.47 \times 10^4) + (9.27 \times 10^6) \Rightarrow (0.498 \times 10^6) + (0.0247 \times 10^6) + (9.27 \times 10^6)$

$\Rightarrow (0.498 + 0.0247 + 9.27) \times 10^6 \Rightarrow 9.793 \times 10^6 \Rightarrow 9.79 \times 10^6\ \Omega$

67. $P = IV \Rightarrow (3.75 \times 10^{-3}\ A)(7.24 \times 10^{-4}\ V) \Rightarrow (3.75 \times 7.24) \times (10^{-3+(-4)}) \Rightarrow 27.15 \times 10^{-7} \Rightarrow 2.72 \times 10^{-6}$ W

69. $C = 8.26 \times 10^{-6} + 1.38 \times 10^{-7} + 5.93 \times 10^{-5} \Rightarrow 0.826 \times 10^{-5} + 0.0138 \times 10^{-5} + 5.93 \times 10^{-5} \Rightarrow 6.7698 \times 10^{-5}$
$= 6.77 \times 10^{-5}$ F

Exercise 8 ◊ Units of Measure
Conversion of Units

1. 152 inches to ft: $152 \text{ in.} = 152\left(\dfrac{1 \text{ ft}}{12 \text{ in}}\right) = \dfrac{152}{12} \Rightarrow 12.666 \text{ ft} \Rightarrow 12.7 \text{ ft}$

3. 762.0 ft to in.: $762.0 \text{ ft} = 762\left(\dfrac{12 \text{ in.}}{1 \text{ ft}}\right) \Rightarrow 9144 \text{ in.}$

5. 29 tons to lb: $29 \text{ tons} = 29\left(\dfrac{2000 \text{ lb}}{1 \text{ ton}}\right) \Rightarrow 29(2000) \Rightarrow 58{,}000 \text{ lb}$

7. 89,600 lb to tons: $89{,}600 \text{ lb} = 89{,}600\left(\dfrac{1 \text{ ton}}{2000 \text{ lb}}\right) \Rightarrow \dfrac{89{,}600}{2000} \Rightarrow 44.8 \text{ ton}$

Converting Between Metric Units

9. 364,000 m to km: $364{,}000 \text{ m} = 364{,}000\left(\dfrac{1 \text{ km}}{1000 \text{ m}}\right) \Rightarrow \dfrac{364{,}000}{1000} \Rightarrow 364 \text{ km}$

11. 735,900 g to kg: $735{,}900 \text{ g} = 735{,}900\left(\dfrac{1 \text{ kg}}{1000 \text{ g}}\right) \Rightarrow \dfrac{735{,}900}{1000} \Rightarrow 735.9 \text{ kg}$

13. 6.2×10^9 ohms to megohms: $6.2 \times 10^9 \text{ ohms} = 6.2 \times 10^9\left(\dfrac{1 \text{ megohm}}{10^6 \text{ ohms}}\right) \Rightarrow \dfrac{6.2 \times 10^9}{1 \times 10^6}$

$\Rightarrow \left(\dfrac{6.2}{1}\right) \times 10^{9-6} \Rightarrow 6.2 \times 10^3 \text{ megohms}$

15. 9348 pF to μF: $9348 \text{ pF} = 9348\left(\dfrac{1 \text{ F}}{10^{12} \text{ pF}}\right)\left(\dfrac{10^6 \mu\text{F}}{1 \text{ F}}\right) \Rightarrow \dfrac{9348 \times 10^6}{1 \times 10^{12}} \Rightarrow 9348 \times 10^{-6} \mu\text{F} \Rightarrow 9.348 \times 10^{-3} \mu\text{F}$

Converting Between Customary and Metric Units

17. 364.0 m to ft: $364.0 \text{ m} = 364.0\left(\dfrac{3.281 \text{ ft}}{1 \text{ m}}\right) \Rightarrow 364.0(3.281) \Rightarrow 1194.284 \text{ ft} \Rightarrow 1.194 \times 10^3 \text{ ft}$

19. 7.35 lb to N: $7.35 \text{ lb} = 7.35\left(\dfrac{4.448 \text{ N}}{1 \text{ lb}}\right) \Rightarrow 7.35(4.448) \Rightarrow 32.6928 \text{ N} \Rightarrow 32.7 \text{ N}$

21. 4.66 U.S. gal to L: $4.66 \text{ gal} = 4.66\left(\dfrac{3.785 \text{ L}}{1 \text{ gal}}\right) \Rightarrow 4.66(3.785) \Rightarrow 17.6381 \text{ L} \Rightarrow 17.6 \text{ L}$

23. 3.94 yards to meters: $3.94 \text{ yd} = 3.94\left(\dfrac{0.9144 \text{ m}}{1 \text{ yd}}\right) \Rightarrow 3.94(0.9144) \Rightarrow 3.6027 \text{ m} \Rightarrow 3.60 \text{ m}$

Converting Areas and Volumes

25. 2840 yd² to acres: $2840 \text{ yd}^2 = 2840\left(\dfrac{9 \text{ ft}^2}{1 \text{ yd}^2}\right)\left(\dfrac{1 \text{ acre}}{43{,}560 \text{ ft}^2}\right) \Rightarrow \dfrac{2840(9)}{43{,}560} \Rightarrow 0.58677 \text{ acre} \Rightarrow 0.587 \text{ acre}$

27. 24.8 ft^2 to m^2: $24.8\text{ ft}^2 = 24.8\left(\dfrac{1\text{ m}^2}{10.76\text{ ft}^2}\right) \Rightarrow \dfrac{24.8}{10.76} \Rightarrow 2.3048\text{ m}^2 \Rightarrow 2.30\text{ m}^2$

29. 0.982 km^2 to acres: $0.982\text{ km}^2 = 0.982\left(\dfrac{247.1\text{ acre}}{1\text{ km}^2}\right) \Rightarrow 0.982(247.1) \Rightarrow 242.652\text{ acres} \Rightarrow 243\text{ acres}$

31. 7.360 ft^3 to in.^3: $7.360\text{ ft}^3 = 7.360\left(\dfrac{12\text{ in.}}{1\text{ ft}}\right)^3 \Rightarrow 7.360(12)^3 \Rightarrow 12718.08\text{ in.}^3 \Rightarrow 1.2720\times10^4\text{ in.}^3$

Converting Rates to Other Units

33. 4.86 ft/s to mi/h: $4.86\text{ ft/s} = \dfrac{4.86}{1}\left(\dfrac{3600\text{ s}}{\text{h}}\right)\left(\dfrac{1\text{ mi}}{5280\text{ ft}}\right) \Rightarrow \dfrac{4.86(3600)}{5280} \Rightarrow 3.31\text{ mi/h}$

35. 66.2 m/h to km/h: $66.2\text{ mi/h} = \dfrac{66.2}{1}\left(\dfrac{1\text{ km}}{0.6214\text{ mi}}\right) \Rightarrow \dfrac{66.2}{0.6214} \Rightarrow 107\text{ km/h}$

37. 953 births/yr to birth/wk: $953\text{ births/yr} = \dfrac{953}{1}\left(\dfrac{1\text{ yr}}{52\text{ wks}}\right) \Rightarrow \dfrac{953}{52} \Rightarrow 18.3\text{ births/wk}$

39. \$800/acre to cents/m^2: $\$800/\text{acre} = \dfrac{800}{1}\left(\dfrac{100\text{ cents}}{\$1}\right)\left(\dfrac{1\text{ acre}}{4047\text{ m}^2}\right) \Rightarrow \dfrac{800(100)}{4047} \Rightarrow 19.8¢/\text{m}^2$

41. \$4720/ton to cents/lb: $\$4720/\text{ton} = \dfrac{4720}{1}\left(\dfrac{100\text{ cents}}{\$1}\right)\left(\dfrac{1\text{ ton}}{2000\text{ lb}}\right) \Rightarrow \dfrac{4720(100)}{2000} \Rightarrow 236¢/\text{lb}$

Angle Conversions

43. $87°25' \Rightarrow 25'\left(\dfrac{1\text{ deg}}{60\text{ min}}\right) \Rightarrow \dfrac{25}{60} \Rightarrow 0.4166;\ 87°25' = 87.42°$

45. $72°12'22''$

$$12\left(\frac{1\text{ deg}}{60\text{ min}}\right) = \frac{12}{60} \Rightarrow 0.2$$

$$22\left(\frac{1\text{ deg}}{3600\text{ sec}}\right) = \frac{22}{3600} \Rightarrow 0.00611$$

$$72 + 0.2 + 0.00611 = 72.20611° \Rightarrow 72.2061°$$

47. $275°18'35'' = 275.3097°$

49. $61.339° = 61°20'20''$

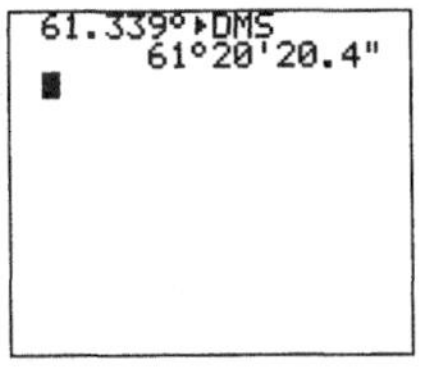

51. $177.344° = 177°20'38''$

53. $128.259° = 128°15'32''$

Substituting Into Formulas

55. $S = \frac{12.0\text{ ft}}{s}(1.30s) + \frac{1}{2}\left(\frac{32.2\text{ ft}}{s^2}\right)(1.30s)^2 \Rightarrow 12(1.30) + \frac{1}{2}(32.2)(1.30s)^2 \Rightarrow 42.8\text{ ft}$

57. $E = \frac{PL}{ae} \Rightarrow \frac{(22{,}500\text{ N})(15.2\text{ m})(1000\text{ mm/m})}{(12.7\text{ m}^2)(2.75\text{ mm})} \Rightarrow 9{,}790{,}000\text{ N/cm}^2$

59. $R = R_1\{1 + a(t - t_1)\} \Rightarrow 775\{1 + 0.00393(80 - 20)°\} \Rightarrow 958\ \Omega$

Applications

61. $15.7(7.481\text{ gal/ft}^3) = 117.4517\text{ gal} \Rightarrow 117\text{ gal}$

63. $6.35(2.205\text{ lb/kg}) = 14.0017\text{ lb} \Rightarrow 14.0\text{ lb}$

65. $3.15\text{ in.} = 3.15(2.54\text{ cm/in.}) \Rightarrow 8.00\text{ cm}$

$1.07\text{ in.} = 1.07(2.54\text{ cm/in.}) \Rightarrow 2.72\text{ cm}$

$31.5\text{mm} = 31.5(1\text{ cm}/10\text{ mm}) \Rightarrow 3.15\text{ cm}$

$36.2\text{ mm} = 36.2(1\text{ cm}/10\text{mm}) \Rightarrow 3.62\text{ cm}$

$0.437\text{ ft} = 0.437(12\text{ in./ft})(2.54\text{ cm/in.}) \Rightarrow 13.3\text{ cm}$

67. $8834\text{ in}^2 = 8834(2.54\text{ cm/in.})^2(1\text{ m}/100\text{ cm})^2 \Rightarrow 5.699\text{ m}^2$

69. $9274\text{ cm}^3 = 9274(1\text{ L}/10^3\text{ cm}^3)(1\text{ gal}/3.785\text{ L}) \Rightarrow 2.450\text{ gal}$

Exercise 9 ◊ Percentage

Conversions

1. $3.72 \Rightarrow 372\%$ (move the decimal two places to the right and affix the percent symbol (%))

3. $0.0055 \Rightarrow 0.55\%$

5. $\frac{2}{5} = 0.40 \Rightarrow 40\%$ (take 2 and divide by 5 to obtain a decimal; then proceed as in 1 – 4)

7. $\frac{7}{10} = 0.70 \Rightarrow 70\%$

9. $23\% \Rightarrow 0.23$ (write the fraction with 100 in the denominator and the percent in the numerator; remove the percent sign and reduce the fraction)

11. $287\frac{1}{2}\% = 287.5\% \Rightarrow \frac{287.5}{100} \Rightarrow 2.875$

13. $37.5\% = \frac{37.5}{100} \Rightarrow \frac{3}{8}$

15. $150\% = \frac{150}{100} \Rightarrow \frac{3}{2}$ or $1\frac{1}{2}$

Finding the Amount

17. 41.1% of 255 tons $\Rightarrow A = BP \Rightarrow A = 0.411(255) \Rightarrow 104.805 = 105$ tons

19. 33.3% of 662 $\Rightarrow 0.333(662) = 22\bar{0}$ kg

21. 35.0% of 343 L $\Rightarrow 0.350(343) = 12\bar{0}$ L

23. R to be added $\Rightarrow 7250\ \Omega\ (0.150) = 1090\ \Omega$

25. Tax credit $\Rightarrow$ (\$1100) (0.42) + (\$5500 - \$1100) (0.25) = \$1562

Finding the Base

27. 86.5 is 16.7% $\Rightarrow \dfrac{85.6}{0.167} = 517.96 \Rightarrow 518$

29. 1.22 is 1.86% $\Rightarrow \dfrac{1.22}{0.0186} = 65.5913 \Rightarrow 65.6$

31. 66.6 is 66.6% $\Rightarrow \dfrac{66.6}{0.666} = 100$

33. Let y = range on earlier electric vehicles

$$y + 0.495y = 161 \text{ km}$$
$$1.495y = 161$$
$$y = \frac{161}{1.495} \Rightarrow 108 \text{ km}$$

35. Let A = the amount spent on solar panels

$$0.35A = \$1560$$
$$A = \frac{1560}{0.35}$$
$$A = \$4457.14$$

Find the Percentage Rate

37. 26.8 is 12.3 $\Rightarrow \dfrac{12.3}{26.8} = 0.458955 \Rightarrow 45.9\%$

39. 44.8 is 8.27 $\Rightarrow \dfrac{8.27}{44.8} = 0.184598 \Rightarrow 18.5\%$

41. 455 h is 152 h $\Rightarrow \dfrac{152}{455} = 0.33406 \Rightarrow 33.4\%$

43. Percent of total capacity $\Rightarrow \left(\dfrac{5840}{50{,}500}\right)(100) = 11.5643\% \Rightarrow 11.6\%$

45. Ripple percentage $\Rightarrow \left(\dfrac{0.75 \text{ V}}{51 \text{ V}}\right)(100) = 1.5\%$

Percent Change

47. from 29.3 to 57.6 $\Rightarrow$ percent change $= \left(\dfrac{57.6 - 29.3}{29.3}\right)(100) = 96.58703\% \Rightarrow 96.6\%$

49. from 227 to 298 $\Rightarrow$ percent change $= \left(\dfrac{298 - 227}{227}\right)(100) = 31.3\%$

51. Percent change in temperature $\Rightarrow \left(\dfrac{21.0°\text{C} - 19.0°\text{C}}{19.0°\text{C}}\right)(100) \Rightarrow 10.526 = 10.5\%$

53. $\left(\dfrac{37.325 - 35.5}{35.5}\right)(100) = 5.14\%$

Percent Efficiency

55. $\left(\dfrac{12.4 \text{ hp}}{18.5 \text{ hp}}\right)(100) = 67.0\%$

57. $\left(\dfrac{1\text{ hp}}{550\text{ ft lb/sec}}\right)\left[\dfrac{(72\text{ ft})(10{,}100\text{ lb/h})\left(\dfrac{\text{h}}{3600\text{ sec}}\right)}{0.50\text{ hp}}\right](100)=\dfrac{\dfrac{(72)(11{,}000)}{3600}}{(550)(0.50)}(100)\Rightarrow 73.4545\%\Rightarrow 73\%$

Percent Error

59. Percent error $=\left(\dfrac{128.0\text{-}125.0}{128.0}\right)(100)=2.3\%$

61. Low voltage = 125.0 V – (0.10) (125.0 V) ⇒ 112.5 V

High voltage – 125.0 V + (1.50) (125.0 V) ⇒ 312.5 V

Percent Concentration

63. Percent concentration of alcohol $=\left(\dfrac{75.0\text{ L}}{75.0\text{ L}+125\text{ L}}\right)(100)=37.5\%$

65. 0.055(455 L) = 25 L

CHAPTER 1 REVIEW PROBLEMS

1. 1.435 – 7.21 + 93.24 – 4.1116 = 83.3534 ⇒ 83.35 (round to the fewest number of significant digits)

3. 21.8(3.775 × 1.07) = 88.0556 ⇒ 88.1

5. Reciprocal of 2.89 = $\dfrac{1}{2.89}=0.34602\Rightarrow 0.346$

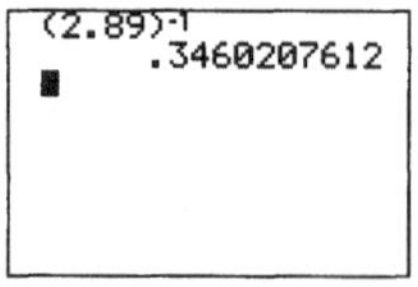

7. $(9.73)^2 = 94.6729 \Rightarrow 94.7$

9. $\sqrt{29.8} = 5.45893 \Rightarrow 5.46$ (round to three significant digits)

11. $(91.2 - 88.6)^2 = 6.76 \Rightarrow 6.8$

13. $y=3(-2.88)^2-2(-2.88)=30.6432\Rightarrow 30.6$

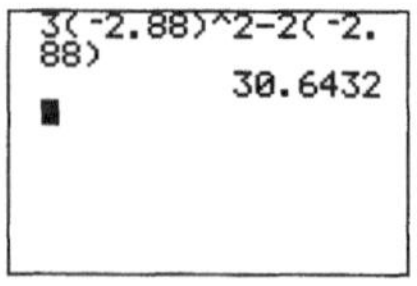

15. $y=2(7.72)-3(3.14)+5(2.27)=17.37\Rightarrow 17.4$

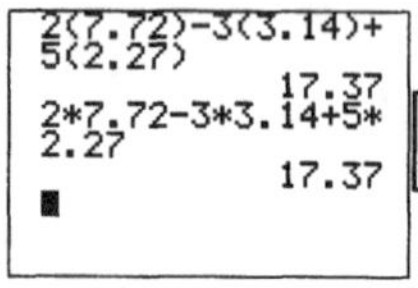

explicit multiplication

The TI-83 computes implicit and explicit multiplication

17. (a) 179.2 ⇒ 179 (b) 1.076 ⇒ 1.08 (c) 4.8550 ⇒ 4.86 (d) 45,725 ⇒ 45,700

19. Percent efficiency $=\left(\dfrac{1310\text{ W}}{2.5\text{ hp}}\right)\left(\dfrac{1\text{ hp}}{746\text{ W}}\right)(100)=70.24128\%\Rightarrow 70.2\%$

21. $(8.34 \times 10^5) + (2.85 \times 10^6)$ -$5.29 \times 10^5) = 3.16 \times 10^6$

23. $\left(7.4\times10^{5}\ \text{J/m}^2\text{h}\right)(15\ \text{acres})\left(\dfrac{4047\ \text{m}^2}{\text{acre}}\right)\left(\dfrac{\text{h}}{3600\ \text{s}}\right)\left(\dfrac{\text{kW}}{10^3\ \text{J/s}}\right)=12{,}478.25\Rightarrow 12{,}000\ \text{kW}$ (round to two significant digits)

25. 15 gal(3.785 L/gal) = 56.8 L; $\left(\dfrac{2.0\ \text{L}}{2.0\ \text{L}+56.8\ \text{L}}\right)(100)=3.40136\%\Rightarrow 3.4\%$

27. 700 billion barrels of oil $\Rightarrow 7\times10^{11}$ barrels

29. $-39.2\div-0.003826=10{,}245.6874\Rightarrow 10{,}200$ (round to three significant digits)

31. $(8.24\times10^{-3})\div(1.98\times10^{7})=4.161616\times10^{-10}\Rightarrow 4.16\times10^{-10}$

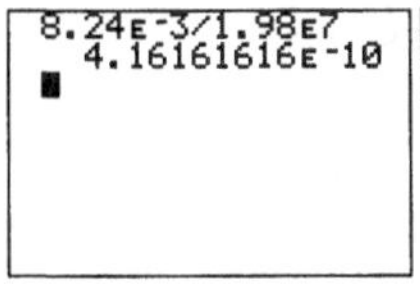

Be sure to be in SCI mode!

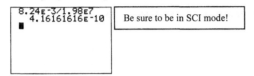

33. $\sqrt[5]{82.8}=2.41883\Rightarrow 2.42$

35. $-\dfrac{2}{3}\ \boxed{<}\ -0.660$

37. $0.492(4827)=2374.884\Rightarrow 2370$ (round to the fewest significant digits)

39. Reciprocal of $-0.582=\dfrac{1}{-0.582}=-1.7182\Rightarrow -1.72$

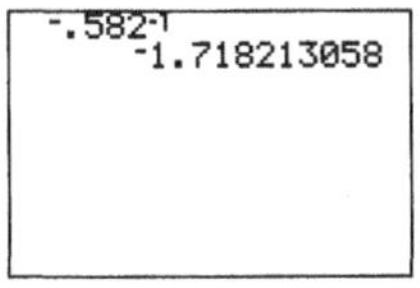

41. $\dfrac{405}{628}=0.644904\Rightarrow 64.5\%$ of the former consumption

43. 49.3 lb(4.448 N/lb) = 219.2864 N $\Rightarrow$ 219 N (round to the fewest significant digits, three)

45. $\left(\dfrac{284\ \text{m}}{1\ \text{min}}\right)(5.25\ \text{s})\left(\dfrac{1\ \text{min}}{60\ \text{s}}\right)+\dfrac{1}{2}\left(\dfrac{9.807\ \text{m}}{\text{s}^2}\right)(5.25\ \text{s})^2=160.0027188$

$$160.0027188\left(\frac{3.281\ \text{ft}}{\text{m}}\right)=524.9689\ \text{ft}\Rightarrow 525\ \text{ft}$$

47. $\dfrac{8460}{38{,}400}=0.2203125\Rightarrow 22.0\%$ (round to the fewest significant digits, three)

49. Let p equal the population: $p+0.125p=8118$

$$1.125p=8118$$
$$p=\frac{8118}{1.125}$$
$$p=7216$$

51. $4.928+2.847-2.836=4.939$

53. 2.84(38.4 = 109.056 $\Rightarrow$ 109 (Why round to three significant digits?)

55. Reciprocal of 4.82 = $\dfrac{1}{4.82}=0.20746\Rightarrow 0.207$

57. $(3.84)^2=14.7456\Rightarrow 14.7$

59. $\sqrt{38.4}=6.19677\Rightarrow 6.20$

61. $(3)^2 - 3(3) + 2 = 2$

63. $83.43 \Rightarrow 83.4$

65. 36.82 in.(2.54 cm/in.) = 93.5228 ⇒ 93.52

67. $\sqrt[3]{746} = 9.06942 \Rightarrow 9.07$

69. $\left(\frac{746}{992}\right)(100) = 75.20161\% \Rightarrow 75.2\%$

71. $73.7 \times 10^{-3} = 0.0737$

73. $6.128 + 8.3470 - 7.23612 = 7.23888 \Rightarrow 7.239$ (round to the least number of significant digits, three)

75. $7.184(16.8) = 120.6912 \Rightarrow 121$

77. Reciprocal of 0.825 ⇒ $\frac{1}{0.825} = 1.212121 \Rightarrow 1.21$

Chapter 2: Introduction to Algebra

CHAPTER 2

Exercise 1 ◊ Algebraic Expressions

Mathematical Expressions

1. $x + 2y$: algebraic expression; contains only algebraic symbols and operations

3. $3 \sin x$: not an algebraic expression; it is transcendental

Algebraic Expressions

5. $5xy - 2x$: not a literal expression; letters at end of alphabet generally represent variables

7. $2az - 3bx$: literal expression; letters at beginning of alphabet generally represent constants

Terms

9. $x^3 - 2x$: 2 terms

11. $ax^2 + bx + c$: 3 term3; terms

Factors

13. $3ax$: 3, a, x

15. $7x^2y^3$: 7, x $(x \cdot x = x^2)$, y $(y \cdot y \cdot y = y^3)$

Coefficient

17. $6x^2$: 6 is the coefficient

19. $-x$: -1 is the coefficient

21. $2ax^5$: $2a$ is the coefficient

Degree

23. $4y^2$: second degree

25. $5x^2y^3$: fifth degree $(2 + 3 = 5)$

27. $5 - xy$: second

29. $2xy^2 + xy - 4$: third degree $(2 + 1 = 3)$

Exercise 2 ◊ Adding and Subtracting Polynomials

Combining Like Terms

1. $8y + 2y \Rightarrow (8 + 2)y = 10y$

3. $5a + 2a = 7a$

5. $8ab - 2ab = 6ab$

7. $2.8x + 3.2x = 6.0x$

9. $53.5a + 21.6a = 75.1a$

11. $8.5ab - 21.2ab = -12.7ab$

13. $5x - 8x + 2x = -x$

15. $9a - 4a + 2a = 7a$

17. $2ab - 8ab - 2ab = -8ab$

19. $5.2x - 2.8x + 1.2x = 3.6x$

21. $53.6m - 12.2m + 31.3m = 72.7m$

23. $38.3xy - 33.9xy + 58.2xy = 62.6xy$

25. $3x + 5x - 4x + 2x = 6x$

27. $5a - 9a - 3a - 2a = -5a$

29. $9.3x + 5.2x - 1.8x + 1.2x = 13.9x$

31. $11.8m - 43.6m - 32.2m + 31.3m = -32.7m$

33. $6x - 3x + 7x - 4x + 2x = 8x$

35. $9ab - 6ab + 5ab - 4ab - 2ab = 2ab$

37. $32.5m - 11.8m - 23.6m - 32.2m + 11.3m = -23.8m$

39. $(x + 2) + (3x - 4) \Rightarrow (1 + 3)x + (2 - 4) = 4x - 2$

41. $(22.7ab + 21.2) + (83.5 - 48.2ab) \Rightarrow (22.7 - 48.2)ab + (21.2 + 83.5) = -25.5ab + 104.7$

43. $(8.33 + 1.05y - (2.44y + 1.12) + (2.88y - 1.74) = 1.49y + 5.47$

Instructions Given Verbally

45. $(a - c + b) + (b + c - a) = (a - a) + (b + b) + (c - c) \Rightarrow 2b$

47. $(8b - 10c + 3a - d) - (5a + 7d - 4b + 6c) = 8b - 10c + 3a - d - 5a - 7d + 4b - 6c \Rightarrow -2a + 12b - 16c - 8d$

49. $(24by^5 - 14bx^4) + (-72bx^5 + 2by^5 - 3bx^4) + (9bx^4 + 23by^4 - 21by^5)$

$= 24by^5 + 2by^5 - 21by^5 - 14bx^4 - 3bx^4 + 9bx^4 + 23by^4 - 72bx^5 = 5by^5 - 8bx^4 - 72bx^5 + 23by^4$

Challenge Problems

51. $(8.33a + 1.15y - 2.4b) - (2.44y + 5.0b - 1.12a) + (3.8b + 2.88y - 1.74a) \Rightarrow (8.33a + 1.12a - 1.74a) + (-2.4b - 5.0b + 3.8b)$

$+ (1.15y - 2.44y + 2.88y) = 7.71a - 3.6b + 1.59y$

53. $(4x + 2y) + (3w - 4y) - (5x + 5w) + (3x + 2y) = 2x - 2w$

55. $(55.2xy + 54.2x) - (28.3x + xy) - (12.5xy + 44.7y) + (2.18xy - 11.6x) = 43.9xy + 14.3x - 44.7y$

Applications

57. $0.12x + 0.08(5000 - x) = 0.12x + 400 - 0.08x \Rightarrow 0.04x + 400$

59. $\pi r^2 + 2\pi rh + \pi r^2 = 2\pi r^2 + 2\pi rh \Rightarrow 2\pi r(r + h)$

Exercise 3 ◊ Laws of Exponents

1. $3^3 \Rightarrow 3 \cdot 3 \cdot 3 = 27$

3. $(-2)^4 \Rightarrow (-2)\,(-2)\,(-2)\,(-2) = 16$

5. $(0.001)^3 \Rightarrow (0.001)\,(0.001)\,(0.001) = 1 \times 10^{-9}$

Multiplying Powers

7. $(w^3)(w^2) \Rightarrow w^{3+2} = w^5$

9. $(p^3)(p^4) \Rightarrow p^{3+4} = p^7$

11. $(y^b)\,(y^3) \Rightarrow y^{b+3}$

13. $(10^5)\,(10^9) \Rightarrow 10^{5+9} = 10^{14}$

15. $(z^{11})\,(z^2) \Rightarrow z^{11+2} = z^{13}$

Dividing Powers

17. $\frac{a^6}{a^4} \Rightarrow \frac{a \cdot a \cdot a \cdot a \cdot a \cdot a}{a \cdot a \cdot a \cdot a} \Rightarrow \frac{\not{a} \cdot \not{a} \cdot \not{a} \cdot \not{a} \cdot a \cdot a}{\not{a} \cdot \not{a} \cdot \not{a} \cdot \not{a}} = a^2$ or $\frac{a^6}{a^4} \Rightarrow a^{6-4} = a^2$

19. $\frac{10^6}{10^2} \Rightarrow 10^{6-2} = 10^4$

21. $\dfrac{x^{-6}}{y} \Rightarrow \dfrac{\frac{1}{x^6}}{y} \Rightarrow \dfrac{1}{x^6} \div y \Rightarrow \dfrac{1}{x^6} \cdot \dfrac{1}{y} = \dfrac{1}{x^6 y}$

23. $\dfrac{y^{a+1}}{y^{a-2}} \Rightarrow y^{(a+1)-(a-2)} \Rightarrow y^{a+1-a+2} = y^3$

Power Raised to a Power

25. $(x^2)^2 \Rightarrow (x \cdot x)^2 \Rightarrow (x \cdot x)(x \cdot x) \Rightarrow x \cdot x \cdot x \cdot x = x^4$ or $(x^2)^2 \Rightarrow x^{2 \cdot 2} = x^4$

27. $(m^5)^2 \Rightarrow m^{5 \cdot 2} = m^{10}$

29. $(a^2)^4 \Rightarrow a^{2 \cdot 4} = a^8$

31. $(z^c)^a \Rightarrow z^{c \cdot a} = z^{ac}$

33. $(a^{x-1})^3 \Rightarrow a^{3(x-1)} = a^{3x-3}$

Product Raised to a Power

35. $(3y)^2 \Rightarrow 3^2 y^2 = 9y^2$

37. $(abp^2)^4 \Rightarrow a^{1 \cdot 4} \cdot b^{1 \cdot 4} \cdot p^{2 \cdot 4} = \Rightarrow a^4 b^4 p^8$

39. $(3a)^2 \Rightarrow 3^2 a^2 = 9a^2$

41. $(2xyz)^5 \Rightarrow 2^5 x^5 y^5 z^5 = 32\, x^5 y^5 z^5$

Quotient Raised to a Power

43. $\left(\dfrac{2}{3}\right)^3 \Rightarrow \dfrac{2^{1 \cdot 3}}{3^{1 \cdot 3}} \Rightarrow \dfrac{2^3}{3^3} \Rightarrow \dfrac{2 \cdot 2 \cdot 2}{3 \cdot 3 \cdot 3} = \dfrac{8}{27}$

45. $\left(\dfrac{a}{b}\right)^3 = \dfrac{a^3}{b^3}$

47. $\left(\dfrac{2ab^3}{3c^2 d}\right)^3 \Rightarrow \dfrac{2^3 a^3 b^{3 \cdot 3}}{3^3 c^{2 \cdot 3} d^3} \Rightarrow \dfrac{2^3 a^3 b^9}{3^3 c^6 d^3} = \dfrac{8a^3 b^9}{27c^6 d^3}$

Zero Exponent

49. $\left(2x^2 - 8x + 32\right)^0 = 1$

51. $\dfrac{82}{y^0} \Rightarrow \dfrac{82}{1} = 82$

53. $\dfrac{\left(z^{-n}\right)\left(z^2\right)}{z^{2-n}} \Rightarrow \dfrac{z^{2-n}}{z^{2-n}} \Rightarrow z^{(2-n)-(2-n)} \Rightarrow z^{2-2-n+n} \Rightarrow z^0 = 1$

Negative Exponent

55. $ay^{-1} \Rightarrow a\left(\dfrac{1}{y}\right) = \dfrac{a}{y}$

57. $(-b)^{-3} = -\dfrac{1}{b^3}$

59. $ab^{-5}c^{-2} \Rightarrow a\left(\dfrac{1}{b^5}\right)\left(\dfrac{1}{c^2}\right) \Rightarrow \left(\dfrac{a}{1}\right)\left(\dfrac{1}{b^5}\right)\left(\dfrac{1}{c^2}\right) = \dfrac{a}{b^5 c^2}$

61. $\left(\frac{a}{b}\right)^{-4} \Rightarrow \frac{a^{-4}}{b^{-4}} \Rightarrow \frac{\frac{1}{a^4}}{\frac{1}{b^4}} \Rightarrow \left(\frac{1}{a^4}\right)\left(\frac{b^4}{1}\right) = \frac{b^4}{a^4}$

An alternative way to solve $\left(\frac{a}{b}\right)^{-4}$ based on the definition of the negative exponent

$$\left(\frac{a}{b}\right)^{-4} \Rightarrow \frac{1}{\left(\frac{a}{b}\right)^4} \Rightarrow \frac{1}{\frac{a^4}{b^4}} \Rightarrow 1 \div \frac{a^4}{b^4} \Rightarrow \left(\frac{1}{1}\right)\left(\frac{b^4}{a^4}\right) = \frac{b^4}{a^4}$$

Solving $\left(\frac{4x^3}{3y^2}\right)^{-3}$ based on the definition of the negative exponent

$$\left(\frac{4x^3}{3y^2}\right)^{-3} \Rightarrow \frac{1}{\left(\frac{4x^3}{3y^2}\right)^3} \Rightarrow \frac{1}{\frac{4^3x^9}{3^3y^6}} \Rightarrow \frac{1}{\frac{64x^9}{27y^6}} \Rightarrow 1 \div \frac{64x^9}{27y^6} \Rightarrow \left(\frac{1}{1}\right)\left(\frac{27y^6}{64x^9}\right) = \frac{27y^6}{64x^9}$$

63. $\frac{1}{a} = a^{-1}$

65. $\frac{c^2}{d^3} \Rightarrow c^2\left(\frac{1}{d^3}\right) \Rightarrow c^2d^{-3}$

67. $\frac{y^2}{x^{-4}} \Rightarrow \frac{y^2}{\frac{1}{x^4}} \Rightarrow \left(\frac{y^2}{1}\right)\left(\frac{x^4}{1}\right) = x^4y^2$

Challenge Problems

69. $\left[2x^2\left(\frac{y^3}{w^2}\right)\right]^2 \Rightarrow 2^2x^{2\cdot 2}\left(\frac{y^{3\cdot 2}}{w^{2\cdot 2}}\right) \Rightarrow 4x^4\left(\frac{y^6}{w^4}\right) \Rightarrow \frac{4x^4}{1}\cdot\frac{y^6}{w^4} = \frac{4x^4y^6}{w^4}$

71. $\left[\left(\frac{ax}{bz}\right)^2\left(\frac{bx}{cz}\right)^3\right]^2 \Rightarrow \left(\frac{ax}{bz}\right)^{2\cdot 2}\left(\frac{bx}{cz}\right)^{3\cdot 2} \Rightarrow \left(\frac{a^4x^4}{b^4z^4}\right)\left(\frac{b^6x^6}{c^6z^6}\right) \Rightarrow \frac{a^4b^{6-4}x^{4+6}}{c^6z^{4+6}} = \frac{a^4b^2x^{10}}{c^6z^{10}}$

Applications

73. $(w)\,(2w)\,(3w) \Rightarrow (2\cdot 3)\,(w\cdot w\cdot w) = 6w^3$

75. $\left(\frac{i}{3}\right)^2 R \Rightarrow \left(\frac{i^2}{3^2}\right)R \Rightarrow \left(\frac{i^2}{9}\right)R \Rightarrow \left(\frac{i^2}{9}\right)\left(\frac{R}{1}\right) = \frac{i^2R}{9}$

Exercise 4 ◊ Multiplying a Monomial by a Monomial

1. $(x^2)\,(x^4) \Rightarrow x^{2+4} = x^6$
3. $(x^2)\,(-x^3) \Rightarrow (-1)(x^{2+3}) = -x^5$
5. $(2a)\,(3b^2) \Rightarrow 6ab^2$
7. $(5m^2n)(\,3mn^2) \Rightarrow 5\cdot 3m^{2+1}n^{1+2} = 15m^3n^3$
9. $(12.5a)(3.26a^2) \Rightarrow (12.5)(3.26)a^{1+2} \Rightarrow 40.75a^3 = 40.8\,a^3$
11. $(3.73xy)(1.77xy^2) \Rightarrow 6.6021x^{1+1}y^{1+2} = 6.60\,x^2y^3$
13. $(2ab)(3b^n) \Rightarrow 2\cdot 3ab^{1+n} = 6ab^{n+1}$
15. $(1.55a^m)(2.36a^n) \Rightarrow (1.55)(2.36)a^{m+n} \Rightarrow 3.658\,a^{m+n} = 3.66\,a^{m+n}$

17. $(2w^2)(5w)(3w^2) \Rightarrow 2 \cdot 5 \cdot 3w^{2+1+2} = 30w^5$

19. $(4a^2b)(2ab)(3b^n) \Rightarrow 4 \cdot 2 \cdot 3a^{2+1}b^{1+1+n} = 24a^3b^{n+2}$

Challenge Problems

21. $(1.84wx^2y)(2.44w^2xy^3)(1.65wx^3y)(2.33w^2xy) \Rightarrow 17.2602672w^{1+2+1+2}x^{2+1+3+1}y^{1+3+1+1} = 17.3w^6x^7y^6$

23. $(3.82abc)(a^xbc^2)(1.55a^2bc^x)(ab^3c) \Rightarrow 5.921a^{1+x+2+1}b^{1+1+1+3}c^{1+2+x+1} = 5.92a^{x+4}b^6c^{x+4}$

Applications

25. $2(1.41x)(3.75x) \Rightarrow 10.575x^{1+1} = 10.6x^2$

Exercise 5 ◊ Multiplying a Monomial and a Multinomial

1. $-(x+2) \Rightarrow (-1)(x) + (-1)(2) = -x - 2$

3. $3 - (-x + 1) \Rightarrow 3 - (-1)(-x) + (-1)(1) \Rightarrow 3 + x - 1 = x + 2$

5. $a + (b + a) \Rightarrow a + b + a = 2a + b$

7. $x + (x - y) \Rightarrow x + x - y = 2x - y$

9. $x(b + 2) \Rightarrow x(b) + x(2) = bx + 2x$

11. $x(x - 5) \Rightarrow x(x) - x(5) \Rightarrow x^{1+1} - 5x = x^2 - 5x$

13. $2.03x(1.27x - 2.36) \Rightarrow 2.03x^{1+1}(1.27) - 2.03x(2.36) \Rightarrow 2.5781x^2 - 4.7905x = 2.58x^2 - 4.79x$

15. $b^4(b^2 + 8) \Rightarrow b^{4+2} + b^4(8) = b^6 + 8b^4$

17. $-(a + 3.92) - (a - 4.14) \Rightarrow -a - 3.92 - a + 4.14 = 0.22 - 2a$

19. $(2x + 5) + (x - 2) \Rightarrow 2x + 5 + x - 2 = 3x + 3$

21. $(2x + 6a) - (4x - a) \Rightarrow 2x + 6a - 4x + a = 7a - 2x$

23. $(7262x + 1.26a) - (2844x - 8.23a) \Rightarrow 7262x + 1.26a - 2844x + 8.23a = 4418x + 9.49a$

25. $(y - z - b) - (b + y + z) \Rightarrow y - z - b - b - y - z = -2b - 2z$

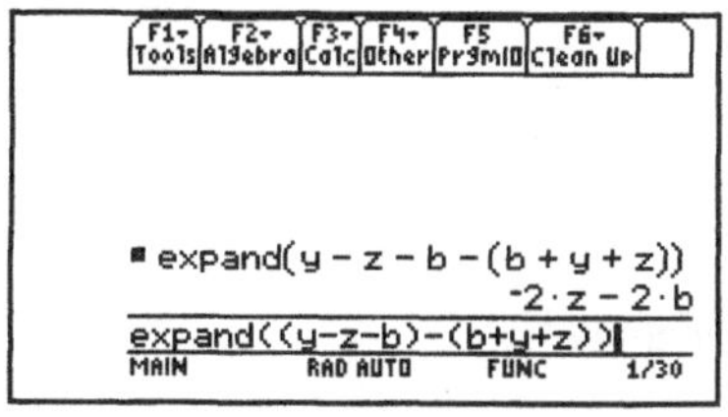

27. $(2z + 5c - 3a) - (6a + 2c - 4z) \Rightarrow 2z + 5c - 3a - 6a - 2c + 4z = 6z + 3c - 9a$

29. $\{[3 - (x + 7)] - 3x\} - (x + 7) \Rightarrow (3 - x - 7 - 3x) - x - 7 \Rightarrow 3 - x - 7 - 3x - x - 7 = -5x - 11$

Challenge Problems

31. $2ab(9a^2 + 6ab - 3b^2) \Rightarrow 2(9)a^{1+2}b + 2(6)a^{1+1}b^{1+1} - 2(3)ab^{1+2} = 18a^3b + 12a^2b^2 - 6ab^3$

33. $-5.16xy(1.23x^2y - 5.83xy^2 + 4.27x^2y^2 - 2.94xy)$

$\Rightarrow -5.16(1.23)x^{1+2}y^{1+1} - (-5.16)(5.83)x^{1+1}y^{1+2} + (-5.16)(4.27)x^{1+2}y^{1+2} - (-5.16)(2.94)x^{1+1}y^{1+1}$

$\Rightarrow -6.3468x^3y^2 + 30.0828x^2y^3 - 22.0332x^3y^3 + 15.1704x^2y^2$

$= -6.35x^3y^2 + 30.1x^2y^3 - 22.0x^3y^3 + 15.2x^2y^2$

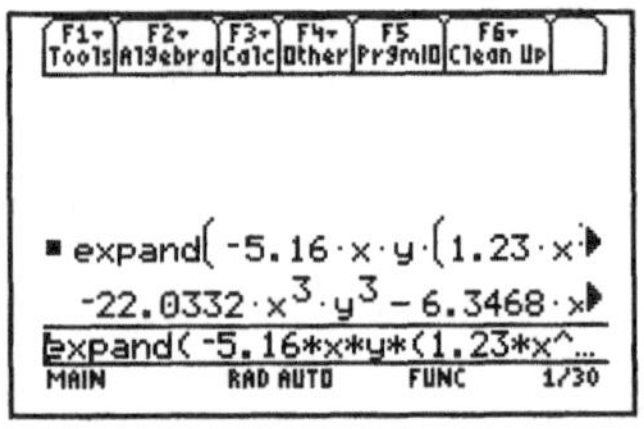

35. $6p-\{3p+[2q-((5p+4q)+p)-(3p+2)]-2p\}\Rightarrow 6p-\{3p+2q-5p-4q-p-3p-2-2p\}$

$\Rightarrow 6p-(-8p-2q-2)\Rightarrow 6p+8p+2q+2=14p+2q+2$

37. $(23y^2+4y^3-12)-(11y^3-8y^2+y)\Rightarrow 23y^2+4y^3-12-11y^3+8y^2-y=-7y^3+31y^2-y-12$

39. $(18y^2-12xy)-(6y^2+xy-a)\Rightarrow 18y^2-12xy-6y^2-xy+a=12y^2-13xy+a$

41. $(-6x-z)-\{3y+[7x-(3z+8y+x)]\}\Rightarrow -6x-z-\{3y+7x-3z-8y-x\}\Rightarrow -6x-z-3y-7x+3z+8y+x$

$=-12x+5y+2z$

Applications

43. $(R_1+R_2)-(R_3-R_4)-(R_5+R_6-R_7)=R_1+R_2-R_3+R_4-R_5-R_6+R_7$

Exercise 6 ◊ Multiplying a Binomial by a Binomial

1. $(x+y)(x+z)\Rightarrow x(x)+x(z)+y(x)+y(z)\Rightarrow x^{1+1}+xz+xy+yz=x^2+xz+xy+yz$

3. $(4m+n)(2m^2-n)\Rightarrow 4m(2m^2)+4m(-n)+n(2m^2)+n(-n)\Rightarrow 8m^3-4mn+2m^2n-n^2=8m^3+2m^2n-4mn-n^2$

5. $(2x-y)(x+y)\Rightarrow 2x(x)+2x(y)-y(x)-y(y)=2x^2+xy-y^2$

By the distributive property:

$$\begin{aligned}(2x-y)(x+y)&=2x(x+y)-y(x+y)\\&=2x(x)+2x(y)-y(x)-y(y)\\&=2x^2+2xy-xy-y^2\\&=2x^2+xy-y^2.\end{aligned}$$

7. $(4xy^2-3a^3b)(3xy^2+4a^3b)\Rightarrow 4xy^2(3xy^2)+4xy^2(4a^3b)-3a^3b(3xy^2)-3a^3b(4a^3b)$

$\Rightarrow 12x^2y^4+16xy^2a^3b-9xy^2a^3b-12a^6b^2=12x^2y^4+7a^3bxy^2-12a^6b^2$

9. $(a-7x)(2a+3x)\Rightarrow a(2a)+a(3x)-7x(2a)-7x(3x)\Rightarrow 2a^2+3ax-14ax-21x^2=2a^2-11ax-21x^2$

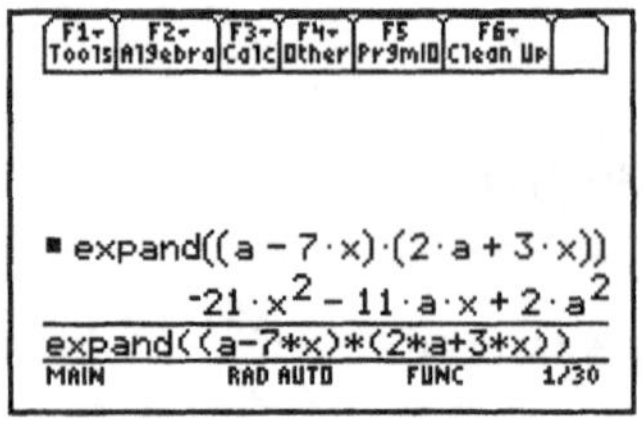

11. $(ax-5b)(ax+5b)\Rightarrow ax(ax)+ax(5b)-5b(ax)-5b(5b)=a^2x^2-25b^2$

By the distributive property:

$$\begin{aligned}(ax-5b)(ax+5b) &= ax(ax+5b)-5b(ax+5b)\\ &= a^2x^2+5abx-5abx-25b^2\\ &= a^2x^2-25b^2\end{aligned}$$

13. $(2.93x-1.11y)(x+y) \Rightarrow 2.93x(x)+2.93x(y)-1.11y(x)-1.11y(y) \Rightarrow 2.93x^2+2.93xy-1.11xy-1.11y^2$

$\Rightarrow 2.93x^2+1.82xy-1.11y^2$

15. $(4.03y^2-3.92a^3b)(3.26y^2+4.73a^3b) \Rightarrow 4.03y^2(3.26y^2)+4.03y^2(4.73a^3b)-3.92a^3b(3.26y^2)-3.92a^3b(4.73a^3b)$

$\Rightarrow 13.1378y^4+19.0619a^3by^2-12.7792a^3by^2-18.5416a^6b^2 \Rightarrow 13.1378y^4+6.2827a^3by^2-18.5416a^6b^2$

$=13.1y^4+6.28a^3by^2-18.5a^6b^2$

Applications

17. Area of rectangle = length × width $\Rightarrow (L+2)(W-3) \Rightarrow L(W)+L(-3)+2(W)+2(-3)=LW-3L+2W-6$

Exercise 7 ◊ Multiplying a Multinomial by a Multinomial

1. $(x-3)(x+4-y) \Rightarrow x(x)+x(4)+x(-y)-3(x)-3(4)-3(-y) \Rightarrow x^2+4x-xy-3x-12+3y$

$=x^2-xy+x+3y-12$

3. $(w^2+w-5)(4w-2) \Rightarrow w^2(4w)+w^2(-2)+w(4w)+w(-2)-5(4w)-5(-2)$

$\Rightarrow 4w^3-2w^2+4w^2-2w-20w+10=4w^3-2w^2-22w+10$

5. $(x+3.88)(x^3-2.15x-6.03) \Rightarrow x(x^3)+x(-2.15x)+x(-6.03)+3.88(x^3)+3.88(-2.15x)+3.88(-6.03)$

$\Rightarrow x^4-2.15x^2-6.03x+3.88x^3-8.342x-23.3964 \Rightarrow x^4+3.88x^3-2.15x^2-14.372x-23.3964$

$=x^4+3.88x^3-2.15x^2-14.4x-23.4$

7. $(2x^2-6xy+3y^2)(3x+3y) \Rightarrow 2x^2(3x)+2x^2(3y)-6xy(3x)-6xy(3y)+3y^2(3x)+3y^2(3y)$

$\Rightarrow 6x^3+6x^2y-18x^2y-18xy^2+9xy^2+9y^3=6x^3-12x^2y-9xy^2+9y^3$

9. $(a^2+2a-2)(a+1) \Rightarrow a^2(a)+a^2(1)+2a(a)+2a(1)-2(a)-2(1) \Rightarrow a^3+a^2+2a^2+2a-2a-2$

$=a^3+3a^2-2$

11. $(c^2-cm+cn+mn)(c-m) \Rightarrow c^2(c)+c^2(-m)-cm(c)-cm(-m)+cn(c)+cn(-m)+mn(c)+mn(-m)$

$\Rightarrow c^3-c^2m-c^2m+cm^2+c^2n-cmn+cmn-m^2n=c^3-2c^2m+c^2n+cm^2-m^2n$

Challenge Problems

13. $(x-y-z)(x+y+z) \Rightarrow x(x)+x(y)+x(z)-y(x)-y(y)-y(z)-z(x)-z(y)-z(z)$

$\Rightarrow x^2+xy+xz-xy-y^2-yz-xz-yz-z^2=x^2-y^2-2yz-z^2$

15. $(5x-y+2x)(4x-y+6) \Rightarrow 5x(4x)+5x(-y)+5x(6)-y(4x)-y(-y)-y(6)+2x(4x)+2x(-y)+2x(6)$

$\Rightarrow 20x^2-5xy+30x-4xy+y^2-6y+8x^2-2xy+12x=28x^2-11xy+42x+y^2-6y$

17. $(a^2-5.93a+31.4)(a^2-5.37a+4.03) \Rightarrow a^2(a^2)+a^2(-5.37a)+a^2(4.03)-5.93a(a^2)$

$-5.93a(-5.37a)-5.93a(4.03)+31.4(a^2)+31.4(-5.37a)+31.4(4.03)$

$\Rightarrow a^4-5.37a^3+4.03a^2-5.93a^3+31.8441a^2-23.8979a+31.4a^2-168.618a+126.542$

$$\Rightarrow a^4 - 11.3a^3 + 67.2741a^2 - 192.5159a + 126.542 = a^4 - 11.3a^3 + 67.3a^2 - 193a + 127$$

19. $(am - ym + yx)(am + ym - yx) \Rightarrow am(am) + am(my) + am(-xy) - my(am) - my(my) - my(-xy)$

$$+xy(am) + xy(my) + xy(-xy)$$

$$\Rightarrow a^2m^2 + am^2y - amxy - am^2y - m^2y^2 + mxy^2 + amxy + mxy^2 - x^2y^2 = a^2m^2 - m^2y^2 + 2mxy^2 - x^2y^2$$

Application

21. $(w + x + a)(l + y - b) = wl + wy - wb + xl + xy - xb + al + ay - ab$

Exercise 8 ◊ Raising a Multinomial to a Power

1. $(x + y)^2 \Rightarrow (x + y)(x + y) \Rightarrow x^2 + xy + xy + y^2 = x^2 + 2xy + y^2$

3. $(a - d)^2 \Rightarrow a^2 - ad - ad + d^2 = a^2 - 2ad + d^2$

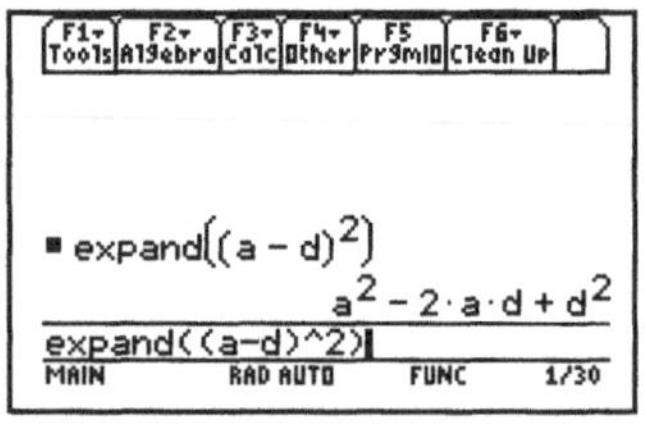

5. $(B + D)^2 \Rightarrow B^2 + BD + BD + D^2 = B^2 + 2BD + D^2$

7. $(4.92y + 3.12z)^2 \Rightarrow (4.92y)^2 + 4.92y(3.12z) + 3.12z(4.92y) + (3.12z)^2$

$$\Rightarrow 24.2064y^2 + 15.3504yz + 15.3504yz + 9.7344z^2 = 24.2y^2 + 30.7yz + 9.73z^2$$

9. $(5n + 6x)^2 \Rightarrow 25n^2 + 30nx + 30nx + 36x^2 = 25n^2 + 60nx + 36x^2$

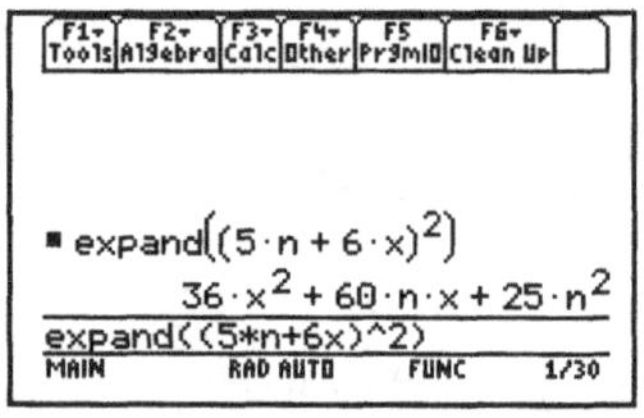

11. $(1 - w)^2 \Rightarrow 1 - 2[(1)(w)] + w^2 \Rightarrow 1 - 2w + w^2 = w^2 - 2w + 1$

13. $(b^3 - 13)^2 \Rightarrow (b^3)^2 - 2[(b^3)(13)] + (13)^2 = b^6 - 26b^3 + 169$

15. $(3.88x^2 - 1.33)^2 \Rightarrow (3.88x^2)^2 - 2[(3.88x^2)(1.33)] + (1.33)^2 = 15.1x^4 - 10.3x^2 + 1.77$

17. $(x + y + z)^2 \Rightarrow (x + y + z)(x + y + z) \Rightarrow x^2 + xy + xz + xy + y^2 + yz + xz + yz + z^2 = x^2 + 2xy + 2xz + y^2 + 2yz + z^2$

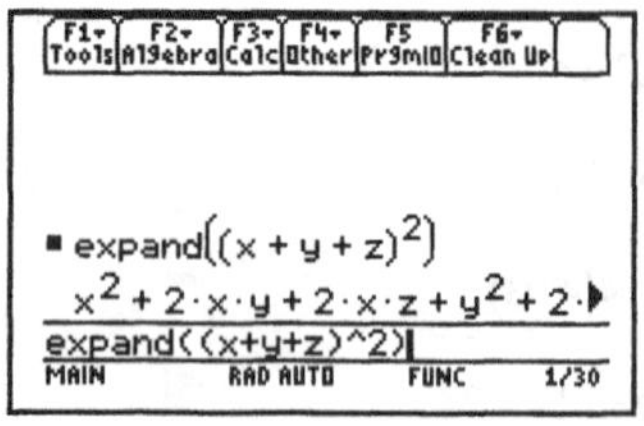

19. $(a+b-1)^2 \Rightarrow a^2+ab-a+ab+b^2-b-a-b+1=a^2+2ab-2a+b^2-2b+1$

21. $\left(c^2-cd+d^2\right)^2=c^4-2c^3d+3c^2d^2-2cd^3+d^4$

Challenge Problems

23. $(x-y)^3 \Rightarrow \left(x^2-2xy+y^2\right)(x-y) \Rightarrow x^2(x)-x^2(y)-2xy(x)-2xy(-y)+y^2(x)+y^2(-y)$

$=x^3-3x^2y+3xy^2-y^3$

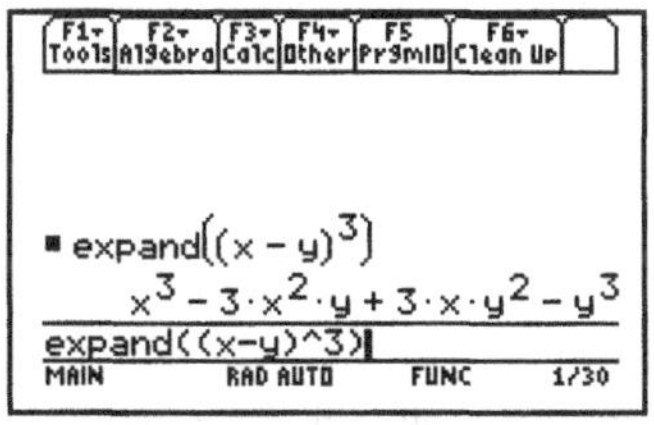

25. $(3.02m+2.16n)^3 \Rightarrow \left(9.1204m^2+13.0464mn+4.6656n^2\right)(3.02m+2.16n)$

$\Rightarrow 9.1204m^2(3.02m)+9.1204m^2(2.16n)$

$\Rightarrow 13.0464mn(3.02m)+13.0464mn(2.16n)+4.6656n(3.02m)+4.6656n(2.16n)$

$=27.5m^3+59.1m^2n+42.3mn^2+10.1n^3$

27. $(c+d)^3 \Rightarrow \left(c^2+2cd+d^2\right)(c+d) \Rightarrow c^2(c)+c^2(d)+2cd(c)+2cd(d)+d^2(c)+d^2(d)=c^3+3c^2d+3cd^2+d^3$

29. $\left(3xy^2+2x^2y\right)^3 \Rightarrow \left(9x^2y^4+12x^3y^3+4x^4y^2\right)\left(3xy^2+2x^2y\right) \Rightarrow 9x^2y^4\left(3xy^2\right)+9x^2y^4\left(2x^2y\right)+12x^3y^3\left(3xy^2\right)+12x^3y^3\left(2x^2y\right)$

$+4x^4y^2\left(3xy^2\right)+4x^4y^2\left(2x^2y\right)$

$\Rightarrow 27x^3y^6+18x^4y^5+36x^4y^5+24x^5y^4+12x^5y^4+8x^6y^3$

$\Rightarrow 27x^3y^6+54x^4y^5+36x^5y^4+8x^6y^3=8x^6y^3+36x^5y^4+54x^4y^5+27x^3y^6$

Applications

31. $(x+2)^2 \Rightarrow (x+2)(x+2)=x^2+4x+4$

33. $\frac{4}{3}\pi(r-2)^3 \Rightarrow \frac{4}{3}\pi\left[\left(r^2-4r+4\right)(r-2)\right] \Rightarrow \frac{4}{3}\pi\left(r^3-2r^2-4r^2+8r+4r-8\right) \Rightarrow \frac{4}{3}\pi\left(r^3-6r^2+12r-8\right)$

$\Rightarrow \left(\frac{4}{3}\pi\right)r^3-\left(\frac{4}{3}\pi\right)6r^2+\left(\frac{4}{3}\pi\right)12r-\left(\frac{4}{3}\pi\right)8=4.19r^3-25.10r^2+50.30r-33.50$

Exercise 9 ◊ Dividing a Monomial by a Monomial

1. $\frac{x^7}{x^4} \Rightarrow x^{7-4}=x^3$

3. $\frac{5xyz}{xy} \Rightarrow 5\left(\frac{xyz}{xy}\right) \Rightarrow 5x^{1-1}y^{1-1}z=5z$

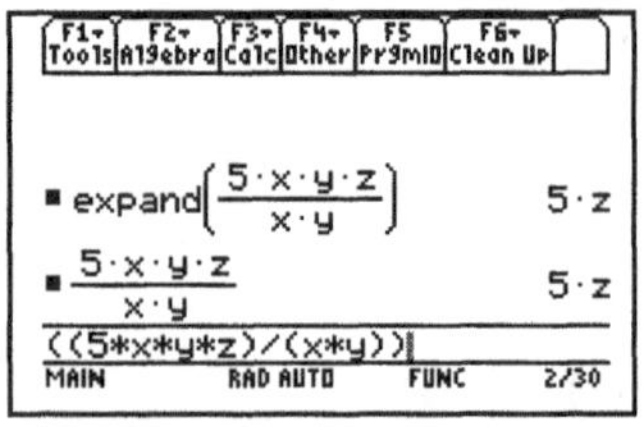

5. $\dfrac{4a^2d}{-2ad} \Rightarrow \left(\dfrac{4}{-2}\right)a^{2-1}d^{1-1} = -2a$

7. $\dfrac{31.4ab^9}{2.66ab^8} \Rightarrow \left(\dfrac{31.4}{2.66}\right)a^{1-1}b^{9-8} \Rightarrow 11.8045a^0b = 11.8b$

9. $\dfrac{8.3ad}{3.26a} \Rightarrow \left(\dfrac{8.3}{3.26}\right)a^{1-1}d \Rightarrow 2.54601a^0d = 2.55d$

11. $42p^5q^4r^2 \div 7p^3qr \Rightarrow 6p^{5-3}q^{4-1}r^{2-1} = 6p^2q^3r$ (divide the coefficients, subtract the exponents of the same base)

13. $-32m^2nx \div 4mx = -8mn$

15. $-36a^4b^2c \div 9ab = -4a^3bc$

17. $\dfrac{-24m^3n^3z}{4m^3z} = -6n^3$ (divide the coefficients, subtract the exponents of the same base)

19. $\dfrac{25a^4bcxyz}{5a^2bcxz} = 5a^2y$

21. $32a^2bc \div (-8ab) = -4ac$ (divide the coefficients, subtract the exponents of the same base)

23. $-36a^2by^2 \div 12a^2y = -3by$

25. $\dfrac{64x^2y^2}{8xy} = 8xy$

27. $\dfrac{x^2y^6z^2}{x^2z^2} = y^6$

29. $\dfrac{a^x}{a^y} = a^{x-y}$ (the laws of exponents remain the same)

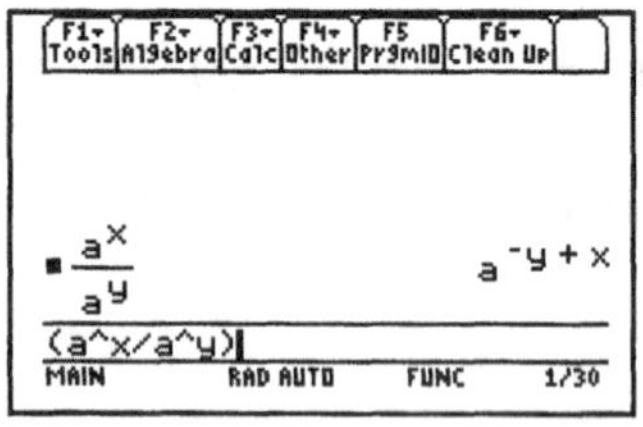

31. $-35a^3b^2z \div 7ab^2 = -5a^2z$

33. $95abc \div 5a^2b^3c \Rightarrow \left(\dfrac{95}{5}\right)a^{1-2}b^{1-3}c^{1-1} \Rightarrow 19a^{-1}b^{-2}c^0 \Rightarrow \dfrac{19}{1}\left(\dfrac{1}{a}\right)\left(\dfrac{1}{b^2}\right) = \dfrac{19}{ab^2}$

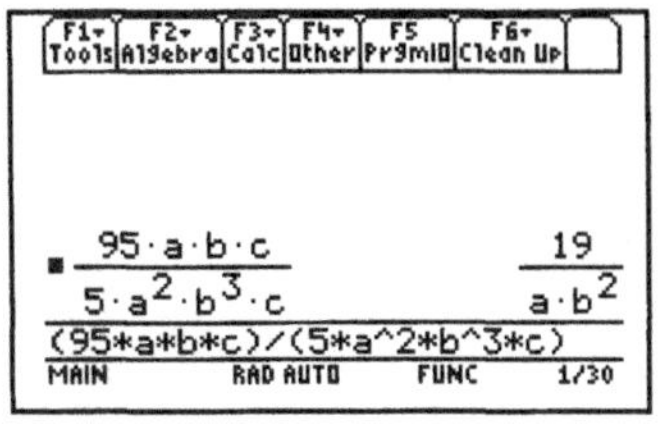

Application

35. $\dfrac{\frac{4}{3}\pi r^3}{4\pi r^2} \Rightarrow \left(\dfrac{\frac{4}{3}}{4}\right)\left(\dfrac{\pi}{\pi}\right)\left(\dfrac{r^3}{r^2}\right) \Rightarrow \dfrac{1}{3}r^{3-2} \Rightarrow \dfrac{1}{3}r = \dfrac{r}{3}$

Exercise 10 ◊ Dividing a Polynomial by a Monomial

1. $\dfrac{15x^3 + 3x^2}{x} \Rightarrow \dfrac{15x^3}{x} + \dfrac{3x^2}{x} \Rightarrow 15x^{3-1} + 3x^{2-1} = 15x^2 + 3x$

3. $\dfrac{36d^5 - 6d^2}{3d} \Rightarrow \dfrac{36d^5}{3d} - \dfrac{6d^2}{3d} = 12d^4 - 2d$

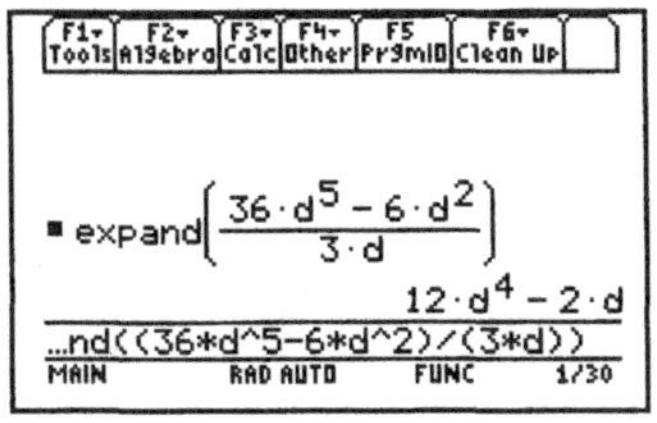

5. $\dfrac{48c^4 + 36c^5}{12c^2} \Rightarrow \dfrac{48c^4}{12c^2} + \dfrac{36c^5}{12c^2} = 4c^2 + 3c^3$

7. $\dfrac{39p^2 + 52p^3}{-13p^2} \Rightarrow \dfrac{39p^2}{-13p^2} + \dfrac{52p^3}{-13p^2} = -3 - 4p$

9. $\dfrac{-25x^3 - 15x^2}{5x} \Rightarrow \dfrac{-25x^3}{5x} - \dfrac{15x^2}{5x} = -5x^2 - 3x$

11. $\dfrac{-8.27bm^3n^2 - 3.22bm^2n}{-1.04m^2n} \Rightarrow \dfrac{-8.27bm^3n^2}{-1.04m^2n} - \dfrac{3.22bm^2n}{-1.04m^2n} \Rightarrow 7.9519b^{3-2}m^{3-2}n^{2-1} - \left(-3.0962bm^{2-2}n^{1-1}\right) = 7.95bmn + 3.10b$

13. $\dfrac{-21.5a^3b^2 + 31.2a^2b^3}{-14.8ab^2} \Rightarrow \dfrac{-21.5}{-14.8}a^{3-1}b^{2-2} + \dfrac{31.2}{-14.8}a^{2-1}b^{3-2} \Rightarrow 1.4527a^2 + (-2.1081ab) = 1.45a^2 - 2.10ab$

15. $\dfrac{x^2y^3z - xy^4z^2}{-xy^2z} \Rightarrow \dfrac{x^2y^3z}{-xy^2z} - \dfrac{xy^4z^2}{-xy^2z} = -xy + y^2z$

17. $\dfrac{m^2n^2 + m^3n^2 - m^2n^3}{-mn} \Rightarrow \dfrac{m^2n^2}{-mn} + \dfrac{m^3n^2}{-mn} - \dfrac{m^2n^3}{-mn} \Rightarrow -mn - m^2n + mn^2 = mn^2 - mn - m^2n$

19. $\dfrac{x^3y^3 - x^4y + xy^4}{-xy} \Rightarrow \dfrac{x^3y^3}{-xy} - \dfrac{x^4y}{-xy} + \dfrac{xy^4}{-xy} = -x^2y^2 + x^3 - y^3$

21. $\dfrac{c^3 - 4c^2d^2 + d^3}{cd^2} \Rightarrow \dfrac{c^3}{cd^2} - \dfrac{4c^2d^2}{cd^2} + \dfrac{d^3}{cd^2} \Rightarrow \dfrac{c^2}{d^2} - 4c + \dfrac{d}{c}$

Challenge Problems

23. $\dfrac{a^4+2a^2b^2-b^4}{a^2b^2} \Rightarrow \dfrac{a^4}{a^2b^2}+\dfrac{2a^2b^2}{a^2b^2}-\dfrac{b^4}{a^2b^2}=\dfrac{a^2}{b^2}+2-\dfrac{b^2}{a^2}$

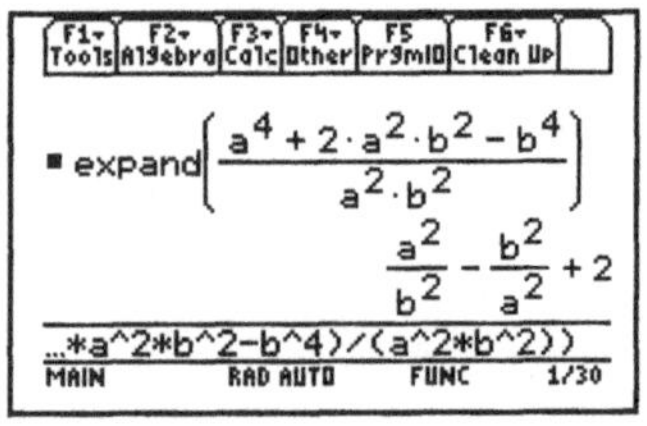

25. $\dfrac{4x^3z+2xz^2-3z^4}{-xz} \Rightarrow \dfrac{4x^3z}{-xz}+\dfrac{2xz^2}{-xz}-\dfrac{3z^4}{-xz}=-4x^2-2z+\dfrac{3z^3}{x}$

27. $\dfrac{p^3q^3+pq^2r^3-p^2r^4}{-p^2r} \Rightarrow \dfrac{p^3q^3}{-p^2r}+\dfrac{pq^2r^3}{-p^2r}-\dfrac{p^2r^4}{-p^2r}=-\dfrac{pq^3}{r}-\dfrac{q^2r^2}{p}+r^3$

29. $\dfrac{4c^4d+3c^2d^3-cd^5}{-cd^2} \Rightarrow \dfrac{4c^4d}{-cd^2}+\dfrac{3c^2d^3}{-cd^2}-\dfrac{cd^5}{-cd^2}=-\dfrac{4c^3}{d}-3cd+d^3$

31. $\dfrac{8b^2c^4+4b^2c-12b^3c^3}{-b^3c^2} \Rightarrow \dfrac{8b^2c^4}{-b^3c^2}+\dfrac{4b^2c}{-b^3c^2}-\dfrac{12b^3c^3}{-b^3c^2}=-\dfrac{8c^2}{b}-\dfrac{4}{bc}+12c$

Exercise 11 ◊ Dividing a Polynomial by a Polynomial

1. $a^2+15a+56 \div a+7$

$$\begin{array}{r} a+8 \\ a+7\overline{)a^2+15a+56} \\ -(a^2+\ \ 7a)\ \downarrow \\ \hline 8a+56 \\ -(8a+56) \\ \hline 0 \end{array}$$

Thus, $a^2+15a+56 \div a+7 = a+8$

Some problems may also be done by factoring (which you will see in Chapter 10)

$$\frac{a^2+15a+56}{a+7} \Rightarrow \frac{(a+8)(a+7)}{a+7} \Rightarrow \frac{(a+8)\cancel{(a+7)}}{\cancel{a+7}} = a+8$$

3. $a^2+a-56 \div a-7$

$$\begin{array}{r} a+8 \\ a-7\overline{)a^2+\ \ a-56} \\ -(a^2-7a)\ \downarrow \\ \hline 8a-56 \\ -(8a-56) \\ \hline 0 \end{array}$$

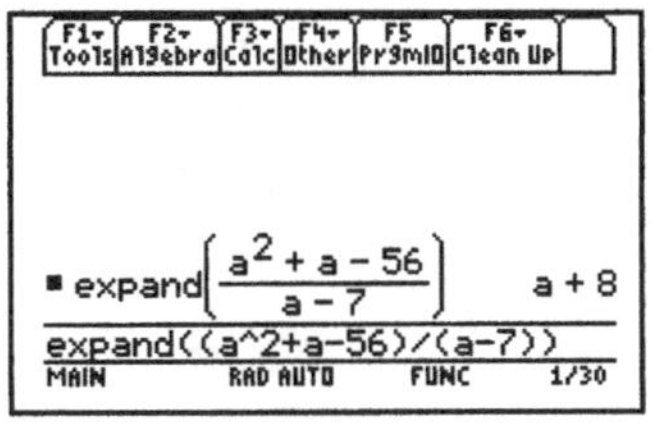

5. $2x^2 + 11x + 5 \div 2x + 1$

$$\begin{array}{r} x + 5 \\ 2x+1 \overline{\smash{\big)}\, 2x^2 + 11x + 5} \\ -\underline{\left(2x^2 + x\right)} \downarrow \\ 10x + 5 \\ -\underline{\left(10x + 5\right)} \\ 0 \end{array}$$

7. $a^2 - 15a + 56 \div a - 7$

$$\begin{array}{r} a - 8 \\ a-7 \overline{\smash{\big)}\, a^2 - 15a + 56} \\ -\underline{\left(a^2 - 7a\right)} \downarrow \\ -8a + 56 \\ -\underline{\left(-8a + 56\right)} \\ 0 \end{array}$$

9. $3x^2 - 4x - 4 \div 2 - x$

$$\begin{array}{r} -3x - 2 \\ 2-x \overline{\smash{\big)}\, 3x^2 - 4x - 4} \\ -\underline{\left(3x^2 - 6x\right)} \downarrow \\ 2x - 4 \\ -\underline{\left(2x - 4\right)} \\ 0 \end{array}$$

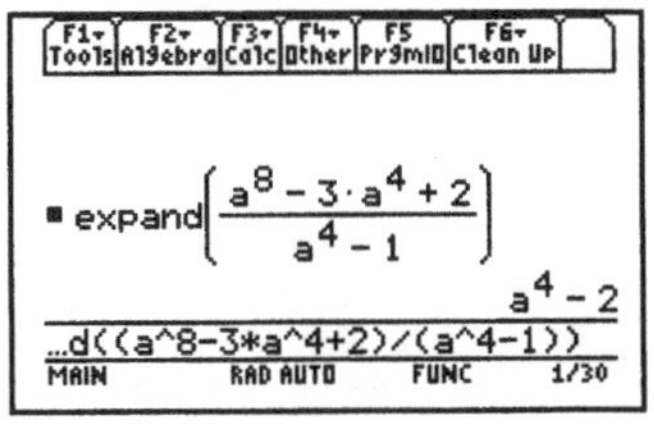

11. $a^3 - 8a - 3 \div a - 3$

$$
\begin{array}{r}
a^2+3a+1 \\
a-3\overline{\big)a^3+0a^2-8a-3} \\
-\left(a^3-3a^2\right)\downarrow\quad\downarrow \\
\hline
3a^2-8a\quad\downarrow \\
-\left(3a^2-9a\right)\downarrow \\
\hline
a-3 \\
-\left(a-3\right) \\
\hline
0
\end{array}
$$

Challenge Problems

13. $x^2-4x+3\div x+2$

$$
\begin{array}{r}
x-6 \\
x+2\overline{\big)x^2-4x+3} \\
-\left(x^2+2x\right)\downarrow \\
\hline
-6x+\ 3 \\
-\left(-6x-12\right) \\
\hline
15
\end{array}
$$

Thus $x^2-4x+3\div x+2=x-6+\dfrac{15}{x+2}$

15. $4+2x-5x^2\div 3-x$

$$
\begin{array}{r}
5x+13 \\
-x+3\overline{\big)-5x^2+\ 2x+\ 4} \\
-\left(-5x^2+15x\right)\ \downarrow \\
\hline
-13x+\ 4 \\
-\left(-13x+39\right) \\
\hline
-35
\end{array}
\quad \text{(write the terms in descending order of the powers)}
$$

Thus $4+2x-5x^2\div 3-x=5x+13-\dfrac{35}{3-x}$

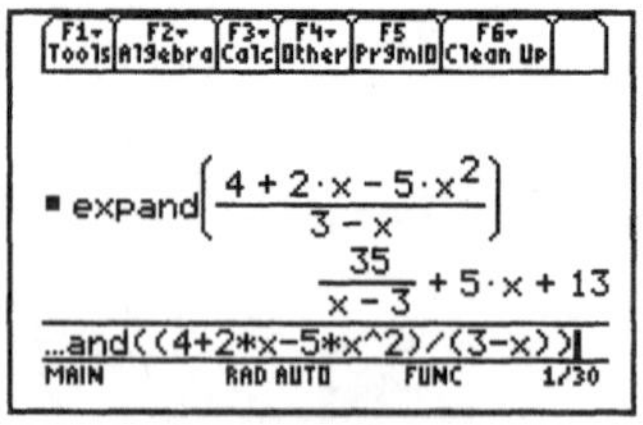

CHAPTER 2 REVIEW PROBLEMS

1. $\left(b^4+b^2x^3+x^4\right)\left(b^2-x^2\right)\Rightarrow b^4\left(b^2\right)+b^4\left(-x^2\right)+b^2x^3\left(b^2\right)+b^2x^3\left(-x^2\right)+x^4\left(b^2\right)+x^4\left(-x^2\right)$

$=b^6-b^4x^2+b^4x^3-b^2x^5+b^2x^4-x^6$

3. $\left(7.28\times10^4\right)^3\Rightarrow(7.28)^3\times\left(10^4\right)^3\Rightarrow 385.82835\times10^{12}=3.86\times10^{14}$

5. $(3x-m)\left(x^2+m^2\right)(3x-m)\Rightarrow(3x-m)(3x-m)\left(x^2+m^2\right)\Rightarrow\left(9x^2-6mx+m^2\right)\left(x^2+m^2\right)$

$\Rightarrow 9x^2\left(x^2\right)+9x^2\left(m^2\right)-6mx\left(x^2\right)-6mx\left(m^2\right)+m^2\left(x^2\right)+m^2\left(m^2\right)\Rightarrow 9x^4+10x^2m^2-6mx^3-6m^3x+m^4$

7. $(2x+1)^3 \Rightarrow (2x+1)(2x+1)(2x+1) \Rightarrow (4x^2+4x+1)(2x+1) \Rightarrow 4x^2(2x)+4x^2(1)+4x(2x)+4x(1)+1(2x)+1(1)$

$\Rightarrow 8x^3+4x^2+8x^2+4x+2x+1 = 8x^3+12x^2+6x+1$

9. $(3ax^2)(2ax^3) \Rightarrow (3)(2)(a^{1+1})(x^{2+3}) = 6a^2x^5$

11. $\dfrac{a^2x-abx-acx}{ax} \Rightarrow \dfrac{a^2x}{ax}-\dfrac{abx}{ax}-\dfrac{acx}{ax} \Rightarrow a^{2-1}x^{1-1}-a^{1-1}bx^{1-1}-a^{1-1}cx^{1-1} = a-b-c$

13. $(4a-3b)^2 \Rightarrow (4a-3b)(4a-3b) \Rightarrow 16a^2-12ab-12ab+9b^2 = 16a^2-24ab+9b^2$

15. $(xy-2)(xy-4) \Rightarrow xy(xy)+xy(-4)-2(xy)-2(-4) \Rightarrow x^2y^2-4xy-2xy+8 = x^2y^2-6xy+8$

17. $(3x+2y)^2 \Rightarrow (3x+2y)(3x+2y) \Rightarrow 9x^2+6xy+6xy+4y^2 = 9x^2+12xy+4y^2$

19. $\dfrac{a^3b^2-a^2b^5-a^4b^2}{a^2b} \Rightarrow \dfrac{a^3b^2}{a^2b}-\dfrac{a^2b^5}{a^2b}-\dfrac{a^4b^2}{a^2b} = ab-b^4-a^2b$

21. $(2m-c)(2m+c)(4m^2+c^2) \Rightarrow (4m^2-c^2)(4m^2+c^2) = 16m^4-c^4$

23. $\dfrac{2a^6}{a^4} \Rightarrow 2a^{6-4} = 2a^2$

25. $y-3[y-2(4-y)] \Rightarrow y-3[y-8+2y] \Rightarrow y-3[3y-8] \Rightarrow y-9y+24 = -8y+24$

27. $(2x^2+xy-2y^2)(3x+3y) \Rightarrow 2x^2(3x)+2x^2(3y)+xy(3x)+xy(3y)-2y^2(3x)-2y^2(3y)$

$= 6x^3+9x^2y-3xy^2-6y^3$

29. $-2[w-3(2w-1)]+3w \Rightarrow -2[w-6w+3]+3w \Rightarrow -2[-5w+3]+3w \Rightarrow 10w-6+3w = 13w-6$

31. $(a^4-2a^3c+4a^2c^2-8ac^3+16c^4)(a+2c) \Rightarrow a^4(a)+a^4(2c)-2a^3c(a)-2a^3c(2c)+4a^2c^2(a)+4a^2c^2(2c)$

$-8ac^3(a)-8ac^3(2c)+16c^4(a)+16c^4(2c) = a^5+32c^5$

33. $\dfrac{(a-c)^m}{(a-c)^2} = (a-c)^{m-2}$

35. $\dfrac{-x^2y-xy^2}{-xy} \Rightarrow \dfrac{-x^2y}{-xy}-\dfrac{xy^2}{-xy} = x+y$

37. $(b-3)^3 \Rightarrow (b-3)(b-3)(b-3) \Rightarrow (b^2-6b+9)(b-3) \Rightarrow b^2(b)+b^2(-3)-6b(b)-6b(-3)+9(b)+9(-3)$

$\Rightarrow b^3-3b^2-6b^2+18b+9b-27 = b^3-9b^2+27b-27$

39. $\dfrac{2x^{-2}y^3}{4x^{-4}y^6} \Rightarrow \dfrac{2\left(\frac{1}{x^2}\right)y^3}{4\left(\frac{1}{x^4}\right)y^6} \Rightarrow \dfrac{\frac{2y^3}{x^2}}{\frac{4y^6}{x^4}} \Rightarrow \dfrac{2y^3}{x^2} \div \dfrac{4y^6}{x^4} \Rightarrow \left(\dfrac{2y^3}{x^2}\right)\left(\dfrac{x^4}{4y^6}\right) \Rightarrow \dfrac{2x^4y^3}{4x^2y^6} = \dfrac{x^2}{2y^3}$

41. $(x-2)(x+4) \Rightarrow x^2+4x-2x-8 = x^2+2x-8$

43. $\dfrac{a^2b^2-2ab-3ab^3}{ab} \Rightarrow \dfrac{a^2b^2}{ab}-\dfrac{2ab}{ab}-\dfrac{3ab^3}{ab} = ab-2-3b^2$

45. $\dfrac{6a^3x^2-15a^4x^2+30a^3x^3}{-3a^3x^2} \Rightarrow \dfrac{6a^3x^2}{-3a^3x^2}-\dfrac{15a^4x^2}{-3a^3x^2}+\dfrac{30a^3x^3}{-3a^3x^2} = -2+5a-10x$

47. $(2.83\times10^{3})^{2}\Rightarrow(2.83)^{2}\times10^{3\cdot2}\Rightarrow8.0089\times10^{6}=8.01\times10^{6}$

49. $(2xy^{3})(5x^{2}y)\Rightarrow(2)(5)x^{1+2}y^{3+1}=10x^{3}y^{4}$

51. $(x-1)(x^{2}+4x)\Rightarrow x^{3}+4x^{2}-x^{2}-4x=x^{3}+3x^{2}-4x$

53. $(1.33\times10^{4})^{2}\Rightarrow(1.33)^{2}\times10^{4\cdot2}\Rightarrow1.7689\times10^{8}=1.77\times10^{8}$

$\Rightarrow y+1-y^{2}-4y-4=-y^{2}-3y-3$

55. $(2ab^{2})(3a^{2}b)\Rightarrow(2)(3)a^{1+2}b^{2+1}=6a^{3}b^{3}$

57. $x^{8}+x^{4}+1\div x^{4}-x$

$$
\begin{array}{r}
x^{4}+x+1 \\
x^{4}-x\,\big)\overline{x^{8}+0x^{5}+x^{4}+1} \\
-(x^{8}-x^{5}) \\
\hline
x^{5}+x^{4} \\
-(x^{5}-x^{2}) \\
\hline
x^{4}+x^{2} \\
-(x^{4}-x) \\
\hline
x^{2}+x+1
\end{array}
$$

Thus $x^{8}+x^{4}+1\div x^{4}-x=x^{4}+x+1+\dfrac{x^{2}+x+1}{x^{4}-x}$

59. $0.85x+0.72(750-x)\Rightarrow0.85+540-0.72x=0.13x+540$

61. $16.1\left(\frac{t}{2}\right)^{2}\Rightarrow16.1\left(\frac{t}{2}\right)\left(\frac{t}{2}\right)\Rightarrow16.1\left(\frac{t^{2}}{4}\right)\Rightarrow\frac{16.1t^{2}}{4}\Rightarrow4.025t^{2}=4.02t^{2}$ ft

Chapter 3: Simple Equations and Word Problems

Exercise 3.1 • Solving Simple Equations

Solve and check each equation. Treat the constants in these equations as exact numbers. Leave your answers in fractional, rather than decimal, form. [See Examples 5 and 6]

1.
$$x+9=16$$
$$x+9-9=16-9$$
$$x=7$$

3.
$$30+5x=20x$$
$$30=20x-5x$$
$$30=15x$$
$$\frac{15x}{15}=\frac{30}{15}$$
$$x=2$$

5.
$$4t+9=11t-3t$$
$$4t=8t-9$$
$$4t-8t=-9$$
$$-4t=-9$$
$$t=\frac{-9}{-4}$$
$$t=\frac{9}{4}$$

7.
$$x+9=5$$
$$x=5-9$$
$$x=-4$$

9.
$$5x+8=9x$$
$$8=9x-5x$$
$$8=4x$$
$$x=2$$

11.
$$10-5y=1-y$$
$$-5y+y=1-10$$
$$-4y=-9$$
$$y=\frac{9}{4}$$

13.
$$x-5=6$$
$$x=6+5$$
$$x=11$$

15.
$$7x-3=5x+1$$
$$7x-5x=1+3$$
$$2x=4$$
$$x=2$$

17.
$$y-4=0$$
$$y=4$$

19.
$$4m-5=10m-2$$
$$4m-10m=-2+5$$
$$-6m=3$$
$$m=-\frac{1}{2}$$

21.
$$6x+4=3x+19$$
$$6x-3x=19-4$$
$$3x=15$$
$$x=5$$

23.
$$17-14x=8-11x$$
$$-14x+11x=8-17$$
$$-3x=-9$$
$$x=3$$

25.
$$5x-10+13=18$$
$$5x+3=18$$
$$5x=18-3$$
$$5x=15$$
$$x=3$$

27.
$$49-5y=3y-7$$
$$-5y-3y=-7-49$$
$$-8y=-56$$
$$y=7$$

29.
$$4x-111=21+14x$$
$$4x-14x=21+111$$
$$-10x=132$$
$$x=-\frac{66}{5}$$

31.
$$16+2z=11-5z$$
$$2z+5z=11-16$$
$$7z=-5$$
$$z=-\frac{5}{7}$$

Equations Having Symbols of Grouping [See Example 7]

33.
$$\begin{aligned} 3(y-5)+2(3+y)&=21\\ 3y-15+6+2y&=21\\ 5y-9&=21\\ 5y&=21+9\\ 5y&=30\\ y&=6 \end{aligned}$$

35.
$$\begin{aligned} 4(y-5)&=2(y-10)\\ 4y-20&=2y-20\\ 4y-2y&=-20+20\\ 2y&=0\\ y&=0 \end{aligned}$$

37.
$$\begin{aligned} 3(15-t)&=3(5+t)\\ 45-3t&=15+3t\\ -3t-3t&=15-45\\ -6t&=-30\\ t&=5 \end{aligned}$$

39.
$$\begin{aligned} 3(2x+1)&=7\\ 6x+3&=7\\ 6x&=7-3\\ 6x&=4\\ x&=\frac{2}{3} \end{aligned}$$

41.
$$\begin{aligned} 5x+6&=6(x-3)+2\\ 5x+6&=6x-18+2\\ 5x+6&=6x-16\\ 5x-6x&=-16-6\\ -x&=-22\\ x&=22 \end{aligned}$$

43.
$$\begin{aligned} 19y-60&=4(y+15)\\ 19y-60&=4y+60\\ 19y-4y&=60+60\\ 15y&=120\\ y&=8 \end{aligned}$$

45.
$$\begin{aligned} 5+3(x-2)&=58\\ 5+3x-6&=58\\ 3x-1&=58\\ 3x&=59\\ x&=\frac{59}{3} \end{aligned}$$

47.
$$\begin{aligned} 4(x-5)-2(6x+3)&=22\\ 4x-20-12x-6&=22\\ -8x-26&=22\\ -8x&=22+26\\ -8x&=48\\ x&=-6 \end{aligned}$$

49.
$$\begin{aligned} 2(4w-3)&=3(5w+2)\\ 8w-6&=15w+6\\ 8w-15w&=6+6\\ -7w&=12\\ w&=-\frac{12}{7} \end{aligned}$$

51.
$$\begin{aligned} 5a&=-2(13-9a)\\ 5a&=-26+18a\\ 5a-18a&=-26\\ -13a&=-26\\ a&=2 \end{aligned}$$

53.
$$\begin{aligned} 6x+5&=7(x-3)\\ 6x+5&=7x-21\\ 6x-7x&=-21-5\\ -x&=-26\\ x&=26 \end{aligned}$$

Simple Fractional Equations [See Examples 8, 9, 10, and 12]

55. $\frac{1}{3a}=7 \quad [a \neq 0]$

$3a\left(\frac{1}{3a}\right)=3a(7)$ [the LCD is $3a$]

$1=21a$

$a=\frac{1}{21}$

57. $\frac{x}{4}-3=11$ [the LCD is 4]

$4\left(\frac{x}{4}-3\right)=4\cdot 11$

$x-12=44$

$x=56$

59. $\frac{5x}{4}=2x-3$ [the LCD is 4]

$4\left(\frac{5x}{4}\right)=4(2x-3)$

$5x=8x-12$

$-3x=-12$

$x=4$

61. $\frac{2x-4}{7}=\frac{2-2x}{4}$ [the LCD is 28]

$28\left(\frac{2x-4}{7}\right)=28\left(\frac{2-2x}{4}\right)$

$4(2x-4)=7(2-2x)$

$8x-16=14-14x$

$8x+14x=14+16$

$22x=30$

$x=\frac{15}{11}$

Equations With Approximate Numbers [See Example 11]
Solve for x. Round your answer to the proper number of significant digits.

63. $24.8x-28.4=0$

$24.8x=28.4$

$x=\frac{28.4}{24.8}=1.15$

rounded to three significant digits

65. $3.82=29.3+3.28x$

$3.28x=3.82-29.3$

$3.28x=-25.48$

$x=\frac{-25.48}{3.28}=-7.77$

rounded to three significant digits

67. $382x+827=625-846x$

$382x+846x=625-827$

$1228x=-202$

$x=\frac{-202}{1228}=-0.164$

rounded to three significant digits

69. $9.38(5.82+x)=23.8$

$54.5916+9.38x=23.8$

$9.38x=-30.7916$

$x=\frac{-30.7916}{9.38}=-3.28$

rounded to three significant digits

71. $92.1(x-2.34)=82.7(x-2.83)$

$92.1x-215.514=82.7x-234.041$

$92.1x-82.7x=-234.041+215.514$

$9.40x=-18.527$

$x=\frac{-18.527}{9.40}=-1.97$

rounded to three significant digits

Simple Literal Equations [See Examples 13 & 14]
Solve for x.

73. $6+bx=b-3$

$$bx=b-9$$
$$x=\frac{b-9}{b}$$

75. $ax+4=b-3$

$$ax=b-7$$
$$x=\frac{b-7}{a}$$

Challenge Problems

77. $6-3(2x+4)-2x=7x+4(5-2x)-8$

$$6-6x-12-2x=7x+20-8x-8$$
$$-6-8x=12-x$$
$$-8x+x=12+6$$
$$-7x=18$$
$$x=-\frac{18}{7}$$

79. $3x-2+x(3-x)=(x-3)(x+2)-x(2x+1)$

$$3x-2+3x-x^2=x^2-x-6-2x^2-x$$
$$6x-2=-2x-6$$
$$6x+2x=-6+2$$
$$8x=-4$$
$$x=-\frac{1}{2}$$

81. $x+(1+x)(3x+4)=(2x+3)(2x-1)-x(x-2)$

$$x+3x^2+7x+4=4x^2+4x-3-x^2+2x$$
$$8x+4=6x-3$$
$$2x=-7$$
$$x=-\frac{7}{2}$$

83. $7x+6(2-x)+3=4+3(6+x)$

$$7x+12-6x+3=4+18+3x$$
$$x+15=22+3x$$
$$-2x=7$$
$$x=-\frac{7}{2}$$

85. $7r-2r(2r-3)-2=2r^2-(r-2)(3+6r)-8$

$$7r-4r^2+6r-2=2r^2-6r^2+9r+6-8$$
$$13r-2=9r-2$$
$$4r=0$$
$$r=0$$

Applications

87. $0.0525x+0.0284(3.25-x)=0.0415(3.25)$

$$0.0525x+0.0923-0.0284x=0.134875$$
$$0.0241x=0.042575$$
$$x=1.77\text{ tons}$$

89. $\frac{17}{12}(4.0+x)+\frac{11}{5.0}x=25$

$$60\left[\frac{17}{12}(4.0+x)+\frac{11}{5.0}x\right]=60[25]$$
$$85(4.0+x)+132x=1500$$
$$340+85x+132x=1500$$
$$217x=1160$$
$$x=5.3\text{ months}$$

Exercise 3.2 • Solving Word Problems - Number Puzzles

Identify an unknown and rewrite each expression as an algebraic expression.

1. If the unknown is x, then the expression is $3x+10$.

3. If the first number is x, then the second number is $x+42$ or $x-42$.

5. If the unknown is x, then the fraction is $\frac{x}{6x+4}$.

7. If x is number of gallons of mixture, then $0.11x$ is the number of gallons of antifreeze.

9. Let x be the length of a rectangle and w be the width.

a) $w+w+x+x=64$
$$2w=64-2x$$
$$w=\frac{64-2x}{2}$$
$$w=32-x$$

b) $xw=320$
$$w=\frac{320}{x}$$

Solve each problem for the required quantity. [See Examples 25 and 27]

11. Let the number be x.
Solution: $16+x=3x$
$$16=2x$$
$$x=8$$

13. Let the number be x.
Solution: $45+x=6x$
$$45=5x$$
$$x=9$$

15. Let the number be x.
Solution: $3+7x=9x-7$
$$3+7=9x-7x$$
$$10=2x$$
$$x=5$$

Exercise 3.3 • Uniform Motion Applications

rate × time = distance

1. Let t represent the time since the planes departed.

	Rate (mi/h)	Time (h)	Distance (mi)
Plane 1	252	t	$252t$
Plane 2	266	t	$266t$

The planes need to be 1750 miles apart which will be the total combined distance the planes traveled
$$252t+266t=1750$$
$$518t=1750$$
$$t=\frac{1750}{518}=3.38\text{ h}$$
The planes take 3.38 hours to be 1750 miles apart.

3. Let t represent the time for the express train.

	Rate (km/h)	Time (h)	Distance (km)
Train 1	22.5	$t+4.25$	$22.5(t+4.25)$
Express Train	85.5	t	$85.5t$

The distance is the same for both trains.

$$22.5(t+4.25)=85.5t$$
$$22.5t+95.625=85.5t$$
$$95.625=85.5t-22.5t$$
$$63.0t=95.625$$
$$t=\frac{95.625}{63.0}=1.52\text{ h}$$

The express travels approximately 1.52 hours and the distance traveled by both trains is $85.5(1.52)=130\text{ km}$.

5. Let t represent the time since the submarines left the same spot.

	Rate (km/day)	Time (day)	Distance (km)
Sub 1	115	t	$115t$
Sub 2	182	t	$182t$

The distance between the submarines will be 1470 km.

$$115t+182t=1470$$
$$297t=1470$$
$$t=\frac{1470}{297}=4.95\text{ days}$$

It will take the submarines 4.95 days to be 1470 km apart.

7. Let t represent the time the ships take to reach the oil slick.

	Rate (mi/day)	Time (day)	Distance (mi)
Ships	525	t	$525t$
Oil Slick	10.5	$t+2$	$10.5(t+2)$

The distance between the ships and the oil slick is 354 miles.

$$525t+10.5(t+2)=354$$
$$525t+10.5t+21=354$$
$$535.5t=333$$
$$t=\frac{333}{535.5}=0.622\text{days}$$

It will take the ships 0.622 days to reach the oil slick. The oil slick will be $525(0.622)=327$ miles from the beach.

9. Let t represent the time since Spacecraft B passed over Houston.

	Rate (km/h)	Time (h)	Distance (km)
Spacecraft A	275	$t+1.25$	$275(t+1.25)$
Spacecraft B	444	t	$444t$

The distance each of the spacecrafts travels is the same.

$$275(t+1.25)=444t$$
$$275t+343.75=444t$$
$$343.75=169t$$
$$t=\frac{343.75}{169}=2.03\text{ h}$$

In order to find the time the decimal part of the hour (0.03) must be converted to minutes: $.03(60\text{ min/h})=1.8\text{ min.}\approx 2\text{ min.}$

The time at which Spacecraft B overtakes Spacecraft A is 3:17 P.M. (1:15 + 2:02).
The distance from Houston is $444(2.03)=901\text{ km}$.

Exercise 3.4 • Financial Applications

1. Let $x=$ number of technicians.
Then $17-x=$ number of helpers.
$$210x+185(17-x)=3345$$
$$210x+3145-185x=3345$$
$$25x=200$$
$$x=8$$
There were 8 technicians employed on the project.

3. Let $x=$ amount before taxes.
$$(1-0.27)x=895000$$
$$0.73x=895000$$
$$x=\frac{895000}{0.73}=1226027$$
The company would need to make $1,226,027 before taxes.

5. Let $x=$ amount of water (in 1000 gallons).
Old rate is $1.95x$.
New rate is $1.16x+45$.
$$1.95x=1.16x+45$$
$$0.79x=45$$
$$x=\frac{45}{0.79}=57$$
57,000 gallons must be purchased.

7. Let $x=$ amount for the skis.
Then $4x=$ amount for the boots.
$$x+4x=210$$
$$5x=210$$
$$x=\frac{210}{5}=42$$
The skis sold for $42 and the boots sold for 4(42)=$168.

9. Let $x=$ annual income
$\frac{1}{4}x=$ amount for board
$\frac{1}{12}x=$ amount for clothes
$\frac{1}{2}x=$ amount for other expenses
$$x-\frac{1}{4}x-\frac{1}{12}x-\frac{1}{2}x=10000$$
$$12\left(x-\frac{1}{4}x-\frac{1}{12}x-\frac{1}{2}x\right)=12(10000)$$
$$12x-3x-x-6x=120000$$
$$2x=120000$$
$$x=60000$$
She has an annual income of $60,000.

11. Let $x=$ amount of lumber bought.
$$(1-0.07)x=4285$$
$$0.93x=4285$$
$$x=\frac{4285}{0.93}=\$4608$$
The carpenter needs to buy $4608 worth of wood.

13. Let $x=$ amount invested at 6.75%.
Then $173924-x=$ amount invested at 8.24%.
$$0.0675x+0.0824(173924-x)=13824$$
$$0.0675x+14331-0.0824x=13824$$
$$-0.0149x=-507$$
$$x=\frac{-507}{-0.0149}=34027$$
There is $34,027 invested at 6.75% and $173,924-$34,027=$139,897 invested at 8.24%.

15. Let $x=$ amount for the computer.
Then $1.5x=$ amount for the printer.
$$x+1.5x=995$$
$$2.5x=995$$
$$x=\frac{995}{2.5}=398$$
The computer sold for $398, and the printer sold for 1.5($398)=$597.

Exercise 3.5 • Mixture Applications

1. Let x represent the amount of the 12% mixture.

	% Alcohol	Amount (gallons)	Amount of Alcohol (gallons)
5% Alcohol	5	252	$0.05(252)$
12% Alcohol	12	x	$0.12x$
Final Mix	9	$252+x$	$0.09(252+x)$

Then adding the two original mixture amounts and setting equal to the final mixture gives the equation solved below.

$$0.05(252)+0.12x=0.09(252+x)$$
$$12.6+0.12x=22.68+0.09x$$
$$0.03x=10.08$$
$$x=\frac{10.08}{0.03}=336 \text{ gal}$$

336 gallons of the 5% mixture must be added to the 12% mixture.

3. Let x represent the amount of nickel silver with 18% zinc.
Then $706-x$ is the amount of nickel silver with 31% zinc.

	% Zinc	Zinc (kg)	Amount of Zinc (kg)
18% Zinc	18	x	$0.18x$
31% Zinc	31	$706-x$	$0.31(706-x)$
Final Mix	22	706	$0.22(706)$

The resulting equation is solved below.

$$0.18x+0.31(706-x)=0.22(706)$$
$$0.18x+218.86-0.31x=155.32$$
$$-0.13x=-63.54$$
$$x=\frac{-63.54}{-0.13}=489 \text{ kg}$$

The final nickel steel will be made up by mixing 489 kg of the 18% variety with $706-489=217$ kg of the 31% variety.

5. Let b represent the amount of brass containing 63% copper.

	% Copper	Amount (kg)	Amount of Copper (kg)
63% Copper	63	b	$0.63b$
72% Copper	72	1120	$0.72(1120)$
Final Bronze	67	$b+1120$	$0.67(b+1120)$

The resulting equation is solved below.

$$0.63b+0.72(1120)=0.67(b+1120)$$
$$0.63b+806.4=0.67b+750.4$$
$$56.0=0.04b$$
$$b=\frac{56.0}{0.04}=1400 \text{ kg}$$

1400 kg of the brass containing 63% copper must be added to the 72% mixture.

7. Let a represent the amount of coolant that is replaced with pure antifreeze.

	% Antifreeze	Amount (liters)	Amount of Antifreeze (liters)
15% Antifreeze	15	$11-a$	$0.15(11-a)$
Pure Antifreeze	100	a	$1.00a$
Final Coolant	25	11	$0.25(11)$

The resulting equation is solved below.

$$0.15(11-a)+1.00a=0.25(11)$$
$$1.65-0.15a+1.00a=2.75$$
$$0.85a=1.10$$
$$a=\frac{1.10}{0.85}=1.29 \text{ liters}$$

1.29 liters of the 15% antifreeze coolant will have to be replaced with pure antifreeze.

9. Let p represent the amount of water allowed to evaporate from the paint that contains 20% solids suspended in water.

	% Water	Amount (lb)	Amount of Water (lb)
20% Solids	80	315	$0.80(315)$
Water	100	p	$1.00p$
Final Paint	75	$315-p$	$0.75(315-p)$

The resulting equation is solved below.

$$0.80(315)-1.00p=0.75(315-p)$$
$$252-1.00p=236.25-0.75p$$
$$15.75=0.25p$$
$$p=\frac{15.75}{0.25}=63.0 \text{ lb}$$

63 pounds of water must evaporate.

11. Let x represent the amount of sand.

	% Sand	Amount (lb)	Amount of Sand (lb)
29% Sand	29	642	$0.29(642)$
Pure Sand	100	x	$1.00x$
Final Concrete	35	$642+x$	$0.35(642+x)$

The resulting equation is solved below.

$$0.29(642)+1.00x=0.35(642+x)$$
$$186.18+1.00x=224.7+0.35x$$
$$0.65x=38.52$$
$$x=\frac{38.52}{0.65}=59.3 \text{ lb}$$

59.3 pounds of sand must be added to the concrete.

Exercise 3.6 • Static Applications

1.

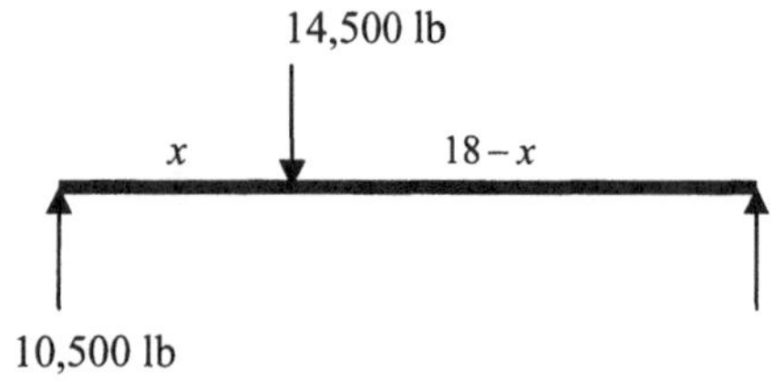

a) Take moments about the right end

$$14500(18-x)=10500(18)$$
$$261000-14500x=189000$$
$$72000=14500x$$
$$x=\frac{72000}{14500}=4.97 \text{ ft}$$

b) Using the figure from problem 1,

$$R+10500=14500$$
$$R=4000 \text{ lb}$$

3. Let x be the distance from the balance point to the right end.

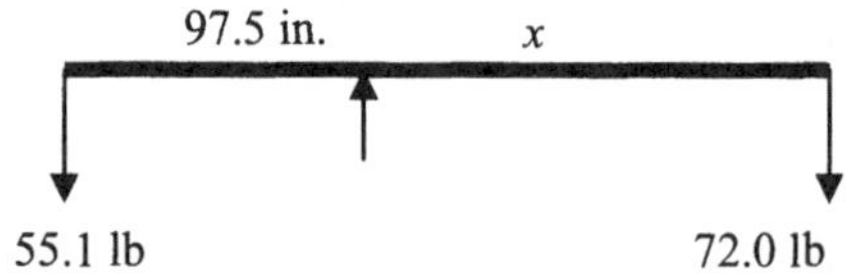

$$55.1(97.5)=72.0x$$
$$x=\frac{55.1(97.5)}{72.0}$$
$$x=74.6$$

Length of Bar $=97.5+74.6=172$ in.

5.

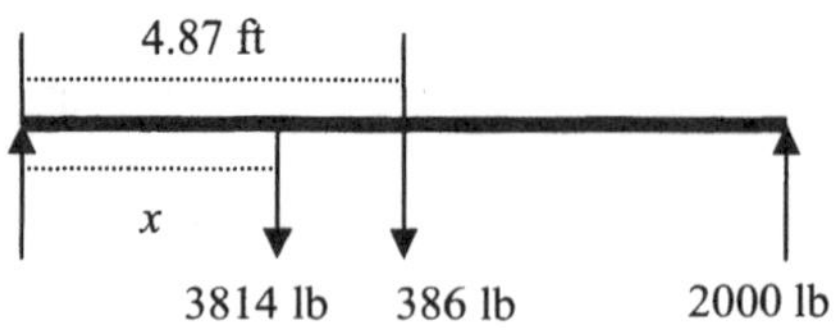

$$3814x+386(4.87)=2000(9.74)$$
$$3814x+1879.82=19480$$
$$3814x=17600.18$$
$$x=4.61 \text{ ft}$$

7.

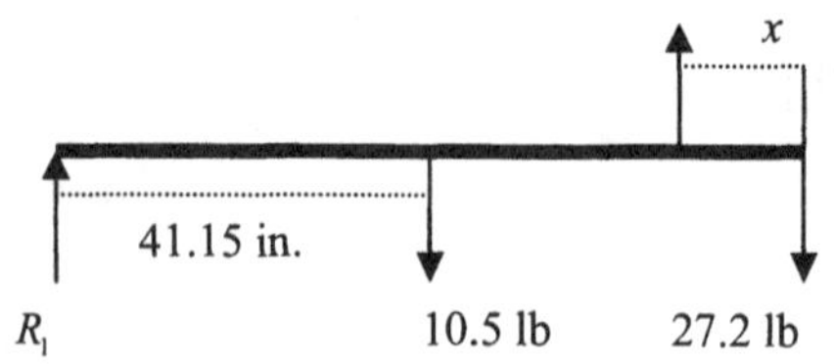

The tension can be found by adding the two weights.

$T=10.5+27.2=37.7$ lb

Then find x by solving the following equation.

$$10.5(41.15)+27.2(82.3)=37.7(82.3-x)$$
$$2670.635=3102.71-37.7x$$
$$-432.075=-37.7x$$
$$x=11.5 \text{ in.}$$

Exercise 3.7 • Work, Fluid Flow, and Energy Flow Applications

Work Problems [See Example 40]

[*amount of work done = rate of work × time worked*]

1. Let t be the time the job takes for all three working together.

   Laborer 1: rate = $\frac{1}{5}$ job per day

   Laborer 2: rate = $\frac{1}{6}$ job per day

   Laborer 3: rate = $\frac{1}{8}$ job per day

   The amount of work done by each laborer is the laborer rate times the time worked (t days). The total work done by the 3 laborers is 1 job.

   $$\frac{1}{5}t+\frac{1}{6}t+\frac{1}{8}t=1 \quad \text{[the LCD is 120]}$$
   $$24t+20t+15t=120$$
   $$59t=120$$
   $$t=\frac{120}{59}\approx 2 \text{ days}$$

3. Let t be the time it takes the two working together to roof the house.

   Roofer 1: rate = $\frac{1}{10}$ house per day

   Roofer 2: rate = $\frac{1}{14}$ house per day

   Total work done by the 2 roofers is 1 roof.

   $$\frac{1}{10}t+\frac{1}{14}t=1 \quad \text{[the LCD is 70]}$$
   $$7t+5t=70$$
   $$12t=70$$
   $$t=\frac{70}{12}=\frac{35}{6} \text{ or } 5\frac{5}{6} \text{ days}$$

5. Let t be the time it takes the new machine to produce a box of parts.

   Old Machine: rate = $\frac{1}{3.3}$ boxes per hour

   New Machine: rate = $\frac{1}{t}$ boxes per hour

   Together: rate = $\frac{1}{1.4}$ boxes per hour

   Set up the equation to find the work done in 1 hour.

   $$\frac{1}{3.3}\cdot 1+\frac{1}{t}\cdot 1=\frac{1}{1.4}\cdot 1 \quad \text{[Multiply both sides by } 4.62t\text{]}$$
   $$1.4t+4.62=3.3t$$
   $$4.62=1.9t$$
   $$t=\frac{4.62}{1.9}\approx 2.4 \text{ hours}$$

Fluid Flow Problems

[*amount of flow = flow rate × duration of flow*]

7. Let t be the time the job takes if both pipes are flowing.

   Pipe 1: rate = $\frac{1}{8.0}$ tank per hour

   Pipe 2: rate = $\frac{1}{9.0}$ tank per hour

   Together the pipes work on 1 tank.

   $$\frac{1}{8.0}t+\frac{1}{9.0}t=1 \quad \text{[the LCD is 72]}$$
   $$9.0t+8.0t=72$$
   $$17.0t=72$$
   $$t=\frac{72}{17.0}\approx 4.2 \text{ hours}$$

9. Let t be the time that the large pipe spent filling the tank.
Then $t-5$ is the time that the small pipe spent filling the tank.

Large Pipe: rate = $\frac{1}{18.0}$ tank per hour

Small Pipe: rate = $\frac{1}{24.0}$ tank per hour

Together the pipes work on 1 tank.

$$\frac{1}{18.0}t+\frac{1}{24.0}(t-5)=1 \quad \text{[the LCD is 72]}$$
$$4.0t+3.0(t-5)=72$$
$$4.0t+3.0t-15.0=72$$
$$7.0t=87$$
$$t=\frac{87}{7.0}\approx 12.4 \text{ hours}$$

Energy Flow Problems [See Example 41]
[*amount of energy = rate of energy flow × time of flow*]

11. Let t be the time the coal stockpile is used up after the second boiler begins operation.

$$\frac{1500}{4.0}(t+3)+\frac{2300}{3.0}t=10000 \quad \text{[the LCD is 12]}$$
$$4500(t+3)+9200t=120000$$
$$4500t+13500+9200t=120000$$
$$13700t=106500$$
$$t\approx 7.8 \text{ weeks}$$

13. Let t be the number of winters needed for the insulated house to use 1250 gal of oil. Then figure the amount of energy consumed in 1 winter.

$$\frac{2100}{3}\cdot 1+\frac{1250}{t}\cdot 1=\frac{1850}{2}\cdot 1 \quad \text{[the LCD is } 6t\text{]}$$
$$4200t+7500=5550t$$
$$7500=1350t$$
$$t\approx 5.6 \text{ winters}$$

15. Let t be the time needed for the new panel to collect 35,000 Btu. Then figure the amount of energy consumed in 1 hour.

$$\frac{9000}{7.0}\cdot 1+\frac{35000}{t}\cdot 1=\frac{35000}{5.0}\cdot 1 \quad \text{[the LCD is } 35t\text{]}$$
$$45000t+1225000=245000t$$
$$1225000=200000t$$
$$t\approx 6.1 \text{ hours}$$

Chapter 3 • Review Problems
Solve each equation.

1. $2x-(3+4x-3x+5)=4$
$$2x-(8+x)=4$$
$$2x-8-x=4$$
$$x=12$$

3. $3(x-2)+2(x-3)+(x-4)=3x-1$
$$3x-6+2x-6+x-4=3x-1$$
$$6x-16=3x-1$$
$$3x=15$$
$$x=5$$

5. $(2x-5)-(x-4)+(x-3)=x-4$
$$2x-5-x+4+x-3=x-4$$
$$2x-4=x-4$$
$$x=0$$

7. $3z-(z+10)-(z-3)=14-z$
$$3z-z-10-z+3=14-z$$
$$2z=21$$
$$z=\frac{21}{2} \text{ or } 10\frac{1}{2}$$

9. $3x+4(3x-5)=12-x$
$$3x+12x-20=12-x$$
$$16x=32$$
$$x=2$$

11. $x^2-2x-3=x^2-3x+1$
$$x=4$$

13. $7x-5-(6-8x)+2=3x-7+106$
$$7x-3-6+8x=3x+99$$
$$12x=108$$
$$x=9$$

15. $\frac{4}{n}=3 \quad [n \neq 0]$
$$4=3n$$
$$n=\frac{4}{3}$$

17. $5.90x-2.80=2.40x+3.40$
$$3.50x=6.20$$
$$x=1.77$$

19. $\frac{x-4.80}{1.50}=6.20x$
$$x-4.80=9.30x$$
$$-4.80=8.30x$$
$$x=-0.578$$

21. $3(x+10)+4(x+20)+5x-170=15$
$$3x+30+4x+80+5x-170=15$$
$$12x=75$$
$$x=\frac{25}{4}$$

23. $5x+3-(2x-2)+(1-x)=6(9-x)$
$$5x+3-2x+2+1-x=54-6x$$
$$8x=48$$
$$x=6$$

25. $4(x-3)=21+7(2x-1)$
$$4x-12=21+14x-7$$
$$-10x=26$$
$$x=-\frac{13}{5}$$

27. $5(x-1)-3(x+3)=12$
$$5x-5-3x-9=12$$
$$2x=26$$
$$x=13$$

29. $5x+(2x-3)+(x+9)=x+1$
$$7x=-5$$
$$x=-\frac{5}{7}$$

31. $2(x-7)=11+2(4x-5)$
$$2x-14=11+8x-10$$
$$-6x=15$$
$$x=-\frac{5}{2}$$

33. $8x-(x-5)+(3x+2)=x-14$
$$8x-x+5+3x+2=x-14$$
$$9x=-21$$
$$x=-\frac{7}{3}$$

Solve for x.

35. $bx+8=2$

$bx=-6$

$x=-\frac{6}{b}$

37. $2(x-3)=4+a$

$2x-6=4+a$

$x=\frac{a+10}{2}$

39. $8-ax=c-5a$

$-ax=c-5a-8$

$x=\frac{5a-c+8}{a}$

Applications

41. Let $x=$ amount of shale from 18.0 gal/ton mine

$25000-x=$ amount of shale from 30.0 gal/ton mine

$18.0x+30.0(25000-x)=23.0(25000)$

$18.0x+750000-30.0x=575000$

$-12.0x=-175000$

$x=14{,}600$ tons from 18.0 mine

$25000-x=10{,}400$ tons from 30.0 mine

43. Let $A=$ amount owed to A

$A+4A+8A+6A=570$

$19A=570$

$A=\$30$

45. Let $2n+1$ be the first odd number. Then $2n+3$ is the second odd number, $2n+5$ is the third odd number, and $2n+7$ is the fourth odd number

$(2n+1)(2n+5)=(2n+3)(2n+7)-64$

$4n^2+12n+5=4n^2+20n+21-64$

$-8n=-48$

$n=6$

$2n+1=13$

$2n+3=15$

$2n+5=17$

$2n+7=19$

47. Taking moments about the left end gives

$485(5.31)+728(9.26)=R_2(15.4)$

$R_2=\frac{485(5.31)+728(9.26)}{15.4}$

$R_2=605$ lb

Then summing forces in the vertical direction,

$R_1+605=485+728$

$R_1=608$ lb

49. Let $x=$ amount of olive oil costing \$4.86

$4.86x+2.75(136)=3.00(x+136)$

$4.86x+374=3.00x+408$

$1.86x=34$

$x=18.3$ liters of olive oil

51. Let $I=$ taxable income

$5700+0.28(I-38100)=8126$

$5700+0.28I-10668=8126$

$0.28I=13094$

$I=\$46{,}764$

53. Let $w=$ income

$w-0.25w=60000$

$w=\frac{60000}{0.75}=\$80{,}000$

55. Let $x=$ amount of steel with 1.15% chromium and $8-x=$ amount of steel with 1.5% chromium.

$$0.0115x+0.015(8-x)=0.0125(8)$$
$$0.0115x+0.12-0.015x=0.1$$
$$-0.0035x=-0.02$$
$$x=5.71 \text{ tons}$$

57. Let $t=$ amount of pure tin.

$$1.00t+0.105(2.75)=0.160(t+2.75)$$
$$1.00t+0.28875=0.160t+0.44$$
$$0.840t=0.15125$$
$$t=0.180 \text{ ton}$$

59. Let $x=$ amount of 12.5% zinc

$$0.125x+0.164(248)=0.150(x+248)$$
$$0.125x+40.672=0.150x+37.2$$
$$3.472=0.025x$$
$$x=139 \text{ lb}$$

61. Let $x=$ amount of antifreeze in final mixture

$$x=0.134(12.5-5.50)+5.50$$
$$x=6.44 \text{ liters}$$

63. Let $t=$ number of technicians
$15-t=$ number of helpers

$$325t+212(15-t)=3971$$
$$325t+3180-212t=3971$$
$$113t=791$$
$$t=7 \text{ technicians}$$

65. Let $x=$ amount brass with 75.5% copper
Then $5250-x=$ amount with 86.3% copper

$$0.755x+0.863(5250-x)=0.800(5250)$$
$$0.755x+4530.75-0.863x=4200$$
$$-0.108x=-330.75$$
$$x=3060 \text{ kg}$$

67. Let $t=$ time for the submerged trip
Then $8.25-t=$ time for surface trip

$$31.5t=42.3(8.25-t)$$
$$31.5t=348.975-42.3t$$
$$73.8t=348.975$$
$$t=4.73 \text{ hours}$$
$$d=31.5(4.73)=149 \text{ km}$$

69. Let $d=$ distance traveled without rockets
Then $75300-d=$ distance traveled with rockets
Total time is 63.5 hours.

$$\frac{d}{1280}+\frac{75300-d}{950}=63.5$$
$$950d+1280(75300-d)=63.5(1216000)$$
$$950d+96384000-1280d=77216000$$
$$-330d=-19168000$$
$$d=58{,}100 \text{ km}$$

58,100 km is the distance traveled without rockets, but it will also be the distance from the start that the retro-rockets were fired.

71. Let $t=$ the time for the launch to catch the ship
Then $t+8.25=$ the time of the original ship

$$24.5t=35.2(t-8.25)$$
$$24.5t=35.2t-290.4$$
$$10.7t=290.4$$
$$t=27.1 \text{ hours}$$
$$d=24.5(27.1)=665 \text{ km}$$

Chapter 4: Functions and Graphs

Exercise 4.1 • Functions

Functions vs. Relations [See Examples 1 and 2]

1. Is a function since only one y value for each x value.

3. Not a function since there are two y values for some x values. (When $x = 2$, y has the values of 2 and -2.)

5. Not a function since there are two y values for some x values. (When $x = 1$, y has the values of $\sqrt{2}$ and $-\sqrt{2}$.)

7. This relation is a function, since each x value only has one y value.

Implicit and Explicit Forms [See Examples 5 and 6]

9. Explicit

11. Implicit

Dependent and Independent Variables

13. x is independent; y is dependent.

15. x and y are independent; w is dependent.

17. x and y are independent; z is dependent.

Changing from Implicit to Explicit Form [See Examples 12 and 13]

19. $2x - y + 4 = 0$
$$y = 2x + 4$$

21. $2(3x + y) = x + 2$
$$6x + 2y = x + 2$$
$$2y = x - 6x + 2$$
$$y = \frac{2 - 5x}{2}$$

Substituting into Functions [See Examples 14 and 15]

23. $f(x) = 5x + 1$
$$f(1) = 5(1) + 1 = 6$$

25. $f(x) = 5 - 13x$
$$f(2) = 5 - 13(2) = 5 - 26 = -21$$

27. $h(x) = x^3 - 2x + 1$
$$h(2.55) = (2.55)^3 - 2(2.55) + 1 = 12.5$$

29. $f(x) = x^2 - 9$
$$f(-2) = (-2)^2 - 9 = 4 - 9 = -5$$

Applications [See Example 16]

31. $f(t) = 55.0t + \frac{1}{2}(32.2)t^2$
$$f(10.0) = 55.0(10.0) + 16.1(10.0)^2 = 2160 \text{ ft}$$
$$f(15.0) = 55.0(15.0) + 16.1(15.0)^2 = 4450 \text{ ft}$$
$$f(20.0) = 55.0(20.0) + 16.1(20.0)^2 = 7540 \text{ ft}$$

33. $f(r) = 0.000030r^2(80 - r)$
$$f(10) = 0.000030(10)^2(80 - (10)) = 0.21 \text{ in.}$$
$$f(15) = 0.000030(15)^2(80 - (15)) = 0.44 \text{ in.}$$

35. $f(I)=I^2 25.6$

$f(1.59)=(1.59)^2 25.6=64.7$ W

$f(2.37)=(2.37)^2 25.6=144$ W

$f(3.17)=(3.17)^2 25.6=257$ W

Exercise 4.2 • More on Functions

A function as a verbal statement [See Examples 25 and 26]

1. $y=x^3$

3. $y=x+2x^2$

5. $y=\frac{2}{3}(x-4)$

[See Example 28]

7. $c=\sqrt{a^2+b^2}$

9. $P=RI^2$

11. $s=0.65w+2.25$

Substituting into a Function [See Examples 30-32]

13. $f(x)=2x^2+4 \Rightarrow f(a)=2a^2+4$

15. $f(x)=5x+1 \Rightarrow f(a+b)=5(a+b)+1=5a+5b+1$

17. $f(x,y)=3x+2y^2-4 \Rightarrow f(2,3)=3(2)+2(3)^2-4=6+18-4=20$

19. $g(a,b)=2b-3a^2 \Rightarrow g(4,-2)=2(-2)-3(4)^2=-4-48=-52$

Manipulating Functions [See Examples 33 and 34]

21. $y=5x+3$

$5x=y-3$

$x=\frac{y-3}{5}$

23. $y=\frac{1}{x}-\frac{1}{5}$

$5xy=5-x$

$5xy+x=5$

$x(5y+1)=5$

$x=\frac{5}{5y+1}$

25. $5p-q=q-p^2$

$p^2+5p=2q$

$q=\frac{p^2+5p}{2}$ or $\frac{p(p+5)}{2}$

27. $E=\frac{PL}{ae}$

$aeE=PL$

$e=\frac{PL}{aE}$

Composite Functions [See Examples 36-38]

29. If $g(x)=x^2-1$ and $f(x)=3+x$, then $f[g(x)]=3+(x^2-1)=x^2+2$

31. If $g(x)=x-4$ and $f(x)=x^2$, then $f[g(x)]=(x-4)^2$ or $x^2-8x+16$

If $g(x)=x^3$ and $f(x)=4-3x$, then

33. $g[f(x)]=(4-3x)^3$

35. $g[f(3)]=[4-3(3)]^3=[-5]^3=-125$

Inverse Functions [See Examples 41-43]

37. $y=5(2x-3)+4x=10x-15+4x$
$y=14x-15$
$x=\dfrac{y+15}{14}$
Interchanging x and y gives
$y=f^{-1}(x)=\dfrac{x+15}{14}$

39. $y=(1+2x)+2(3x-1)=1+2x+6x-2$
$y=8x-1$
$x=\dfrac{y+1}{8}$
Interchanging x and y gives $y=f^{-1}(x)=\dfrac{x+1}{8}$

41. $y=2(4x-3)-3x=8x-6-3x$
$y=5x-6$
$x=\dfrac{y+6}{5}$
Interchanging x and y gives
$y=f^{-1}(x)=\dfrac{x+6}{5}$

43. $y=3(x-2)-4(x+3)=3x-6-4x-12$
$y=-x-18$
$x=\dfrac{y+18}{-1}=-y-18$
Interchanging x and y gives
$y=f^{-1}(x)=-x-18$

Finding Domain and Range [See Examples 44-47]

45. Domain $=\{-10,-7,0,5,10\}$, Range $=\{3,7,10,20\}$

47. The domain cannot include any value that results in a negative value under the square root
Domain: $x\geq 7$, Range: $y\geq 0$

49. The denominator cannot be 0.
Domain: $x\neq 0$, Range: $-\infty<y<\infty$

51. The part under the square root cannot be negative and since it is in the denominator it cannot be 0.
Domain: $x<1$, Range: $y>0$

53. There cannot be a negative number under the square root.
Domain: $x\geq 1$, Range: $y\geq 0$

Chapter 4 • Review Problems

1. (a) This relation is a function.
(b) This relation is not a function. (When $x=0$, y has the values of 5 and -5.)
(c) This relation is not a function. (When $x=2$, y has the values of 2 and -2.)

3. $S=4\pi r^2$

5. (a) Explicit: y is independent and w is dependent.
(b) Implicit

7. $x^2+y^2+2w=3 \Rightarrow w=\dfrac{3-x^2-y^2}{2}$

9. If $y=3x^2+2z$ and $z=2x^2$, then $y=f(x)=3x^2+2(2x^2)=3x^2+4x^2=7x^2$

11. $f(x)=9-3x \Rightarrow 2f(3)+3f(1)-4f(2)=2[9-3(3)]+3[9-3(1)]-4[9-3(2)]=2(0)+3(6)-4(3)=6$

13. $f(x,y,z)=x^2+3xy+z^3 \Rightarrow f(3,2,1)=(3)^2+3(3)(2)+(1)^3=9+18+1=28$

15. If $f(x)=5x$, $g(x)=\frac{1}{x}$, and $h(x)=x^3$, then $\frac{5h(1)-2g(3)}{3f(2)}=\frac{5(1^3)-2\left(\frac{1}{3}\right)}{3[5(2)]}=\frac{15-2}{90}=\frac{13}{90}$

17. y is 7 less than 5 times the cube of x.

19. $y=6-3x^2 \Rightarrow 3x^2=6-y \Rightarrow x=\pm\sqrt{\frac{6-y}{3}}$

21. $y=3x+4(2-x)=3x+8-4x \Rightarrow y=8-x \Rightarrow x=8-y$. Interchanging x and y gives $y=f^{-1}(x)=8-x$

23. If $g(x)=x+x^2$ and $f(x)=5x$, then $g[f(x)]=(5x)+(5x)^2=25x^2+5x$

25. $g_3(w)=1.74+9.25w \Rightarrow g_3(1.44)=1.74+9.25(1.44)=15.1$

27. $g(k,l,m)=\frac{k+3m}{2l} \Rightarrow g(4.62,1.39,7.26)=\frac{4.62+3(7.26)}{2(1.39)}=9.50$

29. $L=f(t)=112.00(1+0.0655t)$ in. $\Rightarrow f(112)=112.00(1+0.0655[112])=934$ in.
$f(176)=112.00(1+0.0655[176])=1400$ in.
$f(195)=112.00(1+0.0655[195])=1540$ in.

31. $P=f(I)=I^2 325 \Rightarrow f(11.2)=325(11.2)^2=40{,}800$ W
$f(15.3)=325(15.3)^2=76{,}100$ W
$f(21.8)=325(21.8)^2=154{,}000$ W

33. $g(x)=5x^2$ and $f(x)=7-2x$

a) $f[g(x)]=7-2[5x^2]=7-10x^2$

b) $g[f(x)]=5[7-2x]^2=5[49-28x+4x^2]=20x^2-140x+245$

c) $f[g(5)]=7-10(5)^2=7-10(25)=7-250=-243$

d) $g[f(5)]=20(5)^2-140(5)+245=20(25)-700+245=500-700+245=45$

Chapter 5: Graphs

Exercise 5.1 • Rectangular Coordinates

1. Fourth
3. Second
5. Fourth
7. First and Fourth
9. $x = 7$

Graphing Point Pairs

11. $E(-1.8,-0.7); F(-1.4,-1.4); G(1.4,-0.6); H(2.5,-1.8)$

13.

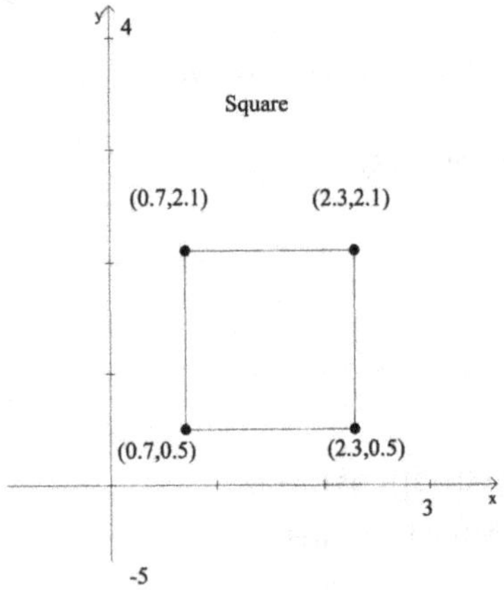

15.

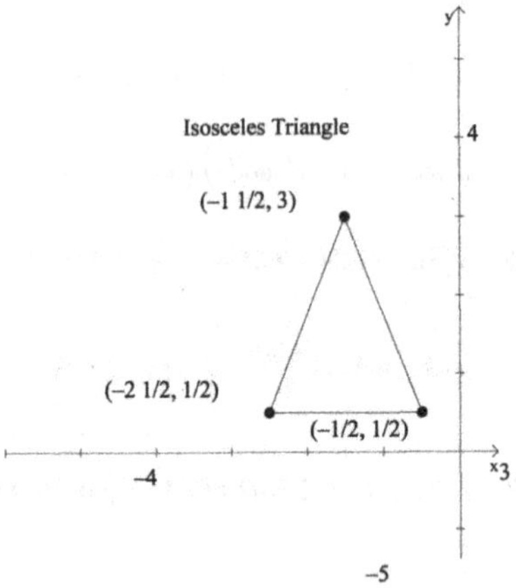

17. $(4,-5)$

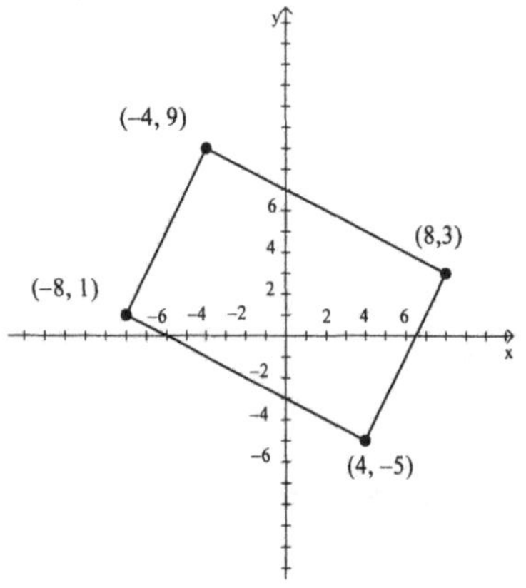

Graphing a Table of Point Pairs

19.

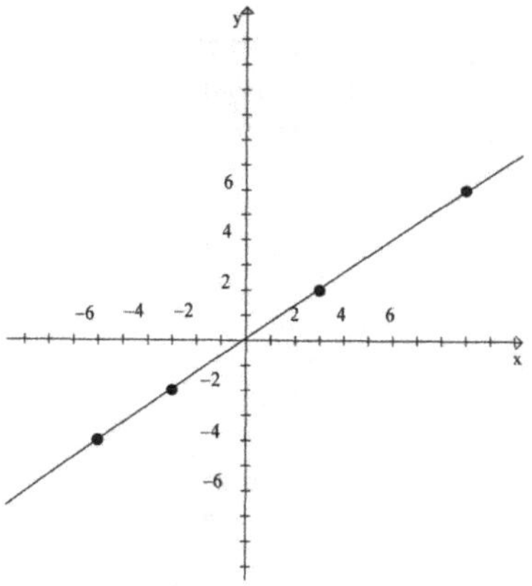

21.

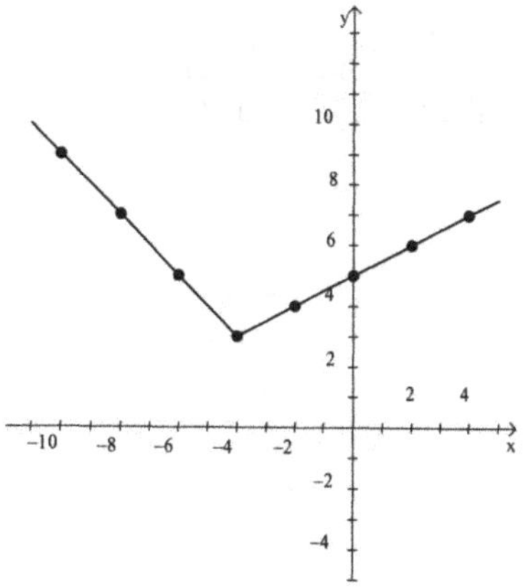

Graphing Empirical Data

23.

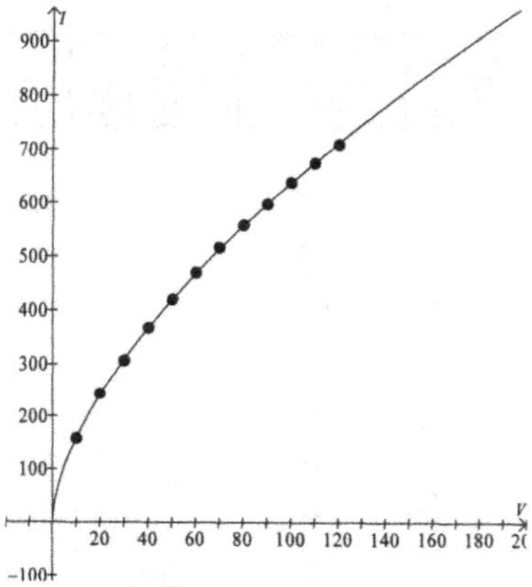

25.

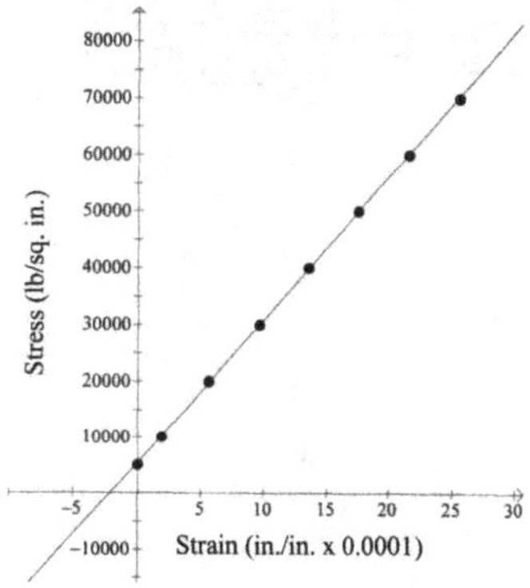

Exercise 5.2 • Graphing an Equation

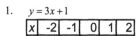

1. $y = 3x + 1$

x	-2	-1	0	1	2
y	-5	-2	1	4	7

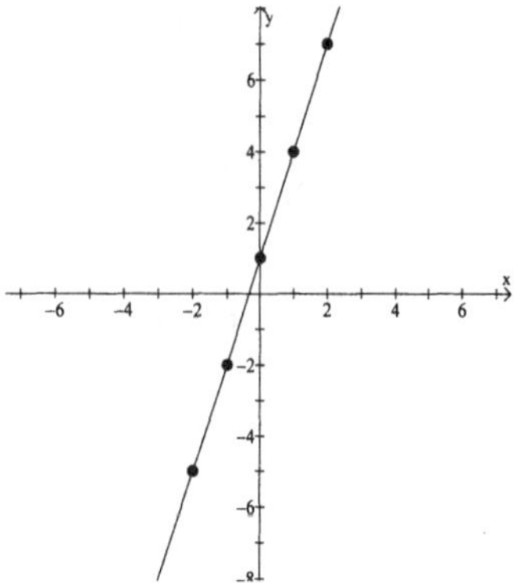

3. $y = 3 - 2x$

x	-2	-1	0	1	2
y	7	5	3	1	-1

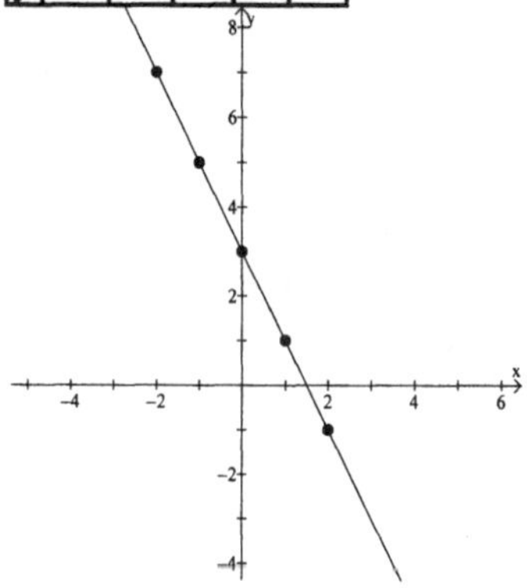

5. $y = x^2$

x	-3	-2	-1	0	1	2	3
y	9	4	1	0	1	4	9

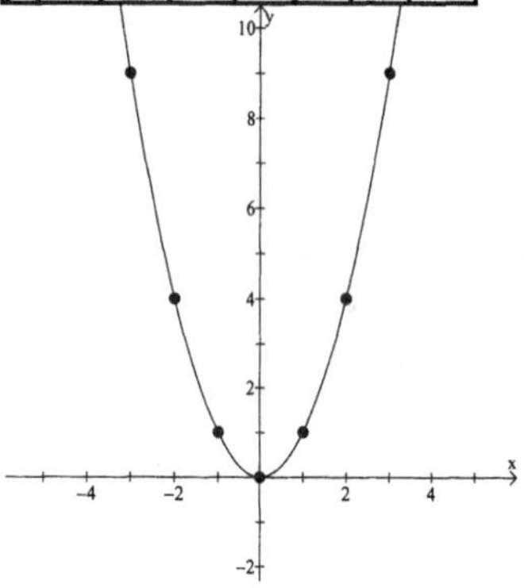

7. $y = \dfrac{x^2}{x+3}$

x	-3	-2	-1	0	1	2	3
y	und	4	0.5	0	0.25	0.8	1.5

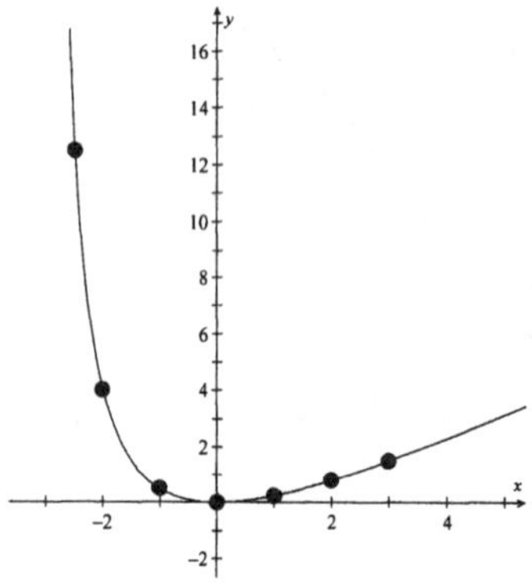

9. $y = x^2 - 1$

x	-3	-2	-1	0	1	2	3
y	8	3	0	-1	0	3	8

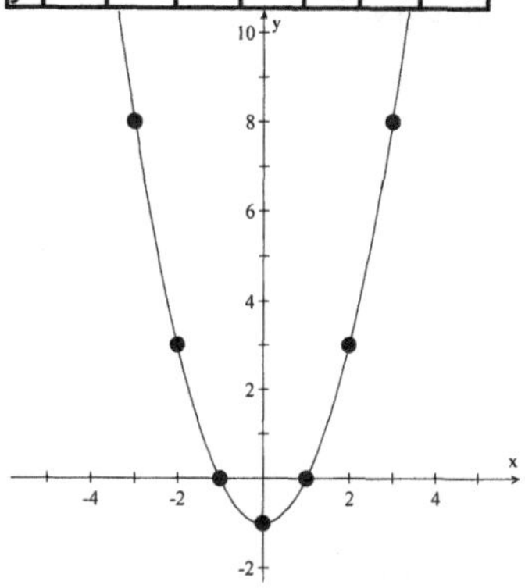

11. $y = x^3$

-3	-2	-1	0	1	2	3
-27	-8	-1	0	1	8	27

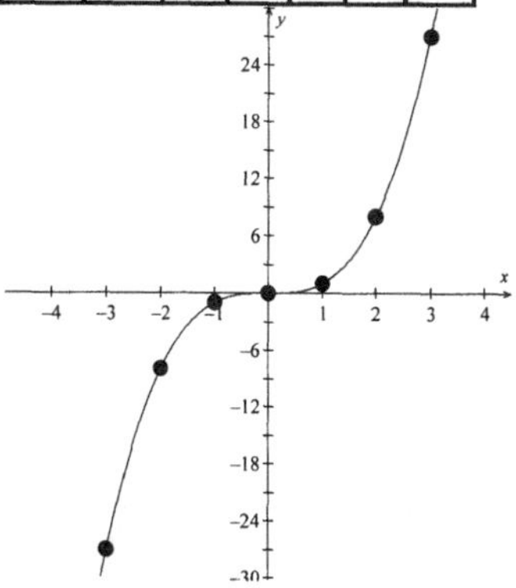

Graphing an Equation Given in Implicit Form

13. $x + y - 5 = 0$

$y = 5 - x$

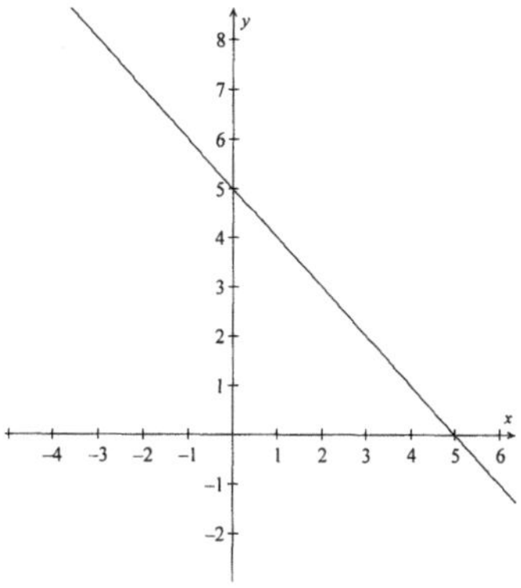

15. $x^2 + y - 4 = 0$

$y = 4 - x^2$

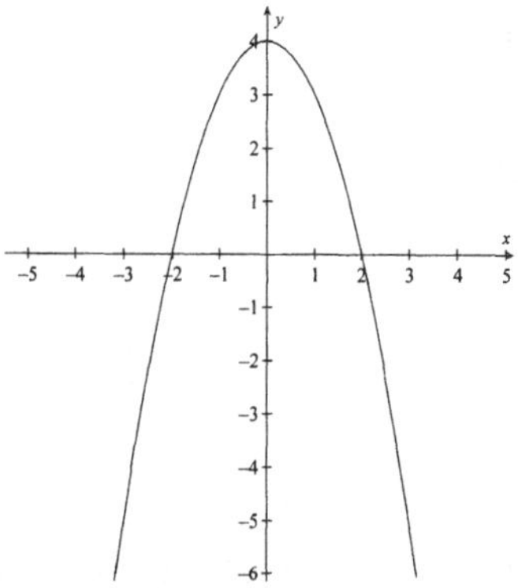

Graphing a Relation

17. $y = \pm 2\sqrt{x}$

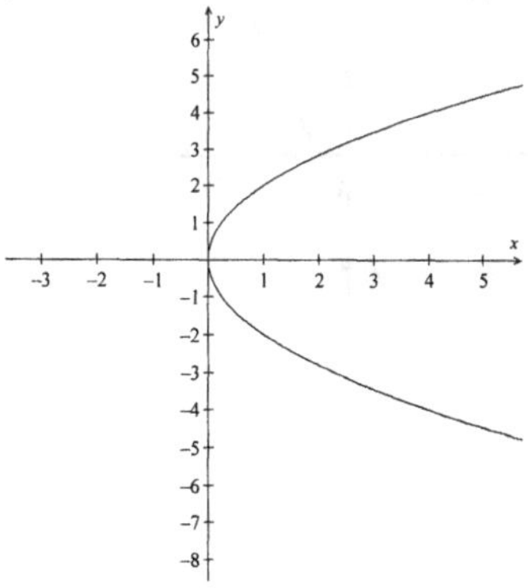

19. $y = 4 \pm \sqrt{2x}$

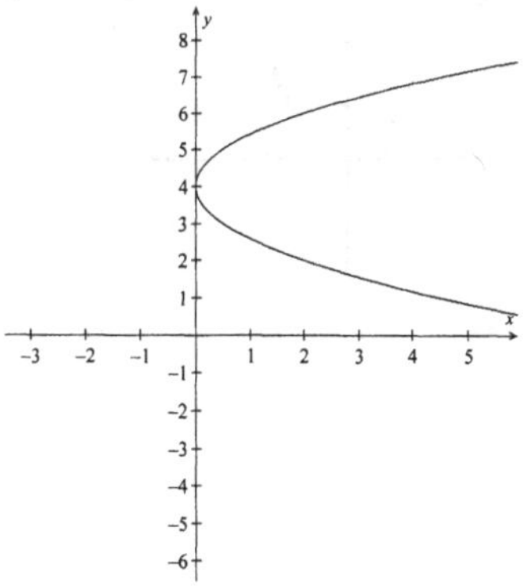

Applications

21. $y = 15600 - 1600t$

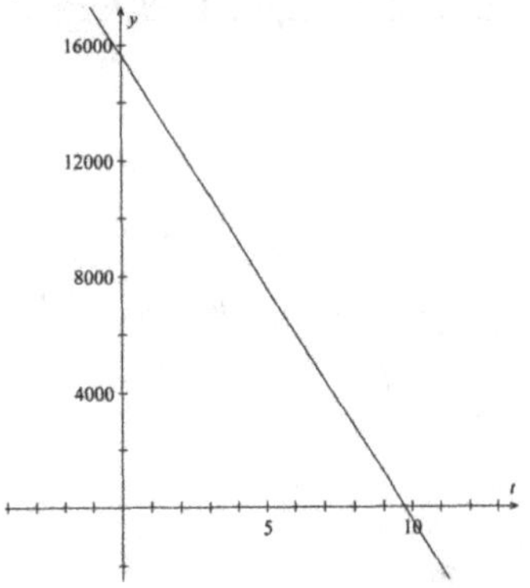

23. $R = \dfrac{2580R_1}{R_1 + 2580}$

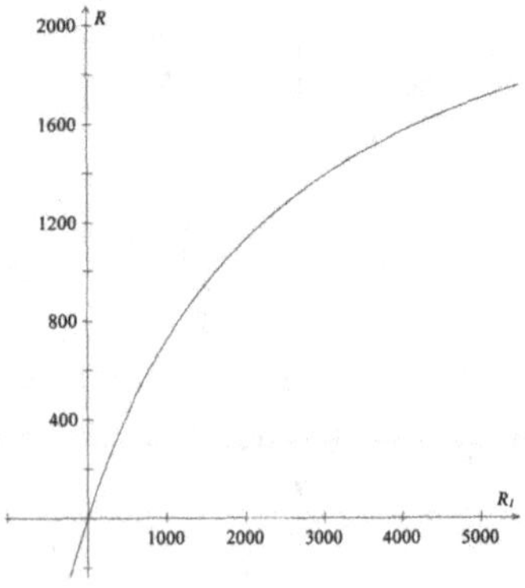

25. $Z = \sqrt{5280^2 + X^2}$

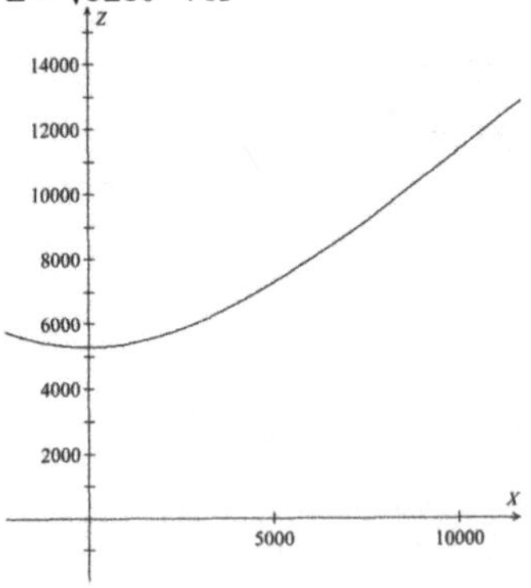

Exercise 5.3 • Graphing Functions by Calculator

Graph each function. Set the viewing window for x and y from -5 to 5.

1. $y = x^2$

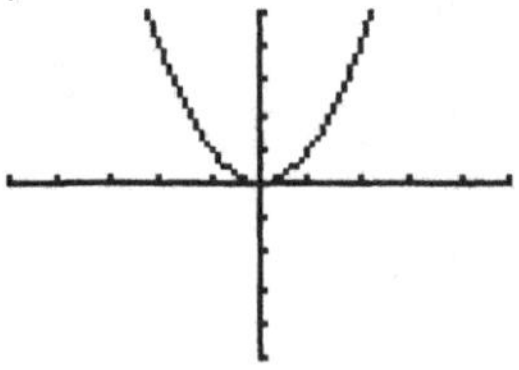

3. $y = 3x - 2$

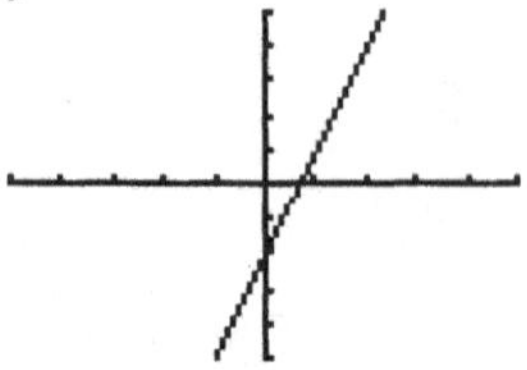

5. $y = x^2 - 2x - 1$

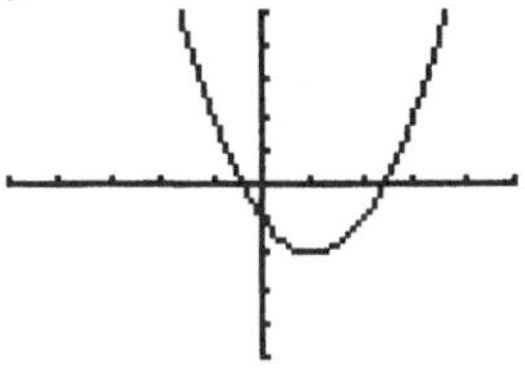

7. $y = x^3 - x$

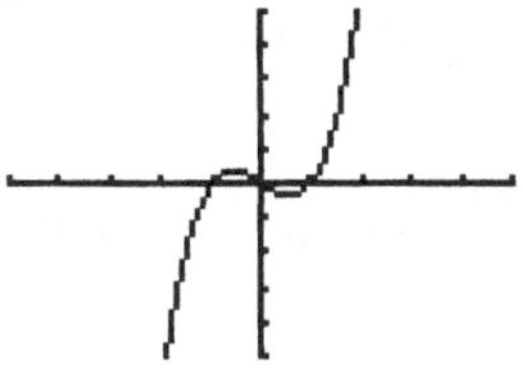

9. $y = x^3 - 2$

11. $y = 1.74x^2 - 2.35x + 1.84$

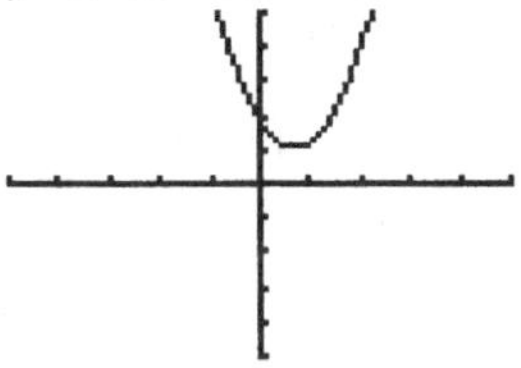

13. $y - 2x^2 + 3x = 3$

$y = 2x^2 + 3x + 3$

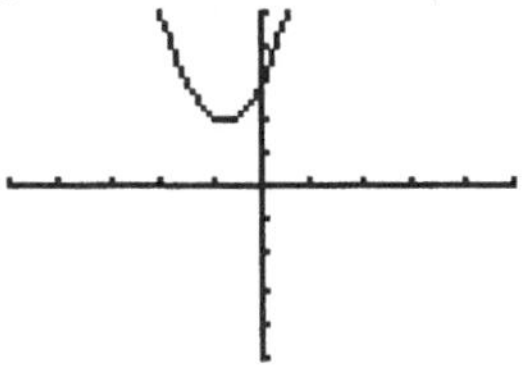

Graph each function. Resize the viewing window or use the ZOOM feature. Locate any zeros, maximum points, or minimum points.

15. $y = 7x^2 + 9x - 14$

Window:
Xmin = -5, Xmax = 5, Xscl = 1
Ymin = -20, Ymax = 5, Yscl = 2

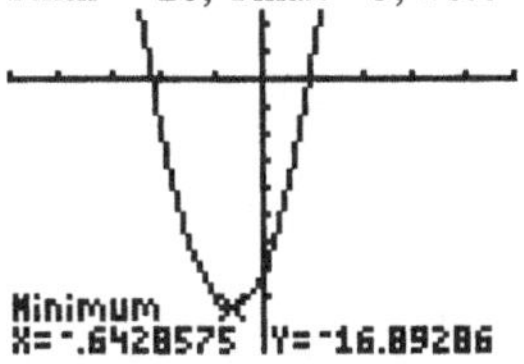

Zeros at $x = -2.20,\ 0.91$

Minimum at $(-0.64, -16.9)$

17. $y = 5x^4 + 13x^2 - 31$

Window:
Xmin = -5, Xmax = 5, Xscl = 1
Ymin = -40, Ymax = 5, Yscl = 5

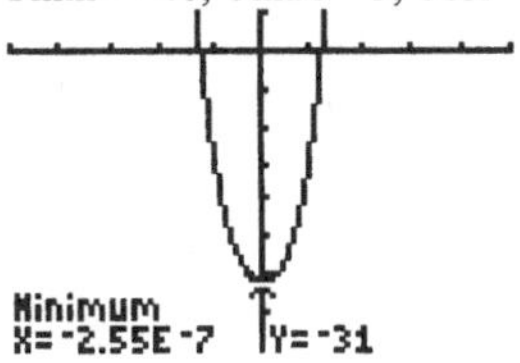

Zeros at $x = -1.23,\ 1.23$

Minimum at $(0, -31)$

Exercise 5.4 • The Straight Line

Slope

1. $m = \frac{\text{rise}}{\text{run}} = \frac{4}{2} = 2$

3. $m = \frac{\text{rise}}{\text{run}} = \frac{-4.25}{5.33} = -0.797$

5. $m = \frac{y_2 - y_1}{x_2 - x_1} = \frac{7-4}{5-2} = \frac{3}{3} = 1$

7. $m = \frac{y_2 - y_1}{x_2 - x_1} = \frac{-6.22 - 5.11}{5.23 - (-2.84)} = \frac{-11.33}{8.07} = -1.40$

Graphing the Straight Line

9. $m = 3 \quad y - \text{intercept} = (0, -5)$

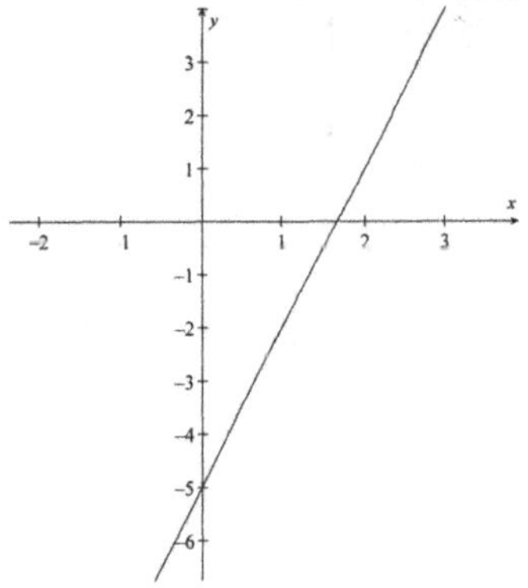

11. $m = -\frac{1}{2} \quad y - \text{intercept} = \left(0, -\frac{1}{4}\right)$

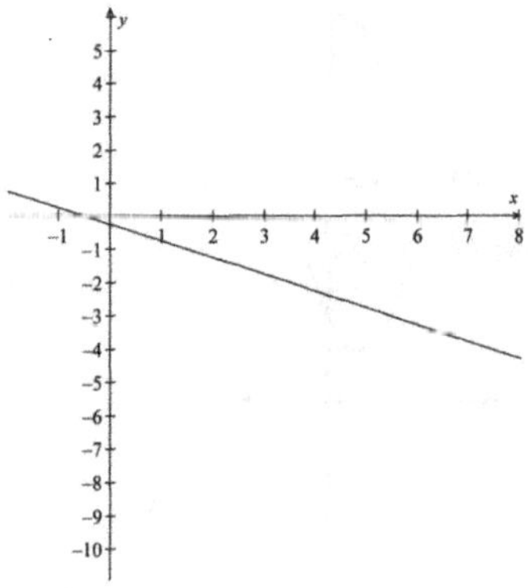

Writing the Equation Given y Intercept and Slope

13. $y = 4x - 3$

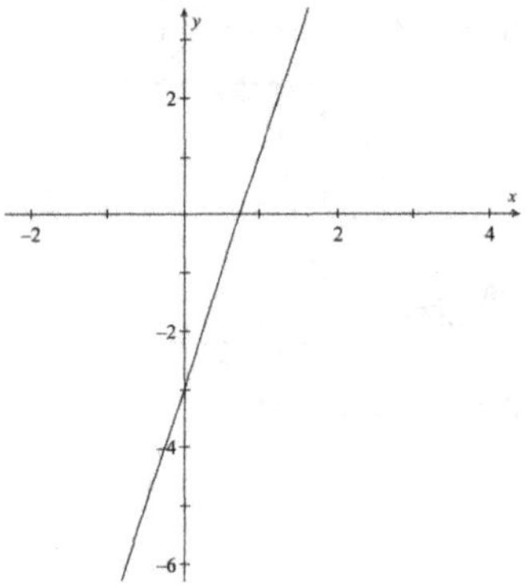

15. $y = 3x - 1$

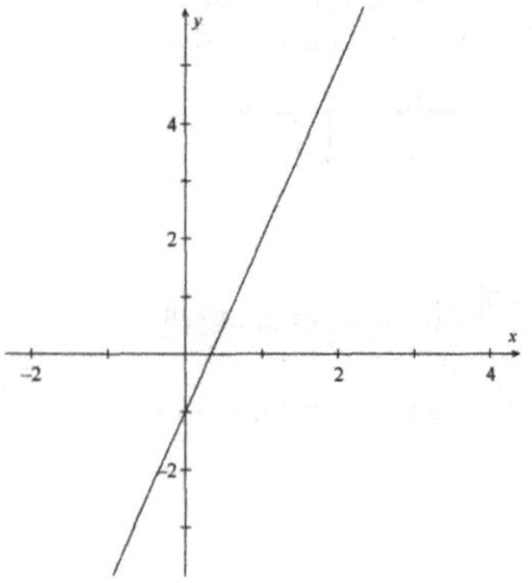

17. $y = 2.30x - 1.50$

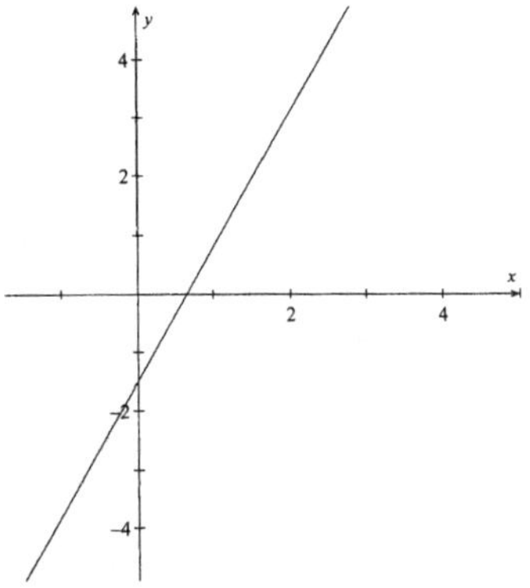

Writing the Equation given Two Points

19. Find the slope:

$$m = \frac{y_2 - y_1}{x_2 - x_1} = \frac{4-3}{-1-2} = -\frac{1}{3}$$

Solve for y:

$$\frac{y-3}{x-2} = -\frac{1}{3}$$

$$3(y-3) = -1(x-2)$$

$$3y - 9 = -x + 2$$

$$3y = -x + 11$$

$$y = -\frac{x}{3} + \frac{11}{3}$$

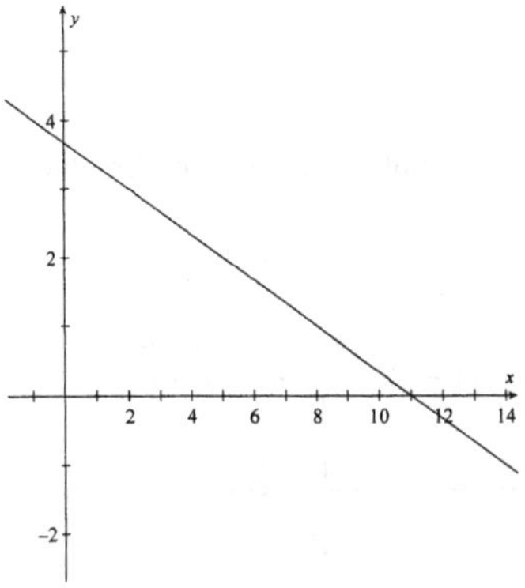

21. Find the slope:

$$m = \frac{y_2 - y_1}{x_2 - x_1} = \frac{3.24 - 2.43}{-2.11 - 1.22} = \frac{0.81}{-3.33} = -0.243$$

Solve for y:

$$\frac{y - 2.43}{x - 1.22} = -0.243$$

$$y - 2.43 = -0.243(x - 1.22)$$

$$y = -0.243x + 0.296 + 2.43$$

$$y = -0.243x + 2.73$$

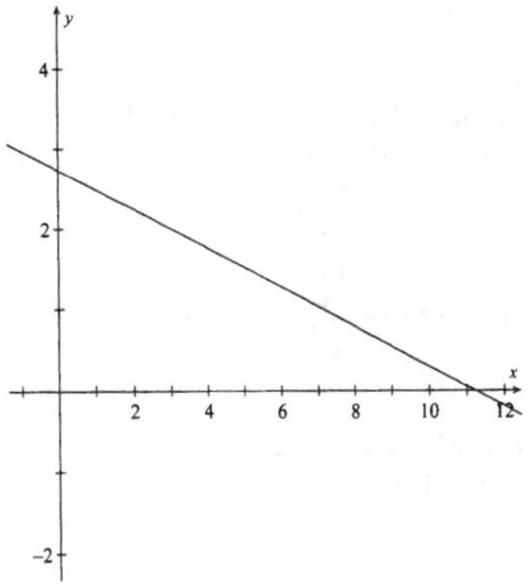

Applications

23. $m = \frac{\text{rise}}{\text{run}} = \frac{9.3}{15.5} = \frac{3}{5}$

25. $F = kL - kL_0$, $L_0 = 4.85$ in., $k = 18.5$ ft/in., $L = 10.5$ ft

$F = 18.5L - 18.5(4.85) \Rightarrow F = 18.5L - 89.7$

$F(10.5) = 18.5(10.5) - 89.7 = 105$ lb

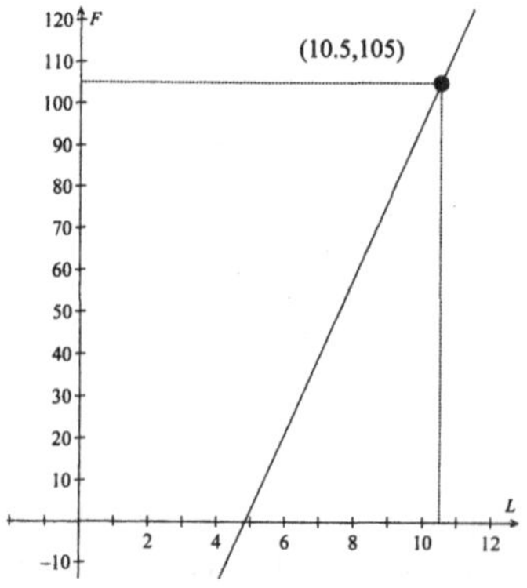

Exercise 5.5 • Solving an Equation Graphically

1. $2.4x^3 - 7.2x^2 - 3.3 = 0$
Window:
Xmin = -10, Xmax = 10, Xscl = 1
Ymin = -20, Ymax = 10, Yscl = 2

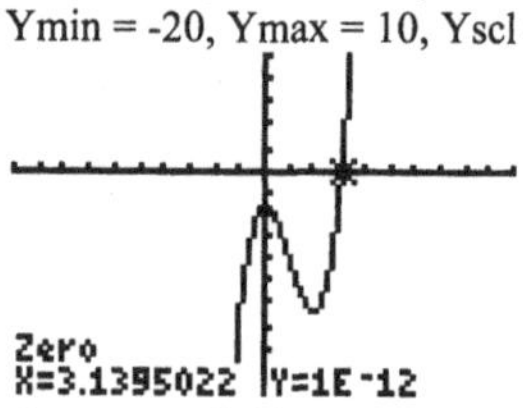

Zero at $x = 3.14$

3. $x^3 - 25x^2 + 48x + 19 = 0$
Window:
Xmin = -10, Xmax = 10, Xscl = 1
Ymin = -20, Ymax = 50, Yscl = 10

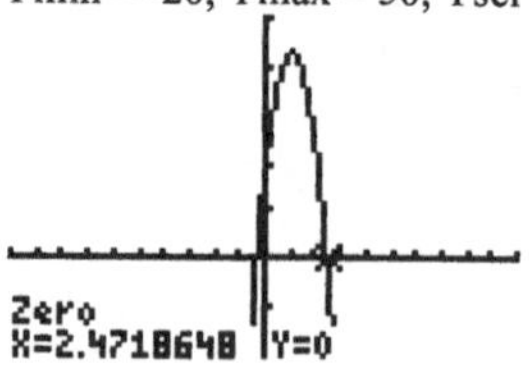

Zeros at $x = -0.34,\ 2.47$

5. $6.4x^4 - 3.8x - 5.5 = 0$
Window:
Xmin = -10, Xmax = 10, Xscl = 1
Ymin = -10, Ymax = 10, Yscl = 1

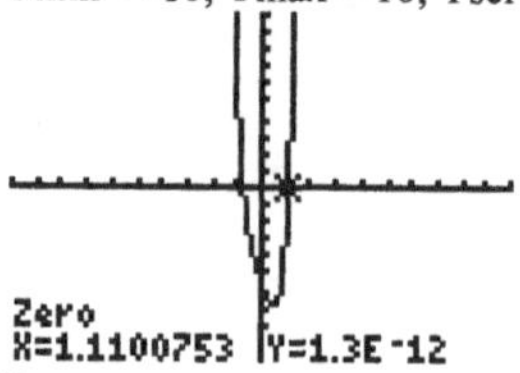

Zeros at $x = -0.79,\ 1.11$

Applications

1. $108x-72.5(x+1.25)=0$

Window:
Xmin = -10, Xmax = 10, Xscl = 1
Ymin = -10, Ymax = 10, Yscl = 1

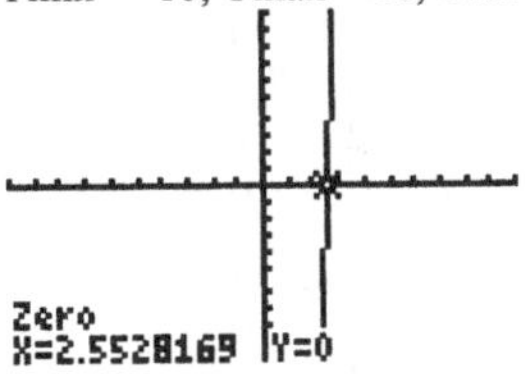

3. $0.0525x+0.0284(3.25-x)-0.0415(3.25)=0$

Window:
Xmin = -10, Xmax = 10, Xscl = 1
Ymin = -1, Ymax = 1, Yscl = 1

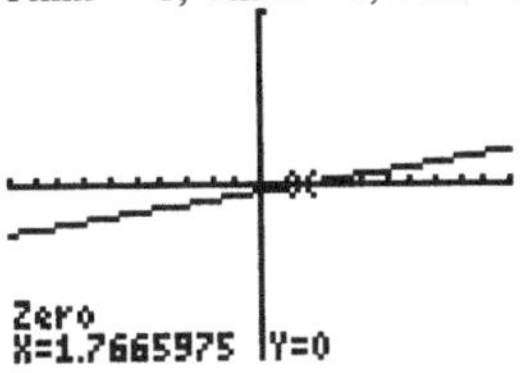

Chapter 5 • Review Problems

1.

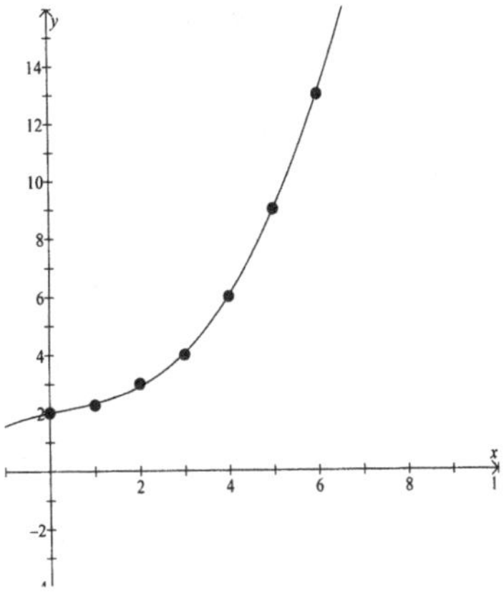

3. $y=3x-2x^2$

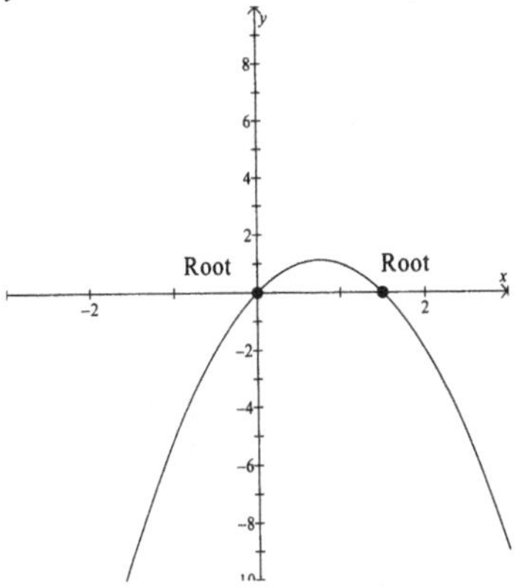

5. a) $m=\dfrac{\text{rise}}{\text{run}}=\dfrac{8}{2}=4$

b) $m=\dfrac{y_2-y_1}{x_2-x_1}=\dfrac{7-5}{2-(-3)}=\dfrac{2}{5}$

7. Find the slope:

$$m=\frac{y_2-y_1}{x_2-x_1}=\frac{-3-5}{1-(-2)}=-\frac{8}{3}$$

Solve for y:

$$\frac{y-5}{x-(-2)}=-\frac{8}{3}$$

$$3(y-5)=-8(x+2)$$

$$3y-15=-8x-16$$

$$3y=-8x-1$$

$$y=-\frac{8}{3}x-\frac{1}{3}$$

9. $y = 5x^2 + 24x - 12$

Xmin = -10, Xmax = 10, Xscl = 1
Ymin = -50, Ymax = 10, Yscl = 5

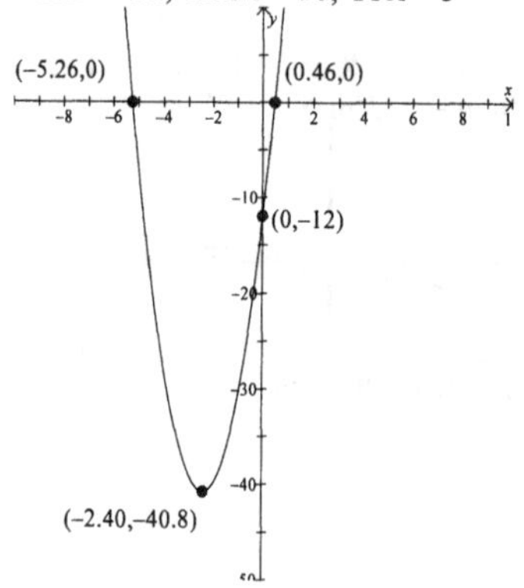

11. $y = 6x^3 + 3x^2 - 14x - 21$

Xmin = -5, Xmax = 5, Xscl = 1
Ymin = -40, Ymax = 10, Yscl = 5

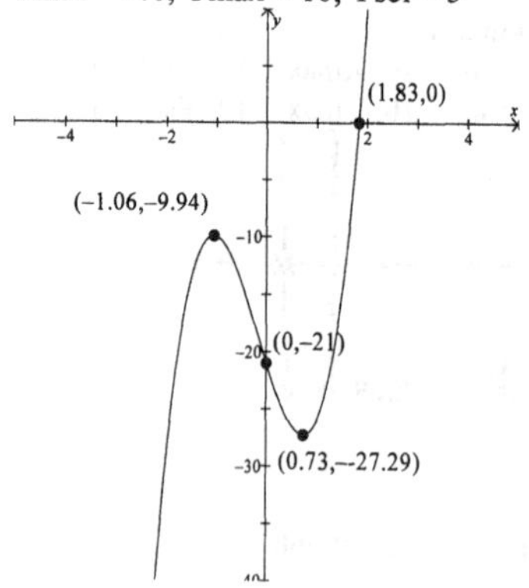

13. $y = -2x + 4$

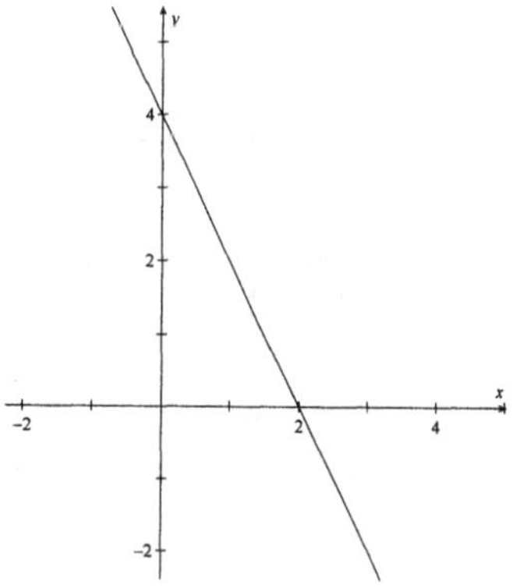

15. $m = 12 \quad y - \text{intercept} = (0,-4)$

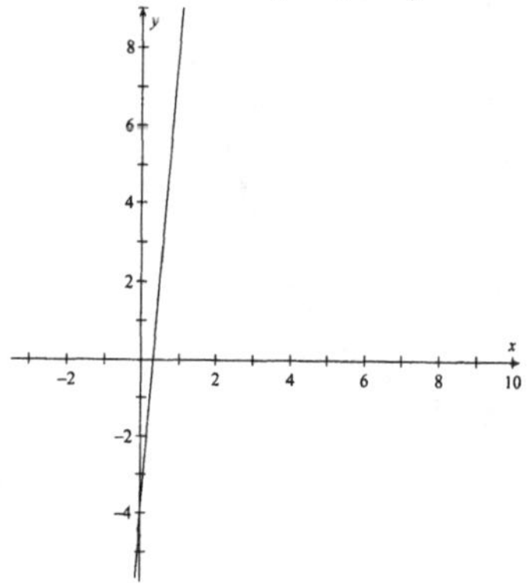

17. Find the slope:

$$m = \frac{R_2 - R_1}{t_2 - t_1} = \frac{12.7 - 10.4}{91.1 - 10.5} = \frac{2.3}{80.6} = 0.029$$

Solve for R:

$$\frac{R - 10.4}{t - 10.5} = 0.029$$

$$R - 10.4 = 0.029(t - 10.5)$$

$$R = 0.029t - 0.305 + 10.4$$

$$R = 0.029t + 10.1$$

$$R(42.7) = 0.029(42.7) + 10.1 = 11.3\,\Omega$$

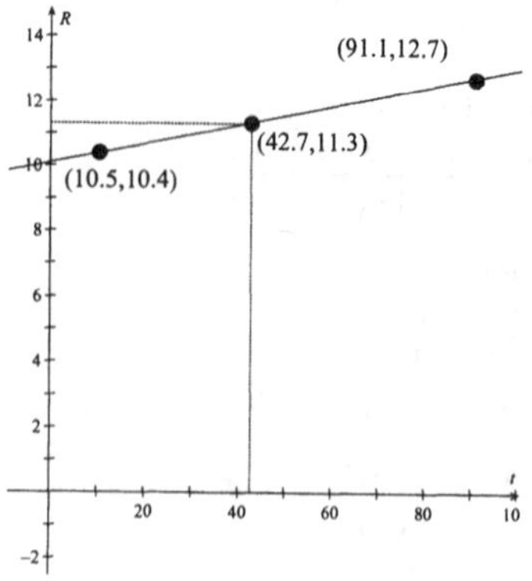

Chapter 6: Geometry

Exercise 6.1 • Straight Lines and Angles

1. (a) $\theta = 90^\circ - 27.2^\circ = 62.8^\circ$ since the angles form a right angle.

   (b) $\theta = 180^\circ - 115.4^\circ = 64.6^\circ$ since the angles form a straight angle.

   (c) $\theta = 37.7^\circ$ since the angles are opposite angles.

3. The three horizontal lines are parallel. Therefore, corresponding segments are proportional:
   $\frac{x}{5.43} = \frac{2.01}{2.16} \Rightarrow x = \frac{2.01 \times 5.43}{2.16} = 5.05$

5. The ground and the horizontal sheet of glass are parallel. Then $A = C = 46.3^\circ$ and $B = D = 180^\circ - 46.3^\circ = 134^\circ$

Exercise 6.2 • Triangles

1. Since θ is the third side of a triangle $\theta = 180^\circ - \left(61^\circ + 42^\circ\right) = 77^\circ$.

   Since ϕ and θ form a straight angle $\phi = 180^\circ - \theta = 103^\circ$

3. $A = \frac{bh}{2} = \frac{25.3 \times 18.6}{2} = 235$ sq. units

5. Since the triangles are similar, the corresponding sides are proportional. Thus
   $\frac{41.8}{32.7} = \frac{a}{21.6} \Rightarrow a = \frac{21.6 \times 41.8}{32.7} = 27.6$ and $\frac{41.8}{32.7} = \frac{b}{45.5} \Rightarrow b = \frac{45.5 \times 41.8}{32.7} = 58.2$

7. Since $a^2 + b^2 = c^2 \Rightarrow a^2 = c^2 - b^2$, then $a = \sqrt{(65.1)^2 - (30.6)^2} = 57.5$

Applications

9. First, the area of the land is $A = \frac{828 \text{ ft} \times 412 \text{ ft}}{2} = 170568 \text{ ft}^2$
   Since there are 43,560 ft^2 in an acre, the triangular piece of land contains 3.916 acres. At \$1125 per acre the cost is $3.916 \times \$1125 = \$4405$ (to the nearest dollar).

11. The ladder forms a right triangle with the wall with the ladder being the hypotenuse, so
    $\text{height} = \sqrt{(39.0 \text{ ft})^2 - (15.0 \text{ ft})^2} = 36.0 \text{ ft}$.

13. First find the diagonal distance of the floor in the room which is the hypotenuse of a right triangle with legs being the width and length of the floor. Thus the diagonal across the floor $= \sqrt{(16.0 \text{ ft})^2 + (20.0 \text{ ft})^2} = 25.6 \text{ ft}$.
    The diagonal distance from one of the lower corners to the opposite upper corner is the hypotenuse of a right triangle with legs being the diagonal across the floor (25.6 ft) and the height of the room (9.0 ft). Hence, the diagonal across the room $= \sqrt{(25.6 \text{ ft})^2 + (9.0 \text{ ft})^2} = 28.3 \text{ ft}$.

15. Let d be the length of the walk which is the hypotenuse of a right triangle. $d = \sqrt{(125\text{ m})^2 + (233\text{ m})^2} = 264\text{ m}$

17. Construct a right triangle from the center of the cone to the tip of the cone. The base is $\frac{1}{2}$ the diameter and the hypotenuse is the slant height. Then, the height (h) is solved as $h = \sqrt{(21.8\text{ in.})^2 - (9.2\text{ in.})^2} = 19.8\text{ in.}$

19. First find half the perimeter: $s = \dfrac{2877\text{ ft} + 1874\text{ ft} + 2025\text{ ft}}{2} = 3388\text{ ft}$. Using Hero's formula, the area of the land is $A = \sqrt{3388(3388-2877)(3388-2025)(3388-1874)} = 1890135.8\text{ ft}^2\left(\dfrac{1\text{ acre}}{43{,}560\text{ ft}^2}\right) = 43.39\text{ acres}$.

21. The triangle formed is a $30°$-$60°$-$90°$ right triangle with the height of the triangle being half the difference in diameters or $h = \dfrac{35\text{ mm} - 17\text{ mm}}{2} = 9\text{ mm}$. The base ($x$) of the triangle is $\frac{1}{2}$ the hypotenuse ($c = 2x$). Thus, $(2x)^2 = (9\text{ mm})^2 + x^2 \Rightarrow 3x^2 = 81 \Rightarrow x^2 = \dfrac{81}{3} = 27 \Rightarrow x = \sqrt{27} = 5.2\text{ mm}$.

23. Think of the octagon as a square with corners cut out. The measure of each interior angle of the octagon is $\dfrac{(8-2)180°}{8} = 135°$ which means the corners cut out of the square will be $45°$-$45°$-$90°$ triangles. Take the length of the sides of the octagon to be x which is also the hypotenuse of the cut out triangle. The other sides of the triangle will then be $\dfrac{x}{\sqrt{2}}$. Thus one side of the square which is 16.0 in. is also made up of the x of one side of the octagon and two of the $\dfrac{x}{\sqrt{2}}$ sides of the cut out triangles. This leads to the following solution:

$$\frac{x}{\sqrt{2}} + x + \frac{x}{\sqrt{2}} = 16.0\text{ in.} \Rightarrow \frac{2x}{\sqrt{2}} + \frac{x\sqrt{2}}{\sqrt{2}} = 16 \Rightarrow \frac{x(2+\sqrt{2})}{\sqrt{2}} = 16 \Rightarrow x = 16\frac{\sqrt{2}}{2+\sqrt{2}} = 6.63\text{ in.}$$

25. The triangle formed is a $45°$-$45°$-$90°$ right triangle with legs measuring $\frac{1}{2}(3.25\text{ in.}) = 1.625\text{ in.}$ and PQ the hypotenuse. The length of $PQ = \sqrt{(1.625\text{ in.})^2 + (1.625\text{ in.})^2} = 2.30\text{ in.}$

27. Since the triangles are similar, the corresponding sides are proportional. Thus $\dfrac{x}{2.16\text{ m}} = \dfrac{5.16\text{ m}}{4.25\text{ m}} \Rightarrow x = \dfrac{5.16 \times 2.16}{4.25} = 2.62\text{ m}$

Exercise 6.3 • Quadrilaterals

1. (a) Area $= 5.83\text{ in.} \times 5.83\text{ in.} = 34.0\text{ in.}^2$ Perimeter $= 4 \times 5.83\text{ in.} = 23.3\text{ in.}$

   (b) Area $= 4.82\text{ m} \times 4.82\text{ m} = 23.2\text{ m}^2$ Perimeter $= 4 \times 4.82\text{ m} = 19.3\text{ m}$

   (c) Area $= 384\text{ cm} \times 734\text{ cm} = 282{,}000\text{ cm}^2$ Perimeter $= 2 \times 384\text{ cm} + 2 \times 734\text{ cm} = 2240\text{ cm}$

   (d) Area $= 55.4\text{ in.} \times 73.5\text{ in.} = 4070\text{ in.}^2$ Perimeter $= 2 \times 55.4\text{ in.} + 2 \times 73.5\text{ in.} = 258\text{ in.}$

Applications

3. $A = 312 \text{ ft} \times 6.5 \text{ ft} = 2028 \text{ ft}^2 \Rightarrow 2028 \text{ ft}^2 \times \left(\frac{1 \text{ yd}}{3 \text{ ft}}\right)^2 = 225.3 \text{ yd}^2 \Rightarrow 225.3 \text{ yd}^2 \times \$13.50/\text{yd}^2 = \$3042$

5. $A = 6.25 \text{ m} \times 7.18 \text{ m} = 44.875 \text{ m}^2 \Rightarrow 44.875 \text{ m}^2 \times \frac{\$7.75}{\text{m}^2} = \$348$

7. Area of walls = Perimeter × Height $= (2 \times 40 \text{ ft} + 2 \times 36 \text{ ft}) \times 22 \text{ ft} = 3344 \text{ ft}^2$

   Area of ceiling $= 40 \text{ ft} \times 36 \text{ ft} = 1440 \text{ ft}^2$

   Total area with correction for doors and windows $= 3344 \text{ ft}^2 + 1440 \text{ ft}^2 - 1375 \text{ ft}^2 = 3409 \text{ ft}^2$

   Cost $= 3409 \text{ ft}^2 \times \left(\frac{\text{yd}}{3 \text{ ft}}\right)^2 = 378.8 \text{ yd}^2 \times \$8.50/\text{yd}^2 = \$3220$

9. Area of sides = Perimeter × Height $= (2 \times 68 \text{ in.} + 2 \times 54 \text{ in.}) \times 48 \text{ in.} = 11{,}712 \text{ in.}^2$

   Area of bottom = Length × Width $= 68 \text{ in.} \times 54 \text{ in.} = 3{,}672 \text{ in.}^2$

   Total area $= 11{,}712 \text{ in.}^2 + 3672 \text{ in.}^2 = 15{,}384 \text{ in.}^2$

   $15384 \text{ in.}^2 \times \left(\frac{\text{ft}}{12 \text{ in.}}\right)^2 = 106.8 \text{ ft}^2 \Rightarrow 106.8 \text{ ft}^2 \times 5.2 \text{ lb}/\text{ft}^2 = 555.4 \text{ lb} \Rightarrow 555.4 \text{ lb} \times \$1.55/\text{lb} = \$861$

11. Area of wall $= 13 \text{ ft} \times 18 \text{ ft} = 234 \text{ ft}^2$ Area of door $= 4 \text{ ft} \times 9 \text{ ft} = 36 \text{ ft}^2$ Area of window $= 3 \text{ ft} \times 6 \text{ ft} = 18 \text{ ft}^2$

   Total Area $= 234 \text{ ft}^2 - 36 \text{ ft}^2 - 2(18 \text{ ft}^2) = 162 \text{ ft}^2 \Rightarrow 162 \text{ ft}^2 \times \left(\frac{12 \text{ in.}}{\text{ft}}\right)^2 = 23{,}328 \text{ in.}^2$

   Area face of bricks $= \left(8\tfrac{1}{2} \text{ in.} + \tfrac{1}{2} \text{ in.}\right) \times \left(2\tfrac{1}{4} \text{ in.} + \tfrac{1}{2} \text{ in.}\right) = 24.75 \text{ in.}^2$

   Number of Bricks $= \frac{23{,}328 \text{ in.}^2}{24.75 \text{ in.}^2/\text{brick}} = 942.5 \approx 1000$ bricks

13. The sum of the interior angles for a n-sided polygon is $(n-2)180°$.
   The miter angle is half the angle between the pieces to be joined

   a) Angles of equilaterial triangle $= \frac{(3-2)180°}{3} = \frac{180°}{3} = 60°$ Miter angle $= \frac{60°}{2} = 30°$

   b) Angles of square or rectangle frame $= \frac{(4-2)180°}{4} = \frac{360°}{4} = 90°$ Miter angle $= \frac{90°}{2} = 45°$

   c) Angles of hexagonal window $= \frac{(6-2)180°}{6} = \frac{720°}{6} = 120°$ Miter angle $= \frac{120°}{2} = 60°$

   d) Angles of octagonal wall clock $= \frac{(8-2)180°}{8} = \frac{1080°}{8} = 135°$ Miter angle $= \frac{135°}{2} = 67.5°$

Exercise 6.4 • The Circle

1. $C = 2\pi(4.82\text{ cm}) = 30.3\text{ cm}$ $\qquad A = \pi(4.82\text{ cm})^2 = 73.0\text{ cm}^2$

3. $C = 2\pi r \Rightarrow r = \dfrac{C}{2\pi} = \dfrac{74.8\text{ in.}}{2\pi} = 11.9\text{ in.}$ $\qquad A = \pi(11.9\text{ in.})^2 = 445\text{ in.}^2$

5. $A = \pi r^2 \Rightarrow r = \sqrt{\dfrac{A}{\pi}} = \sqrt{\dfrac{39.5\text{ ft}^2}{\pi}} = 3.55\text{ ft}$ $\qquad C = \pi(3.55\text{ ft}) = 22.3\text{ ft}$

7. The circumscribed square has a side length equal to the diameter of the circle, 155cm.
The inscribed square has a diagonal equal to the diameter of the circle and this diagonal is the hypotenuse of the right triangles formed by cutting the inscribed square from corner to opposite corner.
$(155\text{ cm})^2 = (\text{side})^2 + (\text{side})^2 \Rightarrow 24{,}025\text{ cm}^2 = 2(\text{side})^2 \Rightarrow \text{side} = \sqrt{24025/2} = 109.6\text{ cm}$
$\text{Difference} = 155\text{ cm} - 109.6\text{ cm} = 45.4\text{ cm}$

9. Use the diagram below

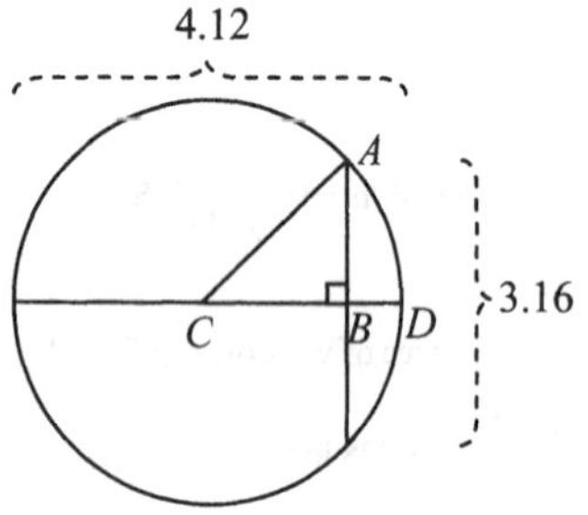

$r = \dfrac{4.12}{2} = 2.06$ half the chord length $= \dfrac{3.16}{2} = 1.58$ $\qquad CD = CA = 2.06 \qquad AB = 1.58$

$CB = \sqrt{(CA)^2 - (AB)^2} = \sqrt{(2.06)^2 - (1.58)^2} = 1.322$ and $x = CD - CB = 2.06 - 1.322 = 0.738$

11. $PQ = \sqrt{(105)^2 - (95)^2} = 44.7$

Applications

13. $A = \pi\left(\dfrac{d}{2}\right)^2 \Rightarrow d = 2\sqrt{\dfrac{A}{\pi}} = 2\sqrt{\dfrac{196\text{ in.}^2}{\pi}} = 15.8\text{ in.}$

15. $C = \pi(33.0\text{ m}) = 104\text{ m}$

17. $C = \pi d \Rightarrow d = \dfrac{C}{\pi} = \dfrac{95.0\text{ in.}}{\pi} = 30.2\text{ in.} \Rightarrow \dfrac{30.2\text{ in.}}{2} = 15.1\text{ in.}$ since cutting from both sides means a tree with a diameter twice as long as chain-saw bar can be cut.

19. $C = \pi(15.0\text{ cm}) = 47.1\text{ cm}$; hypotenuse $= \sqrt{(48\text{ cm})^2 + (52\text{ cm})^2} = 71\text{ cm}$
belt length = sum of three sides + circumference of pulley wheel $= 48\text{ cm} + 52\text{ cm} + 71\text{ cm} + 47.1\text{ cm} = 218\text{ cm}$

21. A 30°-60°-90° right triangle can be constructed by bisecting the screw thread thus forming a 30° angle at the base of the screw head with the hypotenuse *c* running from the base of the screw thread to the center of the wire. The short side, *b*, of the triangle is thus the radius of the wire or 0.0500 in and the hypotenuse is twice that or 0.1000 in. Then $D = T - 2(c+b) = 1.500 - 2(0.1000 + 0.0500) = 1.200$ in.

23. $C = \pi(78.5\text{ cm}) = 247\text{ cm}$

25. A right triangle can be formed by taking a line from the lower left angle of the equilateral triangle to the center of the circle. Then a line can be dropped from the center to the tangent point on the bottom side of the equilateral triangle. The triangle formed is a 30°-60°-90° right triangle with the leg opposite the 60° angle equal to $\frac{1}{2}(2.84\text{ m}) = 1.42\text{ m}$. The leg opposite the 30° angle is the radius of the circle which is equal to $\frac{1.42\text{ m}}{\sqrt{3}} = 0.820\text{ m}$.

27. Since DE and AC are intersecting chords of a circle,
$AB \times BC = DB \times BE \Rightarrow AB \times (7.38\text{ ft}) = (4.77\text{ ft}) \times (5.45\text{ ft})$

$$AB = \frac{4.77 \times 5.45}{7.38} = 3.52\text{ ft}$$

29. Using the idea of intersecting chords in a circle,

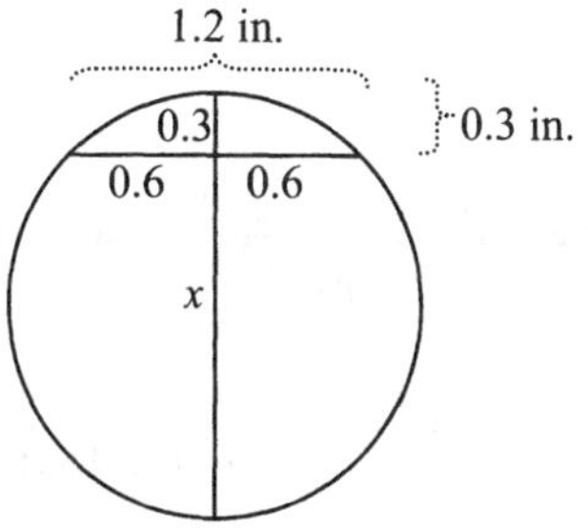

$$(0.3\text{ in.})x = (0.6\text{ in.})^2 \Rightarrow x = \frac{0.6^2}{0.3} = 1.2\text{ in.} \Rightarrow \text{radius} = \frac{1}{2}(x + 0.3\text{ in.}) = \frac{1.5\text{ in.}}{2} = 0.75\text{ in.}$$

Exercise 6.5 • Polyhedra

The Prism

1. $V = 5.63\text{ in.}^2 \times 4.72\text{ in.} = 26.6\text{ in.}^3$

3. $s = \frac{306\text{ in.} + 552\text{ in.} + 772\text{ in.}}{2} = 815\text{ in.}$

$\text{area of base} = \sqrt{815(815-306)(815-552)(815-772)} = 68494\text{ in.}^2$

$V = (\text{area of base}) \times (\text{altitude}) = 68494\text{ in.}^2 \times 925\text{ in.} = 6.34 \times 10^7\text{ in.}^3$

Prism Applications

5. Bases are rectangles measuring 12 in. by 18 in. or 1 ft by 1.5 ft.

$$\text{Lateral area} = (\text{perimeter of base})\times(\text{altitude}) = [2(1\text{ ft})+2(1.5\text{ ft})]\times(8\text{ ft}) = 40\text{ ft}^2$$

$$\text{Area of base} = (1\text{ ft})\times(1.5\text{ ft}) = 1.5\text{ ft}^2$$

$$\text{Volume} = (\text{area of base})\times(\text{altitude}) = 1.5\text{ ft}^2\times 8\text{ ft} = 12\text{ ft}^3$$

Rectangular Parallelepiped and Cube

9. a) $\text{Surface area} = 6(3.75\text{ in.})^2 = 84.4\text{ in.}^2$ $\quad$ $\text{Volume} = (3.75\text{ in.})^3 = 52.7\text{ in.}^3$

b) $\text{Surface area} = 6(26.3\text{ cm})^2 = 4150\text{ cm}^2$ $\quad$ $\text{Volume} = (26.3\text{ cm})^3 = 18{,}200\text{ cm}^3$

c) $\text{Surface area} = 6(2.24\text{ ft})^2 = 30.1\text{ ft}^2$ $\quad$ $\text{Volume} = (2.24\text{ ft})^3 = 11.2\text{ ft}^3$

Rectangular Parallelepiped and Cube Applications

11. $V(\text{one cut}) = 1.00\text{ in.}\times 1.00\text{ in.}\times 0.0050\text{ in.} = 0.0050\text{ in.}^3$

$$\text{cuts in } 0.5\text{ in.}^3 = \frac{0.50\text{ in.}^3}{0.0050\ \text{in.}^3/\text{cut}} = 100\text{ cuts}$$

13. $$\text{Volume} = \left(2.0\text{ mi}\times\frac{1760\text{ yd}}{1\text{ mi}}\right)\times\left(35\text{ ft}\times\frac{1\text{ yd}}{3\text{ ft}}\right)\times\left(3.0\text{ in.}\times\frac{1\text{ yd}}{36\text{ in.}}\right) = 3422\text{ yd}^3 \Rightarrow \frac{3422\text{ yd}^3}{8.0\ \text{yd}^3/\text{load}} = 428\text{ loads}$$

15. $$\text{Volume} = \frac{2.5}{12}\text{ ft}\times\frac{15}{12}\text{ ft}\times 18\text{ ft} = 4.6875\text{ ft}^3 \qquad \text{board feet} = \frac{4.6875\text{ ft}^3}{\frac{1}{12}\text{ ft}^3} = 56\frac{1}{4}\text{ board feet}$$

17. $$\text{Volume} = \frac{48}{3}\text{ yd}\times\frac{36}{3}\text{ yd}\times\frac{12}{3}\text{ yd} = 770\text{ yd}^3$$

19. $$\text{Volume} = 4\text{ ft}\times 4\text{ ft}\times 8\text{ ft} = 128\text{ ft}^3$$

21. $$\text{Truck loads} = \frac{128\text{ ft}^3}{37\text{ ft}^3} = 3\frac{1}{2}\text{ loads}$$

23. The gravel is loose and has a density 100 lb/ft^3 as compared to the packed gravel in the previous problem with a density 110 lb/ft^3.

$$\text{Loose volume} = (\text{packed volume})\frac{(\text{packed density})}{(\text{loose density})} = (4.2\text{ yd}^3)\left(\frac{110\text{ lb/ft}^3}{100\text{ lb/ft}^3}\right) = 4.6\text{ yd}^3$$

The Pyramid

27. $\text{Volume} = \left(\frac{h}{3}\right)(\text{area of base}) = \left(\frac{7.93\text{ in.}}{3}\right)(6.83\text{ in.})^2 = 123\text{ in.}^3$

$s = \text{slant height} = \sqrt{\left(\frac{6.83\text{ in.}}{2}\right)^2 + (7.93\text{ in.})^2} = 8.63\text{ in.}$

$\text{Lateral area} = \left(\frac{s}{2}\right)(\text{perimeter of base}) = \left(\frac{8.63\text{ in.}}{2}\right)(4\times 6.83\text{ in.}) = 118\text{ in.}^2$

29. $\text{Total area} = 4\left(\frac{1}{2}\right)(11.0\text{ in.})(4.00\text{ in.}) + (4.00\text{ in.})^2 = 104\text{ in.}^2$

31. a) $\text{Volume} = \left(\frac{h}{3}\right)(\text{area of base}) = \left(\frac{24.5\text{ ft}}{3}\right)(22.0\text{ ft})^2 = 3950\text{ ft}^3$

b) $s = \text{slant height} = \sqrt{\left(\frac{22.0\text{ ft}}{2}\right)^2 + (24.5\text{ ft})^2} = 26.9\text{ ft}$

$\text{Lateral area} = \left(\frac{s}{2}\right)(\text{perimeter of base}) = \left(\frac{26.9\text{ ft}}{2}\right)(4\times 22.0\text{ ft}) = 1180\text{ ft}^2$

33. The run of the hip rafter is the diagonal to the center of the room. Using Pythagorean Theorem gives the run of hip rafter $= \sqrt{(20.0\text{ ft})^2 + (25.0\text{ ft})^2} = 32.0\text{ ft}$. The rise of the hip rafter is 15.0 ft. Use Pythagorean with the rise and run of the hip rafter to get the length of the hip rafter, length of hip rafter $= \sqrt{(32.0\text{ ft})^2 + (15.0\text{ ft})^2} = 35.3\text{ ft}$.

Exercise 6.6 • Cylinder, Cone, and Sphere

Cylinder

1. a) $\text{Lateral area} = (\text{perimeter of base})(\text{altitude}) = (\pi\times 34.5\text{ in.})(26.4\text{ in.}) = 2860\text{ in.}^2$

$\text{Total area} = \text{Lateral area} + \text{area of bases} = 2860\text{ in.}^2 + 2\left(\pi\times\left(\frac{34.5\text{ in.}}{2}\right)^2\right) = 4730\text{ in.}^2$

$\text{Volume} = (\text{area of base})(\text{altitude}) = \left(\pi\times\left(\frac{34.5\text{ in.}}{2}\right)^2\right)(26.4\text{ in.}) = 24{,}700\text{ in.}^3$

b) $\text{Lateral area} = (\text{perimeter of base})(\text{altitude}) = (\pi\times 134\text{ cm})(226\text{ cm}) = 95{,}100\text{ cm}^2$

$\text{Total area} = \text{Lateral area} + \text{area of bases} = 95{,}100\text{ cm}^2 + 2\left(\pi\times\left(\frac{134\text{ cm}}{2}\right)^2\right) = 123{,}000\text{ cm}^2$

$\text{Volume} = (\text{area of base})(\text{altitude}) = \left(\pi\times\left(\frac{134\text{ cm}}{2}\right)^2\right)(226\text{ cm}) = 3{,}190{,}000\text{ cm}^3$

c) $\text{Lateral area} = (\text{perimeter of base})(\text{altitude}) = (\pi \times 3.65\text{ m})(2.76\text{ m}) = 31.6\text{ m}^2$

$$\text{Total area} = \text{Lateral area} + \text{area of bases} = 31.6\text{ m}^2 + 2\left(\pi \times \left(\frac{3.65\text{ m}}{2}\right)^2\right) = 52.5\text{ m}^2$$

$$\text{Volume} = (\text{area of base})(\text{altitude}) = \left(\pi \times \left(\frac{3.65\text{ m}}{2}\right)^2\right)(2.76\text{ m}) = 28.9\text{ m}^3$$

3. $\text{Volume} = (\text{area of base})(\text{altitude}) = (18.2\text{ in.}^2)\left(\frac{11.2\text{ in.}}{2} \times \sqrt{3}\right) = 175\text{ in.}^3$

Cylinder Applications

5. $V_{\text{bushing}} = V_{\text{outside}} - V_{\text{inside}} = \left(\left(\pi \times \left(\frac{18.0\text{ mm}}{2}\right)^2\right) \times 25.0\text{ mm}\right) - \left(\left(\pi \times \left(\frac{12.0\text{ mm}}{2}\right)^2\right) \times 25.0\text{ mm}\right) = 3530\text{ mm}^3$

7. $\text{Volume of four cylinders} = \frac{\pi \times (82.0\text{ mm})^2 \times 95.0\text{ mm}}{4} \times 4 = 2{,}006{,}787\text{ mm}^3$

$$2{,}002{,}787\text{ mm}^3 \times \frac{1\text{ cm}^3}{1000\text{ mm}^3} \times \frac{1\text{ liter}}{1000\text{ cm}^3} = 2.01\text{ liters}$$

9. $\text{Volume} = \pi\left(\left(\frac{2.55\text{ m}}{2}\right)^2 - \left(\frac{2.00\text{ m}}{2}\right)^2\right)(7.54\text{ m}) = 14.8\text{ m}^3$

11. $\text{Volume} = \pi\left(\frac{4.50\text{ ft}}{2}\right)^2(7.20\text{ ft}) = 114.5\text{ ft}^3 \Rightarrow 114.5\text{ ft}^3\left(\frac{7.48\text{ gal}}{1\text{ ft}^3}\right) = 857\text{ gal}$

Cone

13. a) $\text{slant height} = \sqrt{\left(\frac{9.80\text{ in.}}{2}\right)^2 + (12.6\text{ in.})^2} = 13.5\text{ in.}$

$$\text{Lateral area} = \frac{s}{2}(\text{circumference of base}) = \frac{13.5\text{ in.}}{2}(\pi \times 9.80\text{ in.}) = 208\text{ in.}^2$$

$$\text{Total area} = \text{Lateral area} + \text{area of base} = 208\text{ in.}^2 + \pi\left(\frac{9.80\text{ in.}}{2}\right)^2 = 283\text{ in.}^2$$

$$\text{Volume} = \frac{h}{3}(\text{area of base}) = \frac{12.6\text{ in.}}{3}\left(\pi \times \left(\frac{9.80\text{ in.}}{2}\right)^2\right) = 317\text{ in.}^3$$

b) $\frac{1.50\text{ m}-0.8\text{ m}}{2}=0.35\text{ m} \Rightarrow \text{slant height}=\sqrt{(0.35\text{ m})^2+(1.20\text{ m})^2}=1.25\text{ m}$

$\text{Lateral area}=\frac{s}{2}(\text{sum of base circumferences})=\frac{1.25\text{ m}}{2}(\pi\times 1.50\text{ m}+\pi\times 0.80\text{ m})=4.52\text{ m}^2$

$\text{Total area}=\text{Lateral area}+\text{area of bases}=4.52\text{ m}^2+\pi\left(\frac{1.50\text{ m}}{2}\right)^2+\pi\left(\frac{0.8\text{ m}}{2}\right)^2=6.79\text{ m}^2$

$A_1=\pi\left(\frac{0.8\text{ m}}{2}\right)^2=0.503\text{ m}^2 \qquad A_2=\pi\left(\frac{1.50\text{ m}}{2}\right)^2=1.77\text{ m}^2$

$\text{Volume}=\frac{h}{3}\left(A_1+A_2+\sqrt{A_1A_2}\right)=\frac{1.20\text{ m}}{3}\left(0.503\text{ m}^2+1.77\text{ m}^2+\sqrt{0.503\text{ m}^2\times 1.77\text{ m}^2}\right)=1.29\text{ m}^3$

15. $\text{Volume}=\frac{24.0\text{ cm}}{3}\left(\pi\left(\frac{30.0\text{ cm}}{2}\right)^2\right)=5650\text{ cm}^2$

17. $\text{Volume}=\frac{h}{3}(\text{area of base})=\frac{12.5\text{ ft}}{3}\left(\pi\times\left(\frac{14.2\text{ ft}}{2}\right)^2\right)=660\text{ ft}^3$

19. $A_1=\pi\left(\frac{1\text{ ft}}{2}\right)^2=0.785\text{ ft}^2$ and $A_2=\pi\left(\frac{\frac{15}{12}\text{ ft}}{2}\right)^2=1.23\text{ ft}^2$

$\text{Volume}=\frac{12.0\text{ ft}}{3}\left(0.785\text{ ft}^2+1.23\text{ ft}^2+\sqrt{0.785\text{ ft}^2\times 1.23\text{ ft}^2}\right)=12.0\text{ ft}^3$

Sphere

23. The volume $V=\frac{4}{3}\pi r^3$, then $r=\sqrt[3]{\frac{3V}{4\pi}}=\sqrt[3]{\frac{3\times 5.88}{4\pi}}=1.12$ and $A=4\pi(1.12)^2=15.8$.

25. The surface area is $A=4\pi r^2$, then $r=\sqrt{\frac{A}{4\pi}}=\sqrt{\frac{46.0\text{ cm}^2}{4\pi}}=1.91\text{ cm}$ and $V=\frac{4}{3}\pi(1.91\text{ cm})^3=29.2\text{ cm}^3$.

Sphere Applications

27. Note, the formula for the surface area of a sphere is 4 times the formula for the area of a circle

29. (a) $r=\sqrt[3]{\frac{3V}{4\pi}}=\sqrt[3]{\frac{3\times 9000\text{ m}^3}{4\pi}}=12.9\text{ m}$

(b) $\text{Weight}=\text{Surface Area}\times\text{Density}=4\pi(12.9\text{ m})^2\times\left(2.00\ {}^{\text{kg}}\!/_{\text{m}^2}\right)=4180\text{ kg}$

Chapter 6 • Review Problems

1. Here is another $30°$-$60°$-$90°$ special triangle problem and the length of the side opposite the 30^0 angle is ½ the hypotenuse. Since this side is 12 miles (in one minute) the hypotenuse (the path the rocket is following) is 24 miles and the speed of the rocket is $\left(24\,\frac{\text{mi}}{\text{min}}\right)\left(\frac{60\text{ min}}{1\text{ hr}}\right) = 1440$ miles per hour .

3. The volume of the water raised in the cylindrical tank is 4.00 m diameter by 1.00 m depth is 4π m^3. Finding the radius of a sphere given the volume is $r = \sqrt[3]{\frac{3V}{4\pi}} = \sqrt[3]{\frac{3\left(4\pi \text{ m}^3\right)}{4\pi}} = \sqrt[3]{3} = 1.44\text{ m}$ and $d = 2r = 2.88\text{ m}$

5. $c = \sqrt{(12\text{ m})^2 + (5\text{ m})^2} = 13\text{ m}$

7. $s = \frac{573+638+972}{2} = 1091$ $\qquad A = \sqrt{1091(1091-573)(1091-638)(1091-972)} = 175{,}000\text{ units}^2$

9. r = 4.00 in., d = 8.00 in., and C = 8π in. The belt goes around ½ of C for each pulley plus twice the distances between the centers or distance = π 8.00 in. + 18.00 in. = 43.1 in.

11. $\frac{21.0\text{ m}}{36.0\text{ m}} = \frac{x}{42.0\text{ m}} \Rightarrow x = \frac{42.0\times 21.0}{36.0} = 24.5\text{ m}$ then the length of the shorter side is $42\text{ m} - 24.5\text{ m} = 17.5\text{ m}$.

13. Sum of interior angles = $(n-2)180° = (5-2)180° = 540° \Rightarrow 540° - \left(38° + 96° + 112° + 133°\right) = 161°$.

15. $V = \frac{86\text{ mm}}{3}\left(\pi\left(\frac{49\text{ mm}}{2}\right)^2 + \pi\left(\frac{63\text{ mm}}{2}\right)^2 + \sqrt{\left(\pi\left(\frac{49\text{ mm}}{2}\right)^2\right)\left(\pi\left(\frac{63\text{ mm}}{2}\right)^2\right)}\right) = 213\text{ cm}^3$

17. $A = \frac{1}{2}(38.4\text{ in.})(53.8\text{ in.}) = 1030\text{ in.}^2$

19. $A = 39.2\text{ m}\times 29.3\text{ m} = 1150\text{ m}^2$

21. $A = 4\pi\left(\frac{39.2\text{ in.}}{2}\right)^2 = 4830\text{ in.}^2$

23. $V = 35.8\text{ in.}\times 37.4\text{ in.}\times 73.4\text{ in.} = 98{,}300\text{ in.}^3$

25. $r^2 = (r - 1.16\text{ cm})^2 + (4.125\text{ cm})^2 = r^2 - 2.32r + 1.16^2 + 4.125^2 \Rightarrow r = \frac{1.16^2 + 4.125^2}{2.32} = 7.91\text{ cm}$

Chapter 7: Right Triangles and Vectors

Exercise 7.1 • The Trigonometric Functions

Sine, Cosine, and Tangent

1. (a) $\sin\theta=\dfrac{4.75}{7.51}=0.6325$ $\cos\theta=\dfrac{5.82}{7.51}=0.7750$ $\tan\theta=\dfrac{4.75}{5.82}=0.8162$

   (b) $\sin\theta=\dfrac{17.1}{30.5}=0.5607$ $\cos\theta=\dfrac{25.3}{30.5}=0.8295$ $\tan\theta=\dfrac{17.1}{25.3}=0.6759$

   (c) $\sin\theta=\dfrac{116}{218}=0.5321$ $\cos\theta=\dfrac{185}{218}=0.8486$ $\tan\theta=\dfrac{116}{185}=0.6270$

   (d) $\sin\theta=\dfrac{1.87}{3.66}=0.5109$ $\cos\theta=\dfrac{3.15}{3.66}=0.8607$ $\tan\theta=\dfrac{1.87}{3.15}=0.5937$

3. (a) $\sin 49.3^\circ=0.7581$ $\cos 49.3^\circ=0.6521$ $\tan 49.3^\circ=1.1626$

   (b) $\sin 38.9^\circ=0.6280$ $\cos 38.9^\circ=0.7782$ $\tan 38.9^\circ=0.8069$

   (c) $\sin 18.3^\circ=0.3140$ $\cos 18.3^\circ=0.9494$ $\tan 18.3^\circ=0.3307$

   (d) $\sin 2.07^\circ=0.0361$ $\cos 2.07^\circ=0.9993$ $\tan 2.07^\circ=0.0361$

   (e) $\sin 85.3^\circ=0.9966$ $\cos 85.3^\circ=0.0819$ $\tan 85.3^\circ=12.1632$

   (f) $\sin 28.7^\circ=0.4802$ $\cos 28.7^\circ=0.8771$ $\tan 28.7^\circ=0.5475$

   (g) $\sin 73.7^\circ=0.9598$ $\cos 73.7^\circ=0.2807$ $\tan 73.7^\circ=3.4197$

   (h) $\sin 43.9^\circ=0.6934$ $\cos 43.9^\circ=0.7206$ $\tan 43.9^\circ=0.9623$

   (i) $\sin 3.345^\circ=0.0583$ $\cos 3.345^\circ=0.9983$ $\tan 3.345^\circ=0.0584$

   (j) $\sin 58.49^\circ=0.8525$ $\cos 58.49^\circ=0.5226$ $\tan 58.49^\circ=1.6312$

   (k) $\sin 78.37^\circ=0.9795$ $\cos 78.37^\circ=0.2016$ $\tan 78.37^\circ=4.8587$

   (l) $\sin 22.05^\circ=0.3754$ $\cos 22.05^\circ=0.9269$ $\tan 22.05^\circ=0.4050$

   (m) $\sin 89^\circ 43'=0.9940$ $\cos 89^\circ 43'=0.1094$ $\tan 89^\circ 43'=9.0821$

   (n) $\sin 78^\circ 27'=0.9798$ $\cos 78^\circ 27'=0.2002$ $\tan 78^\circ 27'=4.8933$

   (o) $\sin 33^\circ 47'=0.5561$ $\cos 33^\circ 47'=0.8311$ $\tan 33^\circ 47'=0.6690$

   (p) $\sin 63^\circ 29'=0.8948$ $\cos 63^\circ 29'=0.4465$ $\tan 63^\circ 29'=2.0042$

Finding the Angle

5. $D=\tan^{-1}(1.53)=56.8^\circ$

7. $K=\cos^{-1}(0.77)=39.6^\circ$

9. $E=\cos^{-1}(0.847)=32.1^\circ$

11. $\arccos 0.862=30.5^\circ$

13. $\sin^{-1}(0.175)=10.1^\circ$

15. $\arctan 4.26=76.8^\circ$

Applications

17. $\theta = \tan^{-1}\left(\frac{9}{12}\right) = 36.9^\circ$

19. $\theta = \cos^{-1}\left(\frac{10.6}{12.4}\right) = 31.3^\circ$

Exercise 7.2 • Solution of Right Triangles

Right Triangles When One Side and One Angle are Known

1. $B = 90^\circ - 42.9^\circ = 47.1^\circ$ $\quad b = \frac{155}{\tan 42.9^\circ} = 167$ $\quad c = \frac{155}{\sin 42.9^\circ} = 228$

3. $A = 90^\circ - 31.9^\circ = 58.1^\circ$ $\quad b = \frac{1.74}{\tan 58.1^\circ} = 1.08$ $\quad c = \frac{1.74}{\sin 58.1^\circ} = 2.05$

5. $B = 90^\circ - 64.7^\circ = 25.3^\circ$ $\quad b = \frac{284}{\tan 64.7^\circ} = 134$ $\quad c = \frac{284}{\sin 64.7^\circ} = 314$

7. $A = 90^\circ - 55.2^\circ = 34.8^\circ$ $\quad a = 9.26 \tan 34.8^\circ = 6.44$ $\quad c = \frac{9.25}{\cos 34.8^\circ} = 11.3$

9. $B = 90^\circ - 31.4^\circ = 58.6^\circ$ $\quad a = 82.4 \tan 31.4^\circ = 50.3$ $\quad c = \frac{82.4}{\cos 31.4^\circ} = 96.5$

Right Triangles When Two Sides Are Known

11. $A = \tan^{-1}\left(\frac{382}{274}\right) = 54.3^\circ$ $\quad B = 90^\circ - 54.3^\circ = 35.7^\circ$ $\quad c = \sqrt{382^2 + 274^2} = 470$

13. $A = \cos^{-1}\left(\frac{3.97}{4.86}\right) = 35.2^\circ$ $\quad B = 90^\circ - 35.2^\circ = 54.8^\circ$ $\quad a = \sqrt{4.86^2 - 3.97^2} = 2.80$

15. $A = \sin^{-1}\left(\frac{27.4}{37.5}\right) = 46.9^\circ$ $\quad B = 90^\circ - 46.9^\circ = 43.1^\circ$ $\quad b = \sqrt{37.5^2 - 27.4^2} = 25.6$

17. $A = \sin^{-1}\left(\frac{41.3}{63.7}\right) = 40.4^\circ$ $\quad B = 90^\circ - 40.4^\circ = 49.6^\circ$ $\quad b = \sqrt{63.7^2 - 41.3^2} = 48.5$

19. $A = \cos^{-1}\left(\frac{228}{473}\right) = 61.2^\circ$ $\quad B = 90^\circ - 61.2^\circ = 28.8^\circ$ $\quad a = \sqrt{473^2 - 228^2} = 414$

Exercise 7.3 • Applications of the Right Triangle

Measuring Inaccessible Distances

1. $\tan 57.6^\circ = \frac{a}{255 \text{ m}} \Rightarrow a = (255 \text{ m}) \times \tan 57.6^\circ = 402 \text{ m}$

3. $\tan 28.7^\circ = \dfrac{156\text{ ft}}{b} \Rightarrow b = \dfrac{156\text{ ft}}{\tan 28.7^\circ} = 285\text{ ft}$

5. $\tan 47.6^\circ = \dfrac{PQ}{59.3\text{ m}} \Rightarrow PQ = (59.3\text{ m}) \times \tan 47.6^\circ = 64.9\text{ m}$

7. $\tan 28.0^\circ = \dfrac{b}{75.0\text{ yd}} \Rightarrow b = (75.0\text{ yd}) \times \tan 28.0^\circ = 39.9\text{ yd}$

9. $\tan 6.7^\circ = \dfrac{15\text{ m}}{w} \Rightarrow w = \dfrac{15\text{ m}}{\tan 6.7^\circ} = 128\text{ m}$

Navigation

11. $\tan 43^\circ 15' = \dfrac{Q-\text{Ship}}{15.6\text{ km}} \Rightarrow Q-\text{Ship} = (15.6\text{ km}) \times \tan 43^\circ 15' = 14.7\text{ km}$

$\cos 43^\circ 15' = \dfrac{15.6\text{ km}}{P-\text{Ship}} \Rightarrow P-\text{Ship} = \dfrac{15.6\text{ km}}{\cos 43^\circ 15'} = 21.4\text{ km}$

13.

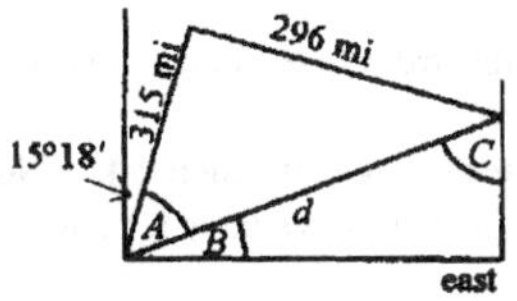

N15°18′E + S74°42′E = 90° thus $d = \sqrt{315^2 + 296^2} = 432$ mi

$\tan A = \dfrac{296}{315} \Rightarrow A = 43^\circ 13' \quad B = 90^\circ - 43^\circ 13' - 15^\circ 18' \quad C = 90^\circ - B = 58^\circ 31'$ or S58°31′W

15. The 255 mi path of the ship is the hypotenuse of a right triangle.

$\cos 46^\circ 12' = \dfrac{\text{N}}{225\text{ mi}} \Rightarrow \text{N} = (225\text{ mi}) \times \cos 46^\circ 12' = 156\text{ mi north}$

$\sin 46^\circ 12' = \dfrac{\text{E}}{225\text{ mi}} \Rightarrow \text{E} = (225\text{ mi}) \times \sin 46^\circ 12' = 162\text{ mi east}$

Structures

17. The base of the right triangle of which AB is the hypotenuse is the hypotenuse of the 45° right triangle below it. The base is found using Pythagorean Theorem: $\sqrt{(12.5\text{ ft})^2 + (12.5\text{ ft})^2} = 17.7\text{ ft}$. Then $AB = \sqrt{(17.7\text{ ft})^2 + (10.3\text{ ft})^2} = 20.5\text{ ft}$. The angle is found by $\theta = \tan^{-1}\dfrac{10.3\text{ ft}}{17.7\text{ ft}} = 30.2^\circ$

19. $\theta = \cos^{-1}\left(\dfrac{65\text{ ft}}{82\text{ ft}}\right) = 37.6^\circ$

21. $A = \tan^{-1}\left(\dfrac{0.975\text{ m}}{0.750\text{ m}}\right) = 52.4^\circ$ $\qquad B = \tan^{-1}\left(\dfrac{0.750\text{ m}}{0.975\text{ m}}\right) = 37.6^\circ$

$AB = \sqrt{(0.975\text{ m})^2 + (0.750\text{ m})^2} = 1.23\text{ m}$

$\text{Area} = \frac{1}{2}(0.975\text{ m})(0.750\text{ m}) = 0.366\text{ m}^2$

Geometry

23. Let the length of the side of the cube = 1, then $AC = \sqrt{1^2 + 1^2} = \sqrt{2}$ and $AB = \sqrt{\left(\sqrt{2}\right)^2 + 1^2} = \sqrt{3}$

then $\theta = \cos^{-1}\dfrac{\sqrt{2}}{\sqrt{3}} = 35.3^\circ$

25. Let the length of the short side = 1 and that of the diagonal = 3. Then, $\sin^{-1}\left(\dfrac{1}{3}\right) = 19.5^\circ$.

27. $\sin 83.2^\circ = \dfrac{h}{255} \Rightarrow h = 255 \times \sin 83.2^\circ = 253;\quad A = 482 \times 253 = 122{,}000\text{ units}^2$

29. Each internal angle of the pentagon is $\dfrac{(5-2)180}{5} = 108^\circ$. Construct lines from two adjacent vertices of the pentagon to the center of the circle. The two base angles of this isosceles triangle are each 54°. Now if we draw a line from this center of the circle to the center of the side of the pentagon, we have a right triangle with a long side equal to the radius of the circle which is 244 in. Then,

$\tan 54^\circ = \dfrac{244\text{ in.}}{s/2} \Rightarrow s/2 = \dfrac{244\text{ in.}}{\tan 54^\circ} = 177.3\text{ in.}$ and $s = 2 \times (177.3\text{ in.}) = 355\text{ in.}$

Shop Trigonometry

31. The hypotenuse of each of the right triangles formed by constructing a vertical line from the x-axis to the center of the hole is 155 mm.

$\cos 60^\circ = \dfrac{x_1}{155\text{ mm}} \Rightarrow x_1 = (155\text{ mm}) \times \cos 60^\circ = 77.5\text{ mm}$

$\cos 30^\circ = \dfrac{x_2}{155\text{ mm}} \Rightarrow x_2 = (155\text{ mm}) \times \cos 30^\circ = 134\text{ mm}$

$\sin 60^\circ = \dfrac{y_1}{155\text{ mm}} \Rightarrow y_1 = (155\text{ mm}) \times \sin 60^\circ = 134\text{ mm}$

$\sin 30^\circ = \dfrac{y_2}{155\text{ mm}} \Rightarrow y_2 = (155\text{ mm}) \times \sin 30^\circ = 77.5\text{ mm}$

33. $\tan\left(\dfrac{3.5^\circ}{2}\right) = \dfrac{x}{9.000\text{ in.}} \Rightarrow x = (9.000\text{ in.}) \times \tan(1.75^\circ) = 0.275\text{ in.}$ and $d = 1.000\text{ in.} + 2(0.275\text{ in.}) = 1.55\text{ in.}$

35. Construct a radius from the center of the pulley to the point of contact of the measuring square with the pulley fragment. This is a tangent line and forms a right angle with the radius. We now have a $30^\circ - 60^\circ - 90^\circ$ special triangle with the base length equal to the radius of the pulley fragment and the hypotenuse equal to the radius

plus the 5.53 in distance from the pulley to the vertex of the measuring square and with $\sin 30^\circ = \frac{1}{2}$:

$$\sin 30^\circ = \frac{r}{r+5.53\text{ in.}} \Rightarrow \frac{1}{2}r + (5.53\text{ in.})\frac{1}{2} = r \Rightarrow r = 5.53\text{ in.}$$

37. This is similar to problem 22 in Section 6.2. Thus,

$$r^2 = \left(\frac{0.750\text{ cm}}{2}\right)^2 + \left(\frac{r}{2}\right)^2 \Rightarrow 4r^2 = 0.5625 + r^2 \Rightarrow 3r^2 = 0.5625 \Rightarrow r = \sqrt{\frac{.5625}{3}} = 0.433\text{ cm}$$

Now, as you can see from the figure, the width of the nut is two times the sides of the equilateral triangles formed above or equal 0.866 cm.

39. a) $\sin 25.60^\circ = \frac{x}{5\text{ in.}} \Rightarrow x = (5\text{ in.}) \times \sin 25.60^\circ = 2.160\text{ in.}$

b) $\sin\theta = \frac{1.554\text{ in.}}{5\text{ in.}} \Rightarrow \theta = \sin^{-1}\left(\frac{1.554\text{ in.}}{5\text{ in.}}\right) = 18.10^\circ$

c) Since the wedge is placed such that its side is parallel to the surface plane the angle will be the same as the angle opposite the page block. Thus, $\sin\theta = \frac{2.215\text{ in.}}{5\text{ in.}} \Rightarrow \theta = \sin^{-1}\left(\frac{2.215\text{ in.}}{5\text{ in.}}\right) = 26.30^\circ$

Exercise 7.4 • Angles in Standard Position

1. $r = \sqrt{2.25^2 + 4.82^2} = 5.32$

$\sin\theta = \frac{y}{r} = \frac{4.82}{5.32} = 0.906$

$\cos\theta = \frac{x}{r} = \frac{2.25}{5.32} = 0.423$

$\tan\theta = \frac{y}{x} = \frac{4.82}{2.25} = 2.14$

$\theta = \tan^{-1}\left(\frac{4.82}{2.25}\right) = 65.0^\circ$

3. $r = \sqrt{3.72^2 + 5.49^2} = 6.63$

$\sin\theta = \frac{y}{r} = \frac{5.49}{6.63} = 0.828$

$\cos\theta = \frac{x}{r} = \frac{3.72}{6.63} = 0.561$

$\tan\theta = \frac{y}{x} = \frac{5.49}{3.72} = 1.48$

$\theta = \tan^{-1}\left(\frac{5.49}{3.72}\right) = 55.8^\circ$

5. $r = \sqrt{1.93^2 + 4.83^2} = 5.20$

$\sin\theta = \frac{y}{r} = \frac{4.83}{5.20} = 0.929$

$\cos\theta = \frac{x}{r} = \frac{1.93}{5.20} = 0.371$

$\tan\theta = \frac{y}{x} = \frac{4.83}{1.93} = 2.50$

$\theta = \tan^{-1}\left(\frac{4.83}{1.93}\right) = 68.2^\circ$

Exercise 7.5 • Introduction to Vectors

Components of a Vector

1. $V_x = \cos 48.3^\circ \times 4.93 = 3.28$ $\quad V_y = \sin 48.3^\circ \times 4.93 = 3.68$

3. $V_x = \cos 58.24^\circ \times 1.884 = 0.9917$ $\quad V_y = \sin 58.24^\circ \times 1.884 = 1.602$

5. $V_x = \cos 45.2^\circ \times 836 = 589$ $\quad V_y = \sin 45.2^\circ \times 836 = 593$

7. $V_x = \cos 64.9^\circ \times 22.7 = 9.63$ $\quad V_y = \sin 64.9^\circ \times 22.7 = 20.6$

9. $V_x = \cos 77.3^\circ \times 18.4 = 4.05$ $\quad V_y = \sin 77.3^\circ \times 18.4 = 17.9$

Resultant of Two Perpendicular Vectors

11. $R = \sqrt{(483)^2 + (382)^2} = 616$ $\quad \theta = \tan^{-1}\dfrac{483}{382} = 51.7^\circ$

13. $R = \sqrt{(7364)^2 + (4837)^2} = 8811$ $\quad \theta = \tan^{-1}\dfrac{7364}{4837} = 56.7^\circ$

15. $R = \sqrt{(1.25)^2 + (2.07)^2} = 2.42$ $\quad \theta = \tan^{-1}\dfrac{1.25}{2.07} = 31.1^\circ$

17. $R = \sqrt{(6.82)^2 + (4.83)^2} = 8.36$ $\quad \theta = \tan^{-1}\dfrac{6.82}{4.83} = 54.7^\circ$

19. $R = \sqrt{(2.27)^2 + (3.97)^2} = 4.57$ $\quad \theta = \tan^{-1}\dfrac{2.27}{3.97} = 29.8^\circ$

Exercise 7.6 • Applications of Vectors

Force Vectors

1. Force is equal to the tangential component: $F_{\text{Tangent}} = (56.5N)(\sin 12.6^\circ) = 12.3N$

3. $W = (624\text{ N})(\cos 25.2^\circ) + (994\text{ N})(\cos 15.5^\circ) = 1520\text{ N}$

5. $\sin\theta = \dfrac{551\text{ N}}{1270\text{ N}} \Rightarrow \theta = \sin^{-1}\left(\dfrac{551\text{ N}}{1270\text{ N}}\right) = 25.7^\circ$

7. $F_{\text{Tangent}} = (18.6\text{ tons})(\sin 15.4^\circ) = 4.94\text{ tons}$

Velocity Vectors

9. $\tan 80°15' = \dfrac{V\text{west}}{20.5 \text{ km/h}} \Rightarrow V\text{west} = (20.5 \text{ km/h})(\tan 80°15') = 119 \text{ km/h}$

11. $V_y = (10.6 \text{ m/min})(\sin 32.5°) = 5.70 \text{ m/min}$ $\qquad$ $\text{Time} = \dfrac{10.0 \text{ m}}{5.70 \text{ m/min}} = 1.75 \text{ min}$

13. $\tan 5°15' = \dfrac{10.6 \text{ mi/h}}{V_{\text{Ship}}} \Rightarrow V_{\text{Ship}} = \dfrac{10.6 \text{ mi/h}}{\tan 5°15'} = 115 \text{ mi/h}$

Impedance Vectors

15. $\tan 72.0° = \dfrac{X}{115\,\Omega} \Rightarrow X = (115\,\Omega)\tan 72.0° = 354\,\Omega$ $\qquad$ $\cos 72.0° = \dfrac{115\,\Omega}{Z} \Rightarrow Z = \dfrac{115\,\Omega}{\cos 72.0°} = 372\,\Omega$

17. $Z = \sqrt{(5.75\,\Omega)^2 + (4.22\,\Omega)^2} = 7.13\,\Omega$ $\qquad$ $\theta = \tan^{-1}\left(\dfrac{5.75\,\Omega}{4.22\,\Omega}\right) = 53.7°$

Chapter 7 • Review Problems

1. $\sin 72.9° = 0.9558$ $\qquad$ $\cos 72.9° = 0.2940$ $\qquad$ $\tan 72.9° = 3.2506$

3. $\sin 60.16° = 0.8674$ $\qquad$ $\cos 60.16° = 0.4976$ $\qquad$ $\tan 60.16° = 1.7433$

5. $\cos\theta = 0.824 \Rightarrow \theta = \cos^{-1} 0.824 = 34.5°$

7. $\arcsin 0.377 = 22.1°$

9. $\arctan 1.43 = 55.0°$

11. $B = 90° - 28.5° = 61.5°$

$\tan 28.5° = \dfrac{a}{3.72} \Rightarrow a = 3.72 \tan 28.5° = 2.02$

$\cos 28.5° = \dfrac{3.72}{c} \Rightarrow c = \dfrac{3.72}{\cos 28.5°} = 4.23$

13. $V_x = 885 \cos 66.3° = 356$ $\qquad$ $V_y = 885 \sin 66.3° = 810$

15. $R = \sqrt{385^2 + 275^2} = 473$ $\qquad$ $\theta = \tan^{-1}\left(\dfrac{275}{385}\right) = 35.5^0$

17. $\tan 22.5° = \dfrac{\text{Steeple}_{\text{Top}}}{125 \text{ ft}} \Rightarrow \text{Steeple}_{\text{Top}} = (125 \text{ ft})\tan 22.5° = 51.78 \text{ ft}$

$\tan 19.6° = \dfrac{\text{Steeple}_{\text{Bottom}}}{125 \text{ ft}} \Rightarrow \text{Steeple}_{\text{Bottom}} = (125 \text{ ft})\tan 19.6° = 44.51 \text{ ft}$

height of the steeple $= 51.75 \text{ ft} - 44.51 \text{ ft} = 7.27 \text{ ft}$

19. $\cos 59.2^\circ = 0.5120$

21. $\tan 52.8^\circ = 1.3175$

23. $\cos 24.7^\circ = 0.9085$

25. $\theta = \tan^{-1}(1.7362) = 60.1^\circ$

27. $\theta = \sin^{-1}(0.7253) = 46.5^\circ$

29. $\theta = \cos^{-1}(0.9475) = 18.6^\circ$

31. $AC = \sqrt{(158.2\text{ ft})^2 - (139.3\text{ ft})^2} = 74.98\text{ ft}$

$B = \cos^{-1}\left(\dfrac{139.3\text{ ft}}{158.2\text{ ft}}\right) = 28.3^\circ = \text{N}28^\circ 18'\text{E}$

Chapter 8: Oblique Triangles and Vectors

Exercise 8.1 • Trigonometric Functions of Any Angle

1. $r=\sqrt{3.00^2+(-5.00)^2}=5.83$

$\sin\theta=\frac{y}{r}=\frac{-5.00}{5.83}=-0.828$ $\cos\theta=\frac{x}{r}=\frac{3.00}{5.83}=0.514$ $\tan\theta=\frac{y}{x}=\frac{-5.00}{3.00}=-1.67$

$\csc\theta=\frac{r}{y}=\frac{5.83}{-5.00}=1.17$ $\sec\theta=\frac{r}{x}=\frac{5.83}{3.00}=1.95$ $\cot\theta=\frac{x}{y}=\frac{3.00}{-5.00}=-0.600$

3. $r=\sqrt{24.0^2+(-7.00)^2}=25.0$

$\sin\theta=\frac{y}{r}=\frac{-7.00}{25.0}=-0.280$ $\cos\theta=\frac{x}{r}=\frac{24.0}{25.0}=0.960$ $\tan\theta=\frac{y}{x}=\frac{-7.00}{24.0}=-0.292$

$\csc\theta=\frac{r}{y}=\frac{25.0}{-7.00}=-3.57$ $\sec\theta=\frac{r}{x}=\frac{25.0}{24.0}=1.04$ $\cot\theta=\frac{x}{y}=\frac{24.0}{-7.00}=-3.43$

5. $r=\sqrt{1.59^2+(-3.11)^2}=3.49$

$\sin\theta=\frac{y}{r}=\frac{-3.11}{3.49}=-0.890$ $\cos\theta=\frac{x}{r}=\frac{1.59}{3.49}=0.455$ $\tan\theta=\frac{y}{x}=\frac{-3.11}{1.59}=-1.96$

$\csc\theta=\frac{r}{y}=\frac{3.49}{-3.11}=-1.12$ $\sec\theta=\frac{r}{x}=\frac{3.49}{1.59}=2.20$ $\cot\theta=\frac{x}{y}=\frac{1.59}{-3.11}=-0.511$

Trigonometric Functions of Any Angle by Calculator

7. $\sin 101^\circ=0.9816$ $\cos 101^\circ=-0.1908$ $\tan 101^\circ=-5.145$

9. $\sin 331^\circ=-0.4848$ $\cos 331^\circ=0.8746$ $\tan 331^\circ=-0.5543$

11. $\sin(-62.85^\circ)=-0.8898$ $\cos(-62.85^\circ)=0.4563$ $\tan(-62.85^\circ)=-1.950$

13. $\sin 486^\circ=0.8090$ $\cos 486^\circ=-0.5878$ $\tan 486^\circ=-1.376$

15. $\sin 114^\circ 23'=0.9108$ $\cos 114^\circ 23'=-0.4128$ $\tan 114^\circ 23'=-2.206$

17. $\sin 412^\circ=0.7780$ $\cos 412^\circ=0.6157$ $\tan 412^\circ=1.280$

Reciprocal Relationships Evaluate to Four Decimal Points

19. $\csc\theta=\frac{1}{\sin\theta}=\frac{1}{0.7352}=1.3602$

21. $\sec\theta=\frac{1}{\cos\theta}=\frac{1}{0.7354}=1.3598$

Cotangent, Secant, and Cosecant by Calculator Evaluate to Four Decimal Points

23. $\cot 153.6^\circ = \dfrac{1}{\tan 153.6^\circ} = -2.0145$

25. $\csc 207.4^\circ = \dfrac{1}{\sin 207.4^\circ} = -2.1730$

27. $\cot 228.7^\circ = \dfrac{1}{\tan 228.7^\circ} = 0.8785$

Cofunctions Express as a Function of the Complementary Angle.

29. $\cos 73^\circ = \sin\left(90^\circ - 73^\circ\right) = \sin 17^\circ$

31. $\sec 85.6^\circ = \csc\left(90^\circ - 85.6^\circ\right) = \csc 4.4^\circ$

33. $\csc 82.7^\circ = \sin\left(90^\circ - 82.7^\circ\right) = \sin 7.3^\circ$

35. $\sin 215^\circ = -0.574$ $\cos 215^\circ = -0{,}718$

Exercise 8.2 • Finding the Angle When the Trigonometric Function is KnownSolution of Right Triangles

Reference Angle

1. $\theta_{\text{ref}} = 180^\circ - 163^\circ = 17^\circ$

3. $\theta_{\text{ref}} = 360^\circ - 305^\circ = 55^\circ$

5. $\theta_{\text{ref}} = 249.3^\circ - 180^\circ = 69.3^\circ$

Algebraic Signs of the Trigonometric Functions

7. IV
9. II
11. IV
13. I or II
15. I or IV
17. negative
19. positive
21. negative

23. $\sin 206^\circ = \dfrac{y}{r} = \dfrac{(-)}{(+)} = \text{negative}$ $\cos 206^\circ = \dfrac{x}{r} = \dfrac{(-)}{(+)} = \text{negative}$ $\tan 206^\circ = \dfrac{y}{x} = \dfrac{(-)}{(-)} = \text{positive}$

25. $\sin\left(-48^\circ\right) = \dfrac{y}{r} = \dfrac{(-)}{(+)} = \text{negative}$ $\cos\left(-48^\circ\right) = \dfrac{x}{r} = \dfrac{(+)}{(+)} = \text{positive}$ $\tan\left(-48^\circ\right) = \dfrac{y}{x} = \dfrac{(-)}{(+)} = \text{negative}$

Finding the Angle by Calculator

27. $\theta_{\text{I}} = \theta_{\text{Ref}} = \sin^{-1} 0.7761 = 50.9^\circ$ and $\theta_{\text{II}} = 180^\circ - 50.9^\circ = 129.1^\circ$

29. $\theta_{\text{I}} = \theta_{\text{Ref}} = \cos^{-1} 0.8372 = 33.2^\circ$ and $\theta_{\text{IV}} = 360^\circ - 33.2^\circ = 326.8^\circ$

31. $\theta_{\text{I}} = \theta_{\text{Ref}} = \tan^{-1} 6.372 = 81.1^\circ$ and $\theta_{\text{III}} = 180^\circ + 81.1^\circ = 261.1^\circ$

33. $\sin^{-1}(-0.6358) = -39.5^\circ$ $\theta_{\text{Ref}} = 39.5^\circ$ $\theta_{\text{III}} = 180^\circ + 39.5^\circ = 219.5^\circ$ and $\theta_{\text{IV}} = 360^\circ - 39.5^\circ = 320.5^\circ$

Inverse of the Cotangent, Secant, and Cosecant

35. $\theta_{\text{I}} = \theta_{\text{Ref}} = \cos^{-1}\left(\dfrac{1}{1.7361}\right) = 54.8^\circ$ and $\theta_{\text{IV}} = 360^\circ - 54.8^\circ = 305.2^\circ$

37. $\sin^{-1}\left(\dfrac{1}{-3.852}\right) = -15.0^\circ$ $\theta_{\text{Ref}} = 15.0^\circ$ $\theta_{\text{III}} = 180^\circ + 15.0^\circ = 195.0^\circ$ and $\theta_{\text{IV}} = 360^\circ - 15.0^\circ = 345.0^\circ$

Exercise 8.3 • Law of Sines

Two Angles and One Side Known (AAS or ASA cases)

1. a) $A = 61.9^\circ, C = 47.0^\circ$, and $a = 7.65$

$B = 180^\circ - (61.9^\circ + 47.0^\circ) = 71.1^\circ$

$$\frac{b}{\sin 71.1^\circ} = \frac{7.65}{\sin 61.9^\circ} \Rightarrow b = \frac{7.65 \sin 71.1^\circ}{\sin 61.9^\circ} = 8.20$$

$$\frac{c}{\sin 47.0^\circ} = \frac{7.65}{\sin 61.9^\circ} \Rightarrow c = \frac{7.65 \sin 47.0^\circ}{\sin 61.9^\circ} = 6.34$$

b) $A = 126^\circ$, $B = 27.0^\circ$, and $\text{b} = 119$

$C = 180^\circ - (126^\circ + 27.0^\circ) = 27.0^\circ$

$$\frac{a}{\sin 126^\circ} = \frac{119}{\sin 27.0^\circ} \Rightarrow a = \frac{119 \sin 126^\circ}{\sin 27.0^\circ} = 212$$

$$\frac{c}{\sin 27.0^\circ} = \frac{119}{\sin 27.0^\circ} \Rightarrow c = \frac{119 \sin 27.0^\circ}{\sin 27.0^\circ} = 119 \quad \text{(An isosceles triangle)}$$

c) $A = 31.6^\circ, C = 44.8^\circ$, and $a = 11.7$

$B = 180^\circ - (31.6^\circ + 44.8^\circ) = 103.6^\circ$

$$\frac{b}{\sin 103.6^\circ} = \frac{11.7}{\sin 31.6^\circ} \Rightarrow b = \frac{11.7 \sin 103.6^\circ}{\sin 31.6^\circ} = 21.7$$

$$\frac{c}{\sin 44.8^\circ} = \frac{11.7}{\sin 31.6^\circ} \Rightarrow c = \frac{11.7 \sin 44.8^\circ}{\sin 31.6^\circ} = 15.7$$

d) $A = 15.0^\circ, C = 72.0^\circ$, and $\text{c} = 375$

$B = 180^\circ - (15.0^\circ + 72.0^\circ) = 93.0^\circ$

$$\frac{a}{\sin 15.0^\circ} = \frac{375}{\sin 72.0^\circ} \Rightarrow a = \frac{375 \sin 15.0^\circ}{\sin 72.0^\circ} = 102$$

$$\frac{b}{\sin 93.0^\circ} = \frac{375}{\sin 72.0^\circ} \Rightarrow b = \frac{375 \sin 93.0^\circ}{\sin 72.0^\circ} = 394$$

3. $A = 24.14°$, $B = 38.27°$, and $a = 5562$

$$C = 180° - (24.14° + 38.27°) = 117.59°$$

$$\frac{b}{\sin 38.27°} = \frac{5562}{\sin 24.14°} \Rightarrow b = \frac{5562 \sin 38.27°}{\sin 24.14°} = 8423$$

$$\frac{c}{\sin 117.59°} = \frac{5562}{\sin 24.14°} \Rightarrow c = \frac{5562 \sin 117.59°}{\sin 24.14°} = 12{,}050$$

5. $A = 44.47°, C = 63.88°$, and $c = 1.065$

$$B = 180° - (44.47° + 63.88°) = 71.65°$$

$$\frac{a}{\sin 44.47°} = \frac{1.065}{\sin 63.88°} \Rightarrow a = \frac{1.065 \sin 44.47°}{\sin 63.88°} = 0.8309$$

$$\frac{b}{\sin 71.65°} = \frac{1.065}{\sin 63.88°} \Rightarrow b = \frac{1.065 \sin 71.65°}{\sin 63.88°} = 1.126$$

Two Sides and One Angle Known (SSA or ambiguous case)

7. a) $A = 46.3°$, $a = 304$, and $b = 228$

$$\frac{\sin B}{228} = \frac{\sin 46.3°}{304} \Rightarrow B = \sin^{-1}\left(\frac{228 \sin 46.3°}{304}\right) = 32.8°$$

$$C = 180° - (46.3° + 32.8°) = 101°$$

$$\frac{c}{\sin 101°} = \frac{304}{\sin 46.3°} \Rightarrow c = \frac{304 \sin 101°}{\sin 46.3°} = 413$$

b) $A = 65.9°$, $a = 1.59$, and $c = 1.46$

$$\frac{\sin C}{1.46} = \frac{\sin 65.9°}{1.59} \Rightarrow C = \sin^{-1}\left(\frac{1.46 \sin 65.9°}{1.59}\right) = 57.0°$$

$$B = 180° - (65.9° + 57.0°) = 57.1°$$

$$\frac{b}{\sin 57.1°} = \frac{1.59}{\sin 65.9°} \Rightarrow b = \frac{1.59 \sin 57.1°}{\sin 65.9°} = 1.46$$

c) $A = 43.0°$, $a = 21.0$, and $b = 15.0$

$$\frac{\sin B}{15.0} = \frac{\sin 43.0°}{21.0} \Rightarrow B = \sin^{-1}\left(\frac{15.0 \sin 43.0°}{21.0}\right) = 29.2°$$

$$C = 180° - (43.0° + 29.2°) = 107.8°$$

$$\frac{c}{\sin 107.8°} = \frac{21.0}{\sin 43.0°} \Rightarrow c = \frac{21.0 \sin 107.8°}{\sin 43.0°} = 29.3$$

9. $C = 61.7^\circ$, $b = 284$, and $c = 382$

$$\frac{\sin B}{284} = \frac{\sin 61.7^\circ}{382} \Rightarrow B = \sin^{-1}\left(\frac{284\sin 61.7^\circ}{382}\right) = 40.9^\circ$$

Other possible $B' = 180^\circ - 40.9^\circ = 139.1^\circ$. However $139.1^\circ + 61.7^\circ > 180^\circ$, so B' is not a possible solution.

$A = 180^\circ - \left(40.9^\circ + 61.7^\circ\right) = 77.4^\circ$

$$\frac{a}{\sin 77.4^\circ} = \frac{382}{\sin 61.7^\circ} \Rightarrow a = \frac{382\sin 77.4^\circ}{\sin 61.7^\circ} = 423$$

11. $A = 45.6^\circ$, $a = 7.83$, and $c = 10.4$

$$\frac{\sin C}{10.4} = \frac{\sin 45.6^\circ}{7.83} \Rightarrow C = \sin^{-1}\left(\frac{10.4\sin 45.6^\circ}{7.83}\right) = 71.6^\circ$$

Other possible $C' = 180^\circ - 71.6^\circ = 108.4^\circ$. However $108.4^\circ + 45.6^\circ < 180^\circ$, so C' is also a solution.

$B = 180^\circ - \left(45.6^\circ + 71.6^\circ\right) = 62.8^\circ$

$$\frac{b}{\sin 62.8^\circ} = \frac{7.83}{\sin 45.6^\circ} \Rightarrow b = \frac{7.83\sin 62.8^\circ}{\sin 45.6^\circ} = 9.75$$

$B' = 180^\circ - \left(45.6^\circ + 108.4^\circ\right) = 26.0^\circ$

$$\frac{b'}{\sin 26.0^\circ} = \frac{7.83}{\sin 45.6^\circ} \Rightarrow b' = \frac{7.83\sin 26.0^\circ}{\sin 45.6^\circ} = 4.80$$

An Application

13. First construct a right triangle by making a side in line with the telephone pole. The angle opposite that pole will be $A + 11.6^\circ$. Since this is a right triangle $48.0^\circ + \left(A + 11.6^\circ\right) = 90^\circ \Rightarrow A = 90^\circ - 48.0^\circ - 11.6^\circ = 30.4^\circ$.

The other angle at the base and left of the pole then is $L = 180^\circ - \left(30.4^\circ + 48.0^\circ\right) = 101.6^\circ$. Then the length of the left wire (l) can be found $\frac{l}{\sin 101.6^\circ} = \frac{185}{\sin 30.4^\circ} \Rightarrow l = \frac{185\sin 101.6^\circ}{\sin 30.4^\circ} = 358$ ft

To find the length of the right wire (r), the angle at the base and right of the pole will simply be $R = 180^\circ - 101.6^\circ = 78.4^\circ$. Then the angle opposite the pole on the right is $C = 180^\circ - \left(48.0^\circ + 78.4^\circ\right) = 53.6^\circ$.

Which gives $\frac{r}{\sin 78.4^\circ} = \frac{185}{\sin 53.6^\circ} \Rightarrow r = \frac{185\sin 78.4^\circ}{\sin 53.6^\circ} = 225$ ft .

Exercise 8.4 • Law of Cosines

Two Sides and One Angle Known (SAS case)

1. a) $C = 106.0^\circ$, $a = 15.7$, and $b = 11.2$

$$c^2 = a^2 + b^2 - 2ab\cos C \Rightarrow c = \sqrt{(15.7)^2 + (11.2)^2 - 2(15.7)(11.2)\cos 106.0^\circ} = 21.7$$

$$\frac{\sin A}{15.7} = \frac{\sin 106.0^\circ}{21.7} \Rightarrow A = \sin^{-1}\left(\frac{15.7\sin 106.0^\circ}{21.7}\right) = 44.1^\circ$$

$B = 180^\circ - \left(44.1^\circ + 106.0^\circ\right) = 29.9^\circ$

b) $B = 51.4°$, $a = 1.95$, and $c = 1.46$

$$b^2 = a^2 + c^2 - 2ac\cos B \;\Rightarrow\; b = \sqrt{(1.95)^2 + (1.46)^2 - 2(1.95)(1.46)\cos 51.4°} = 1.54$$

$$\frac{\sin A}{1.95} = \frac{\sin 51.4°}{1.54} \;\Rightarrow\; A = \sin^{-1}\left(\frac{1.95\sin 51.4°}{1.54}\right) = 81.7°$$

$$C = 180° - \left(81.7° + 51.4°\right) = 46.9°$$

c) $A = 63.8°$, $b = 18.3$, and $c = 21.7$

$$a^2 = b^2 + c^2 - 2bc\cos A \;\Rightarrow\; a = \sqrt{(18.3)^2 + (21.7)^2 - 2(18.3)(21.7)\cos 63.8°} = 21.3$$

$$\frac{\sin B}{18.3} = \frac{\sin 63.8°}{21.3} \;\Rightarrow\; B = \sin^{-1}\left(\frac{18.3\sin 63.8°}{21.3}\right) = 50.4°$$

$$C = 180° - \left(63.8° + 50.4°\right) = 65.8°$$

d) $C = 46.3°$, $a = 728$, and $b = 906$

$$c^2 = a^2 + b^2 - 2ab\cos C \;\Rightarrow\; c = \sqrt{(728)^2 + (906)^2 - 2(728)(906)\cos 46.3°} = 663$$

$$\frac{\sin A}{728} = \frac{\sin 46.3°}{663} \;\Rightarrow\; A = \sin^{-1}\left(\frac{728\sin 46.3°}{663}\right) = 52.5°$$

$$B = 180° - \left(52.5° + 46.3°\right) = 81.2°$$

3. $A = 115°$, $b = 46.8$, and $c = 51.3$

$$a^2 = b^2 + c^2 - 2bc\cos A \;\Rightarrow\; a = \sqrt{(46.8)^2 + (51.3)^2 - 2(46.8)(51.3)\cos 115°} = 82.8$$

$$\frac{\sin B}{46.8} = \frac{\sin 115°}{82.8} \;\Rightarrow\; B = \sin^{-1}\left(\frac{46.8\sin 115°}{82.8}\right) = 30.8°$$

$$C = 180° - \left(115° + 30.8°\right) = 34.2°$$

5. $B = 129°$, $a = 186$, and $c = 179$

$$b^2 = a^2 + c^2 - 2ac\cos B \;\Rightarrow\; b = \sqrt{(186)^2 + (179)^2 - 2(186)(179)\cos 129°} = 329$$

$$\frac{\sin A}{186} = \frac{\sin 129°}{329} \;\Rightarrow\; A = \sin^{-1}\left(\frac{186\sin 129°}{329}\right) = 26.1°$$

$$C = 180° - \left(26.1° + 129°\right) = 24.9°$$

Three Sides Known (SSS case)

7. a) $a = 128,\ b = 152,$ and $c = 70.1$

Find the largest angle first (angle opposite the longest side)

$$b^2 = a^2 + c^2 - 2ac\cos B \ \Rightarrow\ B = \cos^{-1}\left(\frac{a^2 + c^2 - b^2}{2ac}\right) = \cos^{-1}\left(\frac{(128)^2 + (70.1)^2 - (152)^2}{2(128)(70.1)}\right) = 95.8^\circ$$

$$\frac{\sin A}{128} = \frac{\sin 95.8^\circ}{152} \ \Rightarrow\ A = \sin^{-1}\left(\frac{128\sin 95.8^\circ}{152}\right) = 56.9^\circ$$

$$C = 180^\circ - \left(56.9^\circ + 95.8^\circ\right) = 27.3^\circ$$

b) $a = 1.16,\ b = 2.82,$ and $c = 1.95$

Find the largest angle first (angle opposite the longest side)

$$b^2 = a^2 + c^2 - 2ac\cos B \ \Rightarrow\ B = \cos^{-1}\left(\frac{a^2 + c^2 - b^2}{2ac}\right) = \cos^{-1}\left(\frac{(1.16)^2 + (1.95)^2 - (2.82)^2}{2(1.16)(1.95)}\right) = 128^\circ$$

$$\frac{\sin A}{1.16} = \frac{\sin 128^\circ}{2.82} \ \Rightarrow\ A = \sin^{-1}\left(\frac{1.16\sin 128^\circ}{2.82}\right) = 18.9^\circ$$

$$C = 180^\circ - \left(18.9^\circ + 128^\circ\right) = 33.1^\circ$$

c) $a = 157,\ b = 112,$ and $c = 217$

Find the largest angle first (angle opposite the longest side)

$$c^2 = a^2 + b^2 - 2ab\cos C \ \Rightarrow\ C = \cos^{-1}\left(\frac{a^2 + b^2 - c^2}{2ab}\right) = \cos^{-1}\left(\frac{(157)^2 + (112)^2 - (217)^2}{2(157)(112)}\right) = 106^\circ$$

$$\frac{\sin A}{157} = \frac{\sin 106^\circ}{217} \ \Rightarrow\ A = \sin^{-1}\left(\frac{157\sin 106^\circ}{217}\right) = 44.1^\circ$$

$$B = 180^\circ - \left(44.1^\circ + 106^\circ\right) = 29.9^\circ$$

d) $a = 77.3,\ b = 81.4,$ and $c = 48.1$

Find the largest angle first (angle opposite the longest side)

$$b^2 = a^2 + c^2 - 2ac\cos B \ \Rightarrow\ B = \cos^{-1}\left(\frac{a^2 + c^2 - b^2}{2ac}\right) = \cos^{-1}\left(\frac{(77.3)^2 + (48.1)^2 - (81.4)^2}{2(77.3)(48.1)}\right) = 77.1^\circ$$

$$\frac{\sin A}{77.3} = \frac{\sin 77.1^\circ}{81.4} \ \Rightarrow\ A = \sin^{-1}\left(\frac{77.3\sin 77.1^\circ}{81.4}\right) = 67.8^\circ$$

$$C = 180^\circ - \left(67.8^\circ + 77.1^\circ\right) = 35.1^\circ$$

9. $a = 1.475,\ b = 1.836,\ \text{and}\ c = 2.017$
Find the largest angle first (angle opposite the longest side)

$$c^2 = a^2 + b^2 - 2ab\cos C \ \Rightarrow\ C = \cos^{-1}\left(\frac{a^2+b^2-c^2}{2ab}\right) = \cos^{-1}\left(\frac{(1.475)^2+(1.836)^2-(2.017)^2}{2(1.475)(1.836)}\right) = 74.16^\circ$$

$$\frac{\sin A}{1.475} = \frac{\sin 74.16^\circ}{2.017} \ \Rightarrow\ A = \sin^{-1}\left(\frac{1.475\sin 74.16^\circ}{2.017}\right) = 44.71^\circ$$

$$B = 180^\circ - \left(44.71^\circ + 74.16^\circ\right) = 61.13^\circ$$

11. $a = 18.6,\ b = 32.9,\ \text{and}\ c = 17.9$
Find the largest angle first (angle opposite the longest side)

$$b^2 = a^2 + c^2 - 2ac\cos B \ \Rightarrow\ B = \cos^{-1}\left(\frac{a^2+c^2-b^2}{2ac}\right) = \cos^{-1}\left(\frac{(18.6)^2+(17.9)^2-(32.9)^2}{2(18.6)(17.9)}\right) = 129^\circ$$

$$\frac{\sin A}{18.6} = \frac{\sin 129^\circ}{32.9} \ \Rightarrow\ A = \sin^{-1}\left(\frac{18.6\sin 129^\circ}{32.9}\right) = 26.1^\circ$$

$$C = 180^\circ - \left(26.1^\circ + 129^\circ\right) = 24.9^\circ$$

An Application

13. Use Law of Cosines to find the distance PQ

$$PQ = \sqrt{(596\text{ ft})^2 + (718\text{ ft})^2 - 2(596\text{ ft})(718\text{ ft})\cos 69.5^\circ} = 756\text{ ft}$$

Exercise 8.5 • Applications

Determining Inaccessible Distances

1. $c = AB = 88.6\text{ m},\ A = 74.3^\circ,\ \text{and}\ C = 85.4^\circ \qquad a = BC \qquad b = AC$

$$B = 180^\circ - \left(74.3^\circ + 85.4^\circ\right) = 20.3^\circ$$

$$\frac{a}{\sin 74.3^\circ} = \frac{88.6\text{ m}}{\sin 85.4^\circ} \ \Rightarrow\ a = BC = \frac{(88.6\text{ m})\sin 74.3^\circ}{\sin 85.4^\circ} = 85.6\text{ m}$$

$$\frac{b}{\sin 20.3^\circ} = \frac{88.6\text{ m}}{\sin 85.4^\circ} \ \Rightarrow\ b = AC = \frac{(88.6\text{ m})\sin 20.3^\circ}{\sin 85.4^\circ} = 30.8\text{ m}$$

3. Let $a = 115\text{ m},\ b = 187\text{ m},\ \text{and}\ c = 215\text{ m}$
Find the largest angle first (angle opposite the longest side)

$$c^2 = a^2 + b^2 - 2ab\cos C \ \Rightarrow\ C = \cos^{-1}\left(\frac{a^2+b^2-c^2}{2ab}\right) = \cos^{-1}\left(\frac{(115\text{ m})^2+(187\text{ m})^2-(215\text{ m})^2}{2(115\text{ m})(187\text{ m})}\right) = 87.4^\circ$$

$$\frac{\sin A}{115} = \frac{\sin 87.4^\circ}{215} \ \Rightarrow\ A = \sin^{-1}\left(\frac{115\sin 87.4^\circ}{215}\right) = 32.3^\circ$$

$$B = 180^\circ - \left(32.3^\circ + 87.4^\circ\right) = 60.3^\circ$$

Navigation

5. In the figure below, let S be the current location of the ship, H the current location of the helicopter, and M the location they will meet in t hours. Using Law of Sines gives

$$\frac{\sin H}{15.0t \text{ km}} = \frac{\sin 105^\circ}{22.0t \text{ km}} \Rightarrow H = \sin^{-1}\left(\frac{15.0\sin 105^\circ}{22.0}\right) = 41.2^\circ \text{, or } 90^\circ - 41.2^\circ = \text{N } 48.86^\circ \text{ W}.$$

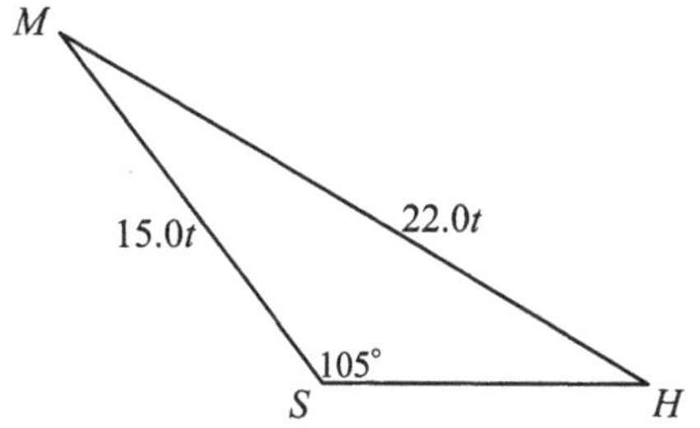

7. As in Problem 5, let S be the current location of the ship, L the current location of the launch, and M the location they will meet in t hours. Using Law of Sines gives

$$\frac{\sin L}{8.60t \text{ mi}} = \frac{\sin\left(90^\circ - 24.25^\circ\right)}{15.4t \text{ mi}} \Rightarrow L = \sin^{-1}\left(\frac{8.60\sin 65.75^\circ}{15.4}\right) = 30.6^\circ \text{, or } 90^\circ - 30.6^\circ = \text{S } 59.4^\circ \text{ E}.$$

9. The distance the plane traveled is $\left(\frac{1}{3}\text{ hr}\right)\left(625\frac{\text{km}}{\text{h}}\right) = 208.3 \text{ km}$. The angles in the formed triangle will be $90^\circ - 31.0^\circ = 59.0^\circ$, $90^\circ + 11.0^\circ = 101.0^\circ$, and $180^\circ - \left(59.0^\circ + 101.0^\circ\right) = 20.0^\circ$. Then using Law of Sines

$$\frac{p}{\sin 101.0^\circ} = \frac{208.3 \text{ km}}{\sin 20.0^\circ} \Rightarrow P = \frac{(208.3 \text{ km})\sin 101.0^\circ}{\sin 20.0^\circ} = 598 \text{ km}.$$

Structures

11. First construct a triangle with the top of the antenna and the two points uphill from the base of the antenna. Let L be the first point on the hill with angle $L = 180^\circ - \left(61^\circ + 8.7^\circ\right) = 110.3^\circ$. Let H be the second point on uphill from the base with angle of and will have a measure of $H = 38.0^\circ + 8.7^\circ = 46.7^\circ$. Then the angle formed at the top of the antenna will be $T = 180^\circ - \left(46.7^\circ + 110.3^\circ\right) = 23.0^\circ$. Use Law of Sines to get the measure of the side opposite angle L: $\frac{l}{\sin 110.3^\circ} = \frac{16.0 \text{ m}}{\sin 23.0^\circ} \Rightarrow l = \frac{(16.0 \text{ m})\sin 110.3^\circ}{\sin 23.0^\circ} = 38.4 \text{ m}$. Then construct a second triangle using side l, the antenna, and the part of the hill from the base of the antenna to the side l. The angle at the base of the antenna will be $90^\circ - 8.70^\circ = 81.3^\circ$. Then using Law of Sines with this triangle to find the height of the antenna, a, gives $\frac{a}{\sin 46.7^\circ} = \frac{38.4 \text{ m}}{\sin 81.3^\circ} \Rightarrow a = \frac{(38.4 \text{ m})\sin 46.7^\circ}{\sin 81.3^\circ} = 28.3 \text{ m}$.

13. Let the top of the mast be B and the bottom be D and the cable ends on the ground be A (downhill point) and C (uphill point).

$m\angle BDC = 90^\circ - 15.6^\circ = 74.4^\circ$ and $C = 180^\circ - \left(42.5^\circ + 74.4^\circ\right) = 63.1^\circ$

$$\frac{BC}{\sin 74.4^\circ} = \frac{71.6\text{ m}}{\sin 63.1^\circ} \Rightarrow BC = \frac{(71.6\text{ m})\sin 74.4^\circ}{\sin 63.1^\circ} = 77.3\text{ m}$$

$m\angle BDA = 180^\circ - 74.4^\circ = 105.6^\circ$ and $A = 180^0 - \left(42.5^0 + 105.6^0\right) = 31.9^0$

$$\frac{AB}{\sin 105.6^\circ} = \frac{71.6\text{ m}}{\sin 31.9^\circ} \Rightarrow AB = \frac{(71.6\text{ m})\sin 105.6^\circ}{\sin 31.9^\circ} = 131.0\text{ m}$$

15. $d = \sqrt{(52.6\text{ ft})^2 + (67.5\text{ ft})^2 - 2(52.6\text{ ft})(67.5\text{ ft})\cos 125^\circ} = 107\text{ ft}$

Mechanisms

17. First find the angle formed at W.

$$\frac{\sin W}{112\text{ mm}} = \frac{\sin 35.7^\circ}{255\text{ mm}} \Rightarrow W = \sin^{-1}\left(\frac{(112\text{ mm})\sin 35.7^\circ}{255\text{ mm}}\right) = 14.85^\circ$$

The angle opposite the distance in question will be $A = 180^\circ - \left(35.7^\circ + 14.85^\circ\right) = 129.45^\circ$.

Find the distance using Law of Sines

$$\frac{x}{\sin 129.45^\circ} = \frac{255\text{ mm}}{\sin 35.7^\circ} \Rightarrow x = \frac{(255\text{ mm})\sin 129.45^\circ}{\sin 35.7^\circ} = 337\text{ mm}$$

19. $AB = \sqrt{(35.6\text{ cm})^2 + (22.8\text{ cm})^2 - 2(35.6\text{ cm})(22.8\text{ cm})\cos 66.3^\circ} = 33.7\text{ cm}$

Geometry

21. $AC = \sqrt{(72.1\text{ in.})^2 + (105\text{ in.})^2 - 2(72.1\text{ in.})(105\text{ in.})\cos 63.0^\circ} = 96.69\text{ in.}$

$$\frac{\sin(m\angle ACD)}{105\text{ in.}} = \frac{\sin 63.0^\circ}{96.69\text{ in.}} \Rightarrow m\angle ACD = \sin^{-1}\left(\frac{(105\text{ in.})\sin 63.0^\circ}{96.69\text{ in.}}\right) = 75.37^\circ$$

$m\angle ACB = 121.0^\circ - 75.37^\circ = 45.63^\circ$

$AB = \sqrt{(43.0\text{ in.})^2 + (96.69\text{ in.})^2 - 2(43.0\text{ in.})(96.69\text{ in.})\cos 45.63^\circ} = 73.4\text{ in.}$

23. The angles opposite the diagonal measure $D = 180^\circ - 22.7^\circ - 15.4^\circ = 141.9^\circ$.
Then use the Law of Sines to find the sides. Let s be the length of the short sides and l be the length of the long sides.

$$\frac{l}{\sin 22.7^\circ} = \frac{125\text{ mm}}{\sin 141.9^\circ} \Rightarrow l = \frac{(125\text{ mm})\sin 22.7^\circ}{\sin 141.9^\circ} = 78.2\text{ mm}$$

$$\frac{s}{\sin 15.4^\circ} = \frac{125\text{ mm}}{\sin 141.9^\circ} \Rightarrow s = \frac{(125\text{ mm})\sin 15.4^\circ}{\sin 141.9^\circ} = 53.8\text{ mm}$$

25. Let D be the midpoint on c.

$$\frac{\sin(m\angle ADC)}{112}=\frac{112\sin 62.3^\circ}{186} \Rightarrow m\angle ADC=\sin^{-1}\left(\frac{112\sin 62.3^\circ}{186}\right)=32.22^\circ$$

$$m\angle DCA=180^\circ-\left(52.3^\circ+32.22^\circ\right)=85.48^\circ$$

$$\frac{AD}{\sin 85.48^\circ}=\frac{186}{\sin 62.3^\circ} \Rightarrow AD=\frac{186\sin 85.48^\circ}{\sin 62.3^\circ}=209.4$$

$$c=2(AD)=2(209.4)=419$$

27. Let x = the proportional constant. Then, $3x+4x+5x=180^\circ \Rightarrow 12x=180^\circ \Rightarrow x=15^\circ$

Let $A=3\left(15^\circ\right)=45^\circ$, $B=4\left(15^\circ\right)=60^\circ$, $C=5\left(15^\circ\right)=75^\circ$, and $a=994$ (shortest side)

$$\frac{b}{\sin 60^\circ}=\frac{994}{\sin 45^\circ} \Rightarrow b=\frac{994\sin 60^\circ}{\sin 45^\circ}=1220$$

$$\frac{c}{\sin 75^\circ}=\frac{994}{\sin 45^\circ} \Rightarrow c=\frac{994\sin 75^\circ}{\sin 45^\circ}=1360$$

29. The angle formed at the intersection at the top of the solar panel with a sun ray is

$$A=180^\circ-\left(18.5^\circ+41.6^\circ\right)=119.9^\circ$$

$$\frac{x}{\sin 119.9^\circ}=\frac{8.00\text{ ft}}{\sin 18.5^\circ} \Rightarrow x=\frac{(8.00\text{ ft})\sin 119.9^\circ}{\sin 18.5^\circ}=21.9\text{ ft}$$

Exercise 8.6 • Applications of Vectors

Resultant of Two Vectors

1. $\theta=180^\circ-21.8^\circ=158.2^\circ$

By the Law of Cosines: $R=\sqrt{(244)^2+(287)^2-2(244)(287)\cos 158.2^\circ}=521$

By the Law of Sines: $\dfrac{\sin\phi}{244}=\dfrac{\sin 158.2^\circ}{521} \Rightarrow \phi=\sin^{-1}\left(\dfrac{244\sin 158.2^\circ}{521}\right)=10.0^\circ$

3. $\theta=180^\circ-55.5^\circ=124.5^\circ$

By the Law of Cosines: $R=\sqrt{(55.9)^2+(42.3)^2-2(55.9)(42.3)\cos 124.5^\circ}=87.1$

By the Law of Sines: $\dfrac{\sin\phi}{55.9}=\dfrac{\sin 124.5^\circ}{87.1} \Rightarrow \phi=\sin^{-1}\left(\dfrac{55.9\sin 124.5^\circ}{87.1}\right)=31.9^\circ$

5. $\theta=180^\circ-100.0^\circ=80.0^\circ$

By the Law of Cosines: $R=\sqrt{(4483)^2+(5829)^2-2(4483)(5829)\cos 80.0^\circ}=6708$

By the Law of Sines: $\dfrac{\sin\phi}{4483}=\dfrac{\sin 80.0^\circ}{6708} \Rightarrow \phi=\sin^{-1}\left(\dfrac{4483\sin 80.0^\circ}{6708}\right)=41.16^\circ$

7. Solve using components:

Vector	$x-$Component	$y-$Component
A	$4.83\cos 18.3^\circ = 4.586$	$4.83\sin 18.3^\circ = 1.52$
B	$5.99\cos 83.5^\circ = 0.678$	$5.99\sin 83.5^\circ = 5.95$
R	$R_x = 5.264$	$R_y = 7.47$

$R = \sqrt{(R_x)^2 + (R_y)^2} = \sqrt{(5.264)^2 + (7.47)^2} = 9.14$ and $\theta = \arctan\left(\frac{R_y}{R_x}\right) = \arctan\left(\frac{7.47}{5.264}\right) = 54.8^\circ$

9. Solve using components:

Vector	$x-$Component	$y-$Component
A	$635\cos 22.7^\circ = 585.8$	$635\sin 22.7^\circ = 245.1$
B	$485\cos 48.8^\circ = 319.5$	$485\sin 48.8^\circ = 364.9$
R	$R_x = 905.3$	$R_y = 610.0$

$R = \sqrt{(R_x)^2 + (R_y)^2} = \sqrt{(905.3)^2 + (610.0)^2} = 1090$ and $\theta = \arctan\left(\frac{R_y}{R_x}\right) = \arctan\left(\frac{610.0}{905.3}\right) = 34.0^\circ$

11. Solve using components:

Vector	$x-$Component	$y-$Component
A	$273\cos 34.0^\circ = 226.3$	$273\sin 34.0^\circ = 152.7$
B	$179\cos 143^\circ = -143.0$	$179\sin 143^\circ = 107.7$
C	$203\cos 225^\circ = -143.5$	$203\sin 225^\circ = -143.5$
D	$138\cos 314^\circ = 95.9$	$138\sin 314^\circ = -99.3$
R	$R_x = 35.7$	$R_y = 17.6$

$R = \sqrt{(R_x)^2 + (R_y)^2} = \sqrt{(35.7)^2 + (17.6)^2} = 39.8$ and $\theta = \arctan\left(\frac{R_y}{R_x}\right) = \arctan\left(\frac{17.6}{35.7}\right) = 26.2^\circ$

Force Vectors

13. $\theta = 180^\circ - 44.6^\circ = 135.4^\circ$

By the Law of Cosines: $R = \sqrt{(18.6\text{ N})^2 + (21.7\text{ N})^2 - 2(18.6\text{ N})(21.7\text{ N})\cos 135.4^\circ} = 37.3\text{ N}$

By the Law of Sines: $\frac{\sin\phi}{18.6\text{ N}} = \frac{\sin 135.4^\circ}{37.3\text{ N}} \Rightarrow \phi = \sin^{-1}\left(\frac{(18.6\text{ N})\sin 135.4^\circ}{37.3\text{ N}}\right) = 20.5^\circ$

15. $\frac{\sin\phi}{125} = \frac{\sin 118.7^\circ}{212} \Rightarrow \phi = \sin^{-1}\frac{125\sin 118.7^\circ}{212} = 31.1^\circ$ $\theta = 180^\circ - (118.7^\circ + 31.1^\circ) = 30.2^\circ$

Direction of the Resultant $= 90^\circ - 30.2^\circ = 59.8^\circ =$ N 59.8° W

The second force F is found by $\frac{F}{\sin 30.2^\circ} = \frac{212\text{ N}}{\sin 118.7^\circ} \Rightarrow F = \frac{(212\text{ N})\sin 30.2^\circ}{\sin 118.7^\circ} = 122\text{ N}$

17. $\theta = 180° - 67.2° = 112.8°$

By the Law of Cosines: $R = \sqrt{(925\text{ N})^2 + (1130\text{ N})^2 - 2(925\text{ N})(1130\text{ N})\cos 112.8°} = 1720\text{ N}$

By the Law of Sines: $\dfrac{\sin\phi}{925\text{ N}} = \dfrac{\sin 112.8°}{1720\text{ N}} \Rightarrow \phi = \sin^{-1}\left(\dfrac{(925\text{ N})\sin 112.8°}{1720\text{ N}}\right) = 29.7°$ from the larger force

19. Use Law of Cosines to find the angle between the resultant and each force.

$$\theta_{\text{Small Force}} = \cos^{-1}\left(\frac{(1120\text{ N})^2 + (2870\text{ N})^2 - (2210\text{ N})^2}{2(1120\text{ N})(2870\text{ N})}\right) = 44.2°$$

$$\theta_{\text{Large Force}} = \cos^{-1}\left(\frac{(2210\text{ N})^2 + (2870\text{ N})^2 - (1120\text{ N})^2}{2(2210\text{ N})(2870\text{ N})}\right) = 20.7°$$

Velocity Vectors

21. $W = \cos^{-1}\left(\dfrac{(35.0\text{ mi/h})^2 + (325\text{ mi/h})^2 - (305\text{ mi/h})^2}{2(35.0\text{ mi/h})(325\text{ mi/h})}\right) = 52.6° \Rightarrow \theta_{\text{Wind}} = 90° - 52.6° = \text{S } 37.4°\text{ E}$

$\dfrac{\sin A}{35.0\text{ mi/h}} = \dfrac{\sin 52.6°}{305\text{ mi/h}} \Rightarrow A = \sin^{-1}\left(\dfrac{(35.0\text{ mi/h})\sin 52.6°}{305\text{ mi/h}}\right) = 5.2^0$ or $\theta_{\text{Airplane}} = 90° - 5.2° = \text{S } 84.8\text{ W}$

23. Let V_A = velocity relative to air, V_G = velocity relative to ground, and V_W = wind velocity = 36.0 km/h

$\dfrac{\sin W}{36.0\text{ km/h}} = \dfrac{\sin 45.0°}{388\text{ km/h}} \Rightarrow W = \sin^{-1}\left(\dfrac{(36.0\text{ km/h})\sin 45.0°}{388\text{ km/h}}\right) = 3.76°$

$G = 180° - (45° + 3.76°) = 131.24°$

$V_G = \sqrt{(36.0\text{ km/h})^2 + (388\text{ km/h})^2 - 2(36.0\text{ km/h})(388\text{ km/h})\cos 131.24°} = 413\text{ km/h}$

Heading $= 131.24° - 90° = \text{N } 41.2°\text{ E}$

25. $V_A = 584$ km/h, $V_W = 58.0$ km/h, and $B = 180° - (48.0° - 12.0°) = 144°$

$V_G = \sqrt{(58.0\text{ km/h})^2 + (584\text{ km/h})^2 - 2(58.0\text{ km/h})(584\text{ km/h})\cos 144°} = 632\text{ km/h}$

$\dfrac{\sin A}{58.0\text{ km/h}} = \dfrac{\sin 144°}{632\text{ km/h}} \Rightarrow A = \sin^{-1}\left(\dfrac{(58.0\text{ km/h})\sin 144°}{632\text{ km/h}}\right) = 3.09°$

Current and Voltage Vectors

27. $\theta = \dfrac{(360° - 2(51.5°))}{2} = 128.5°$

$R = \sqrt{(11.3\text{ A})^2 + (18.4\text{ A})^2 - 2(11.3\text{ A})(18.4\text{ A})\cos 128.5°} = 26.9\text{ A}$

$\dfrac{\sin\phi}{18.4\text{ A}} = \dfrac{\sin 128.5°}{26.9\text{ A}} \Rightarrow \phi = \sin^{-1}\left(\dfrac{(18.4\text{ A})\sin 128.5°}{26.9\text{ A}}\right) = 32.4°$

Chapter 8 • Review Problems

1. $C = 135°$, $a = 44.9$, and $b = 39.1$

$$c^2 = a^2 + b^2 - 2ab\cos C \;\Rightarrow\; c = \sqrt{(44.9)^2 + (39.1)^2 - 2(44.9)(39.1)\cos 135°} = 77.6$$

$$\frac{\sin A}{44.9} = \frac{\sin 135°}{77.6} \;\Rightarrow\; A = \sin^{-1}\left(\frac{44.9\sin 135°}{77.6}\right) = 24.2°$$

$$B = 180° - (24.2° + 135°) = 20.8°$$

3. $B = 38.4°$, $a = 1.84$, and $c = 2.06$

$$b^2 = a^2 + c^2 - 2ac\cos B \;\Rightarrow\; b = \sqrt{(1.84)^2 + (2.06)^2 - 2(1.84)(2.06)\cos 38.4°} = 1.30$$

$$\frac{\sin A}{1.84} = \frac{\sin 38.4°}{1.30} \;\Rightarrow\; A = \sin^{-1}\left(\frac{1.84\sin 38.4°}{1.30}\right) = 61.5°$$

$$C = 180° - (61.5° + 38.4°) = 80.1°$$

5. $A = 132°$, $b = 38.2$, and $c = 51.8$

$$a^2 = b^2 + c^2 - 2bc\cos A \;\Rightarrow\; a = \sqrt{(38.2)^2 + (51.8)^2 - 2(38.2)(51.8)\cos 132°} = 82.4$$

$$\frac{\sin B}{38.2} = \frac{\sin 132°}{82.4} \;\Rightarrow\; B = \sin^{-1}\left(\frac{38.2\sin 132°}{82.4}\right) = 20.2°$$

$$C = 180° - (132° + 20.2°) = 27.8°$$

7. IV

9. II

11. negative

13. negative

15. negative

17. $r = \sqrt{(-3)^2 + (-4)^2} = 5$

$\sin\theta = \dfrac{y}{r} = \dfrac{-4}{5} = -0.800$ $\qquad$ $\cos\theta = \dfrac{x}{r} = \dfrac{-3}{5} = -0.600$ $\qquad$ $\tan\theta = \dfrac{y}{x} = \dfrac{-4}{-3} = 1.33$

19. $\theta = 180° - 58.2° = 121.8°$

By the Law of Cosines: $R = \sqrt{(837)^2 + (527)^2 - 2(837)(527)\cos 121.8°} = 1200$

By the Law of Sines: $\dfrac{\sin\phi}{837} = \dfrac{\sin 121.8°}{1200} \;\Rightarrow\; \phi = \sin^{-1}\left(\dfrac{837\sin 121.8°}{1200}\right) = 36.4°$

21. $\theta = 180° - 155° = 25°$

By the Law of Cosines: $R = \sqrt{(44.9)^2 + (29.4)^2 - 2(44.9)(29.4)\cos 25°} = 22.1$

By the Law of Sines: $\dfrac{\sin\phi}{44.9} = \dfrac{\sin 25°}{22.1} \;\Rightarrow\; \phi = \sin^{-1}\left(\dfrac{44.9\sin 25°}{22.1}\right) = 59.2°$ $\qquad$ $180° - 59.2° = 121°$

23. Let A be the point of the first measurement, B be the point of the second measurement, and C be the lighthouse.

$A = 18^\circ 15' = 18.25^\circ$, $B = 180^\circ - 75^\circ 46' = 104.2^\circ$, and $C = 180^\circ - (18.25^\circ + 104.2^\circ) = 57.55^\circ$

c = distance the boat travelled from A to $B = \left(18.0\,\frac{\text{km}}{\text{h}}\right)(10\text{ min})\left(\frac{1\text{ h}}{60\text{ min}}\right) = 3\text{ km}$

$$\frac{a}{\sin 18.25^\circ} = \frac{3\text{ km}}{\sin 57.55^\circ} \Rightarrow a = \frac{(3\text{ km})\sin 18.25^\circ}{\sin 57.55^\circ} = 1.11\text{ km}$$

25. $\sin 175^\circ = 0.0872$ $\quad\cos 175^\circ = -0.9962$ $\quad\tan 175^\circ = -0.0875$

27. $\sin(127^\circ 22') = 0.7948$ $\quad\cos(127^\circ 22') = -0.6069$ $\quad\tan(127^\circ 22') = -1.3095$

29. $\sin 35^\circ \cos 35^\circ = 0.4698$

31. $(\cos 14^\circ + \sin 14^\circ)^2 = 1.469$

33. Using the hint the angle formed at the intersection of the first two forces is $C = 180^\circ - 37.0^\circ - 28.0^\circ = 115.0^\circ$

$$\frac{\sin\theta}{457\text{ lb}} = \frac{\sin 115^\circ}{638\text{ lb}} \Rightarrow \theta = \sin^{-1}\left(\frac{(457\text{ lb})\sin 115^\circ}{638\text{ lb}}\right) = 40.5^\circ$$

or $37.0^\circ - 40.5^\circ = -3.5^\circ = \text{S } 3.5^\circ \text{ E}$

35. $\tan^{-1}(-1.16) = -49.2^\circ$ $\quad\theta_{\text{Ref}} = 49.2^\circ$ $\quad\theta_{\text{II}} = 180^\circ - 49.2^\circ = 130.8^\circ$ and $\theta_{\text{IV}} = 360^\circ - 49.2^\circ = 310.8^\circ$

37. $\arcsin 0.737 = 47.5^\circ$

39. $\cos^{-1} 0.174 = 80.0^\circ$

41. $R = \sqrt{(273\text{ lb})^2 + (483\text{ lb})^2 - 2(273\text{ lb})(483\text{ lb})\cos 131.8^\circ} = 695\text{ lb}$

$$\frac{\sin A}{273\text{ lb}} = \frac{\sin 131.8^\circ}{695\text{ lb}} \Rightarrow A = \sin^{-1}\left(\frac{(273\text{ lb})\sin 131.8^\circ}{695\text{ lb}}\right) = 17.0^\circ$$

43. $V_x = (523\text{ cm/s})\cos 115.4^\circ = -224\text{ cm/s}$ and $V_y = (523\text{ cm/s})\sin 115.4^\circ = 472\text{ cm/s}$

45. $A = 180^\circ - (74.0^\circ + 48.0^\circ) = 58.0^\circ$

$$\frac{d}{\sin 74.0^\circ} = \frac{175\text{ mm}}{\sin 58.0^\circ} \Rightarrow d = \frac{(175\text{ mm})\sin 74.0^\circ}{\sin 58.0^\circ} = 198\text{ mm}$$

$$\sin 48.0^\circ = \frac{x}{198\text{ mm}} \Rightarrow x = (198\text{ mm})\sin 48.0^\circ = 147\text{ mm}$$

Chapter 9: Systems of Linear Equations

Exercise 9.1 • Systems of Two Linear Equations

Graphical Solution

1.

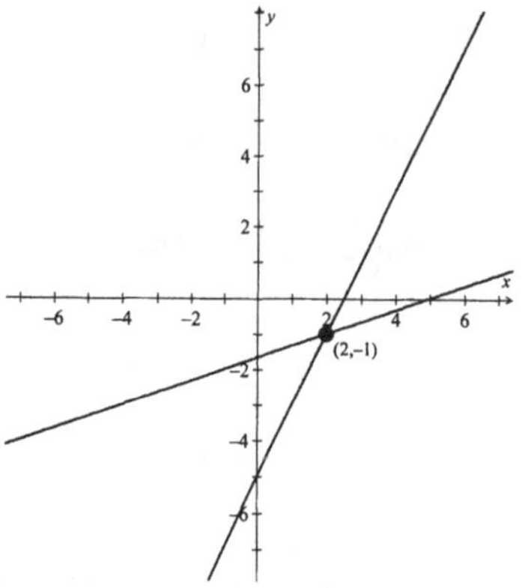

3.

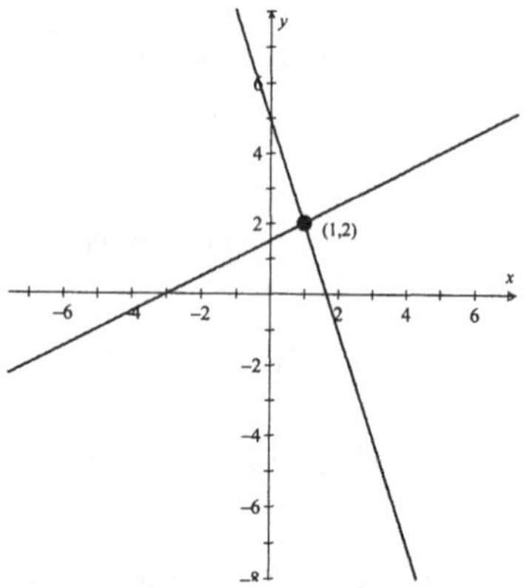

5.

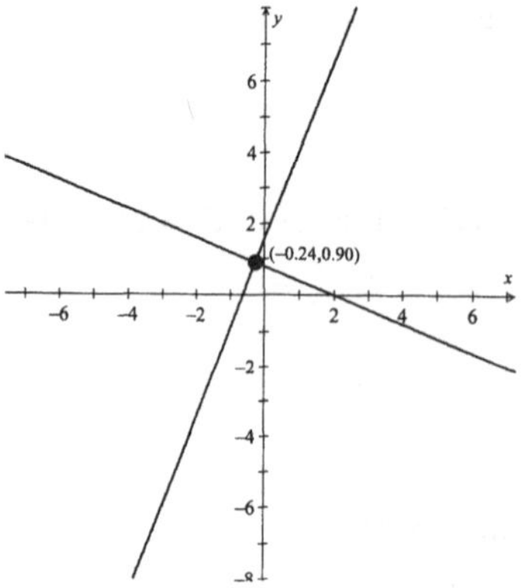

Algebraic Solution

7. $$\begin{array}{l} 2x+y=11 \\ \underline{3x-y=\ 4} \\ 5x \quad =15 \;\Rightarrow\; x=3 \text{ and } y=5 \end{array}$$

9. $$\begin{array}{l} 4x+2y=3 \\ \underline{-4x+\ y=6} \\ 3y=9 \;\Rightarrow\; x=-\dfrac{3}{4} \text{ and } y=3 \end{array}$$

11. $$\begin{array}{l} 3(3x-2y=-15) \;\Rightarrow\; 9x-6y=-45 \\ \underline{5x+6y=\ 3 \;\Rightarrow\; 5x+6y=\ 3} \\ 14x \quad =-42 \;\Rightarrow\; x=-3 \text{ and } y=3 \end{array}$$

13. $$\begin{aligned} -2(x+5y=11) &\Rightarrow -2x-10y=-22 \\ 5(3x+2y=7) &\Rightarrow 15x+10y=35 \\ \hline 13x &= 13 \Rightarrow x=1 \text{ and } y=2 \end{aligned}$$

15. $$\begin{aligned} x=11-4y &\Rightarrow x+4y=11 \\ 2(5x-2y=11) &\Rightarrow 10x-4y=22 \\ \hline 11x &= 33 \Rightarrow x=3 \text{ and } y=2 \end{aligned}$$

17. $$\begin{aligned} 3(7x-4y=81) &\Rightarrow 21x-12y=243 \\ -4(5x-3y=57) &\Rightarrow -20x+12y=-228 \\ \hline x &= 15 \Rightarrow x=15 \text{ and } y=6 \end{aligned}$$

19. $$\begin{aligned} 3x-2y=1 &\Rightarrow 3x-2y=1 \\ 2(2x+y=10) &\Rightarrow 4x+2y=20 \\ \hline 7x &= 21 \Rightarrow x=3 \text{ and } y=4 \end{aligned}$$

21. $$\begin{aligned} 2(y=9-3x) &\Rightarrow 6x+2y=18 \\ x=8-2y &\Rightarrow -x-2y=-8 \\ \hline 5x &= 10 \Rightarrow x=2 \text{ and } y=3 \end{aligned}$$

23. $$\begin{aligned} 4(2x+3y=9) &\Rightarrow 8x+12y=36 \\ -3(5x+4y=5) &\Rightarrow -15x-12y=-15 \\ \hline -7x &= 21 \Rightarrow x=-3 \text{ and } y=5 \end{aligned}$$

25. $$\begin{aligned} 72.7(29.1x-47.6y=42.8) &\Rightarrow 2115.6x-3460.5y=3111.6 \\ 47.6(11.5x+72.7y=25.8) &\Rightarrow 547.4x+3460.5y=1228.1 \\ \hline 2663x &= 4339.7 \Rightarrow x=1.63 \text{ and } y=0.0970 \end{aligned}$$

27. $$\begin{aligned} 4n=18-3m &\Rightarrow 3m+4n=18 \\ -2(m=8-2n) &\Rightarrow -2m-4n=-16 \\ \hline m &= 2 \Rightarrow m=2 \text{ and } n=3 \end{aligned}$$

29. $$\begin{aligned} 3w=13+5z &\Rightarrow 7(3w-5z=13) \Rightarrow -21w+35z=-91 \\ 4w-7z-17=0 &\Rightarrow 5(4w-7z=17) \Rightarrow 20w-35z=85 \\ \hline -w &= -6 \Rightarrow w=6 \text{ and } z=1 \end{aligned}$$

31. $$\begin{aligned} 3.62x=11.7+4.73y &\Rightarrow 7.15(3.62x-4.73y=11.7) \Rightarrow 25.88x-33.82y=83.66 \\ 4.95x-7.15y-12.8=0 &\Rightarrow -4.73(4.95x-7.15y=12.8) \Rightarrow -23.41x+33.82y=-60.54 \\ \hline 2.47x &= 23.12 \\ x=9.36 &\text{ and } y=4.69 \end{aligned}$$

33. $$\begin{array}{lll} 4.17w = 14.7 - 3.72v \Rightarrow & 4.17w + 3.72v = 14.7 \Rightarrow & 4.17w + 3.72v = 14.7 \\ v = 8.11 - 2.73w \Rightarrow & -3.72(2.73w + v = 8.11) \Rightarrow & -10.16w - 3.72v = -30.17 \\ \hline & & -5.99w = -15.47 \\ & & w = 2.58 \text{ and } v = 1.05 \end{array}$$

Exercise 9.2 • Applications

Uniform Motion Applications (Velocity ·Time=Distance)

1. v = speed of the boat $\quad c$ = speed of the current

$$(v+c)1.31 = 18.5 \Rightarrow 3.32(1.31v + 1.31c = 18.5) \Rightarrow 4.35v + 4.35c = 61.42$$
$$(v-c)3.32 = 18.5 \Rightarrow 1.31(3.32v - 3.32c = 18.5) \Rightarrow \underline{4.35v - 4.35c = 24.25}$$
$$8.70v = 85.66$$
$$v = 9.85 \text{ mph and } c = 4.27 \text{ mph}$$

3. v = speed of canoest $\quad c$ = speed of the river

$$\frac{20\text{ mi}}{v+c} = \text{hours down and } \frac{20\text{ mi}}{v-c} = \text{hours up} \Rightarrow \frac{20\text{ mi}}{v+c} + \frac{20\text{ mi}}{v-c} = 8.0$$

Also, $\dfrac{5.0}{v+c} = \dfrac{3.0}{v-c} \Rightarrow 5.0v - 5.0c = 3.0v + 3.0c \Rightarrow v = 4.0c$

Substituting $\dfrac{20.0}{5c} + \dfrac{20.0}{3c} = 8 \Rightarrow c = 1.3\text{ mph and } v = 4.0(1.3\text{ mph}) = 5.3\text{ mph}$

Money Problems

5. $y = a(1+nt) = a + ant$

$$-(3000 = a + an5) \Rightarrow -3000 = -a - 5an$$
$$3100 = a + an6 \Rightarrow \underline{3100 = a + 6an}$$
$$100 = an$$

$3000 = a + (100)5 \Rightarrow a = \$3000 - \$500 = \$2500$

$3000 = 2500 + 2500n5 \Rightarrow 12500n = 500 \text{ and } n = .04 \text{ or } 4\%$

7. Let R = amount in railroad bonds. Then $S = 4400 - R$ = amount in state bonds.

$.062R = .097(4400 - R) \Rightarrow .062R = 426.8 - 0.97R \Rightarrow 0.159R = 426.8 \Rightarrow R = \$2684$

$S = \$4400 - \$2684 = \$1716$

9. $I = ant$

$a(.06)t = \$720 \Rightarrow at = 12000$

$a(.06)(t+3) = \$1800 \Rightarrow at + 3a = \dfrac{1800}{.06} \Rightarrow 12000 + 3a = 30000 \Rightarrow 3a = 18000 \Rightarrow a = \$6000$

$a(6000) = 12000 \Rightarrow a = 2 \text{ years}$

Applications Involving Mixtures

11. Let m = amount of cement. Let s = amount of sand.

$0.05m + 255 = 0.12(m + s + 255) \Rightarrow 0.05m + 255 = 0.12m + 0.12s + 30.6$

$0.08m + s = 0.15(m + s + 255) \Rightarrow 0.08m + s = 0.15m + 0.15s + 38.25$

$$\begin{array}{r} 0.07m + 0.12s = 224.4 \\ \underline{-0.07m + 0.85s = 38.25} \\ 0.97s = 262.65 \Rightarrow s = 271 \text{ lb} \end{array}$$

$0.07m + 0.12(271) = 224.4 \Rightarrow 0.07m = 191.88$ and $m = 2740$ lb

13. Let p = amount of peat. Let v = amount of vermiculate.

$0.12(100) + p = 0.15(100 + p + v) \Rightarrow 12 + p = 15 + 0.15p + 0.15v \Rightarrow 0.85p + 0.15v = 3$

$0.06(100) + v = 0.15(100 + p + v) \Rightarrow 6 + v = 15 + 0.15p + 0.15v \Rightarrow -0.15p + 0.85v = 9$

$300(0.85p + 0.15v = 3) \Rightarrow 255p - 45v = 900$

$1700(-0.15p + 0.85v = 9) \Rightarrow \underline{-255p + 1445v = 15300}$

$1400v = 16200 \Rightarrow v = 11.6$ lb

$255p - 45(11.6) = 900 \Rightarrow 255p = 1422 \Rightarrow p = 5.57$ lb

Statics Applications

15. $T_{1x} - T_{2x} = 0, \quad T_{1y} + T_{2y} = 572$

$\tan 18^0 = \frac{T_{1x}}{T_{1y}} \Rightarrow T_{1x} = 0.3249T_{1y}, \quad \tan 35^0 = \frac{T_{2y}}{T_{2x}} \Rightarrow T_{2y} = 0.7002T_{2x}$

$0.3249T_{1y} - T_{2x} = 0, \quad T_{1y} + 0.7002T_{2x} = 572, \quad T_{1y} + 0.7002 \times 0.3249T_{1y} = 572, \quad T_{1y} = 466.0$

$T_{1x} = 0.3249 \times 466 = 151.4, \quad T_{2x} = T_{1x} = 151.4, \quad T_{2y} = 0.7002 \times 151.4 = 106.0$

$T_1 = \sqrt{151.4^2 + 466^2} = 490$ lb and $T_2 = \sqrt{151.4^2 + 106^2} = 185$ lb

Applications to Work, Fluid Flow, and Energy Flow

17. If h = number of days for helper to complete job then the helper's rate is $\frac{1}{h}$.

If c = number of days for carpenter to complete job then the carpenter's rate is $\frac{1}{c}$.

Since the rate of the carpenter is 1.5 times the helper: $\frac{1}{c} = 1.5\left(\frac{1}{h}\right) = \frac{1.5}{h}$

$15h\left(\frac{1.5}{h} + \frac{1}{h} = \frac{1}{15}\right) \Rightarrow 22.5 + 15 = h \Rightarrow h = 37.5$ days

$\frac{1}{c} = \frac{1.5}{37.5} \Rightarrow c = \frac{37.5}{1.5} = 25$ days

19. Let x = rate of pipe A (gal/h) and y = rate of pipe B (gal/h).

$$-4.8(3.5x+4.5y=117000) \Rightarrow -16.8x-21.6y=-561600$$
$$4.5(5.2x+4.8y=151200) \Rightarrow \underline{23.4x-21.6y=\ 680400}$$
$$6.6x \qquad = 118800 \Rightarrow x=18{,}000 \text{ gal/h}$$
$$3.5(18000)+4.5y=117000 \Rightarrow 4.5y=54000 \Rightarrow y=12{,}000 \text{ gal/h}$$

21. Let h = number of people hydroelectric plant serves with a rate of $\frac{1}{h}$.

Let c = number of people coal-fired plant serves with a rate of $\frac{1}{c}$.

Since the rate of the hydroelectrip plant is 1.75 times the coal-fired plant: $\frac{1}{h}=1.75\left(\frac{1}{c}\right)=\frac{1.75}{c}$

$$255000c\left(\frac{1.75}{c}+\frac{1}{c}=\frac{1}{255000}\right) \Rightarrow 446250+255000=c \Rightarrow c=\text{days}$$
$$\frac{1}{c}=\frac{1.5}{37.5} \Rightarrow c=\frac{37.5}{1.5}=25 \text{ days}$$

Electrical Applications

23.
$$6-736I_1-386I_1+386I_2=0 \Rightarrow 1122I_1-386I_2=6 \Rightarrow 216546I_1-\ 74498I_2=1158$$
$$12-375I_2-386I_2+386I_1 \Rightarrow -386I_1+761I_2=12 \Rightarrow \underline{-216546I_1+426921I_2=6732}$$
$$352423I_2=7890$$
$$I_2=0.0224\text{ A}=22.4\text{ mA}$$
$$1122I_1-386(0.0224)=6 \Rightarrow I_1=13.1\text{ mA}$$

25.
$$31.2=R_1(1+25.4\alpha) \Rightarrow 1787.76=57.3R_1+1455.42R_1\alpha$$
$$35.7=R_1(1+57.3\alpha) \Rightarrow \underline{-906.76=-25.4R_1-1455.42R_1\alpha}$$
$$880.98=31.9R_1 \Rightarrow R_1=27.6\,\Omega$$
$$31.2=(27.6)(1+25.4\alpha) \Rightarrow \alpha=0.00512$$

Miscellaneous Applications

27.
$$\tan 15.8^\circ=\frac{h}{d+346}=0.2830 \Rightarrow h=0.2830d+97.92$$
$$\tan 21.4^\circ=\frac{h}{d}=0.3919 \Rightarrow \underline{-(h=0.3919d)}$$
$$0=-0.1089d+97.92 \Rightarrow d=889\text{ ft and } h=0.3919(889.2)=352\text{ ft}$$

29. $34.8 = 4.28v_0 + a\frac{(4.28)^2}{2} \Rightarrow 4.28v_0 + 9.1592a = 34.8$

$58.3 = 5.57v_0 + a\frac{(5.57)^2}{2} \Rightarrow 5.57v_0 + 15.5125a = 58.3$

$$\begin{array}{rcl} -5.57(4.28v_0 + 9.1592a = 34.8) & \Rightarrow & -23.8396v_0 - 51.0167a = -193.836 \\ 4.28(5.57v_0 + 15.5125a = 58.3) & \Rightarrow & 23.8396v_0 + 66.3935a = 249.524 \\ \hline & & 15.3768a = 55.688 \Rightarrow a = 3.62 \text{ cm/s}^2 \end{array}$$

$4.28v_0 + 9.1592(3.62) = 34.8 \Rightarrow 4.28v_0 = 1.6537 \Rightarrow v_0 = 0.386 \text{ cm/s}^2$

Exercise 9.3 • Other Systems of Equations

Fractional Coefficients

1. $30\left(\frac{x}{5} + \frac{y}{6} = 18\right) \Rightarrow 6x + 5y = 540 \Rightarrow 6x + 5y = 540$

$4\left(\frac{x}{2} - \frac{y}{4} = 21\right) \Rightarrow 5(2x - y = 84) \Rightarrow \underline{10x - 5y = 420}$

$16x = 960 \Rightarrow x = 60$

$2(60) - y = 84 \Rightarrow y = 32$

3. $12\left(\frac{x}{3} + \frac{y}{4} = 8\right) \Rightarrow 4x + 3y = 96$

$x - y = -3 \Rightarrow \underline{3x - 3y = -9}$

$7x = 87 \Rightarrow x = \frac{87}{7}$

$4\left(\frac{87}{7}\right) + 3y = 96 \Rightarrow 3y = \frac{324}{7} \Rightarrow y = \frac{108}{7}$

5. $15\left(\frac{3x}{5} + \frac{2y}{3} = 17\right) \Rightarrow 9(9x + 10y = 255) \Rightarrow 81x + 90y = 2295$

$12\left(\frac{2x}{3} + \frac{3y}{4} = 19\right) \Rightarrow -10(8x + 9y = 228) \Rightarrow \underline{-80x - 90y = -2280}$

$x = 15$

$9(15) + 10y = 255 \Rightarrow 10y = 120 \Rightarrow y = 12$

7. $6\left(\frac{m}{2} + \frac{n}{3} - 3 = 0\right) \Rightarrow 5(3m + 2n = 18) \Rightarrow 15m + 10n = 90$

$10\left(\frac{n}{2} + \frac{m}{5} = \frac{23}{10}\right) \Rightarrow -2(5n + 2m = 23) \Rightarrow \underline{-4m - 10n = -46}$

$11m = 44 \Rightarrow m = 4$

$3(4) + 2n = 18 \Rightarrow 2n = 6 \Rightarrow n = 3$

9. $26.66\left(\frac{r}{6.20}+\frac{s}{4.30}=\frac{1}{3.10}\right) \Rightarrow \frac{1}{4.30}(4.30r-6.20s=8.60) \Rightarrow r-1.44s=2.00$

$16.1\left(\frac{r}{4.60}+\frac{s}{2.30}=\frac{1}{3.50}\right) \Rightarrow -\frac{1}{3.50}(3.50r-7.00s=4.60) \Rightarrow \underline{-r+2s=-1.31}$

$0.56s=0.69 \Rightarrow s=1.23$

$r-1.44(1.23)=2.00 \Rightarrow r=3.77$

Unknowns in the Denominator

11. $xy\left(\frac{1}{x}+\frac{3}{y}=11\right) \Rightarrow -5(3x+y=11xy) \Rightarrow -15x-5y=-55xy$

$xy\left(\frac{5}{x}+\frac{4}{y}=22\right) \Rightarrow 4x+5y=22xy \Rightarrow \underline{4x+5y=22xy}$

$-11x \quad =-33xy \Rightarrow y=\frac{1}{3}$

$\frac{1}{x}+\frac{3}{(1/3)}=11 \Rightarrow \frac{1}{x}+9=11 \Rightarrow \frac{1}{x}=2 \Rightarrow x=\frac{1}{2}$

13. $xy\left(\frac{2}{x}+\frac{4}{y}=14\right) \Rightarrow -3(4x+2y=14xy) \Rightarrow -12x-6y=-42xy$

$xy\left(\frac{6}{x}-\frac{2}{y}=14\right) \Rightarrow -2x+6y=14xy \Rightarrow \underline{-2x+6y=14xy}$

$-14x \quad =-28xy \Rightarrow y=\frac{1}{2}$

$\frac{2}{x}+\frac{4}{1/2}=14 \Rightarrow \frac{2}{x}+8=14 \Rightarrow \frac{2}{x}=6 \Rightarrow x=\frac{1}{3}$

15. $30xy\left(\frac{2}{5x}+\frac{5}{6y}=14\right) \Rightarrow 2(25x+12y=420xy) \Rightarrow 50x+24y=840xy$

$20xy\left(\frac{2}{5x}-\frac{3}{4y}=-5\right) \Rightarrow -3(-15x+8y=-100xy) \Rightarrow \underline{45x-24y=300xy}$

$95x \quad =1140xy \Rightarrow y=\frac{1}{12}$

$\frac{2}{5x}+\frac{5}{6\left(\frac{1}{12}\right)}=14 \Rightarrow \frac{2}{5x}+10=14 \Rightarrow \frac{2}{5x}=4 \Rightarrow x=\frac{2}{5(4)}=\frac{1}{10}$

17. $30wz\left(\frac{1}{5z}+\frac{1}{6w}=18\right) \Rightarrow 2(6w+5z=540wz) \Rightarrow 12w+10z=1080wz$

$8wz\left(\frac{1}{4w}-\frac{1}{2z}+21=0\right) \Rightarrow -5(-4w+2z=-168wz) \Rightarrow \underline{20w-10z=840wz}$

$32w \quad =1920wz \Rightarrow z=\frac{1}{60}$

$\frac{1}{5\left(\frac{1}{60}\right)}+\frac{1}{6w}=18 \Rightarrow 12+\frac{1}{6w}=18 \Rightarrow \frac{1}{6w}=6 \Rightarrow w=\frac{1}{6(6)}=\frac{1}{36}$

Literal Equations

19.
$$\begin{aligned} ax+2by=1 &\Rightarrow \quad ax+2by=1 \\ -2(3ax+by=2) &\Rightarrow \quad \underline{-6ax-2by=-4} \\ &\qquad -5ax \qquad =-3 \Rightarrow x=\frac{3}{5a} \end{aligned}$$

$$a\left(\frac{3}{5a}\right)+2by=1 \Rightarrow \frac{3}{5}+2by=1 \Rightarrow 2by=\frac{2}{5} \Rightarrow y=\frac{1}{5b}$$

21.
$$\begin{aligned} -2(2px+3qy=3) &\Rightarrow -4px-6qy=-6 \\ 3(3px+2qy=4) &\Rightarrow \underline{9px+6qy=12} \\ &\qquad 5px \qquad =6 \Rightarrow x=\frac{6}{5p} \end{aligned}$$

$$2p\left(\frac{6}{5p}\right)+3qy=3 \Rightarrow \frac{12}{5}+3qy=3 \Rightarrow 3qy=\frac{3}{5} \Rightarrow y=\frac{1}{5q}$$

23.
$$\begin{aligned} 3x-2y=a &\Rightarrow 3x-2y=a \\ 2(2x+y=b) &\Rightarrow \underline{4x+2y=2b} \\ &\qquad 7x=a+2b \Rightarrow x=\frac{a+2b}{7} \end{aligned}$$

$$2\left(\frac{a+2b}{7}\right)+y=b \Rightarrow 2a+4b+7y=7b \Rightarrow y=\frac{3b-2a}{7}$$

25.
$$\begin{aligned} m(ax-dy=c) &\Rightarrow \quad amx-mdy=mc \\ -a(mx-ny=c) &\Rightarrow \underline{-amx+any=-ac} \\ &\qquad y(an+md)=mc-ac \Rightarrow y=\frac{mc-ac}{an-dm} \end{aligned}$$

$$ax-d\left(\frac{mc-ac}{an-dm}\right)=c \Rightarrow ax(an-dm)-d(mc-ac)=c(an-dm)$$

$$\Rightarrow xa(an-dm)=acn-cdm+cdm-acd \Rightarrow x=\frac{a(cn-cd)}{a(an-dm)}=\frac{cn-cd}{an-dm}$$

Exercise 9.4 • System of Three Equations

1. $x+z=40 \Rightarrow x=40-z$
$y+z=45 \Rightarrow y=45-z$
$x+y=35 \Rightarrow (40-z)+(45-z)=35 \Rightarrow -2z+85=35 \Rightarrow -2z=-50 \Rightarrow z=25$
$x=40-25=15$ and $y=45-25=20$

3. $3x+y=5 \Rightarrow y=5-3x$
$2(5-3x)-3z=-5 \Rightarrow -3z=-15+6x \Rightarrow z=5-2x$
$x+2(5-2x)=7 \Rightarrow x+10-4x=7 \Rightarrow -3x=-3 \Rightarrow x=1$
$y=5-3(1)=2$ and $z=5-2(1)=3$

5. Adding (2) and (3) gives $x = 5$. Substitute in (1) and (2).

$$y + z = 13$$
$$\underline{-y + z = 1}$$
$$2z = 14 \Rightarrow z = 7$$
$$y + (7) = 13 \Rightarrow y = 13 - 7 = 6$$

7. $-2(x + 2y + 3z = 14) \Rightarrow -2x - 4y - 6z = -28$

$2x + y + 2z = 10 \Rightarrow \underline{2x + y + 2z = 10}$

$-3y - 4z = -18$

$-3(x + 2y + 3z = 14) \Rightarrow -3x - 6y - 9z = -42$

$3x + 4y - 3z = 2 \Rightarrow \underline{3x + 4y - 3z = 2}$

$-2y - 12z = -40$

$2(-3y - 4z = -18) \Rightarrow -6y - 8z = -36$

$-3(-2y - 12z = -40) \Rightarrow \underline{6y + 36z = 120}$

$28z = 84 \Rightarrow z = 3$

$-3y - 4(3) = -18 \Rightarrow -3y = -6 \Rightarrow y = 2$ and $x + 2(2) + 3(3) = 14 \Rightarrow x = 1$

9. $3(x - 2y + 2z = 5) \Rightarrow 3x - 6y + 6z = 15$

$2(5x + 3y + 6z = 57) \Rightarrow \underline{10x + 6y + 12z = 114}$

$13x + 18z = 129$

$x - 2y + 2z = 5$

$\underline{x + 2y + 2z = 21}$

$2x + 4z = 26 \Rightarrow x + 2z = 13$

$13x + 18z = 129 \Rightarrow 13x + 18z = 129$

$-13(x + 2z = 13) \Rightarrow \underline{-13x - 26z = -169}$

$-8z = -40 \Rightarrow z = 5$

$x + 2(5) = 13 \Rightarrow x = 3$ and $3 - 2y + 2(5) = 5 \Rightarrow -2y = -8 \Rightarrow y = 4$

11. $5(5a + b - 4c = -5) \Rightarrow 25a + 5b - 20c = -25$

$3a - 5b - 6c = -20 \Rightarrow \underline{3a - 5b - 6c = -20}$

$28a - 26c = -45$

$3(5a + b - 4c = -5) \Rightarrow 15a + 3b - 12c = -15$

$a - 3b + 8c = -27 \Rightarrow \underline{a - 3b + 8c = -27}$

$16a - 4c = -42 \Rightarrow 8a - 2c = -21$

$28a - 26c = -45 \Rightarrow 28a - 26c = -45$

$-13(8a - 2c = -21) \Rightarrow \underline{-104a + 26c = 273}$

$-76a = 228 \Rightarrow a = -3$

$8(-3) - 2c = -21 \Rightarrow -2c = 3 \Rightarrow c = -\frac{3}{2}$ and $5(-3) + b - 4\left(-\frac{3}{2}\right) = -5 \Rightarrow b = -4$

Fractional Equations

13. Multiply each equation by 3

$3x + y = 15$, $3x + z = 18$, and $3y + z = 27$. Solve by substitution

$y = 15 - 3x$ and $z = 18 - 3x \Rightarrow 3(15 - 3x) + (18 - 3x) = 27$

$\Rightarrow 45 - 9x + 18 - 3x = 27 \Rightarrow -12x = -36 \Rightarrow x = 3$

$y = 15 - 3(3) = 6$ and $z = 18 - 3(3) = 9$

15. Multiply the first equation by 20 to get $2x + 4y + z = 5$ and the third equation by 6 to get $2x + 3y + z = 6$.

$2x + 4y + z = 5 \Rightarrow 2x + 4y + z = 5$

$-(x + y + z = 6) \Rightarrow \underline{-x - y - z = -6}$

$x + 3y = -1$

$-(2x + 4y + z = 5) \Rightarrow -2x - 4y - z = -5$

$2x + 3y + z = 6 \Rightarrow \underline{2x + 3y + z = 6}$

$-y = 1 \Rightarrow y = -1$

$x + 3(-1) = -1 \Rightarrow x = 2$ and $2 + (-1) + z = 6 \Rightarrow z = 5$

Literal Equations

17.
$$\begin{array}{l} x - y \quad\quad = a \\ \underline{\quad y + z = 3a} \\ x \quad + z = 4a \end{array} \qquad \begin{array}{l} x \quad + z = 4a \\ \underline{-x \quad + 5z = 2a} \\ 6z = 6a \Rightarrow z = a \end{array}$$

$x + a = 4a \Rightarrow x = 3a$ and $y + a = 3a \Rightarrow y = 2a$

19. $ax + by = (a+b)c \Rightarrow ax = (a+b)c - by$ and $by + cz = (c+a)b \Rightarrow cz = (c+a)b - by$

Substitute both into the third equation

$$(a+b)c - by + (c+a)b - by = (b+c)a \Rightarrow ac + bc - by + bc + ab - by = ab + ac$$
$$\Rightarrow -2by = -2bc \Rightarrow y = c$$

$ax = (a+b)c - b(c) \Rightarrow ax = ac \Rightarrow x = c$ and $cz = (c+a)b - bc \Rightarrow cz = ab \Rightarrow z = \dfrac{ab}{c}$

Electrical Applications

21.
$$\begin{array}{l} 3I_1 + 2I_2 - 4I_3 = 4 \Rightarrow 3I_1 + 2I_2 - 4I_3 = 4 \\ 2(I_1 - 3I_2 + 2I_3 = -5) \Rightarrow \underline{2I_1 - 6I_2 + 4I_3 = -10} \\ 5I_1 - 4I_2 = -6 \end{array}$$

$$\begin{array}{l} I_1 - 3I_2 + 2I_3 = -5 \Rightarrow I_1 - 3I_2 + 2I_3 = -5 \\ 2(2I_1 + I_2 - I_3 = 3) \Rightarrow \underline{4I_1 + 2I_2 - 2I_3 = 6} \\ 5I_1 - I_2 = 1 \end{array} \qquad \begin{array}{l} 5I_1 - 4I_2 = -6 \Rightarrow 5I_1 - 4I_2 = -6 \\ -(5I_1 - I_2 = 1) \Rightarrow \underline{-5I_1 + I_2 = -1} \\ -3I_2 = -7 \Rightarrow I_2 = \frac{7}{3}\text{ A} \end{array}$$

$5I_1 - \left(\dfrac{7}{3}\right) = 1 \Rightarrow 5I_1 = \dfrac{10}{3} \Rightarrow I_1 = \dfrac{2}{3}\text{ A}$ and $2\left(\dfrac{2}{3}\right) + \dfrac{7}{3} - I_3 = 3 \Rightarrow I_3 = \dfrac{2}{3}\text{ A}$

Statics Applications

23. $F_{1y} - F_{2y} - F_{3y} + 5870 = 0 \Rightarrow F_1 \sin 88.1 - F_2 \sin 57.3 - F_3 \sin 52.8 + 5870 = 0$

$F_{1x} + F_{2x} - F_{3x} = 0 \Rightarrow F_1 \cos 88.1 + F_2 \cos 57.3 - F_3 \cos 52.8 = 0$

$8.00 F_1 \sin 88.1 + 17.0(5870) - 24.0 F_3 \sin 52.8 = 0$

$0.9995F_1 - 0.8415F_2 - 0.7965F_3 = -5870$

$0.03316F_1 + 0.5402F_2 - 0.6046F_3 = 0$

$7.9956F_1 - 19.1167F_3 = -99{,}790 \Rightarrow F_3 = 5220 + 0.4183F_1$

$0.9995F_1 - 0.8415F_2 - 0.7965(5220 + 0.4183F_1) = -5870 \Rightarrow 0.6663F_1 - 0.8415F_2 = -1712$

$0.03316F_1 + 0.5402F_2 - 0.6046(5220 + 0.4183F_1) = 0 \Rightarrow -0.2197F_1 + 0.5402F_2 = 3156$

$1.1959F_2 = 11796 \Rightarrow F_2 = 9863.1 \text{ lb}$

$\frac{1}{0.6663}(0.6663F_1 - 0.8415F_2 = -1712) \Rightarrow F_1 - 1.2629F_2 = -2569$

$\frac{1}{0.2197}(-0.2197F_1 + 0.5402F_2 = 3156) \Rightarrow \underline{-F_1 + 2.4588F_2 = 14365}$

$1.1959F_2 = 11796 \Rightarrow F_2 = 9860 \text{ lb}$

$F_1 - 1.2629(9860) = -2569 \Rightarrow F_1 = 9880 \text{ lb}$ and $F_3 = 5220 + 0.4183(9880) \Rightarrow F_3 = 9350 \text{ lb}$

Chapter 9 • Review Problems

1. $4x + 3y = 27 \Rightarrow 4x + 3y = 27$

$-2(2x - 5y = -19) \Rightarrow \underline{-4x + 10y = 38}$

$13y = 65 \Rightarrow y = 5$ and $4x + 3(5) = 27 \Rightarrow 4x = 12 \Rightarrow x = 3$

3. $xy\left(\frac{15}{x} + \frac{4}{y} = 1\right) \Rightarrow 15y + 4x = xy \Rightarrow 15y + 4x = xy$

$xy\left(\frac{5}{x} - \frac{12}{y} = 7\right) \Rightarrow -3(5y - 12x = 7xy) \Rightarrow \underline{-15y + 36x = -21xy}$

$40x = -20xy \Rightarrow y = -2$

$5(-2) - 12x = 7x(-2) \Rightarrow 2x = 10 \Rightarrow x = 5$

5. $4(5x + 3y - 2z = 5) \Rightarrow 20x + 12y - 8z = 20$

$3(3x - 4y + 3z = 13) \Rightarrow \underline{9x - 12y + 9z = 39}$

$29x + z = 59$

$3(3x - 4y + 3z = 13) \Rightarrow 9x - 12y + 9z = 39$

$2(x + 6y - 4z = -8) \Rightarrow \underline{2x + 12y - 8z = -16}$

$11x + z = 23$

$29x + z = 59 \Rightarrow 29x + z = 59$

$-(11x + z = 23) \Rightarrow \underline{-11x - z = -23}$

$18x = 36 \Rightarrow x = 2$

$11(2) + z = 23 \Rightarrow z = 1$ and $5(2) + 3y - 2(1) = 5 \Rightarrow 3y = -3 \Rightarrow y = -1$

7. (1) $\frac{3}{x+y}+\frac{4}{x-z}=2$; (2) $\frac{6}{x+y}+\frac{5}{y-z}=1$; (3) $\frac{4}{x-z}+\frac{5}{y-z}=2$

Subtracting (3) from (2) $\Rightarrow$ (4) $\frac{6}{x+y}-\frac{4}{x-z}=-1$;

Adding (1) and (4) $\Rightarrow$ $\frac{9}{x+y}=1 \Rightarrow$ (5) $x+y=9$;

Substituting (5) in (1) $\Rightarrow$ $\frac{3}{9}+\frac{4}{x-z}=2 \Rightarrow \frac{4}{x-z}=\frac{5}{3} \Rightarrow$ (6) $x-z=\frac{12}{5}$

Substituting (5) in (2) $\Rightarrow$ $\frac{6}{9}+\frac{5}{y-z}=1 \Rightarrow \frac{5}{y-z}=\frac{1}{3} \Rightarrow$ (7) $y-z=15$

Subtracting (6) from (5) $\Rightarrow$ $y+z=\frac{33}{5}$. Adding this to (7) $\Rightarrow$ $y=\frac{54}{5}$.

$x+\frac{54}{5}=9 \Rightarrow x=-\frac{9}{5}$ and $\frac{54}{5}-z=15 \Rightarrow z=-\frac{21}{5}$

9. $2(2x-3y+14=0) \Rightarrow 4x-6y=-28$

$3(3x+2y=44) \Rightarrow \underline{9x+6y=132}$

$13x=104 \Rightarrow x=8$ and $2(8)-3y=-14 \Rightarrow -3y=-30 \Rightarrow y=10$

11. $6xy\left(\frac{9}{x}+\frac{8}{y}=\frac{43}{6}\right) \Rightarrow 54y+48x=43xy \Rightarrow 54y+48x=43xy$

$6xy\left(\frac{3}{x}+\frac{10}{y}=\frac{29}{6}\right) \Rightarrow -3(18y+60x=29xy) \Rightarrow \underline{-54y-180x=-87xy}$

$-132x=-44xy \Rightarrow y=3$

$54(3)+48x=43x(3) \Rightarrow 81x=162 \Rightarrow x=2$

13. $x+z=b \Rightarrow x=b-z$

$y+z=c \Rightarrow y=c-z$

$x+y=a \Rightarrow (b-z)+(c-z)=a \Rightarrow -2z=a-b-c \Rightarrow z=\frac{b-a+c}{2}$

$x=b-\frac{b-a+c}{2}=\frac{2b-b+a-c}{2}=\frac{b+a-c}{2}$ and $y=c-\frac{b-a+c}{2}=\frac{2c-b+a-c}{2}=\frac{a-b+c}{2}$

15. $21\left(\frac{2x}{7}+\frac{2y}{3}=\frac{16}{3}\right) \Rightarrow 6x+14y=112$

$-14(x+y=12) \Rightarrow \underline{-14x-14y=-168}$

$-8x=56 \Rightarrow x=7$ and $7+y=12 \Rightarrow y=5$

17. $2x-4y+3z=10 \Rightarrow 2x-4y+3z=10$

$4(3x+y-2z=6) \Rightarrow \underline{12x+4y-8z=24}$

$14x-5z=34$

$3(3x+y-2z=6) \Rightarrow 9x+3y-6z=18$

$x-3y-z=20 \Rightarrow \underline{x-3y-z=20}$

$10x-7z=38$

$7(14x-5z=34) \Rightarrow 98x-35z=238$

$-5(10x-7z=38) \Rightarrow \underline{-50x+35z=-190}$

$48x=48 \Rightarrow x=1$

$14(1)-5z=34 \Rightarrow -5z=20 \Rightarrow z=-4$ and $3(1)+y-2(-4)=6 \Rightarrow y=-5$

19. $2(x+3y-z=10) \Rightarrow 2x+6y-2z=20$

$5x-2y+2z=6 \Rightarrow \underline{5x-2y+2z=6}$

$7x+4y=26$

$x+3y-z=10$

$\underline{3x+2y+z=13}$

$4x+5y=23$

$-5(7x+4y=26) \Rightarrow -35x-20y=-130$

$4(4x+5y=23) \Rightarrow \underline{16x+20y=92}$

$-19x=-38 \Rightarrow x=2$

$7(2)+4y=26 \Rightarrow 4y=12 \Rightarrow y=3$ and $3(2)+2(3)+z=13 \Rightarrow z=1$

21. $3(2x-y=9) \Rightarrow 6x-3y=27$

$5x-3y=14 \Rightarrow \underline{-5x+3y=-14}$

$x=13$ and $2(13)-y=9 \Rightarrow y=17$

23. $\frac{A}{B}=\frac{5}{3} \Rightarrow 3A=5B$ and $A=\frac{5}{9}(A+B)+50 \Rightarrow 9A=5A+5B+450 \Rightarrow 4A-5B=450$

Substituting the first equation into the second equation gives: $4A-3A=450 \Rightarrow A=\$450$

$3(450)=5B \Rightarrow B=\$270$

25. $\frac{n+2}{d-2}=1 \Rightarrow n+2=d-2 \Rightarrow \quad n-d=-4$

$\frac{n+d}{d-5}=5 \Rightarrow n+d=5d-25 \Rightarrow \underline{-n+4d=25}$

$3d=21 \Rightarrow d=7$ and $n-7=-4 \Rightarrow n=3$

The fraction is $\frac{3}{7}$.

27. Let l = length, w = width, and lw = area of the rectangle

$(l-4)(w+3)=lw \Rightarrow lw-4w+3l-12=lw \Rightarrow 3l-4w=12$

$(l-4)=(w+3) \Rightarrow -4(l-w=7) \Rightarrow \underline{-4l+4w=-28}$

$-l=-16 \Rightarrow l=16$ units

$16-w=7 \Rightarrow w=9$ units

Original dimensions where 9 units by 16 units.

Chapter 10: Matrices and Determinants

Exercise 10.1 • Definitions

1. A, B, D, E, F, I, J, K

3. C, H

5. I

7. B, I

9. F

11. 6

13. 4×3

15. 2×4

Exercise 10.2 • Solving Systems of Equations by the Unit Matrix Method

Two Equations in Two Unknowns

1. For all problems in this section use a calculator to enter each matrix. Then use the reduced row-echelon form command.

 $(3,5)$

    ```
    [A]
          [[2  1   11]
           [3  -1  4 ]]
    rref([A]
           [[1  0  3]
            [0  1  5]]
    ■
    ```

3. $(-3,3)$

5. $(1,2)$

7. $(3,2)$

9. $(15,6)$

11. $(3,4)$

13. $(2,3)$

15. $(1.63, 0.0971)$

17. $m = 2, n = 3$

19. $w = 6, z = 1$

21. $(-0.462, 2.31)$

Three Equations in Three Unknowns

23. $(5,6,7)$

    ```
    [A]                    rref([A]
      [[1  1   1   18]        [[1 0 0 5]
       [1  -1  1   6 ]         [0 1 0 6]
       [1  1   -1  4 ]]        [0 0 1 7]]
    ■                      ■
    ```

25. $(15,20,25)$

27. $(1,2,3)$

29. $(3,4,5)$

31. $(6,8,10)$

33. $(-2.30, 4.80, 3.09)$

35. $(3,6,9)$

Four Equations in Four Unknowns

37. $x=2, y=3, z=4, w=5$

```
[A]                        rref([A]
  [[1 1 2 1 18]              [[1 0 0 0 2]
    [1 2 1 1 17]               [0 1 0 0 3]
    [1 1 1 2 19]               [0 0 1 0 4]
    [2 1 1 1 16]]              [0 0 0 1 5]]
```

39. $x=-4, y=-3, z=2, w=5$

41. $x=a-c, y=b+c, z=0, w=a-b$

Five Equations in Five Unknowns

43. $x=4, y=5, z=6, w=7, u=8$

```
[A]                          rref([A]
[[1 1 0 0 0 9 ]                [[1 0 0 0 0 4]
  [0 1 1 0 0 11]                 [0 1 0 0 0 5]
  [0 0 1 1 0 13]                 [0 0 1 0 0 6]
  [0 0 0 1 1 15]                 [0 0 0 1 0 7]
  [1 0 0 0 1 12]]                [0 0 0 0 1 8]]
```

45. $x=4, y=5, z=6, v=3, w=2$

Applications

47. Let t = hours of flying time and d = distance covered by plane.

$$\left.\begin{array}{ll} d=226t & \Rightarrow\ d-226t=0 \\ d-125=85.0t & \Rightarrow\ d-85.0t=125 \end{array}\right\} \Rightarrow\ d=200\text{ mi and } t=53.2\text{ min}$$

49. $$\begin{pmatrix} 283 & -274 & 163 & 352 \\ 428 & 163 & 373 & 169 \\ 338 & -112 & -227 & 825 \end{pmatrix} \Rightarrow\ I_1=1.54, I_2=-0.377, I_3=-1.15$$

51. $$\begin{pmatrix} 57.2 & 92.5 & -23.0 & -11.4 & 38.2 \\ 95.3 & -14.9 & 39.0 & 59.9 & 29.3 \\ 66.3 & 81.4 & -91.5 & 33.4 & -73.6 \\ 38.2 & -46.6 & 30.1 & 93.2 & 55.7 \end{pmatrix} \Rightarrow\ I_1=-1.01, I_2=1.69, I_3=2.01, I_4=1.20$$

53. Assembly: $4.50A+5.25B+6.55C+7.75D=15157$ hr

Burn-In: $12A+12B+24C+36D=43680$ hr

Inspection: $1.25A+1.75B+2.25C+3.75D=4824$ hr

Testing: $2.75A+3.00B+3.75C+4.25D=8928$ hr

$$\begin{pmatrix} 4.50 & 5.25 & 6.55 & 7.75 & 15157 \\ 12 & 12 & 24 & 36 & 43680 \\ 1.25 & 1.75 & 2.25 & 3.75 & 4824 \\ 2.75 & 3.00 & 3.75 & 4.25 & 8928 \end{pmatrix} \Rightarrow\ A=1597, B=774, C=453, D=121$$

Exercise 10.3 • Second-Order Determinants

Value of a Determinant

1. $\begin{vmatrix} 3 & 2 \\ 1 & -4 \end{vmatrix} = 3(-4) - 2(1) = -14$

3. $\begin{vmatrix} 0 & 5 \\ -3 & -4 \end{vmatrix} = 0(-4) - 5(-3) = 15$

5. $\begin{vmatrix} -4 & 5 \\ 7 & -2 \end{vmatrix} = -4(-2) - 5(7) = -27$

7. $\begin{vmatrix} 4.82 & 2.73 \\ 2.97 & 5.28 \end{vmatrix} = 4.82(5.28) - 2.73(2.97) = 17.3$

9. $\begin{vmatrix} -\frac{2}{3} & \frac{2}{5} \\ -\frac{1}{3} & \frac{4}{5} \end{vmatrix} = -\frac{2}{3}\left(\frac{4}{5}\right) - \frac{2}{5}\left(-\frac{1}{3}\right) = -\frac{2}{5}$

11. $\begin{vmatrix} a & b \\ c & d \end{vmatrix} = a(d) - b(c) = ad - bc$

Solving a System of Two Linear Equations by Determinants

13. $4x + 2y = 3$

$-4x + y = 6$

$\Delta = \begin{vmatrix} 4 & 2 \\ -4 & 1 \end{vmatrix} = 4(1) - (-4)(2) = 12$

$x = \dfrac{\begin{vmatrix} 3 & 2 \\ 6 & 1 \end{vmatrix}}{\Delta} = \dfrac{3(1) - 6(2)}{12} = -\dfrac{9}{12} = -\dfrac{3}{4}$ and $y = \dfrac{\begin{vmatrix} 4 & 3 \\ -4 & 6 \end{vmatrix}}{\Delta} = \dfrac{4(6) - (-4)(3)}{12} = \dfrac{36}{12} = 3$

15. $3x - 2y = -15$

$5x + 6y = 3$

$\Delta = \begin{vmatrix} 3 & -2 \\ 5 & 6 \end{vmatrix} = 3(6) - 5(-2) = 28$

$x = \dfrac{\begin{vmatrix} -15 & -2 \\ 3 & 6 \end{vmatrix}}{\Delta} = \dfrac{-15(6) - 3(-2)}{28} = \dfrac{-84}{28} = -3$ and $y = \dfrac{\begin{vmatrix} 3 & -15 \\ 5 & 3 \end{vmatrix}}{\Delta} = \dfrac{3(3) - 5(-15)}{28} = \dfrac{84}{28} = 3$

17. $x + 5y = 11$

$3x + 2y = 7$

$\Delta = \begin{vmatrix} 1 & 5 \\ 3 & 2 \end{vmatrix} = 1(2) - 3(5) = -13$

$x = \dfrac{\begin{vmatrix} 11 & 5 \\ 7 & 2 \end{vmatrix}}{\Delta} = \dfrac{11(2) - 7(5)}{-13} = \dfrac{-13}{-13} = 1$ and $y = \dfrac{\begin{vmatrix} 1 & 11 \\ 3 & 7 \end{vmatrix}}{\Delta} = \dfrac{1(7) - 3(11)}{-13} = \dfrac{-26}{-13} = 2$

19. $x = 11 - 4y \Rightarrow x + 4y = 11$

$5x - 2y = 11$

$$\Delta = \begin{vmatrix} 1 & 4 \\ 5 & -2 \end{vmatrix} = 1(-2) - 5(4) = -22$$

$$x = \frac{\begin{vmatrix} 11 & 4 \\ 11 & -2 \end{vmatrix}}{\Delta} = \frac{11(-2) - 11(4)}{-22} = \frac{-66}{-22} = 3 \text{ and } y = \frac{\begin{vmatrix} 1 & 11 \\ 5 & 11 \end{vmatrix}}{\Delta} = \frac{1(11) - 5(11)}{-22} = \frac{-44}{-22} = 2$$

21. $\frac{x}{3} + \frac{y}{4} = 8$

$x - y = -3$

$$\Delta = \begin{vmatrix} \frac{1}{3} & \frac{1}{4} \\ 1 & -1 \end{vmatrix} = \frac{1}{3}(-1) - 1\left(\frac{1}{4}\right) = -\frac{4}{12} - \frac{3}{12} = -\frac{7}{12}$$

$$x = \frac{\begin{vmatrix} 8 & \frac{1}{4} \\ -3 & -1 \end{vmatrix}}{\Delta} = \frac{8(-1) - (-3)\left(\frac{1}{4}\right)}{-\frac{7}{12}} = \frac{-8 + \frac{3}{4}}{-\frac{7}{12}} = \frac{-\frac{29}{4}}{-\frac{7}{12}} = \frac{87}{7} \text{ and } y = \frac{\begin{vmatrix} \frac{1}{3} & \frac{1}{4} \\ 1 & -1 \end{vmatrix}}{\Delta} = \frac{\frac{1}{3}(-3) - 1(8)}{-\frac{7}{12}} = \frac{-9}{-\frac{7}{12}} = \frac{108}{7}$$

23. $\frac{3x}{5} + \frac{2y}{3} = 17$

$\frac{2x}{3} + \frac{3y}{4} = 19$

$$\Delta = \begin{vmatrix} \frac{3}{5} & \frac{2}{3} \\ \frac{2}{3} & \frac{3}{4} \end{vmatrix} = \frac{3}{5}\left(\frac{3}{4}\right) - \frac{2}{3}\left(\frac{2}{3}\right) = \frac{9}{20} - \frac{4}{9} = \frac{81 - 80}{180} = \frac{1}{180}$$

$$x = \frac{\begin{vmatrix} 17 & \frac{2}{3} \\ 19 & \frac{3}{4} \end{vmatrix}}{\Delta} = \frac{17\left(\frac{3}{4}\right) - 19\left(\frac{2}{3}\right)}{\frac{1}{180}} = \frac{\frac{51}{4} - \frac{38}{3}}{\frac{1}{180}} = \frac{\frac{1}{12}}{\frac{1}{180}} = 15 \text{ and } y = \frac{\begin{vmatrix} \frac{3}{5} & 17 \\ \frac{2}{3} & 19 \end{vmatrix}}{\Delta} = \frac{\frac{3}{5}(19) - \frac{2}{3}(17)}{\frac{1}{180}} = \frac{\frac{171}{6} - \frac{170}{6}}{\frac{1}{180}} = 30$$

25. $7x - 4y = 81$

$5x - 3y = 57$

$$\Delta = \begin{vmatrix} 7 & -4 \\ 5 & -3 \end{vmatrix} = 7(-3) - 5(-4) = -1$$

$$x = \frac{\begin{vmatrix} 81 & -4 \\ 57 & -3 \end{vmatrix}}{\Delta} = \frac{81(-3) - 57(-4)}{-1} = 15 \text{ and } y = \frac{\begin{vmatrix} 7 & 81 \\ 5 & 57 \end{vmatrix}}{\Delta} = \frac{7(57) - 5(81)}{-1} = 6$$

27. $3x-2y=1$

$2x+y=10$

$$\Delta=\begin{vmatrix}3 & -2\\ 2 & 1\end{vmatrix}=3(1)-2(-2)=7$$

$$x=\frac{\begin{vmatrix}1 & -2\\ 10 & 1\end{vmatrix}}{\Delta}=\frac{1(1)-10(-2)}{7}=\frac{21}{7}=3 \text{ and } y=\frac{\begin{vmatrix}3 & 1\\ 2 & 10\end{vmatrix}}{\Delta}=\frac{3(10)-2(1)}{7}=\frac{28}{7}=4$$

29. $3x+y=9$

$x+2y=8$

$$\Delta=\begin{vmatrix}3 & 1\\ 1 & 2\end{vmatrix}=3(2)-1(1)=5$$

$$x=\frac{\begin{vmatrix}9 & 1\\ 8 & 2\end{vmatrix}}{\Delta}=\frac{9(2)-8(1)}{5}=\frac{10}{5}=2 \text{ and } y=\frac{\begin{vmatrix}3 & 9\\ 1 & 8\end{vmatrix}}{\Delta}=\frac{3(8)-1(9)}{5}=\frac{15}{5}=3$$

31. $2x+3y=9$

$5x+4y=5$

$$\Delta=\begin{vmatrix}2 & 3\\ 5 & 4\end{vmatrix}=2(4)-5(3)=-7$$

$$x=\frac{\begin{vmatrix}9 & 3\\ 5 & 4\end{vmatrix}}{\Delta}=\frac{9(4)-5(3)}{-7}=\frac{21}{-7}=-3 \text{ and } y=\frac{\begin{vmatrix}2 & 9\\ 5 & 5\end{vmatrix}}{\Delta}=\frac{2(5)-5(9)}{-7}=\frac{-35}{-7}=5$$

33. $\frac{m}{2}+\frac{n}{3}=3$

$\frac{m}{5}+\frac{n}{2}=\frac{23}{10}$

$$\Delta=\begin{vmatrix}\frac{1}{2} & \frac{1}{3}\\ \frac{1}{5} & \frac{1}{2}\end{vmatrix}=\frac{1}{2}\left(\frac{1}{2}\right)-\frac{1}{5}\left(\frac{1}{3}\right)=\frac{15-4}{60}=\frac{11}{60}$$

$$m=\frac{\begin{vmatrix}3 & \frac{1}{3}\\ \frac{23}{10} & \frac{1}{2}\end{vmatrix}}{\Delta}=\frac{3\left(\frac{1}{2}\right)-\frac{23}{10}\left(\frac{1}{3}\right)}{\frac{11}{60}}=\frac{\frac{45-23}{30}}{\frac{11}{60}}=\frac{\frac{22}{30}}{\frac{11}{60}}=4 \text{ and } n=\frac{\begin{vmatrix}\frac{1}{2} & 3\\ \frac{1}{5} & \frac{23}{10}\end{vmatrix}}{\Delta}=\frac{\frac{1}{2}\left(\frac{23}{10}\right)-\frac{1}{5}(3)}{\frac{11}{60}}=\frac{\frac{23-12}{20}}{\frac{11}{60}}=\frac{\frac{11}{20}}{\frac{11}{60}}=3$$

35. $3m+4n=18$

$m+2n=8$

$$\Delta=\begin{vmatrix}3 & 4\\ 1 & 2\end{vmatrix}=3(2)-1(4)=2$$

$$m=\frac{\begin{vmatrix}18 & 4\\ 8 & 2\end{vmatrix}}{\Delta}=\frac{18(2)-8(4)}{2}=\frac{4}{2}=2 \text{ and } n=\frac{\begin{vmatrix}3 & 18\\ 1 & 8\end{vmatrix}}{\Delta}=\frac{3(8)-1(18)}{2}=\frac{6}{2}=3$$

37. $3w-5z=13$
$4w-7z=17$
$$\Delta=\begin{vmatrix}3 & -5\\ 4 & -7\end{vmatrix}=3(-7)-4(-5)=-1$$
$$w=\frac{\begin{vmatrix}13 & -5\\ 17 & -7\end{vmatrix}}{\Delta}=\frac{13(-7)-17(-5)}{-1}=\frac{-6}{-1}=6 \text{ and } z=\frac{\begin{vmatrix}3 & 13\\ 4 & 17\end{vmatrix}}{\Delta}=\frac{3(17)-4(13)}{-1}=\frac{-1}{-1}=1$$

39. $3.72v+4.17w=14.7$
$v+2.73w=8.11$
$$\Delta=\begin{vmatrix}3.72 & 4.17\\ 1 & 2.73\end{vmatrix}=3.72(2.73)-1(4.17)=5.9856$$
$$v=\frac{\begin{vmatrix}14.7 & 4.17\\ 8.11 & 2.73\end{vmatrix}}{\Delta}=\frac{14.7(2.73)-4.17(8.11)}{5.9856}=1.05 \text{ and } w=\frac{\begin{vmatrix}3.72 & 14.7\\ 1 & 8.11\end{vmatrix}}{\Delta}=\frac{3.72(8.11)-1(14.7)}{5.9856}=2.58$$

Literal Equations

41. $ax+by=p$
$cx+dy=q$
$$\Delta=\begin{vmatrix}a & b\\ c & d\end{vmatrix}=ad-bc \quad Note: ad-bc\neq 0$$
$$x=\frac{\begin{vmatrix}p & b\\ q & d\end{vmatrix}}{\Delta}=\frac{dp-bq}{ad-bc} \text{ and } y=\frac{\begin{vmatrix}a & p\\ c & q\end{vmatrix}}{\Delta}=\frac{aq-cp}{ad-bc}$$

43. $2x+by=3$
$cx+4y=d$
$$\Delta=\begin{vmatrix}2 & b\\ c & 4\end{vmatrix}=2(4)-b(c)=8-bc \quad Note: bc\neq 8$$
$$x=\frac{\begin{vmatrix}3 & b\\ d & 4\end{vmatrix}}{\Delta}=\frac{3(4)-b(d)}{8-bc}=\frac{12-bd}{8-bc} \text{ and } y=\frac{\begin{vmatrix}2 & 3\\ c & d\end{vmatrix}}{\Delta}=\frac{2(d)-3(c)}{8-bc}=\frac{2d-3c}{8-bc}$$

Exercise 10.4 • Higher Order Determinants

1. $$\begin{vmatrix}1 & 0 & 2\\ 3 & 1 & 0\\ 1 & 2 & 1\end{vmatrix}=1\begin{vmatrix}1 & 0\\ 2 & 1\end{vmatrix}-0+2\begin{vmatrix}3 & 1\\ 1 & 2\end{vmatrix}=1[1(1)-2(0)]+2[3(2)-1(1)]=11$$

3. $$\begin{vmatrix}-3 & 1 & 2\\ 0 & -1 & 5\\ 6 & 0 & 1\end{vmatrix}=-3\begin{vmatrix}-1 & 5\\ 0 & 1\end{vmatrix}-0+6\begin{vmatrix}1 & 2\\ -1 & 5\end{vmatrix}=-3[-1(1)-0(5)]+6[1(5)-2(-1)]=45$$

5. $\begin{vmatrix} 5 & 1 & 2 \\ -3 & 2 & -1 \\ 4 & -3 & 5 \end{vmatrix} = 5\begin{vmatrix} 2 & -1 \\ -3 & 5 \end{vmatrix} - 1\begin{vmatrix} -3 & -1 \\ 4 & 5 \end{vmatrix} + 2\begin{vmatrix} -3 & 2 \\ 4 & -3 \end{vmatrix}$

$= 5[2(5)-(-3)(-1)] - 1[-3(5)-4(-1)] + 2[-3(-3)-4(2)] = 48$

7. $\begin{vmatrix} 2 & 1 & 3 \\ 0 & -2 & 4 \\ 0 & 1 & 5 \end{vmatrix} = 2\begin{vmatrix} -2 & 4 \\ 1 & 5 \end{vmatrix} - 0 + 0 = 2[-2(5)-1(4)] = -28$

9. Determinant by Calculator is 2.

11. Determinant by Calculator is 18.

13. Determinant by Calculator is −66.

Solving a System of Equations by Determinants

The following systems can all be solved using a calculator to figure out the needed determinants. Some of the work is shown below.

15. $x + y + z = 18$

$x - y + z = 6$

$x + y - z = 4$

$\Delta = \begin{vmatrix} 1 & 1 & 1 \\ 1 & -1 & 1 \\ 1 & 1 & -1 \end{vmatrix} = \begin{vmatrix} -1 & 1 \\ 1 & -1 \end{vmatrix} - \begin{vmatrix} 1 & 1 \\ 1 & -1 \end{vmatrix} + \begin{vmatrix} 1 & 1 \\ -1 & 1 \end{vmatrix} = 4$

$x = \dfrac{\begin{vmatrix} 18 & 1 & 1 \\ 6 & -1 & 1 \\ 4 & 1 & -1 \end{vmatrix}}{\Delta} = \dfrac{20}{4} = 5,\quad y = \dfrac{\begin{vmatrix} 1 & 18 & 1 \\ 1 & 6 & 1 \\ 1 & 4 & -1 \end{vmatrix}}{\Delta} = \dfrac{24}{4} = 6,\quad \text{and } z = \dfrac{\begin{vmatrix} 1 & 1 & 18 \\ 1 & -1 & 6 \\ 1 & 1 & 4 \end{vmatrix}}{\Delta} = \dfrac{28}{4} = 7$

17. $x + y = 35$

$x + z = 40$

$y + z = 45$

$\Delta = \begin{vmatrix} 1 & 1 & 0 \\ 1 & 0 & 1 \\ 0 & 1 & 1 \end{vmatrix} = -2$

$x = \dfrac{\begin{vmatrix} 35 & 1 & 0 \\ 40 & 0 & 1 \\ 45 & 1 & 1 \end{vmatrix}}{\Delta} = \dfrac{-30}{-2} = 15,\quad y = \dfrac{\begin{vmatrix} 1 & 35 & 0 \\ 1 & 40 & 1 \\ 0 & 45 & 1 \end{vmatrix}}{\Delta} = \dfrac{-40}{-2} = 20,\quad \text{and } z = \dfrac{\begin{vmatrix} 1 & 1 & 35 \\ 1 & 0 & 40 \\ 0 & 1 & 45 \end{vmatrix}}{\Delta} = \dfrac{-50}{-2} = 2$

19. $(1,2,3)$

21. $(3,4,5)$

23. $(6,8,10)$

25. $(-2.30, 4.80, 3.09)$

27. $(3,6,9)$

29. $x=2, y=3, z=4, w=5$

31. $x=-4, y=-3, z=2, w=5$

33. $x=a-c, y=b+c, z=0, w=a-b$

35. $x=4, y=5, z=6, w=7, u=8$

37. $\Delta=\begin{vmatrix} 1 & 1 & 1 & 1 & 0 \\ 1 & 1 & 1 & 0 & 1 \\ 1 & 1 & 0 & 1 & 1 \\ 1 & 0 & 1 & 1 & 1 \\ 0 & 1 & 1 & 1 & 1 \end{vmatrix}=4$

$v=\frac{12}{4}=3,\ w=\frac{8}{4}=2,\ x=\frac{16}{4}=4,\ y=\frac{20}{4}=5,\ z=\frac{24}{4}=6$

Chapter 10 • Review Problems

1. $\begin{vmatrix} 6 & \frac{1}{2} & -2 \\ 3 & \frac{1}{4} & 4 \\ 2 & -\frac{1}{2} & 3 \end{vmatrix}=-\frac{1}{2}\begin{vmatrix} 3 & 2 \\ 4 & 3 \end{vmatrix}+\frac{1}{4}\begin{vmatrix} 6 & -2 \\ 2 & 3 \end{vmatrix}+\frac{1}{2}\begin{vmatrix} 6 & -2 \\ 3 & 4 \end{vmatrix}=-\frac{1}{2}(9-8)+\frac{1}{4}(18+4)+\frac{1}{2}(24+6)=20$

3. $\begin{vmatrix} 0 & n & m \\ -n & 0 & l \\ -m & -l & 0 \end{vmatrix}=-n\begin{vmatrix} -n & l \\ -m & 0 \end{vmatrix}+m\begin{vmatrix} -n & 0 \\ -m & -l \end{vmatrix}=-n(ml)+m(nl)=0$

5. $\begin{vmatrix} 25 & 23 & 19 \\ 14 & 11 & 9 \\ 21 & 17 & 14 \end{vmatrix}=25\begin{vmatrix} 11 & 9 \\ 17 & 14 \end{vmatrix}-23\begin{vmatrix} 14 & 9 \\ 21 & 14 \end{vmatrix}+19\begin{vmatrix} 14 & 11 \\ 21 & 17 \end{vmatrix}=25(154-153)-23(196-189)+19(238-231)$

$=25(1)-23(7)+19(7)=-3$

7. $\begin{vmatrix} 9 & 13 & 17 \\ 11 & 15 & 19 \\ 17 & 21 & 25 \end{vmatrix}=9\begin{vmatrix} 15 & 19 \\ 21 & 25 \end{vmatrix}-13\begin{vmatrix} 11 & 19 \\ 17 & 25 \end{vmatrix}+17\begin{vmatrix} 11 & 15 \\ 17 & 21 \end{vmatrix}=9(-24)-13(-48)+17(-24)=0$

9. $\begin{vmatrix} 1 & 2 & 3 \\ 3 & 1 & 2 \\ 2 & 3 & 1 \end{vmatrix}=\begin{vmatrix} 1 & 2 \\ 3 & 1 \end{vmatrix}-2\begin{vmatrix} 3 & 2 \\ 2 & 1 \end{vmatrix}+3\begin{vmatrix} 3 & 1 \\ 2 & 3 \end{vmatrix}=-5-2(-1)+3(7)=18$

11. $$\begin{vmatrix}1&2&2&4\\1&4&4&1\\1&1&2&2\\4&8&11&13\end{vmatrix}=\begin{vmatrix}4&4&1\\1&2&2\\8&11&13\end{vmatrix}-2\begin{vmatrix}1&4&1\\1&2&2\\4&11&13\end{vmatrix}+2\begin{vmatrix}1&4&1\\1&1&2\\4&8&13\end{vmatrix}-4\begin{vmatrix}1&4&4\\1&1&2\\4&8&11\end{vmatrix}$$

Expanding 1st row   Expanding 1st row   Expanding 1st row

$$=4\begin{vmatrix}2&2\\11&13\end{vmatrix}-4\begin{vmatrix}1&2\\8&13\end{vmatrix}+\begin{vmatrix}1&2\\8&11\end{vmatrix}-2\left[\begin{vmatrix}2&2\\11&13\end{vmatrix}-4\begin{vmatrix}1&2\\4&13\end{vmatrix}+\begin{vmatrix}1&2\\4&11\end{vmatrix}\right]+2\left[\begin{vmatrix}1&2\\8&13\end{vmatrix}-4\begin{vmatrix}1&2\\4&13\end{vmatrix}+1\begin{vmatrix}1&1\\4&8\end{vmatrix}\right]$$

$$-4\left[\begin{vmatrix}1&2\\8&11\end{vmatrix}-4\begin{vmatrix}1&2\\4&11\end{vmatrix}+4\begin{vmatrix}1&1\\4&8\end{vmatrix}\right]=23-2(-13)+2(-19)-4(-1)=15$$

13. $$\begin{vmatrix}3&1&5&2\\4&10&14&6\\8&9&1&4\\6&15&21&9\end{vmatrix}=3\begin{vmatrix}10&14&6\\9&1&4\\15&21&9\end{vmatrix}-\begin{vmatrix}4&14&6\\8&1&4\\6&21&9\end{vmatrix}+5\begin{vmatrix}4&10&6\\8&9&4\\6&15&9\end{vmatrix}-2\begin{vmatrix}4&10&14\\8&9&1\\6&15&21\end{vmatrix}$$

Expanding 1st row   Expanding 1st row   Expanding 1st row

$$=3\left[10\begin{vmatrix}1&4\\21&9\end{vmatrix}-14\begin{vmatrix}9&4\\15&9\end{vmatrix}+6\begin{vmatrix}9&1\\15&21\end{vmatrix}\right]-\left[4\begin{vmatrix}1&4\\21&9\end{vmatrix}-14\begin{vmatrix}8&4\\6&9\end{vmatrix}+6\begin{vmatrix}8&1\\6&21\end{vmatrix}\right]+5\left[4\begin{vmatrix}9&4\\15&9\end{vmatrix}-10\begin{vmatrix}8&4\\6&9\end{vmatrix}+6\begin{vmatrix}8&9\\6&15\end{vmatrix}\right]$$

Expanding 1st row

$$-2\left[4\begin{vmatrix}9&1\\15&21\end{vmatrix}-10\begin{vmatrix}8&1\\6&21\end{vmatrix}+14\begin{vmatrix}8&9\\6&15\end{vmatrix}\right]$$

$$3(-750-294+1044)-(-300-672+972)+5(84-480+396)-2(696-1620+924)=0$$

15. $$\begin{vmatrix}1&5&2\\4&7&3\\9&8&6\end{vmatrix}=\begin{vmatrix}7&3\\8&6\end{vmatrix}-5\begin{vmatrix}4&3\\9&6\end{vmatrix}+2\begin{vmatrix}4&7\\9&8\end{vmatrix}=18-5(-3)+2(-31)=-29$$

17. $$\begin{vmatrix}6&4&7\\9&0&8\\5&3&2\end{vmatrix}=-4\begin{vmatrix}9&8\\5&2\end{vmatrix}-3\begin{vmatrix}6&7\\9&8\end{vmatrix}=-4(-22)-3(-15)=133$$

Solve by any method.

19. $x+y+z+w=-4; x+2y+3z+4w=0; x+3y+6z+10w=9; x+4y+10z+20w=24$

$$\text{So } \Delta=\begin{vmatrix}1&1&1&1\\1&2&3&4\\1&3&6&10\\1&4&10&20\end{vmatrix}=\begin{vmatrix}2&3&4\\3&6&10\\4&10&20\end{vmatrix}-\begin{vmatrix}1&3&4\\1&6&10\\1&10&20\end{vmatrix}+\begin{vmatrix}1&2&4\\1&3&10\\1&4&20\end{vmatrix}-\begin{vmatrix}1&2&3\\1&3&6\\1&4&10\end{vmatrix}$$

$$=\left[2\begin{vmatrix}6&10\\10&20\end{vmatrix}-1\begin{vmatrix}3&4\\10&20\end{vmatrix}+\begin{vmatrix}3&4\\6&10\end{vmatrix}\right]-\left[\begin{vmatrix}6&10\\10&20\end{vmatrix}-3\begin{vmatrix}3&4\\10&20\end{vmatrix}+4\begin{vmatrix}3&4\\6&10\end{vmatrix}\right]$$

$$+\left[\begin{vmatrix}3&10\\4&20\end{vmatrix}-\begin{vmatrix}2&4\\4&20\end{vmatrix}+\begin{vmatrix}2&4\\3&10\end{vmatrix}\right]-\left[\begin{vmatrix}3&4\\6&10\end{vmatrix}-\begin{vmatrix}2&3\\4&10\end{vmatrix}+\begin{vmatrix}2&3\\3&6\end{vmatrix}\right]=4-6+4-1=1$$

Solving for x: Expanding 1st column

$$\begin{vmatrix}-4&1&1&1\\0&2&3&4\\9&3&6&10\\24&4&10&20\end{vmatrix}=-4\begin{vmatrix}2&3&4\\3&6&10\\4&10&20\end{vmatrix}+9\begin{vmatrix}1&1&1\\2&3&4\\4&10&20\end{vmatrix}-24\begin{vmatrix}1&1&1\\2&3&4\\3&6&10\end{vmatrix}$$

Expanding 1st rows

$$=-4\left[2\begin{vmatrix}6&10\\10&20\end{vmatrix}-3\begin{vmatrix}3&10\\4&20\end{vmatrix}+4\begin{vmatrix}3&6\\4&10\end{vmatrix}\right]+9\left[\begin{vmatrix}3&4\\10&20\end{vmatrix}-\begin{vmatrix}2&4\\4&20\end{vmatrix}+\begin{vmatrix}2&3\\4&10\end{vmatrix}\right]-24\left[\begin{vmatrix}3&4\\6&10\end{vmatrix}-\begin{vmatrix}2&4\\3&10\end{vmatrix}+\begin{vmatrix}2&3\\3&6\end{vmatrix}\right]$$

$=-4(4)+9(4)-24(1)=-4$. So $x={}^{-4}\!/_{\Delta}={}^{-4}\!/_{1}=-4$

Solving for y: the array is $\begin{vmatrix}1&-4&1&1\\1&0&3&4\\1&9&6&10\\1&24&10&20\end{vmatrix}$, and in a similar manner, $y={}^{-3}\!/_{\Delta}={}^{-3}\!/_{1}=-3$

Solving for z: the array is $\begin{vmatrix}1&1&-4&1\\1&2&0&4\\1&3&9&10\\1&4&24&20\end{vmatrix}$, and in a similar manner, $z={}^{2}\!/_{\Delta}={}^{2}\!/_{1}=2$

Solving for w: the array is $\begin{vmatrix}1&1&1&-4\\1&2&3&0\\1&3&6&9\\1&4&10&24\end{vmatrix}$, and in a similar manner, $w={}^{1}\!/_{\Delta}={}^{1}\!/_{1}=1$

21. $4x+3y=27; 2x-5y=-19$

$$\Delta=\begin{vmatrix}4&3\\2&-5\end{vmatrix}=-20-6=-26$$

$$x=\frac{\begin{vmatrix}27&3\\-19&-5\end{vmatrix}}{\Delta}=\frac{-78}{-26}=3 \text{ and } y=\frac{\begin{vmatrix}4&27\\2&-19\end{vmatrix}}{\Delta}=\frac{-130}{-26}=5$$

23. $x + 2y = 10$

$2x - 3y = -1$

$$\Delta = \begin{vmatrix} 1 & 2 \\ 2 & -3 \end{vmatrix} = -7$$

$$x = \frac{\begin{vmatrix} 10 & 2 \\ -1 & -3 \end{vmatrix}}{\Delta} = \frac{-28}{-7} = 4 \text{ and } y = \frac{\begin{vmatrix} 1 & 10 \\ 2 & -1 \end{vmatrix}}{\Delta} = \frac{-21}{-7} = 3$$

25. Solve using matrices on calculator: $2x - 3y = 7$

$5x + 2y = 27$ $(5,\ 1)$

27. Solve using matrices on calculator: $2x + 5y = 29$

$2x - 5y = -21$ $(2,\ 5)$

29. Solve using matrices on calculator: $4x - 5y = 3$

$3x + 5y = 11$ $(2,1)$

31. Solve using matrices on calculator: $x + 2y = 7$

$x + y = 5$ $(3,\ 2)$

33. Solve using matrices on calculator: $3x + 4y = 25$

$4x + 3y = 21$ $\left(9/7,\ 37/7\right)$

35. Solve using matrices on calculator: $5.03a = 8.16 + 5.11b \Rightarrow 5.03a - 5.11b = 8.16$

$3.63b + 7.26a = 28.8 \Rightarrow 7.26a + 3.63b = 28.8$ $(3.19,\ 1.55)$

37. Solve using matrices on calculator: $72.3x + 54.2y + 83.3z = 52.5$

$52.2x - 26.6y + 83.7z = 75.2$

$33.4x + 61.6y + 30.2z = 58.5$ $(-15.1, 3.52,\ 11.4)$

Chapter 11: Factoring and Fractions

Exercise 1 ◊ Common Factors

1. $3y^2 + y^3 = y^2(3 + y)$

3. $x^5 - 2x^4 + 3x^3 = x^3(x^2 - 2x + 3)$

5. $3a + a^2 - 3a^3 = a(3 + a - 3a^2)$

7. $5(x + y) + 15(x + y)^2 = 5(x + y)[1 + 3(x + y)]$

9. $\frac{3}{x} + \frac{2}{x^2} - \frac{5}{x^3} = \left(\frac{1}{x}\right)\left(3 + \frac{2}{x} - \frac{5}{x^2}\right)$

11. $\frac{5m}{2n} + \frac{15m^2}{4n^2} - \frac{25m^3}{8n} = \frac{5m}{2n}\left(1 + \frac{3m}{2n} - \frac{5m^2}{4}\right)$

Challenge Problems

13. $5a^2b + 6a^2c = a^2(5b + 6c)$

15. $4x^2y + cxy^2 + 3xy^3 = xy(4x + cy + 3y^2)$

17. $3a^3y - 6a^2y^2 + 9ay^3 = 3ay(a^2 - 2ay + 3y^2)$

19. $5acd - 2c^2d^2 + bcd = cd(5a - 2cd + b)$

21. $8x^2y^2 + 12x^2z^2 = 4x^2(2y^2 + 3z^2)$

23. $3a^2b + abc - abd = ab(3a + c - d)$

Applications

25. $L_0 + L_0\alpha t = L_0(1 + \alpha t)$

27. $R_1[1 + \alpha(t - t_1)]$

29. $v_0t + \frac{a}{2}t^2 = t\left(v_0 + \frac{at}{2}\right)$

Exercise 2 ◊ Difference of Two Squares

1. $4 - x^2 = (2 - x)(2 + x)$

3. $9a^2 - x^2 = (3a - x)(3a + x)$

5. $4x^2 - 4y^2 = 4(x - y)(x + y)$

7. $x^2 - 9y^2 = (x - 3y)(x + 3y)$

9. $9c^2 - 16d^2 = (3c - 4d)(3c + 4d)$

11. $9y^2 - 1 = (3y - 1)(3y + 1)$

Challenge Problems

13. $m^4 - n^4 \Rightarrow (m^2 - n^2)(m^2 + n^2) = (m - n)(m + n)(m^2 + n^2)$

15. $4m^2 - 9n^4 = (2m - 3n^2)(2m + 3n^2)$

17. $a^{16} - b^8 = (a^2 - b)(a^2 + b)(a^4 + b^2)(a^8 + b^4)$

19. $25x^4 - 16y^6 = (5x^2 - 4y^3)(5x^2 + 4y^3)$

21. $16a^4 - 121 = (4a^2 - 11)(4a^2 + 11)$

23. $25a^4b^4 - 9 = (5a^2b^2 - 3)(5a^2b^2 + 3)$

25. $\frac{1}{a^2} - \frac{1}{b^2} = \left(\frac{1}{a} + \frac{1}{b}\right)\left(\frac{1}{a} - \frac{1}{b}\right)$

27. $\frac{a^2}{x^2} - \frac{b^2}{y^2} = \left(\frac{a}{x} + \frac{b}{y}\right)\left(\frac{a}{x} - \frac{b}{y}\right)$

Applications

29. $\pi r_2^2 - \pi r_1^2 \Rightarrow \pi(r_2^2 - r_1^2) = \pi(r_2 - r_1)(r_2 + r_1)$

31. $4\pi r_1^2 - 4\pi r_1^2 \Rightarrow 4\pi(r_1^2 - r_1^2) = 4\pi(r_1 - r_2)(r_1 + r_2)$

33. $\frac{1}{2}mv_1^2 - \frac{1}{2}mv_2^2 \Rightarrow \frac{1}{2}m(v_1^2 - v_2^2) = \frac{1}{2}m(v_1 - v_2)(v_1 + v_2)$ or $\frac{m(v_1 - v_2)(v_1 + v_2)}{2}$

35. The volume of a hollow cylinder is the volume of the larger cylinder with the volume of the smaller cylinder subtracted. Since the volume of a solid cylinder of radius R is $v = \pi R^2 h$ and we have a hole of radius r contained within, the volume of the hollow cylinder is $v = \pi R^2 h - \pi r^2 h$. This expression factors to $\pi h(R - r)(R + r)$.

Exercise 3 ◊ Factoring Trinomials

Trinomials with a Leading Coefficient of 1

1. $x^2 - 10x + 21 = (x - 7)(x - 3)$

3. $x^2 - 10x + 9 = (x - 9)(x - 1)$

5. $x^2 + 7x - 30 = (x + 10)(x - 3)$

7. $x^2 + 7x + 12 = (x + 4)(x + 3)$

9. $x^2 - 4x - 21 = (x - 7)(x + 3)$

11. $x^2 + 6x + 8 = (x + 4)(x + 2)$

13. $b^2 - 8b + 15 = (b - 5)(b - 3)$

15. $b^2 - b - 12 = (b - 4)(b + 3)$

17. $2y^2 - 26y + 60 = 2(y - 10)(y - 3)$

The General Quadratic Trinomial

19. $4x^2 - 13x + 3 = (4x - 1)(x - 3)$

21. $5x^2 + 11x + 2 = (5x + 1)(x + 2)$

23. $12b^2 - b - 6 = (3b + 2)(4b - 3)$

25. $2a^2 + a - 6 = (2a - 3)(a + 2)$

27. $5x^2 - 38x + 21 = (x-7)(5x-3)$

29. $3x^2 + 6x + 3 = 3(x+1)(x+1)$

31. $3x^2 - x - 2 = (3x+2)(x-1)$

33. $4x^2 - 10x + 6 = 2(2x-3)(x-1)$

35. $4a^2 + 4a - 3 = (2a-1)(2a+3)$

37. $9a^2 - 15a - 14 = (3a-7)(3a+2)$

The Perfect Square Trinomial

39. $x^2 + 4x + 4 = (x+2)^2$

41. $y^2 - 2y + 1 = (y-1)^2$

43. $2y^2 - 12y + 18 \Rightarrow 2(y^2 - 6y + 9) = 2(y-3)^2$

45. $9 + 6x + x^2 = (3+x)^2$

47. $9x^2 + 6x + 1 = (3x+1)^2$

49. $9y^2 - 18y + 9 \Rightarrow 9(y^2 - 2y + 1) = 9(y-1)^2$

51. $16 + 16a + 4a^2 \Rightarrow 4(4 + 4a + a^2) = 4(2+a)^2$

Applications

53. $x^2 - 35x + 300 = (x-15)(x-20)$

55. $R^2 - 400R + 30{,}000 = (R-300)(R-100)$

57. $16t^2 - 82t + 45 = (2t-9)(8t-5)$

59. $m^2 - 1000m + 250{,}000 = (m-500)^2$

Exercise 4 ◊ Other Factorable Expressions

Factoring by Grouping

1. $a^3 + 3a^2 + 4a + 12 = (a^2+4)(a+3)$

3. $x^3 - x^2 + x - 1 = (x-1)(x^2+1)$

5. $x^2 - bx + 3x - 3b = (x-b)(x+3)$

7. $3x - 2y - 6 + xy = (x-2)(3+y)$

9. $x^2 + y^2 + 2xy - 4 = (x+y-2)(x+y+2)$

11. $m^2 - n^2 - 4 + 4n = (m-n+2)(m+n-2)$

Sum or Difference of Two Cubes

13. $64 + x^3 = (4+x)(16 - 4x + x^2)$

15. $2a^3 - 16 = 2(a-2)(a^2 + 2a + 4)$

17. $x^3 - 1 = (x-1)(x^2 + x + 1)$

19. $x^3 + 1 = (x+1)(x^2 - x + 1)$

21. $a^3 + 64 = (a+4)(a^2 - 4a + 16)$

23. $x^3 + 125 = (x+5)(x^2 - 5x + 25)$

25. $216 - 8a^3 \Rightarrow 8(27 - a^3) = 8(3-a)(9 + 3a + a^2)$

Applications

27. $V = \frac{4}{3}\pi r_2^3 - \frac{4}{3}\pi r_1^3 \Rightarrow \frac{4}{3}\pi(r_2^3 - r_1^3) = \frac{4}{3}\pi(r_2 - r_1)(r_2^2 + r_2 r_1 + r_1^2)$ or $\frac{4\pi(r_2 - r_1)(r_2^2 + r_2 r_1 + r_1^2)}{3}$

Exercise 5 ◊ Simplifying Fractions

1. $x \neq 0$

3. $x - 5 = 0; x \neq 5$

5. $x^2 - 3x + 2 = 0;\ (x-1)(x-2) = 0\ \ x \neq 1; x \neq 2$

7. $\frac{-(b-a)}{(b-a)} \Rightarrow -\frac{b-a}{b-a} = -1$

9. $\frac{-(\cancel{b-a})(c-d)}{\cancel{b-a}} \Rightarrow -(c-d) = d - c$

11. $\frac{2\cdot\cancel{7}}{3\cdot\cancel{7}} = \frac{2}{3}$

13. $\frac{75}{35} \Rightarrow \frac{5\cdot 15}{5\cdot 7} = \frac{15}{7}$

15. $\frac{2ab}{6b} \Rightarrow \frac{\cancel{2}a\cancel{b}}{\cancel{2}\cdot 3\cancel{b}} = \frac{a}{3}$

17. $\frac{21m^2p^2}{28mp^4} \Rightarrow \frac{3\cdot 7}{4\cdot 7}m^{2-1}p^{2-4} \Rightarrow \frac{3}{4}mp^{-2} = \frac{3m}{4p^2}$

19. $\frac{x^2-4}{x^3-8} \Rightarrow \frac{(\cancel{x-2})(x+2)}{(\cancel{x-2})(x^2+2x+4)} = \frac{x+2}{x^2+2x+4}$

Challenge Problems

21. $\frac{2m^3n - 2m^2n - 24mn}{6m^3 + 6m^2 - 36m} \Rightarrow \frac{2mn(m^2 - m - 12)}{6m(m^2 + m - 6)} \Rightarrow \frac{2mn(m+3)(m-4)}{6m(m+3)(m-2)} = \frac{n(m-4)}{3(m-2)}$

23. $\frac{2a^2 - 2}{a^2 - 2a + 1} \Rightarrow \frac{2(a^2-1)}{a^2 - 2a + 1} \Rightarrow \frac{2(a-1)(a+1)}{(a-1)^2} = \frac{2(a+1)}{a-1}$

25. $\frac{x^2 - z^2}{x^3 - z^3} \Rightarrow \frac{(x-z)(x+z)}{(x-z)(x^2 + xz + z^2)} = \frac{x+z}{x^2 + xz + z^2}$

27. $\frac{2a^2 - 8}{2a^2 - 2a - 12} \Rightarrow \frac{2(a^2-4)}{2(a^2 - a - 6)} \Rightarrow \frac{2(a+2)(a-2)}{2(a+2)(a-3)} = \frac{a-2}{a-3}$

29. $\dfrac{x^2-1}{2xy+2y} \Rightarrow \dfrac{(x+1)(x-1)}{2y(x+1)} = \dfrac{x-1}{2y}$

31. $\dfrac{2x^4y^4+2}{3x^8y^8-3} \Rightarrow \dfrac{2(x^4y^4+1)}{3(x^8y^8-1)} \Rightarrow \dfrac{2(x^4y^4+1)}{3(x^4y^4+1)(x^4y^4-1)} = \dfrac{2}{3(x^4y^4-1)}$

Exercise 6 ◊ Multiplication and Division of Fractions

1. $\dfrac{1}{3} \times \dfrac{2}{5} = \dfrac{2}{15}$

3. $\dfrac{2}{3} \times \dfrac{9}{7} \Rightarrow \dfrac{2 \cdot 9}{3 \cdot 7} \Rightarrow \dfrac{18}{21} \Rightarrow \dfrac{3 \cdot 6}{3 \cdot 7} = \dfrac{6}{7}$

5. $\dfrac{2}{3} \times 3\dfrac{1}{5} \Rightarrow \dfrac{2}{3} \times \dfrac{16}{5} \Rightarrow \dfrac{32}{15} = 2\dfrac{2}{15}$

7. $3\dfrac{3}{4} \times 2\dfrac{1}{2} \Rightarrow \dfrac{15}{4} \times \dfrac{5}{2} = \dfrac{75}{8}$

9. $3 \times \dfrac{5}{8} \times \dfrac{4}{5} \Rightarrow \dfrac{3 \cdot 5 \cdot 4}{8 \cdot 5} = 1\dfrac{1}{2}$

11. $\dfrac{a^4b^4}{2a^2y^n} \cdot \dfrac{a^2x}{xy^n} \Rightarrow \dfrac{a^{4+2}b^4x}{2a^2xy^{n+n}} \Rightarrow \dfrac{a^6b^4x}{2a^2xy^{2n}} \Rightarrow \dfrac{a^{6-2}b^4x^{1-1}}{2y^{2n}} = \dfrac{a^4b^4}{2y^{2n}}$

13. $\dfrac{x^2-a^2}{xy} \cdot \dfrac{xy}{x+a} \Rightarrow \dfrac{\cancel{xy}\,\cancel{(x+a)}(x-a)}{\cancel{xy}\,\cancel{(x+a)}} = x-a$

15. $\dfrac{x+y}{10} \cdot \dfrac{ax}{3(x+y)} \Rightarrow \dfrac{ax\cancel{(x+y)}}{30\cancel{(x+y)}} = \dfrac{ax}{30}$

17. $\dfrac{7}{9} \div \dfrac{5}{3} \Rightarrow \dfrac{7}{9} \cdot \dfrac{3}{5} \Rightarrow \dfrac{7}{3 \cdot \cancel{3}} \cdot \dfrac{\cancel{3}}{5} \Rightarrow \dfrac{7}{3 \cdot 5} = \dfrac{7}{15}$

19. $\dfrac{7}{8} \div 4 \Rightarrow \dfrac{7}{8} \cdot \dfrac{1}{4} = \dfrac{7}{32}$

21. $24 \div \dfrac{5}{8} \Rightarrow \dfrac{24}{1} \cdot \dfrac{8}{5} = 38\dfrac{2}{5}$

23. $2\dfrac{7}{8} \div 1\dfrac{1}{2} \Rightarrow \dfrac{23}{8} \cdot \dfrac{2}{3} = 1\dfrac{11}{12}$

25. $50 \div 2\dfrac{3}{5} \Rightarrow \dfrac{50}{1} \cdot \dfrac{5}{13} = 19\dfrac{3}{13}$

27. $\dfrac{7x^2y}{3ad} \div \dfrac{2xy^2}{3a^2d} \Rightarrow \dfrac{7x^2y}{3ad} \cdot \dfrac{3a^2d}{2xy^2} \Rightarrow \dfrac{21a^2dx^2y}{6adxy^2} = \dfrac{7ax}{2y}$

29. $\dfrac{5x^2y^3z}{6a^2b^2c} \div \dfrac{10xy^3z^2}{8ab^2c^2} \Rightarrow \dfrac{5x^2y^3z}{6a^2b^2c} \cdot \dfrac{8ab^2c^2}{10xy^3z^2} = \dfrac{2cx}{3az}$

31. $\dfrac{a^2+4a+4}{d+c} \div (a+2) \Rightarrow \dfrac{a^2+4a+4}{d+c} \cdot \dfrac{1}{a+2} \Rightarrow \dfrac{(a+2)^2}{(a+2)(d+c)} = \dfrac{(a+2)}{(d+c)}$

33. $\dfrac{5(x+y)^2}{x-y} \div (x+y) \Rightarrow \dfrac{5(x+y)^2}{x-y} \cdot \dfrac{1}{x+y} = \dfrac{5(x+y)}{(x-y)}$

35. $\dfrac{5xy}{a-x} \div \dfrac{10xy}{a^2-x^2} \Rightarrow \dfrac{5xy}{a-x} \cdot \dfrac{a^2-x^2}{10xy} \Rightarrow \dfrac{5xy}{a-x} \cdot \dfrac{(a+x)(a-x)}{10xy} = \dfrac{a+x}{2}$

37. $\dfrac{a^2-a-2}{5a-1} \div \dfrac{a-2}{10a^2+13a-3} \Rightarrow \dfrac{(a-2)(a+1)}{5a-1} \cdot \dfrac{(5a-1)(2a+3)}{a-2} = (a+1)(2a+3)$

Applications

39. $1\frac{7}{8} \cdot 4 = 7\frac{1}{2}$ in

41. $f/f_1 = f \div f_1 = \dfrac{pq}{p+q} \div \dfrac{pq_1}{p+q_1} \Rightarrow \dfrac{pq}{p+q} \cdot \dfrac{p+q_1}{pq_1} \Rightarrow \dfrac{p^2q+pqq_1}{p^2q_1+pqq_1} \Rightarrow \dfrac{p(pq+qq_1)}{p(pq_1+qq_1)} = \dfrac{pq+qq_1}{pq_1+qq_1}$

43. $\dfrac{F}{\frac{\pi d^2}{4}} \Rightarrow F \cdot \dfrac{4}{\pi d^2} = \dfrac{4F}{\pi d^2}$

45. $\dfrac{F}{\frac{4\pi r^3}{3}D} \Rightarrow F \cdot \dfrac{3}{4\pi r^3 D} = \dfrac{3F}{4\pi D r^3}$

47. $\dfrac{\frac{\pi d^2}{4}}{\frac{ds}{4}} \Rightarrow \dfrac{\pi d^2}{4} \cdot \dfrac{4}{ds} = \dfrac{\pi d}{s}$

Exercise 7 ◊ Adding and Subtracting Fractions
Common Fractions and Mixed Numbers

1. $\frac{3}{5} + \frac{2}{5} \Rightarrow \frac{3+2}{5} \Rightarrow \frac{5}{5} = 1$

3. $\frac{2}{7} + \frac{5}{7} - \frac{6}{7} \Rightarrow \frac{2+5-6}{7} = \frac{1}{7}$

5. $\frac{1}{3} - \frac{7}{3} + \frac{11}{3} \Rightarrow \frac{1-7+11}{3} = \frac{5}{3}$

7. $\frac{1}{2} + \frac{2}{3} \Rightarrow \frac{3}{6} + \frac{4}{6} = \frac{7}{6}$

9. $\frac{3}{4} + \frac{7}{16} \Rightarrow \frac{12}{16} + \frac{7}{16} = \frac{19}{16}$

11. $\frac{5}{9} - \frac{1}{3} + \frac{3}{18} \Rightarrow \frac{10}{18} - \frac{6}{18} + \frac{3}{18} = \frac{7}{18}$

13. $2 + \frac{3}{5} \Rightarrow \frac{10}{5} + \frac{3}{5} = \frac{13}{5}$

Algebraic Fractions

15. $\frac{1}{a} + \frac{5}{a} \Rightarrow \frac{1+5}{a} = \frac{6}{a}$

17. $\frac{2a}{y} + \frac{3}{y} - \frac{a}{y} \Rightarrow \frac{2a+3aa}{y} = \frac{a+3}{y}$

19. $\frac{5x}{2}-\frac{3x}{2}\Rightarrow\frac{5x-3x}{2}\Rightarrow\frac{2x}{2}=x$

21. $\frac{3x}{a-b}+\frac{2x}{b-a}\Rightarrow\frac{3x}{a-b}+\frac{2x}{-(a-b)}\Rightarrow\frac{3x}{a-b}-\frac{2x}{a-b}\Rightarrow\frac{3x-2x}{a-b}=\frac{x}{a-b}$

23. $\frac{3a}{2x}+\frac{2a}{5x}\Rightarrow\frac{3a}{2x}\left(\frac{5}{5}\right)+\frac{2a}{5x}\left(\frac{2}{2}\right)\Rightarrow\frac{15a}{10x}+\frac{4a}{10x}\Rightarrow\frac{15a+4a}{10x}=\frac{19a}{10x}$

25. $\frac{a+b}{3}-\frac{a-b}{2}\Rightarrow\frac{a+b}{3}\left(\frac{2}{2}\right)-\frac{a-b}{2}\left(\frac{3}{3}\right)\Rightarrow\frac{2(a+b)}{6}-\frac{3(a-b)}{6}\Rightarrow\frac{2a+2b-3a+3b}{6}=\frac{5b-a}{6}$

27. $\frac{4}{x-1}-\frac{5}{x+1}\Rightarrow\frac{4}{x-1}\left(\frac{x+1}{x+1}\right)-\frac{5}{x+1}\left(\frac{x-1}{x-1}\right)\Rightarrow\frac{4x+4-5x+1}{x^2-1}=\frac{9-x}{x^2-1}$

Applications

29. $96\frac{3}{16}+1\frac{1}{2}\Rightarrow\frac{1539}{16}+\frac{3}{2}\Rightarrow\frac{1539}{16}+\frac{3}{2}\left(\frac{8}{8}\right)\Rightarrow\frac{1539+24}{16}\Rightarrow\frac{1563}{16}=97\frac{11}{16}$

31. $97\frac{3}{4}-\left(42\frac{3}{8}+38\frac{5}{16}+\frac{5}{32}\right)\Rightarrow\frac{391}{4}-\left(\frac{339}{8}+\frac{609}{16}+\frac{5}{32}\right)=17\frac{5}{32}$

33. 7 miles in 3 days is $\frac{7}{3}$ mi/day. Similarly, 9 miles in 4 days is $\frac{9}{4}$ mi/day. These are both rates of work and so we simply add up the rates; as $\frac{7}{3}+\frac{9}{4}=\frac{55}{12}=4\frac{7}{12}$ mi/day.

35. Add the rates $\frac{1}{15}+\frac{1}{75}\Rightarrow\frac{5}{75}+\frac{1}{75}\Rightarrow\frac{6}{75}=\frac{2}{25}$ per min

37. $\frac{5}{8}+\frac{3}{4}\Rightarrow\frac{5}{8}+\frac{6}{8}=\frac{11}{8}$ machines

39. $\frac{(a+b)h}{2}-\frac{\pi d^2}{4}\Rightarrow\frac{(a+b)h}{2}\left(\frac{2}{2}\right)-\frac{\pi d^2}{4}\Rightarrow\frac{(a+b)2h}{4}-\frac{\pi d^2}{4}=\frac{2h(a+b)-\pi d^2}{4}$

41. $\frac{d}{V}+\frac{d_1}{V_1}+\frac{d_2}{V_2}\Rightarrow\frac{d}{V}\left(\frac{V_1V_2}{V_1V_2}\right)+\frac{d_1}{V_1}\left(\frac{VV_2}{VV_2}\right)+\frac{d_2}{V_2}\left(\frac{VV_1}{VV_1}\right)=\frac{V_1V_2d+VV_2d_1+VV_1d_2}{VV_1V_2}$

Exercise 8 ◊ Complex Fractions

1. $\frac{\frac{2}{3}+\frac{3}{4}}{\frac{1}{5}}\Rightarrow\frac{\frac{17}{12}}{\frac{1}{5}}\Rightarrow\frac{17}{12}\cdot\frac{5}{1}=\frac{85}{12}$

3. $\frac{\frac{1}{2}+\frac{1}{3}+\frac{1}{4}}{3-\frac{4}{5}}\Rightarrow\frac{\frac{13}{12}}{\frac{11}{5}}\Rightarrow\frac{13}{12}\cdot\frac{5}{11}=\frac{65}{132}$

5. $\frac{5-\frac{2}{5}}{6+\frac{1}{3}}\Rightarrow\frac{\frac{23}{5}}{\frac{19}{3}}\Rightarrow\frac{23}{5}\cdot\frac{3}{19}=\frac{69}{95}$

7. $\dfrac{x+\dfrac{y}{4}}{x-\dfrac{y}{3}} \Rightarrow \dfrac{\dfrac{4x}{4}+\dfrac{y}{4}}{\dfrac{3x}{3}-\dfrac{y}{3}} \Rightarrow \dfrac{\dfrac{4x+y}{4}}{\dfrac{3x-y}{3}} \Rightarrow \dfrac{4x+y}{4}\cdot\dfrac{3}{3x-y}=\dfrac{3(4x+y)}{4(3x-y)}$

9. $\dfrac{1+\dfrac{x}{y}}{1-\dfrac{x^2}{y^2}} \Rightarrow \dfrac{\dfrac{x+y}{y}}{\dfrac{x^2-y^2}{y^2}} \Rightarrow \dfrac{x+y}{y}\cdot\dfrac{y^2}{x^2-y^2} \Rightarrow \dfrac{x+y}{y}\cdot\dfrac{y^2}{(x+y)(x-y)}=\dfrac{y}{(y-x)}$

11. $\dfrac{a^2+\dfrac{x}{3}}{4+\dfrac{x}{5}} \Rightarrow \dfrac{\dfrac{3a^2+x}{3}}{\dfrac{20+x}{5}} \Rightarrow \dfrac{3a^2+x}{3}\cdot\dfrac{5}{20+x}=\dfrac{5(3a^2+x)}{3(20+x)}$

13. $\dfrac{x+\dfrac{2d}{3ac}}{x+\dfrac{3d}{2ac}} \Rightarrow \dfrac{\dfrac{3acx+2d}{3ac}}{\dfrac{2acx+3d}{2ac}} \Rightarrow \dfrac{3acx+2d}{3ac}\cdot\dfrac{2ac}{2acx+3d}=\dfrac{2(3acx+2d)}{3(2acx+3d)}$

15. $\dfrac{x^2-\dfrac{y^2}{2}}{\dfrac{x-3y}{2}} \Rightarrow \dfrac{2x^2-y^2}{2}\cdot\dfrac{2}{x-3y}=\dfrac{2x^2-y^2}{x-3y}$

17. $\dfrac{1+\dfrac{1}{x+1}}{1-\dfrac{1}{x-1}} \Rightarrow \dfrac{x+2}{x+1}\cdot\dfrac{x-1}{x-2}=\dfrac{(x+2)(x-1)}{(x+1)(x-2)}$

Applications

19. $\dfrac{d_1+d_2}{\dfrac{d_1V_2+d_2V_1}{V_1V_2}} \Rightarrow d_1+d_2\cdot\dfrac{V_1V_2}{d_1V_2+d_2V_1}=\dfrac{(d_1+d_2)V_1V_2}{d_1V_2+d_2V_1}$

21. $\dfrac{\dfrac{1}{x+h}-\dfrac{1}{x}}{h} \Rightarrow \dfrac{-h}{x(x+h)}\cdot\dfrac{1}{h}=-\dfrac{1}{x(x+h)}$

Exercise 9 ◊ Fractional Equations

1. $2x+\dfrac{x}{3}=28 \Rightarrow 3\left(2x+\dfrac{x}{3}=28\right) \Rightarrow 6x+x=84 \Rightarrow 7x=84 \Rightarrow x=12$

3. $x+\dfrac{x}{5}=24 \Rightarrow 5\left(x+\dfrac{x}{5}=24\right) \Rightarrow 5x+x=120 \Rightarrow 6x=120 \Rightarrow x=20$

5. $3x-\dfrac{x}{7}=40 \Rightarrow 7\left(3x-\dfrac{x}{7}=40\right) \Rightarrow 21x-x=280 \Rightarrow 20x=280 \Rightarrow x=14$

7. $\dfrac{2x}{3}-x=-24 \Rightarrow 3\left(\dfrac{2x}{3}-x=-24\right) \Rightarrow 2x-3x=-72 \Rightarrow -x=-72 \Rightarrow x=72$

9. $\dfrac{x}{2}+\dfrac{x}{3}+\dfrac{x}{4}=26 \Rightarrow 12\left(\dfrac{x}{2}+\dfrac{x}{3}+\dfrac{x}{4}=26\right) \Rightarrow 6x+4x+3x=312 \Rightarrow 13x=312 \Rightarrow x=24$

11. $\dfrac{3x-1}{4}=\dfrac{2x+1}{3} \Rightarrow 12\left(\dfrac{3x-1}{4}\right)=12\left(\dfrac{2x+1}{3}\right) \Rightarrow 3(3x-1)=4(2x+1) \Rightarrow 9x-3=8x+4 \Rightarrow x=7$

13. $2x+\frac{x}{3}-\frac{x}{4}=50 \Rightarrow 12\left(2x+\frac{x}{3}-\frac{x}{4}=50\right) \Rightarrow 25x=600 \Rightarrow x=24$

15. $3x-\frac{x}{6}+\frac{x}{12}=70 \Rightarrow 12\left(3x-\frac{x}{6}+\frac{x}{12}=70\right) \Rightarrow 35x=840 \Rightarrow x=24$

17. $\frac{6x-19}{2}=\frac{2x-11}{3} \Rightarrow 6\left(\frac{6x-19}{2}\right)=6\left(\frac{2x-11}{3}\right) \Rightarrow 3(6x-19)=2(2x-11) \Rightarrow 18x-57=4x-22 \Rightarrow 14x=35 \Rightarrow x=\frac{5}{2}$

19. $\frac{3x-116}{4}+\frac{180-5x}{6}=0 \Rightarrow 12\left(\frac{3x-116}{4}+\frac{180-5x}{6}=0\right) \Rightarrow 3(3x-116)+2(180-5x)$

$\Rightarrow 9x-348+360-10x=0 \Rightarrow -x=-12 \Rightarrow x=12$

21. $\frac{x-1}{8}-\frac{x+1}{18}=1 \Rightarrow 72\left(\frac{x-1}{8}-\frac{x+1}{18}=1\right) \Rightarrow 9(x-1)-4(x+1)=72 \Rightarrow 9x-9-4x-4=72 \Rightarrow 5x=85 \Rightarrow x=17$

23. $\frac{15x}{4}=\frac{9}{4}-\frac{3-x}{2} \Rightarrow 4\left(\frac{15x}{4}=\frac{9}{4}-\frac{3-x}{2}\right) \Rightarrow 15x=9-2(3-x) \Rightarrow 15x=9-6+2x \Rightarrow 13x=3 \Rightarrow x=\frac{3}{13}$

Equations with Unknown in Denominator

25. $\frac{2}{3x}+6=5 \Rightarrow 3x\left(\frac{2}{3x}+6=5\right) \Rightarrow 2+18x=15x \Rightarrow 2=-3x \Rightarrow x=-\frac{2}{3}$

27. $4+\frac{1}{x+3}=8 \Rightarrow (x+3)\left(4+\frac{1}{x+3}=8\right) \Rightarrow 4(x+3)+1=8(x+3) \Rightarrow 4x+12+1=8x+24 \Rightarrow -11=4x \Rightarrow x=-\frac{1}{4}$

29. $\frac{x-3}{x+2}=\frac{x+4}{x-5} \Rightarrow$ LCD is $(x+2)(x-5) \Rightarrow (x+2)(x-5)\left(\frac{x-3}{x+2}\right)=(x+2)(x-5)\left(\frac{x+4}{x-5}\right) \Rightarrow (x-5)(x-3)=(x+2)(x+4)$

$\Rightarrow x^2-8x+15=x^2+6x+8 \Rightarrow 7=14x \Rightarrow x=\frac{1}{2}$

31. $\frac{9}{x^2+x-2}=\frac{7}{x-1}-\frac{3}{x+2} \Rightarrow$ LCD is $(x-1)(x+2) \Rightarrow (x-1)(x+2)\left(\frac{9}{x^2+x-2}=\frac{7}{x-1}-\frac{3}{x+2}\right) \Rightarrow 9=7(x+2)-3(x-1)$

$\Rightarrow 9=7x+14-3x+3 \Rightarrow 9=4x+17 \Rightarrow 4x=-8 \Rightarrow x=-2$

However, when $x=-2$, there is a division by zero error; therefore, there is no solution.

33. $\frac{4}{x^2-1}+\frac{1}{x-1}+\frac{1}{x+1}=0 \Rightarrow$ LCD is $(x-1)(x+1) \Rightarrow (x-1)(x+1)\left(\frac{4}{x^2-1}+\frac{1}{x-1}+\frac{1}{x+1}=0\right) \Rightarrow 4+1(x+1)+1(x-1)=0$

$\Rightarrow 2x+4=0 \Rightarrow 2x=-4 \Rightarrow x=-2$

Applications

35. $\frac{\text{distance}}{\text{travel to right}}+\frac{\text{distance}}{\text{travel to left}}=\text{total time}$

$\frac{12}{10}+\frac{12}{t}=2 \Rightarrow$ LCD is $10t \Rightarrow 10t\left(\frac{12}{10}+\frac{12}{t}=2\right) \Rightarrow 12t+120=20t \Rightarrow 120=8x \Rightarrow t=15$ cm/s

37. Let t = time Mason A worked

Then $2t$ = time Mason B worked and $\frac{1}{2}(t+2t) \Rightarrow \frac{3}{2}t$ = time Mason C worked

Since Mason A's rate is 7.0 m/day, Mason B's rate is 6.0 m/day and Mason C's rate is 5.0 m/day and the total distance is 318 m, then

$7.0t + 6.0(2t) + 5.0\left(\frac{3}{2}t\right) = 318 \Rightarrow 2\left(7.0t + 6.0(2t) + 5.0\left(\frac{3}{2}t\right) = 318\right) \Rightarrow 14.0t + 24.0t + 15.0t = 636 \Rightarrow 53t = 636 \Rightarrow t = 12$ days

Thus, Mason A takes 12 days, Mason B takes 2(12) or 24 days and Mason C takes $\frac{3}{2}(12)$ or 18 days

39. Let r = the rate of the return trip for the shaper. The stroke needs to be converted to feet: $10.5\left(\frac{1 \text{ ft}}{12 \text{ in}}\right) = 0.875$ ft

	Rate (ft/min)	Time (min)	Distance (ft)
Forward cutting	115	$\frac{429(0.875)}{115}$	429(0.875)
Return trip	r	$\frac{429(0.875)}{r}$	429(0.875)

The time to complete 429 cuts is 4.0 minutes.

$\frac{429(0.875)}{115} + \frac{429(0.875)}{r} = 4.0 \Rightarrow \text{LCD is } 115r \Rightarrow 115r\left(\frac{429(0.875)}{115} + \frac{429(0.875)}{r} = 4.0\right)$

$\Rightarrow 429r(0.875) + (429)(115)(0.875) = 4.0(115r) \Rightarrow 375.375r + 43168.125 = 460r$

$\Rightarrow 43168.125 = 84.625r \Rightarrow r = 510.11$ or 510 ft/min

41. Let t = the number of winters needed for the insulated house to use 1250 gallons of oil. Find the amount of energy consumed in one winter.

$\frac{2100}{3} + \frac{1250}{t} = \frac{1850}{2} \Rightarrow \text{LCD is } 6t \Rightarrow 6t\left(\frac{2100}{3} + \frac{1250}{t} = \frac{1850}{2}\right) \Rightarrow 4200t + 7500 = 5550t \Rightarrow 7500 = 1350t$

$\Rightarrow t = 5.555$ or 5.6 winters

Exercise 10 ◊ Literal Equations and Formulas

1. $2ax = bc \Rightarrow \frac{\cancel{2a}x}{\cancel{2a}} = \frac{bc}{2a} \Rightarrow x = \frac{bc}{2a}$

3. $a(x+y) = b(x+z) \Rightarrow ax + ay = bx + bz \Rightarrow ax - bx = bz - ay \Rightarrow x(a-b) = bz - ay \Rightarrow x = \frac{bz - ay}{a-b}$

5. $4acx - 3d^2 = a^2d - d^2x \Rightarrow 4acx + d^2x = a^2d + 3d^2 \Rightarrow x(4ac + d^2) = a^2d + 3d^2 \Rightarrow x = \frac{a^2d + 3d^2}{4ac + d^2}$

7. $a^2x - cd = b - ax + dx \Rightarrow a^2x + ax - dx = b + cd \Rightarrow x(a^2 + a - d) = b + cd \Rightarrow x = \frac{b + cd}{a^2 + a - d}$

9. $\frac{a}{2}(x - 3w) = z \Rightarrow a(x - 3w) = 2z \Rightarrow ax - 3aw = 2z \Rightarrow ax = 2z + 3aw \Rightarrow x = \frac{2z}{a} + 3w$

11. $3x + m = b \Rightarrow 3x = b - m \Rightarrow x = \frac{b - m}{3}$

13. $ax - bx = c + dx - m \Rightarrow ax - bx - dx = c - m \Rightarrow x(a - b - d) = c - m \Rightarrow x = \frac{c - m}{a - b - d}$

15. $ax - ab = cx - bc \Rightarrow ax - cx = ab - bc \Rightarrow x(a - c) = b(a - c) \Rightarrow x = b$

17. $5dw = ab \Rightarrow w = \frac{ab}{5d}$

19. $5mn = mz - 2 \Rightarrow mz = 5mn + 2 \Rightarrow z = \dfrac{5mn+2}{m}$

21. $m(mn+y) = 3 - y \Rightarrow m^2n + my = 3 - y \Rightarrow my + y = 3 - m^2n \Rightarrow y(m+1) = 3 - m^2n \Rightarrow y = \dfrac{3-m^2n}{m+1}$

Literal Fractional Equations

23. $\dfrac{w+x}{x} = w(w+y) \Rightarrow w + x = xw(w+y) \Rightarrow w = xw(w+y) - x \Rightarrow w = x\left[w(w+y)-1\right] \Rightarrow x = \dfrac{w}{w(w+y)-1}$

25. $\dfrac{p-q}{x} = 3p \Rightarrow p - q = 3px \Rightarrow x = \dfrac{p-q}{3p}$

27. $\dfrac{a-x}{5} = \dfrac{b-x}{2} \Rightarrow 10\left(\dfrac{a-x}{5} = \dfrac{b-x}{2}\right) \Rightarrow 2a - 2x = 5b - 5x \Rightarrow 3x = 5b - 2a \Rightarrow x = \dfrac{5b-2a}{3}$

$\Rightarrow x(d-ad) = ac + adb \Rightarrow x = \dfrac{a(c+bd)}{d(a+1)}$

29. $\dfrac{x}{a} - a = \dfrac{a}{c} - \dfrac{x}{c-a} \Rightarrow (ac)(c-a)\left(\dfrac{x}{a} - a = \dfrac{a}{c} - \dfrac{x}{c-a}\right) \Rightarrow c(c-a)x - a^2c(c-a) = a^2(c-a) - acx \Rightarrow cx(c-a) + acx = a^2(c-a) + a^2c(c-a$

$\Rightarrow x\left[c(c-a) + ac\right] = a^2(c-a)(1+c) \Rightarrow x = \dfrac{a^2(c-a)(1+c)}{c^2}$

31. $\dfrac{x-a}{x-b} = \left(\dfrac{2x-a}{2x-b}\right)^2 \Rightarrow \dfrac{x-a}{x-b} = \dfrac{(2x-a)(2x-a)}{(2x-b)(2x-b)} \Rightarrow (x-b)(2x-b)^2\left(\dfrac{x-a}{x-b} = \dfrac{(2x-a)(2x-a)}{(2x-b)(2x-b)}\right) \Rightarrow (x-a)(2x-b)^2 = (x-b)(2x-a)^2$

$\Rightarrow 4x^3 - 4bx^2 + b^2x - 4ax^2 + 4abx - ab^2 = 4x^3 - 4ax^2 + a^2x - 4bx^2 + 4abx - a^2b \Rightarrow b^2x - ab^2 = a^2x - a^2b$

$\Rightarrow b^2x - a^2x = ab^2 - a^2b \Rightarrow x(b^2 - a^2) = ab^2 - a^2b \Rightarrow x = \dfrac{ab^2 - a^2b}{b^2 - a^2} \Rightarrow x = \dfrac{ab(b-a)}{(b-a)(b+a)} \Rightarrow x = \dfrac{ab}{a+b}$

Formulas

33. $C = \dfrac{w^2L^3}{24P^2} \Rightarrow 24CP^2 = w^2L^3 \Rightarrow \dfrac{24CP^2}{w^2} = L^3 \Rightarrow L = \sqrt[3]{\dfrac{24CP^2}{w^2}}$

35. $L = L_0(1 + \alpha\Delta t) \Rightarrow L_0 = \dfrac{L}{1+\alpha\Delta t}$

37. $s = v_0t + \dfrac{1}{2}at^2 \Rightarrow s - v_0t = \dfrac{1}{2}at^2 \Rightarrow 2(s - v_0t) = at^2 \Rightarrow a = \dfrac{2(s-v_0t)}{t^2}$

39. $y = a + ant \Rightarrow y = a(1+nt) \Rightarrow a = \dfrac{y}{1+nt}$

41. $R = R_1\left[1 + \alpha(t - t_1)\right] \Rightarrow R = R_1 + \alpha tR_1 - \alpha t_1R_1 \Rightarrow \alpha t_1R_1 = R_1 + \alpha tR_1 - R \Rightarrow t_1 = \dfrac{R_1(1+\alpha t) - R}{\alpha R_1}$

43. $M = R_1L - F(L-x) \Rightarrow M = R_1L - FL + Fx \Rightarrow M - Fx = R_1L - FL \Rightarrow M - Fx = L(R_1 - F) \Rightarrow L = \dfrac{M-Fx}{R_1 - F}$

45. $x = \dfrac{10m_1 + 25m_2}{m_1 + m_2} \Rightarrow x(m_1 + m_2) = 10m_1 + 25m_2 \Rightarrow xm_1 + xm_2 = 10m_1 + 25m_2 \Rightarrow xm_1 - 10m_1 = 25m_2 - xm_2$

$\Rightarrow m_1(x-10) = 25m_2 - xm_2 \Rightarrow m_1 = \dfrac{m_2(25-x)}{x-10}$

47. $E = mgy + \frac{1}{2}mv^2 \Rightarrow E = m\left(gy + \frac{1}{2}v^2\right) \Rightarrow m = \dfrac{E}{gy + \frac{1}{2}v^2}$

Chapter 11 Review Problems

1. $x^2 - 2x - 15 = (x-5)(x+3)$

3. $x^6 - y^4 = (x^3 - y^2)(x^3 + y^2)$

5. $2x^2 + 3x - 2 = (2x-1)(x+2)$

7. $8x^3 - \frac{y^3}{27} \Rightarrow (2x)^3 - \left(\frac{y}{3}\right)^3 = \left(2x - \frac{y}{3}\right)\left(4x^2 + \frac{2xy}{3} + \frac{y^2}{9}\right)$

9. $2ax^2y^2 - 18a \Rightarrow 2a(x^2y^2 - 9) = 2a(xy-3)(xy+3)$

11. $3a^2 - 2a - 8 = (3a+4)(a-2)$

13. $\frac{2a^2}{12} - \frac{8b^2}{27} \Rightarrow \frac{2}{3}\left(\frac{a^2}{4} - \frac{4b^2}{9}\right) = \frac{2}{3}\left(\frac{a}{2} - \frac{2b}{3}\right)\left(\frac{a}{2} + \frac{2b}{3}\right)$

15. $2x^2 - 20ax + 50a^2 \Rightarrow 2(x^2 - 10ax + 25a^2) = 2(x-5a)^2$

17. $4a^2 - (3a-1)^2 \Rightarrow (2a - (3a-1))(2a + (3a-1)) \Rightarrow (2a - 3a + 1)(5a-1) = (1-a)(5a-1)$

19. $a^2 - 2a - 8 = (a-4)(a+2)$

21. $x^2 - 21x + 110 = (x-10)(x-11)$

23. $3x^2 - 6x - 45 \Rightarrow 3(x^2 - 2x - 15) = 3(x+3)(x-5)$

25. $64m^3 - 27n^3 = (4m-3n)(16m^2 + 12mn + 9n^2)$

27. $15a^2 - 11a - 12 = (3a-4)(5a+3)$

29. $ax - bx + ay - by \Rightarrow (ax - bx) + (ay - by) \Rightarrow x(a-b) + y(a-b) = (a-b)(x+y)$

31. $a(x-3) - b(x+2) = c \Rightarrow ax - 3a - bx - 2b = c \Rightarrow ax - bx = 3a + 2b + c \Rightarrow x(a-b) = 3a + 2b + c \Rightarrow x = \dfrac{3a+2b+c}{a-b}$

33. $\frac{3x-1}{11} - \frac{2-x}{10} = \frac{6}{5} \Rightarrow 110\left(\frac{3x-1}{11} - \frac{2-x}{10} = \frac{6}{5}\right) \Rightarrow 10(3x-1) - 11(2-x) = 22(6) \Rightarrow 30x - 10 - 22 + 11x = 132$

$\Rightarrow 41x = 164 \Rightarrow x = 4$

35. $\frac{2x+1}{4} - \frac{4x-1}{10} + \frac{5}{4} = 0 \Rightarrow 20\left(\frac{2x+1}{4} - \frac{4x-1}{10} + \frac{5}{4} = 0\right) \Rightarrow 5(2x+1) - 2(4x-1) + 5(5) = 20(0) \Rightarrow 10x + 5 - 8x + 2 + 25 = 0$

$\Rightarrow 2x = -32 \Rightarrow x = -16$

37. $mx - n = \dfrac{nx - m}{p} \Rightarrow mpx - np = nx - m \Rightarrow mpx - nx = np - m \Rightarrow x(mp - n) = np - m \Rightarrow x = \dfrac{np - m}{mp - n}$

39. $m(x-a) + n(x-b) + p(x-c) = 0 \Rightarrow mx - ma + nx - nb + px - pc = 0 \Rightarrow x(m+n+p) = am + bn + cp \Rightarrow x = \dfrac{am + bn + cp}{m+n+p}$

41. $\frac{1}{x-5} - \frac{1}{4} = \frac{1}{3} \Rightarrow \text{LCD is} 12(x-5) \Rightarrow 12(x-5)\left(\frac{1}{x-5} - \frac{1}{4} = \frac{1}{3}\right) \Rightarrow 12 - 3(x-5) = 4(x-5) \Rightarrow 12 - 3x + 15 = 4x - 20$

$\Rightarrow 7x = 47; x = \dfrac{47}{7}$

43. $\dfrac{2}{x-2} = \dfrac{5}{2(x-1)} \Rightarrow$ LCD is $2(x-2)(x-1) \Rightarrow 2(x-2)(x-1)\left(\dfrac{2}{x-2} = \dfrac{5}{2(x-1)}\right) \Rightarrow 4(x-1) = 5(x-2) \Rightarrow 4x-4 = 5x-10$

$\Rightarrow -x = -6 \Rightarrow x = 6$

45. $\dfrac{x+2}{x-2} - \dfrac{x-2}{x+2} = \dfrac{x+7}{x^2-4} \Rightarrow$ LCD is $(x-2)(x+2) \Rightarrow (x-2)(x+2)\left(\dfrac{x+2}{x-2} - \dfrac{x-2}{x+2} = \dfrac{x+7}{x^2-4}\right) \Rightarrow (x+2)^2 - (x-2)^2 = x+7$

$\Rightarrow x^2 + 4x + 4 - (x^2 - 4x + 4) = x + 7 \Rightarrow 8x = x + 7 \Rightarrow 7x = 7 \Rightarrow x = 1$

47. $\dfrac{3}{5x^2} - \dfrac{2}{15xy} + \dfrac{1}{6y^2} \Rightarrow \left(\dfrac{6y^2}{6y^2}\right)\dfrac{3}{5x^2} - \left(\dfrac{2xy}{2xy}\right)\dfrac{2}{15xy} + \left(\dfrac{5x^2}{5x^2}\right)\dfrac{1}{6y^2} \Rightarrow \dfrac{18y^2 - 4xy + 5x^2}{30x^2y^2} \Rightarrow \dfrac{5x^2 - 4xy + 18y^2}{30x^2y^2}$

49. $\dfrac{\dfrac{a-1}{6} - \dfrac{2a-7}{2}}{\dfrac{3a}{4} - 3} \Rightarrow \dfrac{\dfrac{20-5a}{6}}{\dfrac{3a-12}{4}} \Rightarrow \dfrac{20-5a}{6} \cdot \dfrac{4}{3a-12} \Rightarrow \dfrac{20(4-a)}{18(a-4)} \Rightarrow \dfrac{20(4-a)}{-18(4-a)} \Rightarrow -\dfrac{20}{18} = -\dfrac{10}{9}$

51. $\left(1 + \dfrac{x+y}{x-y}\right)\left(1 - \dfrac{x-y}{x+y}\right) \Rightarrow \left(\dfrac{x-y+x+y}{x-y}\right)\left(\dfrac{x+y-x+y}{x+y}\right) \Rightarrow \dfrac{(2x)(2y)}{x^2-y^2} \Rightarrow \dfrac{4xy}{x^2-y^2}$

53. $\dfrac{3wx^2y^3}{7axyz} \times \dfrac{4a^3xz}{6aw^2y} \Rightarrow \dfrac{12a^3wx^3y^3z}{42a^2w^2xy^2z} \Rightarrow \dfrac{2ax^2y}{7w}$

55. $\dfrac{b^2 - 5b}{b^2 - 4b - 5} \Rightarrow \dfrac{b(\cancel{b-5})}{(\cancel{b-5})(b+1)} \Rightarrow \dfrac{b}{b+1}$

57. $\dfrac{20(a^3 - c^3)}{4(a^2 + ac + c^2)} \Rightarrow \dfrac{20(a-c)(a^2 + ac + c^2)}{4(a^2 + ac + c^2)} \Rightarrow 5(a-c)$

59. $\dfrac{2a^2 + 17a + 21}{3a^2 + 26a + 35} \Rightarrow \dfrac{(2a+3)(a+7)}{(3a+5)(a+7)} \Rightarrow \dfrac{2a+3}{3a+5}$

61. $\dfrac{\pi}{3}r_1^2h - \dfrac{\pi}{3}r_2^2d \Rightarrow \dfrac{\pi}{3}(r_1^2h - r_2^2d)$

Chapter 12: Quadratic Equations

Exercise 1 ◊ Solving a Quadratic Equation Graphically and by Calculator

Explicit Functions

1. $x^2 - 12x + 28 = 0$

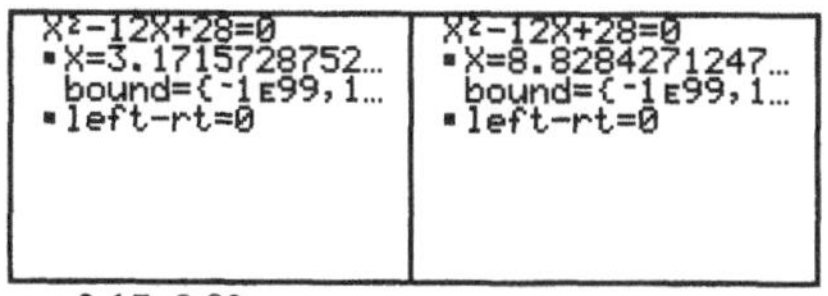

$x = 3.17, 8.83$

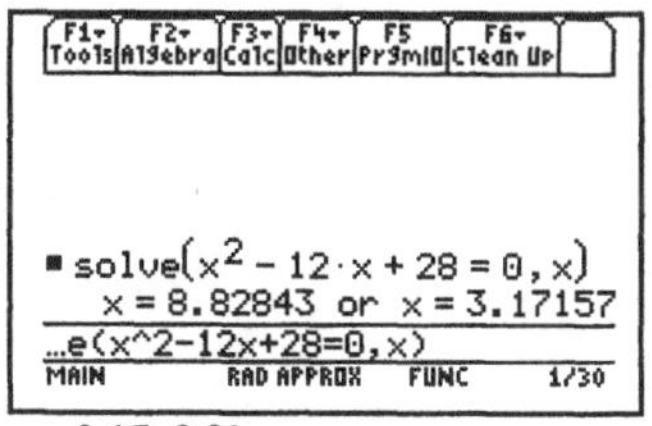

$x = 3.17, 8.83$

3. $x^2 + x - 19 = 0$

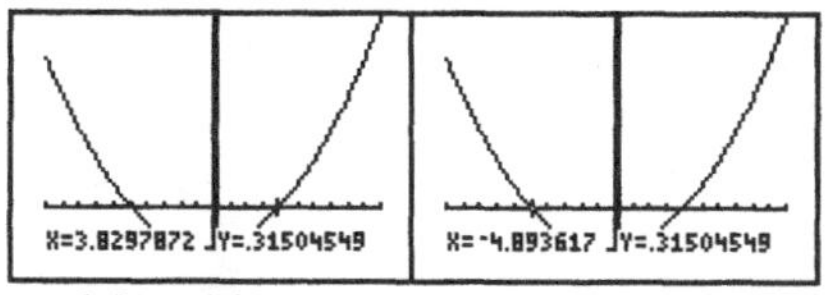

$x = 3.89, -4.89$

5. $3x^2 + 12x - 35 = 0$

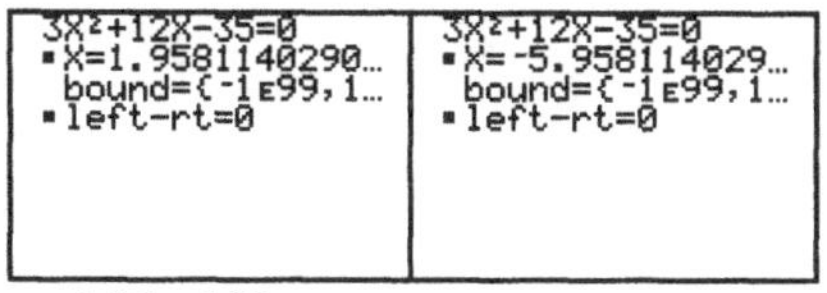

$x = 1.96, -5.96$

7. $36x^2 + 3x - 7 = 0$

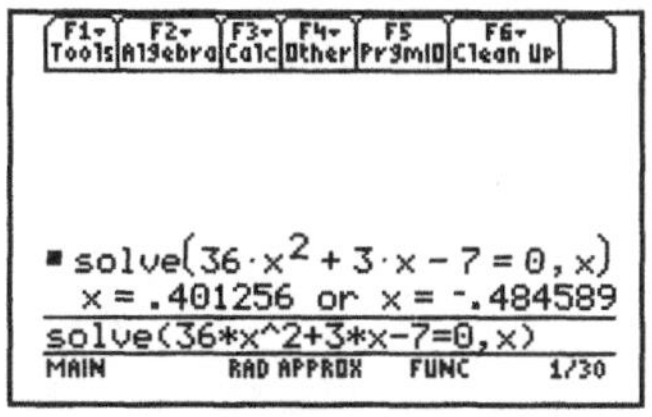

$x = 0.401, -0.485$

9. $49x^2 + 21x - 5 = 0$

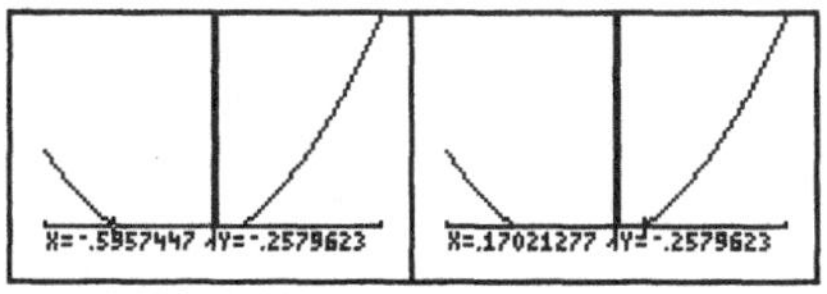

$x = -0.599, 0.170$

11. $3x^2 - 10x + 4 = 0$

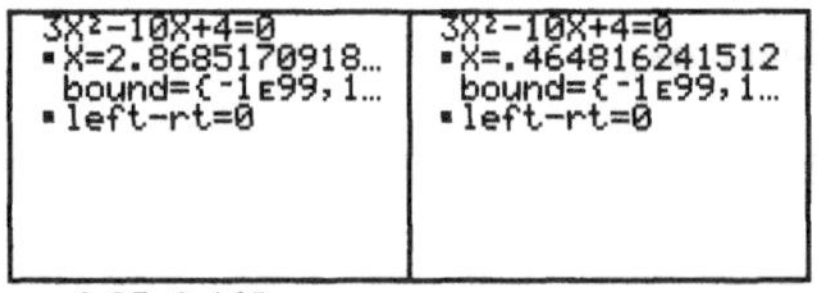

$x = 2.87, 0.465$

Implicit Functions

13. $3x^2 + 5x = 7 \Rightarrow 3x^2 + 5x - 7 = 0$

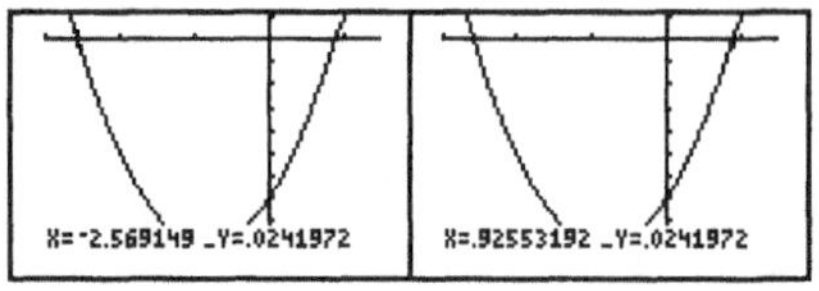

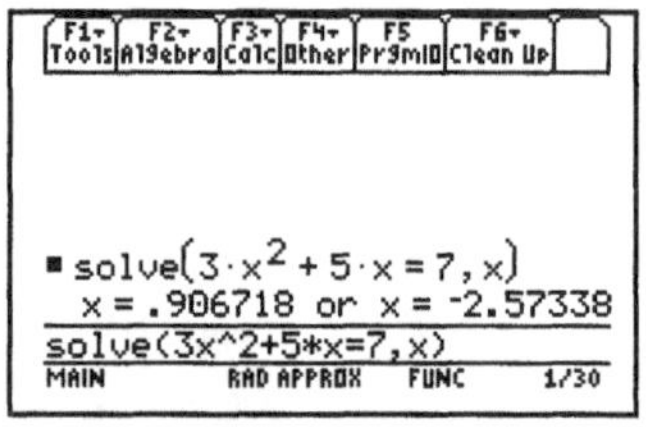

$x = -2.57, 0.907$

15. $x^2 - 4 = 4x + 7 \Rightarrow x^2 - 4 - 4x - 7 \Rightarrow x^2 - 4x - 11 = 0$

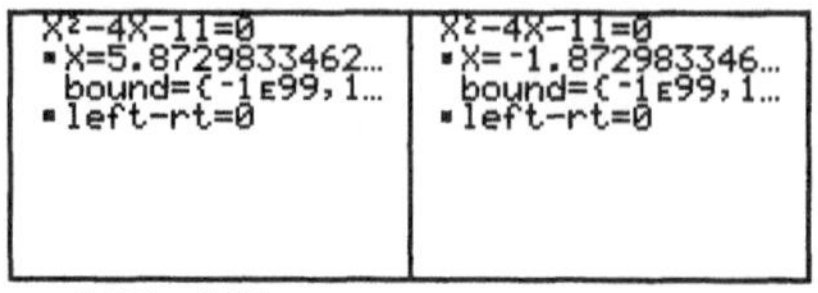

$x = 5.87, -1.87$

17. $6x - 300 = 205 - 3x^2 \Rightarrow 3x^2 - 205 + 6x - 300 \Rightarrow 3x^2 + 6x - 505 = 0$

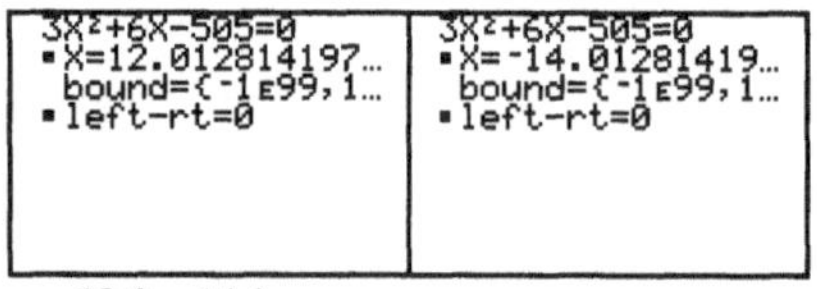

$x = 12.0, -14.0$

19. $2x^2 + 100 = 32x - 11 \Rightarrow 2x^2 + 100 - 32x + 11 = 0 \Rightarrow 2x^2 - 32x + 111 = 0$

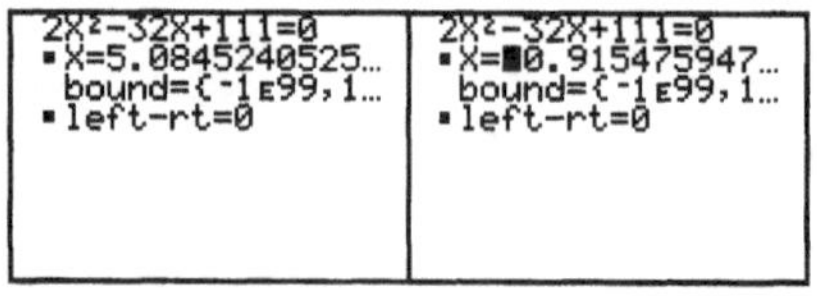

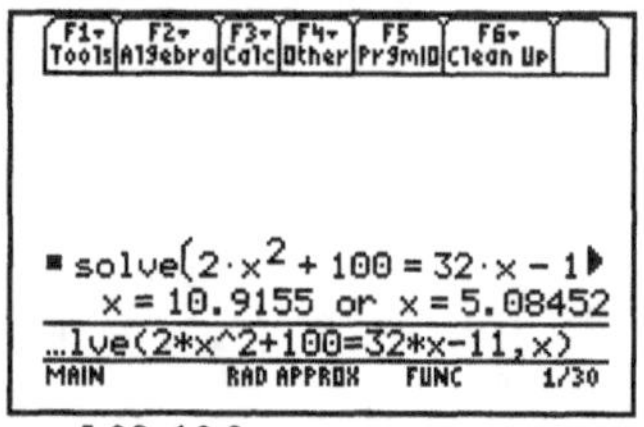

$x = 5.08, 10.9$

Challenge Problems

21. $4.26x + 5.74 = 1.27x^2 + 2.73x \Rightarrow 0 = 1.27x^2 + 2.73x - 4.26x - 5.74 \Rightarrow 1.27x^2 - 1.53x - 5.74 = 0$

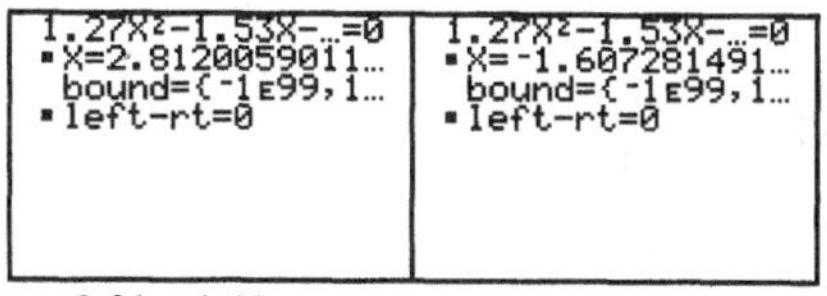

$x = 2.81, -1.61$

23. $x^2 - 6.27x - 14.4 = 3.17 \Rightarrow x^2 - 6.27x - 14.4 - 3.17 = 0 \Rightarrow x^2 - 6.27x - 17.57 = 0$

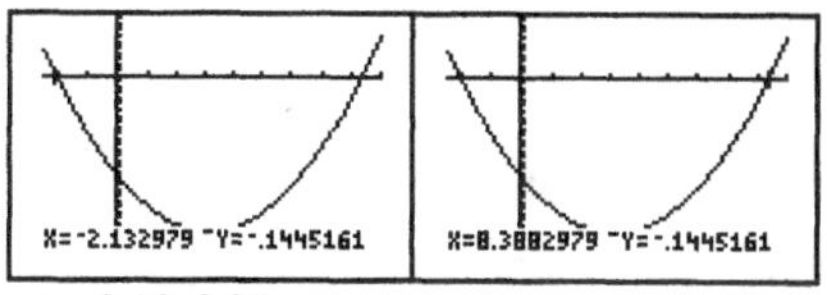

$x = -2.10, 8.37$

25. $x(2x - 3) = 3x(x + 4) - 2 \Rightarrow 2x^2 - 3x = 3x^2 + 12x - 2 \Rightarrow 0 = x^2 + 15x - 2$

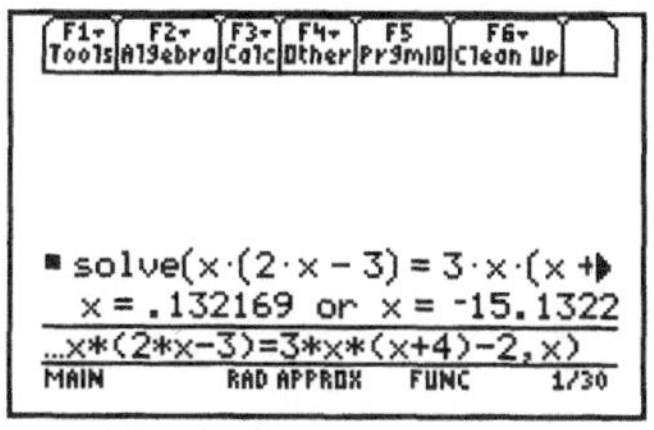

$x = -15.1, 0.132$

27. $(4.20x - 5.80)(7.20x - 9.20) = 8.20x + 9.90 \Rightarrow 30.24x^2 - 38.64x - 41.76x + 53.36 - 8.20x - 9.90 = 0$

$\Rightarrow 30.24x^2 - 88.60x + 43.46 = 0$

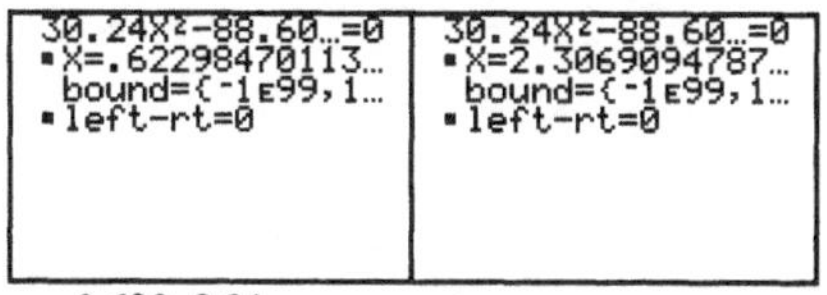

$x = 0.623, 2.31$

An Application

29. $(32.5 - x)x + x(24.6) = 225 \Rightarrow 32.5x - x^2 + 24.6x - 225 = 0 \Rightarrow x^2 - 57.1x + 225 = 0$

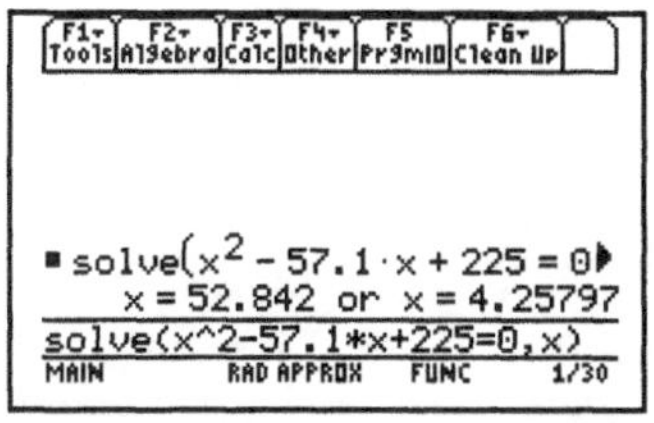

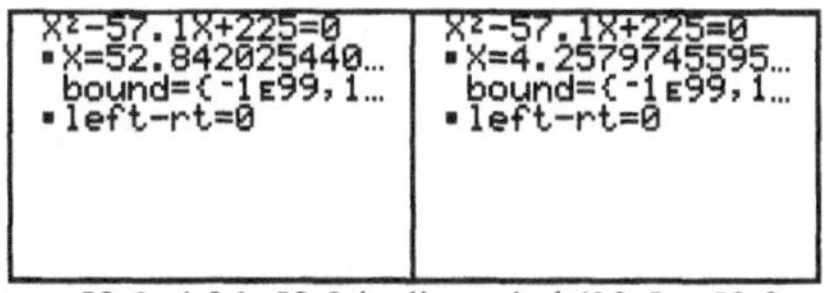

$x = 52.8, 4.26$; 52.8 is discarded ($32.5 - 52.8$ results in a negative distance); thus, $x = 4.26$ in.

Exercise 2 ◊ Solving a Quadratic by Formula

1. $x^2 + 5x - 6 = 0$; where $a = 1, b = 5, c = -6$; $x = \dfrac{-b \pm \sqrt{b^2 - 4ac}}{2a}$

$$x = \frac{-5 \pm \sqrt{(-5)^2 - 4(1)(-6)}}{2(1)} \Rightarrow \frac{-5 + \sqrt{49}}{2}, \frac{-5 - \sqrt{49}}{2} \Rightarrow \frac{-5+7}{2}, \frac{-5-7}{2}; x = 1, -6$$

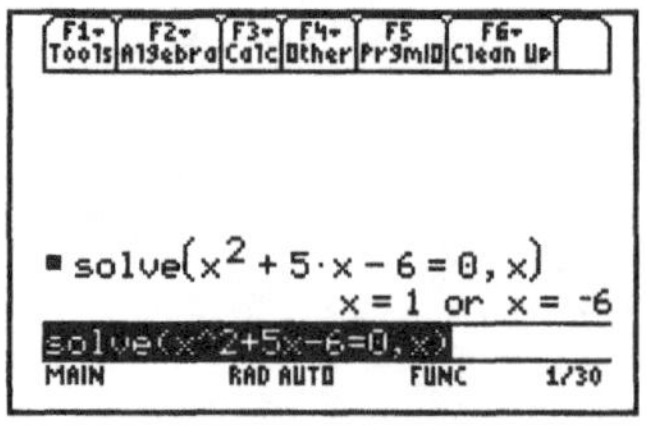

Solving $x^2 + 5x - 6 = 0$ by completing the square

$$x^2 + 5x = 6$$

$$x^2 + 5x + \left[\left(\frac{1}{2}\right)5\right]^2 = 6 + \left[\left(\frac{1}{2}\right)5\right]^2$$

$$x^2 + 5x + \frac{25}{4} = \frac{49}{4}$$

$$\left(x + \frac{5}{2}\right)^2 = \frac{49}{4}$$

$$x + \frac{5}{2} = \pm\frac{7}{2}$$

$$x = -\frac{5}{2} \pm \frac{7}{2}$$

$$x = 1, -6$$

3. $x^2 - 12x + 3 = 0$; where $a = 1, b = -12, c = 3$

$$x = \frac{-(-12) \pm \sqrt{(-12)^2 - 4(1)(3)}}{2(1)} \Rightarrow \frac{12 \pm \sqrt{132}}{2}; x = 11.7, 0.255$$

5. $2x^2 - 15x + 9 = 0$; where $a = 2, b = -15, c = 9$

$$x = \frac{-(-15) \pm \sqrt{(-15)^2 - 4(2)(9)}}{2(2)} \Rightarrow \frac{15 \pm \sqrt{153}}{4}; x = 6.84, 0.657$$

Solving $2x^2 - 15x + 9 = 0$ by completing the square

$$\frac{2x^2}{2} - \frac{15x}{2} + \frac{9}{2} = \frac{0}{2}$$

$$x^2 - \frac{15}{2}x + \frac{9}{2} = 0$$

$$x^2 - \frac{15}{2}x + \left[\left(\frac{1}{2}\right)\left(\frac{15}{2}\right)\right]^2 = -\frac{9}{2} + \left[\left(\frac{1}{2}\right)\left(\frac{15}{2}\right)\right]^2$$

$$x^2 = \frac{15}{2}x + \frac{225}{16} = \frac{153}{16}$$

$$\left(x-\frac{15}{4}\right)^2=\frac{153}{16}$$

$$x-\frac{15}{4}=\pm\frac{\sqrt{153}}{4}$$

$$x=\frac{15}{4}\pm\frac{\sqrt{153}}{4}$$

$x = 6.84, 0.657$

7. $5x^2 - 25x + 4 = 0$; where $a = 5$, $b = -25$, $c = 4$

$$x=\frac{-(-25)\pm\sqrt{(-25)^2-4(5)(4)}}{2(5)}\Rightarrow\frac{25\pm\sqrt{545}}{10};\ x = 4.83,\ 0.165$$

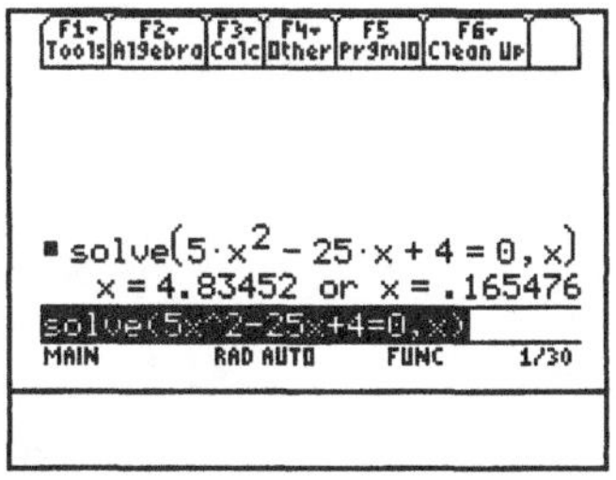

Challenge Problems

9. $1.22x^2 - 11.5x + 9.89 = 0$; where $a = 1.22$, $b = -11.5$, $c = 9.89$

$$x=\frac{-(-11.5)\pm\sqrt{(-11.5)^2-4(1.22)(9.89)}}{2(1.22)}\Rightarrow\frac{11.5\pm\sqrt{83.9868}}{2.44};\ x = 8.46,\ 0.957$$

11. $2.96x^2 - 33.2x + 4.05 = 0$; where $a = 2.96$, $b = -33.2$, $c = 4.05$

$$x=\frac{-(-33.2)\pm\sqrt{(-33.2)^2-4(2.96)(4.05)}}{2(2.96)}\Rightarrow\frac{33.2\pm\sqrt{1054.288}}{5.92};\ x = 11.0,\ -0.123$$

13. $9 + 2x^2 = 25x \Rightarrow 2x^2 - 25x + 9 = 0$; where $a = 2$, $b = -25$, $c = 9$

$$x=\frac{-(-25)\pm\sqrt{(-25)^2-4(2)(9)}}{2(2)}\Rightarrow\frac{25\pm\sqrt{553}}{4};\ x = 12.1,\ 0.370$$

15. $3.25 - 31x = 4.99x - 63.5 \Rightarrow -35.99x = -66.75 \Rightarrow 1.85468 \Rightarrow 1.85$

17. $3.88(x^2 + 7.72) = 6.34x(3.99x - 3.81) \Rightarrow 21.4166x^2 - 24.1554x - 22.6236 = 0$; where $a = 21.4166$, $b = -24.1554$, $c = -22.6236$

$$x=\frac{-(-24.1554)\pm\sqrt{(-24.1544)^2-4(21.4166)(-22.6236)}}{2(21.4166)}\Rightarrow\frac{24.1554\pm\sqrt{2521.517406}}{42.8332};\ x = 1.73,\ -0.608$$

Exercise 3 ◊ Applications

Number Puzzles

1. Let x = the fraction; $\frac{1}{x}$ = the reciprocal; thus $x+\frac{1}{x}=\frac{13}{6}$

$$x+\frac{1}{x}-\frac{13}{6}=0$$

$$6x^2+6-13x=0\ [\text{multiply by } 6x]$$
$$6x^2-13x+6=0$$
$$(2x-3)(3x-2)=0$$
$$x=\frac{3}{2},\frac{2}{3}$$

3. Let y = the smaller number and $y + 7$ = the larger number

$$(y+7)^3=y^3+1267$$
$$y^3+21y^2+147y+343=y^3+1267$$
$$21y^2+147y-924=0$$
$$7y^2+49y-308=0\ [\text{divide by } 3]$$
$$y=\frac{-(49)\pm\sqrt{(49)^2-4(7)(-308)}}{2(7)}\Rightarrow\frac{-49\pm\sqrt{11025}}{14}$$
$y=4,-11$; therefore, $y=4$ and $y+7\Rightarrow 11$

[-11 is discarded; the numbers are all positive as per dierctions]

5. Let r = the first number and $r + 10$ = the second number

$$r^2+(r+10)^2=250$$
$$r^2+r^2+20r+100-250=0$$
$$2r^2+20r-150=0$$
$$r^2+10r-75=0\ [\text{divide by } 2]$$
$$(r+15)(r-5)=0$$
$$r=-15,5\ (r\neq-15)$$
Therefore, $r=5$ and $r+10=15$

Geometry Problems

7. Let $l = w + 2$, $wl = 24$

$$w(w+2)=24$$
$$w^2+2w-24=0$$
$$(w+6)(w-4)=0$$
$w=-6, 4$ [-6 is discarded (cannot have a negative distance)]

Thus $w=4$, $l=w+2=6$; therefore, the dimensions are 4 m $\times$ 6 m

9. Let $l = 2w$ and $(l-6)(w-6)(3) = 648$

$$(2w-6)(w-6)(3)=648\ [\text{substitute } 2w \text{ for } l]$$
$$(2w^2-18w+36)(3)=648$$
$$2w^2-18w+36=216\ [\text{divide by } 3]$$
$$w^2-9w+18=108\ [\text{divide by } 2]$$
$$w^2-9w-90=0$$
$$(w+6)(w-15)=0$$
$w=-6, 15$ [-6 is discarded]

Thus $w=15$; $l=2w=30$; therefore, the dimensions are 15 cm $\times$ 30 cm

11. Let $2w + 2l = 724$ and $wl = 32{,}400$

$$2w+2l=724\Rightarrow 2w=724-2l\Rightarrow w=362-l$$

$(362-l)l = 32{,}400$ [since $wl = 32{,}400$]

$362l - l^2 = 32{,}400$

$l^2 - 362l + 32{,}400 = 0$

$(l-200)(l-162) = 0$

$l = 200, 162$; therefore, the dimensions are 162 m × 200 m

13. Let x = the original side, then x^3 = original volume

$(x-0.175)^3 = x^3 - 2.48$

$x^3 - 0.525x^2 + 0.09188x - 0.005359 = x^3 - 2.48$

$-0.525x^2 + 0.09188x + 2.475 = 0$

$x^2 - 0.1750x - 4.714 = 0$ [divide by -0.525]; where $a = 1, b = -0.1750, c = -4.714$

$$x = \frac{-(-0.1750) \pm \sqrt{(-0.1750)^2 - 4(1)(-4.714)}}{2(1)} \Rightarrow \frac{0.1750 \pm \sqrt{18.8866}}{2}$$

$x = -2.09, 2.26$ [-2.09 is discarded]; therefore, $x = 2.26$ in.

Uniform Motion

15. $D = RT = R_1T_1 = R_2T_2 = 350$ and $T_1 + T_2 = 14.4h$

$R_1T_1 = (R_1 + 8)(14.4 - T_1)$

$R_1T_1 = 14.4R_1 - R_1T_1 + 115.2 - 8T_1$

$2R_1T_1 = 14.4R_1 - 8T_1 + 115.2$

$$T_1 = \frac{350}{R_1}$$

$$2R_1\left(\frac{350}{R_1}\right) = 14.4R_1 - 8\left(\frac{350}{R_1}\right)_1 + 115.2 \text{ [substitute for } T_1]$$

$$14.4R_1 - \frac{2800}{R_1} - 584.8 = 0$$

$14.4R_1^2 - 584.8R_1 - 2800 = 0$ [multiply by R_1]; where $a = 14.4, b = -584.8, c = -2800$

$$R_1 = \frac{-(-584.8) \pm \sqrt{(-584.8)^2 - 4(14.4)(-2800)}}{2(14.4)} \Rightarrow \frac{584.8 \pm \sqrt{503{,}271.04}}{28.8}$$

$R_1 = 44.938, -4.32$ [-4.32 is discarded]; therefore, $R_1 = 44.9$ mi/h

17. Let R = the speed of the local bus, $R + 10$ = the speed of the express bus and t = the time for the local bus

[express bus] $(R+10)(t-1) = 250 \Rightarrow Rt + 10t - R - 10 = 250 \Rightarrow Rt + 10t - R = 260$

[local bus] $Rt = 240 \Rightarrow t = \frac{240}{R}$

Substitute $t = \frac{240}{R}$ into $Rt + 10t - R = 260$

$$R\left(\frac{240}{R}\right) + 10\left(\frac{240}{R}\right) - R = 260$$

$$240 + \frac{2400}{R} - R = 260$$

$240R + 2400 - R^2 = 260R$ [multiply by R]

$R^2 + 20R - 2400 = 0$
$(R+60)(R-40) = 0$
$R = -60, 40$ [-60 is discarded]
Therefore, the rate of the local bus is 40 mi/h, the rate of the express bus is $R + 10$ or 50 mi/h

19. Let R = rate and t = time

[increased rate] $(R+1)(t-1) = 30$

[uniform rate] $Rt = 30 \Rightarrow t = \dfrac{30}{R}$

Substitute $t = \dfrac{30}{R}$ into $(R+1)(t-1) = 30$

$(R+1)\left(\dfrac{30}{R} - 1\right) = 30$

$30 - R + \dfrac{30}{R} - 1 = 30$

$R^2 - 30 + R = 0$ [multiply by R]

$R^2 + R - 30 = 0$

$(R-5)(R+6) = 0$

$R = 5, -6$ [-6 is discarded]; therefore, the rate of travel is 5 km/h

Work Problems

21. Let x = the time for the larger pipe alone to fill the tank and y = the time for the smaller pipe to fill the tank

[together] $\dfrac{1}{x} + \dfrac{1}{y} = \dfrac{1}{8.40}$

[smaller pipe] $y = x + 2.50$

Substitute $y = x + 2.50$ into $\dfrac{1}{x} + \dfrac{1}{y} = \dfrac{1}{8.40}$

$\dfrac{1}{x} + \dfrac{1}{x+2.50} = \dfrac{1}{8.40}$

$8.40(x+2.50) + 8.40x = x(x+2.50)$ [multiply by $(x)(x+2.50)(8.40)$]

$8.40x + 21 + 8.40x = x^2 + 2.50x$

$x^2 - 14.3x - 21 = 0$; where $a = 1, b = -14.3, c = -21$

$$x = \frac{-(-14.3) \pm \sqrt{(-14.3)^2 - 4(1)(-21)}}{2(1)} \Rightarrow \frac{14.3 \pm \sqrt{288.49}}{2}$$

$x = 15.6, -1.34$ [-1.34 is discarded]; therefore, the larger pipe alone takes 15.6 h

23. Let x = the number of days worked and w = her wage

[working less] $(w-10)(x+3) = xw \Rightarrow 3w - 10x - 30 = 0$

[regular wage] $xw = 1800 \Rightarrow w = \dfrac{1800}{x}$

Substitute $w = \dfrac{1800}{x}$ into $3w - 10x - 30 = 0$

$3\left(\dfrac{1800}{x}\right) - 10x - 30 = 0$

$10x^2 + 30x - 5400 = 0$ [multiply by x]; where $a = 10$, $b = 30$, $c = -5400$

$$x = \frac{-(30) \pm \sqrt{(30)^2 - 4(10)(-5400)}}{2(10)} \Rightarrow \frac{-30 \pm \sqrt{216{,}900}}{20}$$

$x = -24.786, 21.786$ [-24.786 is discarded]; therefore, she worked 21.8 days

Simply Supported Beam

25. $l = 25.0$ ft, w = 1550 ft·lb and $M = 112{,}000$ ft·lb

$$M = \frac{1}{2}wlx - \frac{1}{2}wx^2 \Rightarrow 2M = wlx - wx^2 \Rightarrow wx^2 - wlx + 2M = 0;\ \text{where } a = w,\ b = -wl,\ c = 2M$$

$$x = \frac{-(-wl) \pm \sqrt{(-wl)^2 - 4(w)(2M)}}{2(w)} \Rightarrow \frac{wl \pm \sqrt{(wl)^2 - 4(w)(2M)}}{2w}$$

$$x = \frac{(1550)(25.0) \pm \sqrt{[(1550)(25.0)]^2 - 4(1550)(224{,}000)}}{2(1550)}$$

$$x = \frac{38{,}750 \pm \sqrt{1{,}406{,}250{,}000 - 694{,}400{,}000}}{3100}$$

$x = 9.15$ ft and 15.9 ft

Free Falling Body

27. $v_1 = 120$ m/s, $y = 0$ ft and $g = -9.8$ m/s^2

$$\frac{1}{2}at^2 + v_1 t - y = 0;\ \text{where } a = \frac{g}{2},\ b = v_1,\ c = -y$$

$$t = \frac{-(v_1) \pm \sqrt{(v_1)^2 - 4\left(\frac{g}{2}\right)(-y)}}{2\left(\frac{g}{2}\right)} \Rightarrow \frac{-120 \pm \sqrt{(120)^2 - 4\left(\frac{-9.8}{2}\right)(0)}}{2\left(\frac{-9.8}{2}\right)} \Rightarrow \frac{-120 \pm \sqrt{14{,}400 - 0}}{-9.8}$$

$t = -0.1699, 24.506$ [-0.1699 is discarded since time cannot be negative];
therefore, 24.5 s to return to its starting point

Electrical Problems

29. $P = EI - I^2R \Rightarrow RI^2 - EI + P = 0$

$E = 115$V; $R = 100\Omega$; $P = 29.3$ W

$100I^2 - 115I + 29.3 = 0$; where $a = 100$, $b = -115$, $c = 29.3$

$$I = \frac{-(-115) \pm \sqrt{(-115)^2 - 4(100)(29.3)}}{2(100)} \Rightarrow \frac{115 \pm \sqrt{1505}}{200}$$

$I = 0.769$ A, 0.381 A

31. $P = (I_1 + I_2)^2 R \Rightarrow RI_1^2 + 2RI_1I_2 + RI_2^2 - P = 0$; where $a = R, b = 2RI_2, c = RI_2^2 - P$

$R = 100\Omega$, $I_2 = 0.2$ A, $P = 9.0$ W

$$I_1 = \frac{-(2RI_2) \pm \sqrt{(2RI_2)^2 - 4(R)(RI_2^2 - P)}}{2(R)} \Rightarrow \frac{(-2)(100)(0.2) \pm \sqrt{[(2)(100)(0.2)]^2 - 4(100)\left[(100)(0.2)^2 - 9\right]}}{2(100)}$$

$$I_1 = \frac{-40 \pm \sqrt{3600}}{200}$$

$I_1 = -0.5$ A, 0.1 A

33. $V = B \pm B\sqrt{\frac{I}{A}}$; where $A = 4.8$ mA, $B = -2.5$V, $I = 1.5$ mA

$\Rightarrow -2.5 \pm (-2.5)\sqrt{\frac{1.5}{4.8}} = -2.5 \pm (-2.5)(0.559016) \Rightarrow -2.5 \pm (-1.3975) = -3.8975\text{V}, \ -1.1\text{V}$

$[-3.8975\text{ V}$ is discarded since it is beyond the pinch-off voltage of $-2.5\text{ V}]$; therefore, $V = -1.1$ V

CHAPTER 12 REVIEW PROBLEMS

1. $y^2 - 5y - 6 = 0$

$(y - 6)(y + 1) = 0$

$y - 6 = 0 \Rightarrow y = 6$ and $y + 1 = 0 \Rightarrow y = -1$

3. $w^2 - 5w = 0$

$w(w - 5) = 0$

$w = 0$ and $w - 5 = 0 \Rightarrow w = 5$

5. $\frac{r}{3} = \frac{r}{r+5}$

$r^2 + 5r = 3r$ [multiply by $(3)(r+5)$]

$r^2 + 2r = 0$

$r(r+2) = 0$

$r = 0$ and $r + 2 = 0 \Rightarrow r = -2$

7. $3t^2 - 10 = 13t \Rightarrow 3t^2 - 13t - 10 = 0$

$(3t + 2)(t - 5) = 0$

$3t + 2 = 0 \Rightarrow t = -\frac{2}{3}$ and $t - 5 = 0 \Rightarrow t = 5$

9. $\frac{1}{t^2} + 2 = \frac{3}{t}$

$1 + 2t^2 = 3t$ [multiply by t^2]

$2t^2 - 3t + 1 = 0$

$(2t - 1)(t - 1) = 0$

$2t - 1 = 0 \Rightarrow t = \frac{1}{2}$ and $t - 1 = 0 \Rightarrow t = 1$

11. $9 - x^2 = 0$

$(3 - x)(3 + x) = 0$

$3 - x = 0 \Rightarrow x = 3$ and $3 + x = 0 \Rightarrow x = -3$

or $9 - x^2 = 0 \Rightarrow 9 = x^2 \Rightarrow x = \pm 2$ [take square root of both sides]

13. $2x(x + 2) = x(x + 3) + 5$

$2x^2 + 4x = x^2 + 3x + 5$

$x^2 + x - 5 = 0$; where $a = 1$, $b = 1$, $c = -5$

$$x = \frac{-(1) \pm \sqrt{(1)^2 - 4(1)(-5)}}{2(1)} \Rightarrow \frac{-1 \pm \sqrt{21}}{2}; \ x = 1.79, -2.79$$

15. $9y^2 + y = 5 \Rightarrow 9y^2 + y - 5 = 0$; where $a = 9$, $b = 1$, $c = -5$

$$y = \frac{-(1) \pm \sqrt{(1)^2 - 4(9)(-5)}}{2(9)} \Rightarrow \frac{-1 \pm \sqrt{181}}{18}; \; y = 0.692, -0.803$$

17. $\frac{z}{2} = \frac{5}{z}$

$z^2 = 10$ [multiply by $2z$]

$z^2 - 10 = 0$; where $a = 1, b = 0, c = -10$

$$z = \frac{-(0) \pm \sqrt{(0)^2 - 4(1)(-10)}}{2(1)} \Rightarrow \frac{0 \pm \sqrt{40}}{2}; \; z = 3.16, -3.16$$

or $z^2 = 10 \Rightarrow z = \pm\sqrt{10}$ [take square root of both sides]$\Rightarrow z = \pm 3.16$

19. $2x^2 + 3x = 2 \Rightarrow 2x^2 + 3x - 2 = 0$

$(2x - 1)(x + 2) = 0$

$2x - 1 = 0 \Rightarrow x = \frac{1}{2}$ and $x + 2 = 0 \Rightarrow x = -2$

21. $27x = 3x^2 \Rightarrow 3x^2 - 27x = 0$

$3x(x - 9) = 0$

$3x = 0 \Rightarrow x = 0$ and $x - 9 = 0 \Rightarrow x = 9$

23. $5y^2 = 125 \Rightarrow 5y^2 - 125 = 0$

$5(y^2 - 25) = 0 \Rightarrow y^2 - 25 = 0$ [divide by 5]

$(y + 5)(y - 5) = 0$

$y + 5 = 0 \Rightarrow y = -5$ and $y - 5 = 0 \Rightarrow y = 5$

or $5y^2 = 125 \Rightarrow y^2 = 25 \Rightarrow y = \pm 5$ [take square root of both sides]

25. $1.26x^2 - 11.8 = 1.13x \Rightarrow 1.26x^2 - 1.13x - 11.8 = 0$; where $a = 1.26, b = -1.13, c = -11.8$

$$x = \frac{-(-1.13) \pm \sqrt{(-1.13)^2 - 4(1.26)(-11.8)}}{2(1.26)} \Rightarrow \frac{1.13 \pm \sqrt{60.7489}}{2.52}; \; x = 3.54, -2.64$$

27. $5.12y^2 + 8.76y - 9.89 = 0$; where $a = 5.12, b = 8.76, c = -9.89$

$$y = \frac{-(8.76) \pm \sqrt{(8.76)^2 - 4(5.12)(-9.89)}}{2(5.12)} \Rightarrow \frac{-8.76 \pm \sqrt{279.2848}}{10.24}; \; y = 0.777, -2.49$$

29. $27.2w^2 + 43.6w = 45.2 \Rightarrow 27.2w^2 + 43.6w - 45.2 = 0$; where $a = 27.2, b = 43.6, c = -45.2$

$$w = \frac{-(43.6) \pm \sqrt{(43.6)^2 - 4(27.2)(-45.2)}}{2(27.2)} \Rightarrow \frac{-43.6 \pm \sqrt{6818.72}}{54.4}; \; w = 0.716, -2.32$$

31. Let x = number of bags bought and p = price per bag

$xp = 1000 \Rightarrow p = \frac{1000}{x}$

$(x + 5)(p - 0.12) = 1000 \Rightarrow xp + 5p - 0.12x - 0.6 = 1000$; Substituting for $p = \frac{1000}{x}$

$$x\left(\frac{1000}{x}\right) + 5\left(\frac{1000}{x}\right) - 0.12x - 1000.6 = 0$$

$1000x + 5000 - 0.12x^2 - 1000.6x = 0$ [multiply by x]

$0.12x^2 + 0.6x - 5000 = 0$; where $a = 0.12$, $b = 0.6$, $c = -5000$

$$x = \frac{-(0.6) \pm \sqrt{(0.6)^2 - 4(0.12)(-5000)}}{2(0.12)} \Rightarrow \frac{-0.6 \pm \sqrt{2400.36}}{0.24}; \; x = 201.63, -206.639[\text{discarded}]$$

Therefore, $x = 202$ bags [nearest whole bag]

33. $x(2x) = 450$

$2x^2 = 450 \Rightarrow 2x^2 - 450 = 0 \Rightarrow x^2 - 225 = 0$ [divide by 2] $\Rightarrow (x + 15)(x - 15) = 0$

$x = -15, 15$ [-15 is discarded]

Therefore, the dimensions are $15 \times 2(15)$ ft $\Rightarrow 15 \times 30$ ft

35. Let x = his original speed and t = his predicted time $\left[10 \text{ min} = \frac{1}{6} \text{ h}\right]$

$$xt = 3 \Rightarrow t = \frac{3}{x}$$

$$\frac{1 \text{ mile}}{x} + \frac{1}{6} + \frac{2 \text{ miles}}{x+1} = \frac{3}{x}$$

$$6(x+1) + x(x+1) + 2(6x) = \frac{3}{x}[6x(x+1)] \; [\text{multiply by } 6x(x+1)]$$

$$6x + 6 + x^2 + x + 12x = 18x + 18$$

$$x^2 + x - 12 = 0$$

$$(x+4)(x-3) = 0$$

$x + 4 = 0 \Rightarrow x = -4$[discarded] and $x - 3 = 0 \Rightarrow x = 3$; therefore, his original speed was 3 mi/h

37. Let x = the number of seconds per revolution and $\frac{1}{x}$ = the number of revolutions per second

$$\frac{15 \text{ ft}}{\text{rev}}\left(\frac{1.0 \text{ rev}}{x \text{ sec}}\right)\left(\frac{3600 \text{ sec}}{\text{h}}\right) = \frac{15 \text{ ft}}{\text{rev}}\left(\frac{1.0}{x+1}\right)3600 + 14,400$$

$$15(3600)\left(\frac{1}{x} - \frac{1}{x+1}\right) = 14,400$$

$$15\left(\frac{1}{x} - \frac{1}{x+1}\right) = \frac{14,400}{3600}$$

$$\frac{1}{x} - \frac{1}{x+1} = \frac{4}{15}$$

$$15(x+1) - 15x = 4x(x+1)$$

$$15x + 15 - 15x = 4x^2 + 4x$$

$$4x^2 + 4x - 15 = 0$$

$$(2x+5)(2x-3) = 0$$

$2x + 5 = 0 \Rightarrow x = -\frac{5}{2}$[discarded] and $2x - 3 = 0 \Rightarrow x = \frac{3}{2}$; therefore, it takes $1\frac{1}{2}$ seconds for one revolution

39. $w = 26 - 2d$ and $wd = 80$

Substitute $w = 26 - 2d$ into $wd = 80 \Rightarrow (26 - 2d)d = 80$

$26d - 2d^2 = 80$

$-2d^2 + 26d - 80 = 0$

$d^2 - 13d + 40 = 0$ [divide by -2]

$(d - 5)(d - 8) = 0$

$d-5=0 \Rightarrow d=5$ and $d-8=0 \Rightarrow d=8$

For $d=5$, $w=\frac{80}{d} \Rightarrow \frac{80}{5}=16$; therefore the dimensions are 5.0 in. × 16.0 in.

For $d=8$, $w=\frac{80}{d} \Rightarrow \frac{80}{8}=10$; therefore the dimensions are 8.0 in. × 10.0 in.

41. Let y = the woman's rate of travel and t = the woman's time of travel

[faster rate] $(y+1.5)(t-3)=36 \Rightarrow yt-3y+1.5t=40.5$

[normal rate] $yt=36 \Rightarrow t=\frac{36}{y}$

Substitute $t=\frac{36}{y}$ into $yt-3y+1.5t=40.5$

$$y\left(\frac{36}{y}\right)-3y+1.5\left(\frac{36}{y}\right)=40.5$$

$$-3y+\frac{54}{y}=4.5$$

$3y^2+4.5y-54=0$; where $a=3$, $b=4.5$, $c=-54$

$$y=\frac{-(4.5)\pm\sqrt{(4.5)^2-4(3)(-54)}}{2(3)} \Rightarrow \frac{-4.5\pm\sqrt{668.25}}{6};\ y=-5.06[\text{discarded}],\ 3.56$$

Therefore, she traveled 3.56 km/h

43. $2x^2=(x+6)(x+4)$

$2x^2=x^2+10x+24$

$x^2-10x-24=0$

$(x+2)(x-12)=0$

$x+2=0 \Rightarrow x=-2$ [discarded] and $x-12=0 \Rightarrow x=12$; therefore the dimensions are 12 ft × 12 ft

Chapter 13: Exponents and Radicals

Exercise 1 ◊ Integral Exponents

1. $3x^{-1} \Rightarrow 3\left(\frac{1}{x^1}\right)$ [generally, when the exponent is one, it is not written] $\Rightarrow \frac{3}{1}\left(\frac{1}{x}\right) = \frac{3}{x}$

3. $2p^0 \Rightarrow 2(1)\ [p^0 = 1] \Rightarrow 2$

5. $a(2b)^{-2} \Rightarrow a(2^{-2})(b^{-2}) \Rightarrow a\left(\frac{1}{2^2}\right)\left(\frac{1}{b^2}\right) \Rightarrow \frac{a}{1}\left(\frac{1}{4}\right)\left(\frac{1}{b^2}\right) = \frac{a}{4b^2}$

7. $a^3b^{-2} \Rightarrow a^3\left(\frac{1}{b^2}\right) = \frac{a^3}{b^2}$

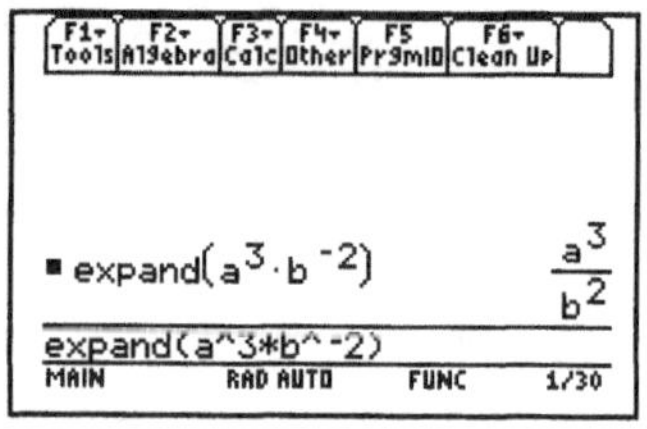

9. $p^3q^{-1} \Rightarrow p^3\left(\frac{1}{q}\right) = \frac{p^3}{q}$

11. $(3m^3n^2p)^0 = 3^{1\cdot 0}m^{3\cdot 0}n^{1\cdot 0}p^{1\cdot 0} \Rightarrow 3^0m^0n^0p^0 \Rightarrow 1\cdot 1\cdot 1\cdot 1 = 1$

13. $\left(4a^3b^2c^6\right)^{-2} \Rightarrow \frac{1}{\left(4a^3b^2c^6\right)^2} = \frac{1}{16a^6b^4c^{12}}$

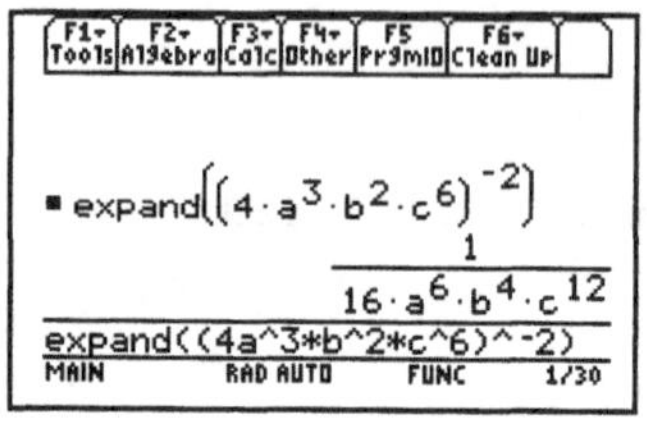

15. $(x+y)^{-1} \Rightarrow \frac{1}{(x+y)^1} = \frac{1}{x+y}$

17. $(m^{-2} - 6n)^{-2} \Rightarrow \frac{1}{\left(m^{-2}-6n\right)^2} \Rightarrow \frac{1}{m^{-4} - 12m^{-2}n + 36n^2} \Rightarrow \frac{1}{\frac{1}{m^4} - \frac{12n}{m^2} + 36n^2} \Rightarrow \frac{1}{\frac{1-12m^2n+36m^4n^2}{m^4}} \Rightarrow \frac{m^4}{\left(1-6m^2n\right)^2}$

19. $2x^{-1} + y^{-2} \Rightarrow 2\left(\frac{1}{x}\right) + \frac{1}{y^2} = \frac{2}{x} + \frac{1}{y^2}$

21. $\left(\frac{3}{b}\right)^{-1} \Rightarrow \frac{3^{-1}}{b^{-1}} \Rightarrow \frac{\frac{1}{3}}{\frac{1}{b}} \Rightarrow \frac{1}{3} \div \frac{1}{b} \Rightarrow \frac{1}{3}\cdot\frac{b}{1} = \frac{b}{3}$

or $\left(\frac{3}{b}\right)^{-1} \Rightarrow \frac{1}{\left(\frac{3}{b}\right)^1} \Rightarrow \frac{1}{\frac{3^1}{b^1}} \Rightarrow 1 \div \frac{3}{b} \Rightarrow 1\cdot\frac{b}{3} = \frac{b}{3}$ or $\left(\frac{3}{b}\right)^{-1} \Rightarrow \left(\frac{b}{3}\right)^1 = \frac{b}{3}$

23. $\left(\dfrac{2x}{3y}\right)^{-2} \Rightarrow \left(\dfrac{3y}{2x}\right)^{2} \Rightarrow \dfrac{3^2y^2}{2^2x^2} = \dfrac{9y^2}{4x^2}$

25. $\left(\dfrac{2px}{3qy}\right)^{-3} \Rightarrow \left(\dfrac{3qy}{2px}\right)^{3} \Rightarrow \dfrac{3^3q^3y^3}{2^3p^3x^3} = \dfrac{27q^3y^3}{8p^3x^3}$

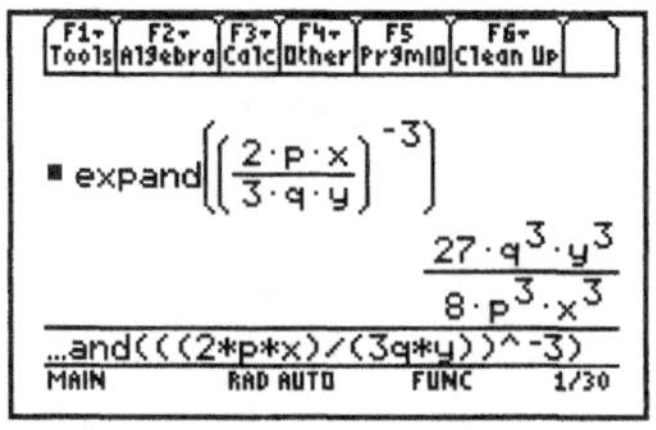

27. $\left(\dfrac{3a^4b^3}{5x^2y}\right)^{2} \Rightarrow \dfrac{3^2a^8b^6}{5^2x^4y^2} = \dfrac{9a^8b^6}{25x^4y^2}$

Challenge Problems

29. $(3m)^{-3} - 2n^{-2} \Rightarrow \dfrac{1}{(3m)^3} - 2\left(\dfrac{1}{n^2}\right) \Rightarrow \dfrac{1}{27m^3} - \dfrac{2}{n^2}$

31. $(x^n + y^m)^2 \Rightarrow (x^n + y^m)(x^n + y^m) \Rightarrow x^{n+n} + x^n y^m + x^n y^m + y^{m+m} \Rightarrow x^{2n} + 2x^n y^m + y^{2m}$

33. $\left(16x^6y^0 \div 8x^4y\right) \div 4xy^6 \Rightarrow \dfrac{16x^6y^0}{8x^4y} \div 4xy^6 \Rightarrow \dfrac{16x^6y^0}{8x^4y} \cdot \dfrac{1}{4xy^6} \Rightarrow \dfrac{(\cancel{4})(\cancel{4})x^6}{(\cancel{4})(2)x^4y} \cdot \dfrac{1}{(\cancel{4})xy^6} = \dfrac{x}{2y^7}$

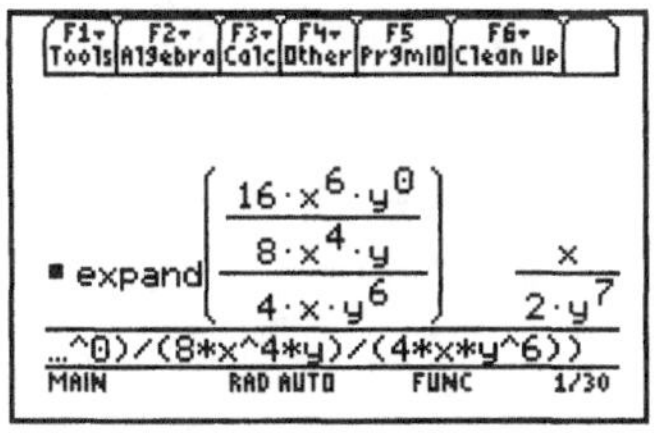

35. $\left(72p^6q^7 \div 9p^4q\right) \div 8pq^6 \Rightarrow \dfrac{72p^6q^7}{9p^4q} \div 8pq^6 \Rightarrow \dfrac{72p^6q^7}{9p^4q} \cdot \dfrac{1}{8pq^6} = p$

37. $\left(\dfrac{-2a^3x^3}{3b^2y}\right)^{2n} \Rightarrow \dfrac{(-2)^{1\cdot 2n}a^{3\cdot 2n}x^{3\cdot 2n}}{3^{1\cdot 2n}b^{2\cdot 2n}y^{1\cdot 2n}} \Rightarrow \dfrac{(-2)^{2n}a^{6n}x^{6n}}{3^{2n}b^{4n}y^{2n}} = \dfrac{4^n a^{6n}x^{6n}}{9^n b^{4n}y^{2n}}$

39. $\left(\dfrac{2p^{-2}z^3}{3q^{-2}x^{-4}}\right)^{-1} \Rightarrow \dfrac{1}{\dfrac{2p^{-2}z^3}{3q^{-2}x^{-4}}} \Rightarrow \dfrac{1}{\dfrac{2\left(\frac{1}{p^2}\right)(z^3)}{3\left(\frac{1}{q^2}\right)\left(\frac{1}{x^4}\right)}} \Rightarrow \dfrac{1}{\dfrac{\frac{2z^3}{p^2}}{\frac{3}{q^2x^4}}} \Rightarrow \dfrac{1}{\dfrac{2z^3}{p^2} \div \dfrac{3}{q^2x^4}} \Rightarrow \dfrac{1}{\dfrac{2z^3}{p^2} \cdot \left(\dfrac{q^2x^4}{3}\right)} \Rightarrow \dfrac{1}{\dfrac{2q^2x^4z^3}{3p^2}}$

$\Rightarrow 1 \div \left(\dfrac{2q^2x^4z^3}{3p^2}\right) \Rightarrow 1 \cdot \dfrac{3p^2}{2q^2x^4z^3} = \dfrac{3p^2}{2q^2x^4z^3}$

41. $\left(\dfrac{5w^2}{2z}\right)^{p} \Rightarrow \dfrac{5^{1\cdot p}w^{2\cdot p}}{2^{1\cdot p}z^{1\cdot p}} = \dfrac{5^p w^{2p}}{2^p z^p}$

43. $\left(\frac{3n^{-2}y^3}{5m^{-3}x^4}\right)^{-2} \Rightarrow \frac{1}{\left(\frac{3n^{-2}y^3}{5m^{-3}x^4}\right)^2} \Rightarrow \frac{1}{\frac{3^2n^{-4}y^6}{5^2m^{-6}x^8}} \Rightarrow \frac{1}{\frac{\frac{9y^6}{n^4}}{\frac{25x^8}{m^6}}} \Rightarrow \frac{1}{\frac{9m^6y^6}{25n^4x^8}} = \frac{25n^4x^8}{9m^6y^6}$

Applications

45. $\frac{1}{R} = \frac{1}{R_1} + \frac{1}{R_2} \Rightarrow R^{-1} = R_1^{-1} + R_2^{-1}$

47. $(2I)^2\left(\frac{R}{2}\right) \Rightarrow 4I^2\left(\frac{R}{2}\right) \Rightarrow \frac{4I^2R}{2} \Rightarrow \frac{(2)(2)I^2R}{2} \Rightarrow \frac{(\not{2})(2)I^2R}{\not{2}} = 2I^2R$

Exercise 2 ◊ Simplification of Radicals

Exponential and Radical Form

1. $a^{1/4} \Rightarrow \sqrt[4]{a^1} = \sqrt[4]{a}$

3. $z^{3/4} \Rightarrow \sqrt[4]{z^3}$

5. $(m-n)^{1/2} \Rightarrow \sqrt[2]{(m-n)^1} = \sqrt{m-n}$

7. $\left(\frac{x}{y}\right)^{-1/3} \Rightarrow \frac{1}{\left(\frac{x}{y}\right)^{1/3}} \Rightarrow \frac{1}{\frac{x^{1/3}}{y^{1/3}}} \Rightarrow 1 \div \frac{x^{1/3}}{y^{1/3}} \Rightarrow 1 \cdot \left(\frac{y^{1/3}}{x^{1/3}}\right) = \sqrt[3]{\frac{y}{x}}$

9. $\sqrt{b} \Rightarrow \sqrt[2]{b^1} \Rightarrow \left(b^1\right)^{1/2} \Rightarrow b^{1\cdot(1/2)} = b^{1/2}$

11. $\sqrt{y^2} \Rightarrow \sqrt[2]{y^2} \Rightarrow y^{2\cdot(1/2)}y^{2/2} \Rightarrow y^1 = |y|$

13. $\sqrt[n]{a+b} \Rightarrow \sqrt[n]{(a+b)^1} = (a+b)^{1/n}$

15. $\sqrt{x^2y^2} \Rightarrow \sqrt[2]{x^2y^2} \Rightarrow \left(x^2y^2\right)^{1/2} \Rightarrow x^{2\cdot(1/2)}y^{2\cdot(1/2)} \Rightarrow x^{2/2}y^{2/2} = |xy|$

Simplifying Radicals

17. $\sqrt{18} \Rightarrow \sqrt{(9)(2)} \Rightarrow \sqrt{9}\cdot\sqrt{2} \Rightarrow \sqrt{3^2}\cdot\sqrt{2} = 3\sqrt{2}$

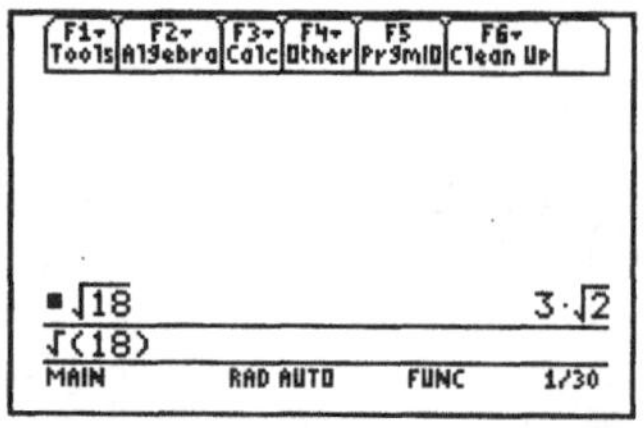

19. $\sqrt{63} \Rightarrow \sqrt{(9)(7)} \Rightarrow \sqrt{9}\cdot\sqrt{7} \Rightarrow \sqrt{3^2}\cdot\sqrt{7} = 3\sqrt{7}$

21. $\sqrt[3]{-56} \Rightarrow \sqrt[3]{(-8)(7)} \Rightarrow \sqrt[3]{(-2)^3}\cdot\sqrt[3]{7} = -2\sqrt[3]{7}$

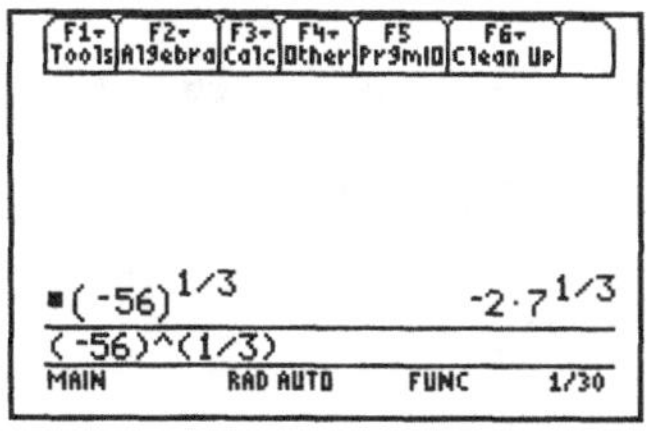

23. $\sqrt{a^3} \Rightarrow \sqrt{(a^2)(a)} \Rightarrow \sqrt{a^2}\cdot\sqrt{a} = a\sqrt{a}$

25. $\sqrt{36x^2y} \Rightarrow \sqrt{6^2}\cdot\sqrt{x^2}\cdot\sqrt{y} = 6x\sqrt{y}$

27. $\sqrt{\frac{3}{7}} \Rightarrow \frac{\sqrt{3}}{\sqrt{7}} \Rightarrow \frac{\sqrt{3}}{\sqrt{7}}\left(\frac{\sqrt{7}}{\sqrt{7}}\right) \Rightarrow \frac{\sqrt{21}}{\sqrt{49}} = \frac{\sqrt{21}}{7}$

29. $\sqrt[3]{\frac{1}{4}} \Rightarrow \frac{\sqrt[3]{1}}{\sqrt[3]{4}} \Rightarrow \frac{\sqrt[3]{1}}{\sqrt[3]{4}}\left(\frac{\sqrt[3]{2}}{\sqrt[3]{2}}\right) \Rightarrow \frac{\sqrt[3]{2}}{\sqrt[3]{8}} = \frac{\sqrt[3]{2}}{2}$

31. $\sqrt[3]{\frac{2}{9}} \Rightarrow \frac{\sqrt[3]{2}}{\sqrt[3]{9}} \Rightarrow \frac{\sqrt[3]{2}}{\sqrt[3]{9}}\left(\frac{\sqrt[3]{3}}{\sqrt[3]{3}}\right) \Rightarrow \frac{\sqrt[3]{6}}{\sqrt[3]{27}} = \frac{\sqrt[3]{6}}{3}$

33. $\sqrt{\frac{1}{2x}} \Rightarrow \frac{\sqrt{1}}{\sqrt{2x}} \Rightarrow \frac{\sqrt{1}}{\sqrt{2x}}\left(\frac{\sqrt{2x}}{\sqrt{2x}}\right) \Rightarrow \frac{\sqrt{2x}}{\sqrt{4x^2}} = \frac{\sqrt{2x}}{2x}$

Challenge Problems

35. $x\sqrt[3]{16x^3y} \Rightarrow x\sqrt[3]{(8)(2)(x^3)(y)} \Rightarrow x\sqrt[3]{(2^3)(2)(x^3)(y)} \Rightarrow x\sqrt[3]{2^3}\cdot\sqrt[3]{2}\cdot\sqrt[3]{x^3}\cdot\sqrt[3]{y} \Rightarrow x\cdot 2\cdot\sqrt[3]{2}\cdot x\cdot\sqrt[3]{y} = 2x^2\sqrt[3]{2y}$

37. $3\sqrt[5]{32xy^{11}} \Rightarrow 3\sqrt[5]{2^5xy^5y^5y} \Rightarrow 3(2)(y)(y)\sqrt[5]{xy} = 6y^2\sqrt[5]{xy}$

39. $\sqrt{a^3 - a^2b} \Rightarrow \sqrt{a^2(a-b)} \Rightarrow \sqrt{a^2}\cdot\sqrt{(a-b)} = a\sqrt{(a-b)}$

41. $\sqrt{9m^3 + 18n} \Rightarrow \sqrt{9(m^3 + 2n)} \Rightarrow \sqrt{9}\cdot\sqrt{m^3 + 2n} \Rightarrow \sqrt{3^2}\cdot\sqrt{m^3 + 2n} = 3\sqrt{m^3 + 2n}$

43. $\sqrt{\frac{3a^2}{5b}} \Rightarrow \frac{\sqrt{3a^2}}{\sqrt{5b}} \Rightarrow \frac{\sqrt{3a^2}}{\sqrt{5b}}\left(\frac{\sqrt{5b}}{\sqrt{5b}}\right) \Rightarrow \frac{\sqrt{15a^2b}}{\sqrt{25b^2}} = \frac{a\sqrt{15b}}{5b}$

45. $\sqrt[3]{\frac{1}{x^2}} \Rightarrow \frac{\sqrt[3]{1}}{\sqrt[3]{x^2}} \Rightarrow \frac{\sqrt[3]{1}}{\sqrt[3]{x^2}}\left(\frac{\sqrt[3]{x}}{\sqrt[3]{x}}\right) \Rightarrow \frac{\sqrt[3]{x}}{\sqrt[3]{x^3}} \Rightarrow \frac{\sqrt[3]{x}}{x}$

47. $\sqrt[6]{\frac{4x^6}{9}} \Rightarrow \frac{\sqrt[6]{4x^6}}{\sqrt[6]{9}} \Rightarrow \frac{\sqrt[6]{4x^6}}{\sqrt[6]{9}}\left(\frac{\sqrt[6]{81}}{\sqrt[6]{81}}\right) \Rightarrow \frac{\sqrt[6]{324x^6}}{\sqrt[6]{729}} \Rightarrow \frac{\sqrt[6]{(18^2)x^6}}{\sqrt[6]{3^6}} \Rightarrow \frac{x\sqrt[6]{18^2}}{3} \Rightarrow \frac{x(18^{2/6})}{3} \Rightarrow \frac{x(18^{1/3})}{3} = \frac{x\sqrt[3]{18}}{3}$

Applications

49. $\omega_n = \sqrt{\frac{kg}{W}} \Rightarrow \left(\frac{kg}{W}\right)^{1/2}$

51. $Z = \sqrt{R^2 + X^2} \Rightarrow (R^2 + X^2)^{1/2}$

53. $c = \sqrt{a^2 + b^2} \Rightarrow \sqrt{a^2 + (3a)^2} \Rightarrow \sqrt{a^2 + 9a^2} \Rightarrow \sqrt{10a^2} \Rightarrow \sqrt{10}\cdot\sqrt{a^2} = a\sqrt{10}$

Addition and Subtraction of Radicals

1. $2\sqrt{24}-\sqrt{54}\Rightarrow 2\sqrt{(4)(6)}-\sqrt{(9)(6)}\Rightarrow 2\sqrt{4}\cdot\sqrt{6}-\sqrt{9}\cdot\sqrt{6}\Rightarrow 2(2)\sqrt{6}-3\sqrt{6}\Rightarrow 4\sqrt{6}-3\sqrt{6}\Rightarrow(4-3)\sqrt{6}=\sqrt{6}$

3. $\sqrt{24}-\sqrt{96}+\sqrt{54}\Rightarrow\sqrt{(4)(6)}-\sqrt{(16)(6)}+\sqrt{(9)(6)}\Rightarrow 2\sqrt{6}-4\sqrt{6}+3\sqrt{6}\Rightarrow(2-4+3)\sqrt{6}=\sqrt{6}$

5. $2\sqrt{50}+\sqrt{72}+3\sqrt{18}\Rightarrow 2\sqrt{(2)(25)}+\sqrt{(36)(2)}+3\sqrt{(9)(2)}\Rightarrow 2(5)\sqrt{2}+6\sqrt{2}+3(3)\sqrt{2}$

   $\Rightarrow 10\sqrt{2}+6\sqrt{2}+9\sqrt{2}=25\sqrt{2}$

7. $2\sqrt[3]{2}-3\sqrt[3]{16}+\sqrt[3]{-54}\Rightarrow 2\sqrt[3]{2}-3\sqrt[3]{(8)(2)}+\sqrt[3]{(-27)(2)}\Rightarrow 2\sqrt[3]{2}-3(2)\sqrt[3]{2}-3\sqrt[3]{2}\Rightarrow 2\sqrt[3]{2}-6\sqrt[3]{2}-3\sqrt[3]{2}=-7\sqrt[3]{2}$

9. $\sqrt[3]{625}-2\sqrt[3]{135}-\sqrt[3]{320}\Rightarrow\sqrt[3]{(125)(5)}-2\sqrt[3]{(27)(5)}-\sqrt[3]{(64)(5)}\Rightarrow 5\sqrt[3]{5}-2(3)\sqrt[3]{5}-4\sqrt[3]{5}=-5\sqrt[3]{5}$

11. $\sqrt[4]{768}-\sqrt[4]{48}-\sqrt[4]{243}\Rightarrow\sqrt[4]{(256)(3)}-\sqrt[4]{(16)(3)}-\sqrt[4]{(81)(3)}\Rightarrow 4\sqrt[4]{3}-2\sqrt[4]{3}-3\sqrt[4]{3}=-\sqrt[4]{3}$

13. $\sqrt{128x^2y}-\sqrt{98x^2y}+\sqrt{162x^2y}\Rightarrow\sqrt{(64)(2)x^2y}-\sqrt{(49)(2)x^2y}+\sqrt{(81)(2)x^2y}\Rightarrow 8x\sqrt{2y}-7x\sqrt{2y}+9x\sqrt{2y}$

    $=10x\sqrt{2y}$

15. $7\sqrt{\dfrac{27}{50}}-3\sqrt{\dfrac{2}{3}}\Rightarrow\dfrac{21\sqrt{3}}{5\sqrt{2}}-\dfrac{3\sqrt{2}}{\sqrt{3}}\Rightarrow\dfrac{21\sqrt{3}}{5\sqrt{2}}\left(\dfrac{\sqrt{2}}{\sqrt{2}}\right)-\dfrac{3\sqrt{2}}{\sqrt{3}}\left(\dfrac{\sqrt{3}}{\sqrt{3}}\right)\Rightarrow\dfrac{21\sqrt{6}}{10}-\sqrt{6}\Rightarrow\dfrac{21\sqrt{6}-10\sqrt{6}}{10}=\dfrac{11\sqrt{6}}{10}$

17. $\sqrt{a^2x}+\sqrt{b^2x}\Rightarrow a\sqrt{x}+b\sqrt{x}=(a+b)\sqrt{x}$

Multiplication of Radicals

19. $(2\sqrt{3})(3\sqrt{8})\Rightarrow(2\cdot 3)(\sqrt{3}\cdot\sqrt{8})\Rightarrow 6\sqrt{24}\Rightarrow 6\sqrt{(4)(6)}\Rightarrow 6\sqrt{4}\cdot\sqrt{6}\Rightarrow 6\cdot 2\sqrt{6}=12\sqrt{6}$

21. $\left(\sqrt{\dfrac{5}{8}}\right)\left(\sqrt{\dfrac{3}{4}}\right)\Rightarrow\sqrt{\dfrac{15}{32}}\Rightarrow\dfrac{\sqrt{15}}{\sqrt{32}}\Rightarrow\dfrac{\sqrt{15}}{\sqrt{(16)(2)}}\Rightarrow\dfrac{\sqrt{15}}{(\sqrt{16})(\sqrt{2})}\Rightarrow\dfrac{\sqrt{15}}{4\sqrt{2}}\Rightarrow\dfrac{\sqrt{15}}{4\sqrt{2}}\left(\dfrac{\sqrt{2}}{\sqrt{2}}\right)\Rightarrow\dfrac{\sqrt{30}}{4(2)}=\dfrac{\sqrt{30}}{8}$

23. $(3\sqrt{3})(2\sqrt[3]{2})\Rightarrow[3(3^{1/2})][2(2^{1/3})]\Rightarrow[3(3^{3/6})][2(2^{2/6})]\Rightarrow(2)(3)\sqrt[6]{(3^3)(2^2)}\Rightarrow 6\sqrt[6]{(27)(4)}=6\sqrt[6]{108}$

25. $(2\sqrt{3})(\sqrt[4]{5})\Rightarrow[2(3^{1/2})][5^{1/4}]\Rightarrow[2(3^{2/4})][5^{1/4}]\Rightarrow 2\sqrt[4]{(3^2)(5)}\Rightarrow 2\sqrt[4]{(9)(5)}=2\sqrt[4]{45}$

27. $(2\sqrt[3]{24})\left(\sqrt[9]{\dfrac{8}{27}}\right)\Rightarrow[2(24^{1/3})]\left[\left(\dfrac{8}{27}\right)^{1/9}\right]\Rightarrow[2(24^{1/3})]\left[\left(\dfrac{2^3}{3^3}\right)^{1/9}\right]\Rightarrow[2(24^{1/3})]\left[\left(\dfrac{2^{1/3}}{3^{1/3}}\right)\right]\Rightarrow[2(24^{1/3})]\left[\left(\dfrac{2}{3}\right)^{1/3}\right]$

    $\Rightarrow\left[2\sqrt[3]{24}\right]\left[\sqrt[3]{\dfrac{2}{3}}\right]\Rightarrow 2\sqrt[3]{(24)\left(\dfrac{2}{3}\right)}\Rightarrow 2\sqrt[3]{\dfrac{48}{3}}\Rightarrow 2\sqrt[3]{\dfrac{(8)(6)}{3}}\Rightarrow(2)(2)\sqrt[3]{2}=4\sqrt[3]{2}$

29. $(3\sqrt[3]{9a^2})(\sqrt[3]{3abc})\Rightarrow 3\sqrt[3]{(9a^2)(3abc)}\Rightarrow 3\sqrt[3]{(9a^3)(3bc)}\Rightarrow(3)(3a)\sqrt[3]{bc}=9a\sqrt[3]{bc}$

31. $\left(\sqrt{\dfrac{a}{b}}\right)\left(\sqrt{\dfrac{c}{d}}\right)\Rightarrow\sqrt{\dfrac{ac}{bd}}\Rightarrow\dfrac{\sqrt{ac}}{\sqrt{bd}}\Rightarrow\dfrac{\sqrt{ac}}{\sqrt{bd}}\left(\dfrac{\sqrt{bd}}{\sqrt{bd}}\right)\Rightarrow\dfrac{\sqrt{acbd}}{\sqrt{b^2d^2}}=\dfrac{\sqrt{abcd}}{bd}$

33. $(\sqrt[3]{x})(\sqrt{y})\Rightarrow(x^{1/3})(y^{1/2})\Rightarrow(x^{2/6})(y^{3/6})=\sqrt[6]{x^2y^3}$

35. $(\sqrt[3]{a^2b})(\sqrt[3]{2a^2b^2})(\sqrt{3a^3b^2})\Rightarrow(\sqrt[3]{2a^4b^3})(\sqrt{3a^3b^2})\Rightarrow(\sqrt[3]{2(a^3)(a)b^3})(\sqrt{3a^3b^2})\Rightarrow(ab\sqrt[3]{2a})(\sqrt{3a^3b^2})$

    $\Rightarrow ab(2a)^{1/3}(3a^3b^2)^{1/2}\Rightarrow ab(2a)^{2/6}(3a^3b^2)^{3/6}\Rightarrow ab\sqrt[6]{4a^2}\left(\sqrt[6]{27a^9b^6}\right)\Rightarrow ab\sqrt[6]{108a^{11}b^6}$

$$\Rightarrow ab\sqrt[6]{108(a^6)(a^5)b^6} = a^2b^2\sqrt[6]{108a^5}$$

Powers

37. $\left(4\sqrt[3]{4x^2}\right)^2 \Rightarrow \left(4\sqrt[3]{4x^2}\right)\left(4\sqrt[3]{4x^2}\right) \Rightarrow 16\sqrt[3]{16x^4} \Rightarrow 16\sqrt[3]{(8)(2)(x^3)(x)} \Rightarrow 16(2)(x)\sqrt[3]{2x} = 32x\sqrt[3]{2x}$

39. $\left(5+4\sqrt{x}\right)^2 \Rightarrow \left(5+4\sqrt{x}\right)\left(5+4\sqrt{x}\right) \Rightarrow 25+20\sqrt{x}+20\sqrt{x}+16\sqrt{x^2} = 25+40\sqrt{x}+16x$

41. $\left(\sqrt{a}+5a\sqrt{b}\right)^2 \Rightarrow \left(\sqrt{a}+5a\sqrt{b}\right)\left(\sqrt{a}+5a\sqrt{b}\right) \Rightarrow \sqrt{a^2}+5a\sqrt{ab}+5a\sqrt{ab}+25a^2\sqrt{b^2} = a+10a\sqrt{ab}+25a^2b$

Division of Radicals

43. $6\sqrt{72} \div 12\sqrt{32} \Rightarrow \dfrac{36\sqrt{2}}{48\sqrt{2}} \Rightarrow \dfrac{36}{48} = \dfrac{3}{4}$

45. $\sqrt{72} \div 2\sqrt[4]{64} \Rightarrow \dfrac{\sqrt{72}}{2\sqrt[4]{64}} \Rightarrow \dfrac{\sqrt{(9)(8)}}{2\sqrt[4]{(16)(4)}} \Rightarrow \dfrac{3\sqrt{8}}{4\sqrt[4]{4}} \Rightarrow \dfrac{3\sqrt{8}}{4\sqrt{2}} \Rightarrow \dfrac{3\sqrt{8}}{4\sqrt{2}}\left(\dfrac{\sqrt{2}}{\sqrt{2}}\right) \Rightarrow \dfrac{3\sqrt{16}}{4\sqrt{4}} \Rightarrow \dfrac{12}{8} = \dfrac{3}{2}$

47. $8\sqrt[3]{ab} \div 4\sqrt{ac} \Rightarrow \dfrac{8\sqrt[3]{ab}}{4\sqrt{ac}} \Rightarrow \dfrac{2(ab)^{1/3}}{(ac)^{1/2}} \Rightarrow \dfrac{2(ab)^{2/6}}{(ac)^{3/6}} \Rightarrow \dfrac{2(ab)^{2/6}}{(ac)^{3/6}}\left(\dfrac{(ac)^{3/6}}{(ac)^{3/6}}\right) \Rightarrow \dfrac{2\left(a^2b^2a^3c^3\right)^{1/6}}{ac} = \dfrac{2\sqrt[6]{a^5b^2c^3}}{ac}$

49. $\left(3+\sqrt{2}\right) \div \left(2-\sqrt{2}\right) \Rightarrow \dfrac{3+\sqrt{2}}{2-\sqrt{2}}\left(\dfrac{2+\sqrt{2}}{2+\sqrt{2}}\right) \Rightarrow \dfrac{6+3\sqrt{2}+2\sqrt{2}+2}{4-2} \Rightarrow \dfrac{8+5\sqrt{2}}{2}$

51. $4\sqrt{x} \div \sqrt{a} \Rightarrow \dfrac{4\sqrt{x}}{\sqrt{a}} \Rightarrow \dfrac{4\sqrt{x}}{\sqrt{a}}\left(\dfrac{\sqrt{a}}{\sqrt{a}}\right) = \dfrac{4\sqrt{ax}}{a}$

53. $12 \div \sqrt[3]{4x^2} \Rightarrow \dfrac{12}{\sqrt[3]{4x^2}} \Rightarrow \dfrac{12}{\sqrt[3]{4x^2}}\left(\dfrac{\sqrt[3]{2x}}{\sqrt[3]{2x}}\right) \Rightarrow \dfrac{12\sqrt[3]{2x}}{\sqrt[3]{8x^3}} \Rightarrow \dfrac{12\sqrt[3]{2x}}{2x} = \dfrac{6\sqrt[3]{2x}}{x}$

Challenge Problems

55. $\sqrt{x^2y} - \sqrt{4a^2y} \Rightarrow x\sqrt{y} - 2a\sqrt{y} = (x-2a)\sqrt{y}$

57. $4\sqrt{3a^2x} - 2a\sqrt{48x} \Rightarrow 4a\sqrt{3x} - 2a(4)\sqrt{3x} \Rightarrow 4a\sqrt{3x} - 8a\sqrt{3x} = -4a\sqrt{3x}$

59. $\sqrt[3]{1250a^3b} + \sqrt[3]{270c^3b} \Rightarrow \sqrt[3]{(125)(10)a^3b} + \sqrt[3]{(27)(10)c^3b} \Rightarrow 5a\sqrt[3]{10b} + 3c\sqrt[3]{10b} = (5a+3c)\sqrt[3]{10b}$

61. $\sqrt[5]{a^{13}b^{11}c^{12}} - 2\sqrt[5]{a^8bc^2} + \sqrt[5]{a^3b^6c^7} \Rightarrow \sqrt[5]{a^5a^5a^3b^5b^5bc^5c^5c^2} - 2\sqrt[5]{a^5a^3bc^2} + \sqrt[5]{a^3b^5bc^5c^2}$

$a^2b^2c^2\sqrt[5]{a^3bc^2} - 2a\sqrt[5]{a^3bc^2} + bc\sqrt[5]{a^3bc^2} = \left(a^2b^2c^2 - 2a + bc\right)\sqrt[5]{a^3bc^2}$

63. $\left(\sqrt{5}-\sqrt{3}\right)\left(2\sqrt{3}\right) \Rightarrow 2\sqrt{3}\left(\sqrt{5}\right) - 2\sqrt{3}\left(\sqrt{3}\right) \Rightarrow 2\sqrt{15} - 2\sqrt{9} \Rightarrow 2\sqrt{15} - 2(3) = 2\sqrt{15} - 6$

65. $\left(\sqrt{x^3 - x^4y}\right)\left(\sqrt{x}\right) \Rightarrow \sqrt{x\left(x^3 - x^4y\right)} \Rightarrow \sqrt{x^4 - x^5y} \Rightarrow \sqrt{x^4(1-xy)} = x^2\sqrt{1-xy}$

67. $\left(a+\sqrt{b}\right)\left(a-\sqrt{b}\right) \Rightarrow a^2 - a\sqrt{b} + a\sqrt{b} - \sqrt{b^2} = a^2 - b$

69. $\left(4\sqrt{x}+2\sqrt{y}\right)\left(4\sqrt{x}-5\sqrt{y}\right) \Rightarrow 16\sqrt{x^2} - 20\sqrt{xy} + 8\sqrt{xy} - 10\sqrt{y^2} = 16x - 12\sqrt{xy} - 10y$

71. $\sqrt{x} \div \left(\sqrt{x}+\sqrt{y}\right) \Rightarrow \dfrac{\sqrt{x}}{\sqrt{x}+\sqrt{y}} \Rightarrow \dfrac{\sqrt{x}}{\sqrt{x}+\sqrt{y}}\left(\dfrac{\sqrt{x}-\sqrt{y}}{\sqrt{x}-\sqrt{y}}\right) = \dfrac{x-\sqrt{xy}}{x-y}$

73. $\left(a+\sqrt{b}\right)\div\left(a-\sqrt{b}\right)\Rightarrow\frac{a+\sqrt{b}}{a-\sqrt{b}}\Rightarrow\frac{a+\sqrt{b}}{a-\sqrt{b}}\left(\frac{a+\sqrt{b}}{a+\sqrt{b}}\right)\Rightarrow\frac{a^2+a\sqrt{b}+a\sqrt{b}+b}{a^2-b}=\frac{a^2+2a\sqrt{b}+b}{a^2-b}$

75. $\left(3\sqrt{m}-\sqrt{2n}\right)\div\left(\sqrt{3n}+\sqrt{m}\right)\Rightarrow\frac{3\sqrt{m}-\sqrt{2n}}{\sqrt{3n}+\sqrt{m}}\Rightarrow\frac{3\sqrt{m}-\sqrt{2n}}{\sqrt{3n}+\sqrt{m}}\left(\frac{\sqrt{3n}-\sqrt{m}}{\sqrt{3n}-\sqrt{m}}\right)=\frac{3\sqrt{3mn}-3m-n\sqrt{6}+2\sqrt{mn}}{3n-m}$

77. $\left(2x\sqrt{3x}\right)^3\Rightarrow\left(2x\sqrt{3x}\right)\left(2x\sqrt{3x}\right)\left(2x\sqrt{3x}\right)\Rightarrow 8x^3\sqrt{27x^3}\Rightarrow 8x^3\sqrt{(9)(3)\left(x^2\right)(x)}\Rightarrow\left(8x^3\right)(3)(x)\sqrt{3x}=24x^4\sqrt{3x}$

Exercise 4 ◊ Radical Equations
Radical Equations

1. $\sqrt{x}=6$

$\left(\sqrt{x}\right)^2=6^2$

$x=36$

3. $\sqrt{7x+8}=6$

$\left(\sqrt{7x+8}\right)^2=6^2$

$7x+8=36$

$7x=28$

$x=4$

5. $\sqrt{2.95x-1.84}=6.23$

$\left(\sqrt{2.95x-1.84}\right)^2=6.23^2$

$2.95x-1.84=38.8129$

$2.95x=40.6529$

$x=13.7806\Rightarrow 13.8$

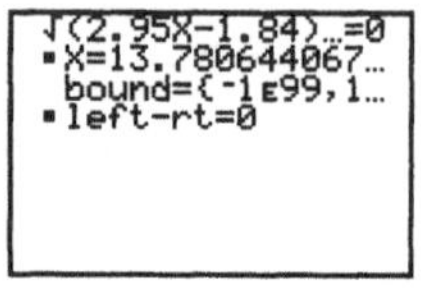

7. $\sqrt{x+1}=\sqrt{2x-7}$

$\left(\sqrt{x+1}\right)^2=\left(\sqrt{2x-7}\right)^2$

$x+1=2x-7$

$-x=-8$

$x=8$

9. $\sqrt{3x+1}=5$

$\left(\sqrt{3x+1}\right)^2=5^2$

$3x+1=25$

$3x=24$

$x=8$

11. $\sqrt{x-3}=\frac{4}{\sqrt{x-3}}$

$x-3=4$ [multiply both sides by $\sqrt{x-3}$]

$x=7$

13. $\sqrt{x^2 - 7.25} = 8.75 - x$

$\left(\sqrt{x^2 - 7.25}\right)^2 = (8.75 - x)^2$

$x^2 - 7.25 = 76.5625 - 17.5x + x^2$

$-7.25 = 76.5625 - 17.5x$

$-83.8125 = -17.5x$

$4.789 = x;\ x = 4.79$

15. $\dfrac{6}{\sqrt{3+x}} = \sqrt{x+3}$

$6 = x + 3 \left[\text{multiply both sides by } \sqrt{3+x}\right]$

$3 = x$

Applications

17. $x + 25.3 + \sqrt{x^2 + (25.3)^2} = 68.4$

$\sqrt{x^2 + (25.3)^2} = 43.1 - x$

$x^2 + 640.09 = 1857.61 - 86.2x + x^2$ [square both sides]

$640.09 = 1857.61 - 86.2x$

$-1217.52 = -86.2x$

$14.12436 = x;\ x = 14.1$

Thus, $x = 14.1$ cm; the hypotenuse $= \sqrt{x^2 + (25.3)^2} \Rightarrow \sqrt{(14.1)^2 + (25.3)^2} = 28.96 \Rightarrow 29$ cm

19. $x + 2.73 + \sqrt{x^2 + (2.73)^2} = 11.4$

$\sqrt{x^2 + (2.73)^2} = 8.67 - x$

$x^2 + 7.4529 = 75.1689 - 17.34x + x^2$ [square both sides]

$7.4529 = 75.1689 - 17.34x$

$-67.716 = -17.34x$

$3.90519 = x;\ x = 3.91$ m; the hypotenuse $\Rightarrow \sqrt{(3.91)^2 + (2.73)^2} = 4.7687 \Rightarrow 4.77$ m

21. $Z = \sqrt{R^2 + \left(\omega L - \dfrac{1}{\omega C}\right)^2}$

$Z^2 = R^2 + \left(\omega L - \dfrac{1}{\omega C}\right)^2$

$Z^2 - R^2 = \left(\omega L - \dfrac{1}{\omega C}\right)^2$

$\pm\sqrt{Z^2 - R^2} = \omega L - \dfrac{1}{\omega C}$

$\omega L \pm \sqrt{Z^2 - R^2} = \dfrac{1}{\omega C}$

$\omega C\left(\omega L \pm \sqrt{Z^2 - R^2}\right) = 1$

$C = \dfrac{1}{\omega\left(\omega L \pm \sqrt{Z^2 - R^2}\right)} \Rightarrow \dfrac{1}{\omega^2 L \pm \omega\sqrt{Z^2 - R^2}}$

CHAPTER 13 REVIEW PROBLEMS

1. $\sqrt{52} \Rightarrow \sqrt{(4)(13)} \Rightarrow \sqrt{2^2}\sqrt{13} = 2\sqrt{13}$

3. $\sqrt[3]{162} \Rightarrow \sqrt[3]{(27)(6)} \Rightarrow \sqrt[3]{3^3}\left(\sqrt[3]{6}\right) = 3\sqrt[3]{6}$

5. $\sqrt[6]{4} \Rightarrow \sqrt[6]{2^2} \Rightarrow 2^{2/6} \Rightarrow 2^{1/3} = \sqrt[3]{2}$

or $\sqrt[6]{4} \Rightarrow \sqrt[3]{\sqrt{4}} \Rightarrow \sqrt[3]{\sqrt{2^2}} \Rightarrow \sqrt[3]{2}$

7. $3\sqrt[4]{81x^5} \Rightarrow 3\sqrt[4]{3^4x^4x} \Rightarrow 3\sqrt[4]{3^4}\sqrt[4]{x^4}\sqrt[4]{x} \Rightarrow 3(3)x\sqrt[4]{x} = 9x\sqrt[4]{x}$

9. $\sqrt[3]{(a-b)^5x^4} \Rightarrow \sqrt[3]{(a-b)^3(a-b)^2x^3x} \Rightarrow \sqrt[3]{(a-b)^3}\sqrt[3]{x^3}\sqrt[3]{(a-b)^2}\sqrt[3]{x} = (a-b)x\sqrt[3]{(a-b)^2x}$

11. $\sqrt{x}\cdot\sqrt{x^3-x^4y} \Rightarrow \sqrt{x}\cdot\sqrt{x^3(1-xy)} \Rightarrow \sqrt{x^4(1-xy)} = x^2\sqrt{1-xy}$

13. $\dfrac{3-2\sqrt{3}}{2-5\sqrt{2}} \Rightarrow \dfrac{3-2\sqrt{3}}{2-5\sqrt{2}}\left(\dfrac{2+5\sqrt{2}}{2+5\sqrt{2}}\right) \Rightarrow \dfrac{6+15\sqrt{2}-4\sqrt{3}-10\sqrt{6}}{4+10\sqrt{2}-10\sqrt{2}-50} \Rightarrow \dfrac{6+15\sqrt{2}-4\sqrt{3}-10\sqrt{6}}{4-50}$

$= -\dfrac{1}{46}\left(6+15\sqrt{2}-4\sqrt{3}-10\sqrt{6}\right)$

15. $\sqrt{98x^2y^2} - \sqrt{128x^2y^2} \Rightarrow \sqrt{(49)(2)x^2y^2} - \sqrt{(64)(2)x^2y^2} \Rightarrow 7xy\sqrt{2} - 8xy\sqrt{2} = -xy\sqrt{2}$

17. $\sqrt{a}\sqrt[3]{b} \Rightarrow a^{1/2}b^{1/3} \Rightarrow a^{3/6}b^{2/6} = \sqrt[6]{a^3b^2}$

19. $\left(3+2\sqrt{x}\right)^2 \Rightarrow \left(3+2\sqrt{x}\right)\left(3+2\sqrt{x}\right) \Rightarrow 9+6\sqrt{x}+6\sqrt{x}+4x = 9+12\sqrt{x}+4x$

21. $3\sqrt{50}-2\sqrt{32} \Rightarrow 3(5)\sqrt{2}-2(4)\sqrt{2} \Rightarrow 15\sqrt{2}-8\sqrt{2} = 7\sqrt{2}$

23. $\sqrt{2ab} \div \sqrt{4ab^2} \Rightarrow \dfrac{\sqrt{2ab}}{\sqrt{4ab^2}} \Rightarrow \dfrac{\sqrt{2ab}}{\sqrt{4ab^2}}\left(\dfrac{\sqrt{a}}{\sqrt{a}}\right) \Rightarrow \dfrac{\sqrt{2a^2b}}{2ab} \Rightarrow \dfrac{a\sqrt{2b}}{2ab} = \dfrac{\sqrt{2b}}{2b}$

25. $3\sqrt{9}\cdot 4\sqrt{8} \Rightarrow (3)(4)\sqrt{9}\sqrt{8} \Rightarrow 12\sqrt{72} \Rightarrow 12\sqrt{36(2)} \Rightarrow 12(6)\sqrt{2} = 72\sqrt{2}$

27. $\sqrt{x+6} = 4$

$x+6 = 16$ [square both sides]

$x = 10$

29. $\sqrt[5]{2x+4} = 2$

$\left(\sqrt[5]{2x+4}\right)^5 = 2^5$

$2x+4 = 32$

$2x = 28$

$x = 14$

31. $\sqrt{x}+\sqrt{x-9.75} = 6.23$

$\sqrt{x-9.75} = 6.23-\sqrt{x}$

$x-9.75 = 38.8129-12.46\sqrt{x}+x$ [square both sides]

$-9.75 = 38.8129-12.46\sqrt{x}$

$-48.5629 = -12.46\sqrt{x}$

$\dfrac{-48.5629}{-12.46} = \sqrt{x}$

$3.8975 = \sqrt{x}$

$15.1905 = x;\ x = 15.2$

33. $\left(x^{n-1}+y^{n-2}\right)\left(x^n+y^{n-1}\right) \Rightarrow x^{n-1+n}+x^{n-1}y^{n-1}+y^{n-2}x^n+y^{n-2+n-1} \Rightarrow x^{2n-1}+(xy)^{n-1}+x^ny^{n-2}+y^{2n-3}$

35. $\left(\frac{9x^4y^3}{6x^3y}\right)^3 \Rightarrow \frac{9^3x^{12}y^9}{6^3x^9y^3} \Rightarrow \frac{729x^{12}y^9}{216x^9y^3} = \frac{27x^3y^6}{8}$

37. $\left(\frac{2x^5y^3}{x^2y}\right)^3 \Rightarrow \frac{2^3x^{15}y^9}{x^6y^3} = 8x^9y^6$

39. $3w^{-2} \Rightarrow 3\left(\frac{1}{w^2}\right) = \frac{3}{w^2}$

41. $(3x)^{-1} \Rightarrow 3^{-1}x^{-1} \Rightarrow \frac{1}{3}\left(\frac{1}{x}\right) = \frac{1}{3x}$

43. $x^{-1} - 2y^{-2} \Rightarrow \frac{1}{x} - 2\left(\frac{1}{y^2}\right) = \frac{1}{x} - \frac{2}{y^2}$

45. $(3x)^{-2} + (2x^2y^{-4})^{-2} \Rightarrow \frac{1}{3^2x^2} + \frac{y^8}{2^2x^4} = \frac{1}{9x^2} + \frac{y^8}{4x^4}$

47. $(p^{a-1} + q^{a-2})(p^a + q^{a-1}) \Rightarrow p^{a-1+a} + p^{a-1}q^{a-1} + q^{a-2}p^a + q^{a-2+a-1} = p^{2a-1} + (pq)^{a-1} + p^aq^{a-2} + q^{2a-3}$

49. $p^2q^{-1} \Rightarrow p^2\left(\frac{1}{q}\right) = \frac{p^2}{q}$

51. $r^2s^{-3} \Rightarrow r^2\left(\frac{1}{s^3}\right) = \frac{r^2}{s^3}$

53. $(3x^3y^2z)^0 \Rightarrow 3^0x^0y^0z^0 \Rightarrow (1)(1)(1)(1) = 1$

55. $V = \frac{4}{3}\pi(3r)^3 \Rightarrow \frac{4}{3}\pi(27r^3) \Rightarrow \frac{4(27)}{3}\pi r^3 = 36\pi r^3$

57. $B = \sqrt{AC} \Rightarrow B = \sqrt{A}\sqrt{B}$

Chapter 14: Radian Measure, Arc Length, and Rotation

Exercise 1 ◊ Radian Measure

1. $47.8° \Rightarrow 47.8°\left(\dfrac{2\pi \text{ rad}}{360°}\right) \Rightarrow \dfrac{47.8°(2\pi \text{ rad})}{360°} \Rightarrow 0.83426 = 0.834 \text{rads}$

3. $35.25° \Rightarrow 35.25°\left(\dfrac{2\pi \text{ rad}}{360°}\right) \Rightarrow \dfrac{35.25°(2\pi \text{ rad})}{360°} \Rightarrow 0.61522 = 0.615 \text{rads}$

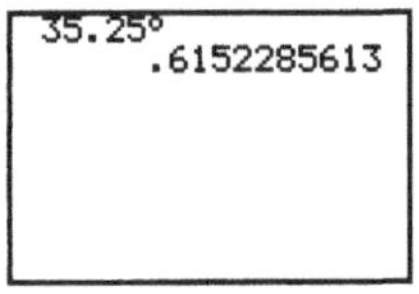

5. First, convert to revs to degree: $1.55 \text{ rev} \Rightarrow 1.55\left(\dfrac{360°}{1}\right) = 558°$

   Now, convert degrees to rads: $558°\left(\dfrac{2\pi \text{ rad}}{360°}\right) \Rightarrow \dfrac{558°(2\pi \text{ rad})}{360°} \Rightarrow 9.7389 = 9.74 \text{rads}$

7. $1.75 \text{ rad} \Rightarrow 1.75 \text{ rad}\left(\dfrac{1 \text{ rev}}{2\pi \text{ rad}}\right) \Rightarrow \dfrac{1.75(1 \text{ rev})}{2\pi \text{ rad}} \Rightarrow 0.27852 = 0.279 \text{ rev}$

9. $3.12 \text{ rad} \Rightarrow 3.12 \text{ rad}\left(\dfrac{1 \text{ rev}}{2\pi \text{ rad}}\right) \Rightarrow \dfrac{3.12(1 \text{ rev})}{2\pi \text{ rad}} \Rightarrow 0.49656 = 0.497 \text{ rev}$

11. $1.12 \text{ rad} \Rightarrow 1.12 \text{ rad}\left(\dfrac{1 \text{ rev}}{2\pi \text{ rad}}\right) \Rightarrow \dfrac{1.12(1 \text{ rev})}{2\pi \text{ rad}} \Rightarrow 0.17825 = 0.178 \text{ rev}$

13. $2.83 \text{ rad} \Rightarrow 2.83 \text{ rad}\left(\dfrac{360°}{2\pi \text{ rad}}\right) \Rightarrow \dfrac{2.83(360°)}{2\pi \text{ rad}} \Rightarrow 162.147 = 162°$

15. $0.372 \text{ rad} \Rightarrow 0.372 \text{ rad}\left(\dfrac{360°}{2\pi \text{ rad}}\right) \Rightarrow \dfrac{0.372(360°)}{2\pi \text{ rad}} \Rightarrow 21.31402 = 21.3°$

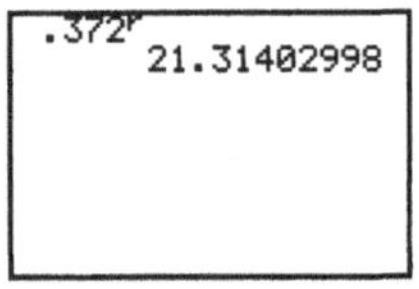

17. $1.14 \text{ rad} \Rightarrow 1.14 \text{ rad}\left(\dfrac{360°}{2\pi \text{ rad}}\right) \Rightarrow \dfrac{1.14(360°)}{2\pi \text{ rad}} \Rightarrow 65.3171 = 65.3°$

19. $60° \Rightarrow 60°\left(\dfrac{\pi \text{ rad}}{180°}\right) \Rightarrow \dfrac{60(\pi \text{ rad})}{180°} \Rightarrow \dfrac{60\pi}{180} = \dfrac{\pi}{3}$

21. $66° \Rightarrow 66°\left(\dfrac{\pi \text{ rad}}{180°}\right) \Rightarrow \dfrac{66(\pi \text{ rad})}{180°} \Rightarrow \dfrac{66\pi}{180} = \dfrac{11\pi}{30}$

23. $126° \Rightarrow 126°\left(\dfrac{\pi \text{ rad}}{180°}\right) \Rightarrow \dfrac{126(\pi \text{ rad})}{180°} \Rightarrow \dfrac{126\pi}{180} = \dfrac{7\pi}{10}$

25. $78° \Rightarrow 78°\left(\dfrac{\pi \text{ rad}}{180°}\right) \Rightarrow \dfrac{78(\pi \text{ rad})}{180°} \Rightarrow \dfrac{78\pi}{180} = \dfrac{13\pi}{30}$

27. $400° \Rightarrow 400°\left(\dfrac{\pi \text{ rad}}{180°}\right) \Rightarrow \dfrac{400(\pi \text{ rad})}{180°} \Rightarrow \dfrac{400\pi}{180} = \dfrac{20\pi}{9}$

29. $81° \Rightarrow 81°\left(\dfrac{\pi \text{ rad}}{180°}\right) \Rightarrow \dfrac{81(\pi \text{ rad})}{180°} \Rightarrow \dfrac{81\pi}{180} = \dfrac{9\pi}{20}$

31. $\dfrac{\pi}{8} \Rightarrow \dfrac{\pi}{8}\left(\dfrac{180°}{\pi \text{ rad}}\right) \Rightarrow \dfrac{180\pi}{8\pi \text{ rad}} = 22\dfrac{1}{2}°$

33. $\dfrac{9\pi}{11} \Rightarrow \dfrac{9\pi}{11}\left(\dfrac{180°}{\pi \text{ rad}}\right) \Rightarrow \dfrac{1620\pi}{11\pi \text{ rad}} \Rightarrow 147.2727° = 147°$

35. $\dfrac{\pi}{9} \Rightarrow \dfrac{\pi}{9}\left(\dfrac{180°}{\pi \text{ rad}}\right) \Rightarrow \dfrac{180\pi}{9\pi \text{ rad}} = 20°$

37. $\dfrac{7\pi}{8} \Rightarrow \dfrac{7\pi}{8}\left(\dfrac{180°}{\pi \text{ rad}}\right) \Rightarrow \dfrac{1260\pi}{8\pi \text{ rad}} \Rightarrow 157.5° = 158°$

39. $\dfrac{2\pi}{15} \Rightarrow \dfrac{2\pi}{15}\left(\dfrac{180°}{\pi \text{ rad}}\right) \Rightarrow \dfrac{360\pi}{15\pi \text{ rad}} = 24°$

41. $\dfrac{\pi}{12} \Rightarrow \dfrac{\pi}{12}\left(\dfrac{180°}{\pi \text{ rad}}\right) \Rightarrow \dfrac{180\pi}{12\pi \text{ rad}} = 15°$

43. $\sin\dfrac{\pi}{3} \Rightarrow 0.86602 = 0.8660$

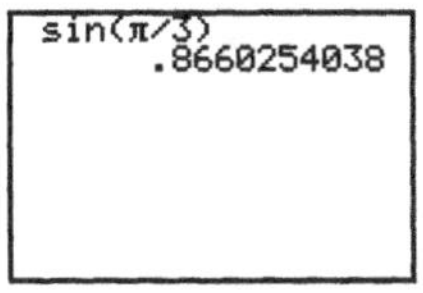

45. $\cos 1.063 \Rightarrow 0.486252 \Rightarrow 0.4863$

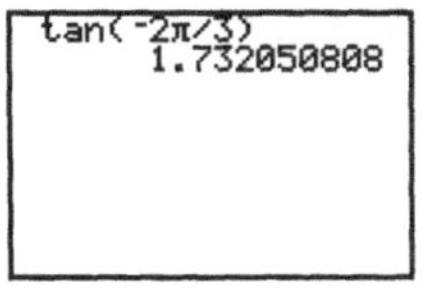

47. $\cos\dfrac{3\pi}{5} \Rightarrow -0.309016 = -0.3090$

49. $\sec 0.355 \Rightarrow (\cos 0.355)^{-1} \Rightarrow 1.0665 = 1.067$

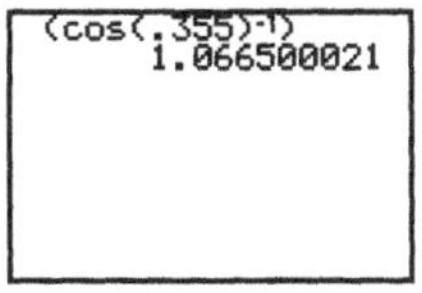

51. $\cot\dfrac{8\pi}{9} \Rightarrow \left(\tan\dfrac{8\pi}{9}\right)^{-1} \Rightarrow -2.74747 = -2.747$

53. $\cos\left(-\dfrac{6\pi}{5}\right) \Rightarrow -0.809016 = -0.8090$

55. $\cos 1.832 \Rightarrow -0.258243 = -0.2582$

57. $\sin 0.6254 \Rightarrow 0.58542 = 0.5854$

59. $\arcsin 0.7263 \Rightarrow \sin^{-1} 0.7263 \Rightarrow 0.812923 = 0.8129$

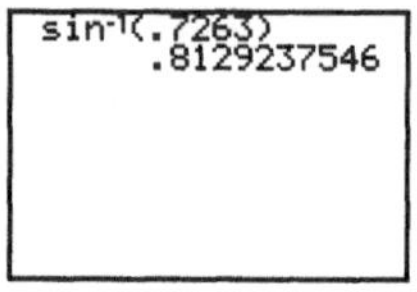

61. $\cos^{-1} 0.2320 \Rightarrow 1.3366 = 1.337$

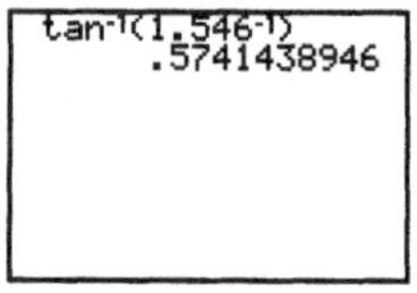

63. $\sin^{-1} 0.2649 \Rightarrow 0.268100 = 0.2681$

65. $\arctan 3.7253 \Rightarrow \tan^{-1} 3.7253 \Rightarrow 1.30854 = 1.309$

67. $\sin^2 \frac{\pi}{6} + \cos\frac{\pi}{6} \Rightarrow \left(\sin\frac{\pi}{6}\right)^2 + \cos\frac{\pi}{6} \Rightarrow 0.25 + 0.866025 \Rightarrow 1.11602 = 1.116$

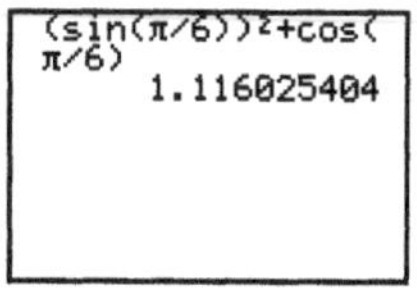

69. $\cos^2 \frac{3\pi}{4} \Rightarrow \left(\cos\frac{3\pi}{4}\right)^2 = 0.5000$

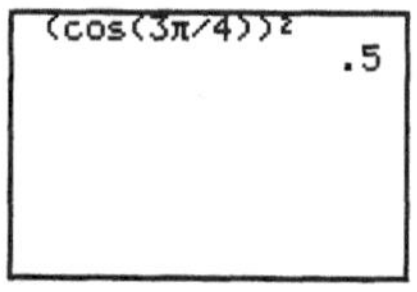

71. $\sin\frac{\pi}{8}\tan\frac{\pi}{8} \Rightarrow (0.382683)(0.414213) \Rightarrow 0.158512 = 0.1585$

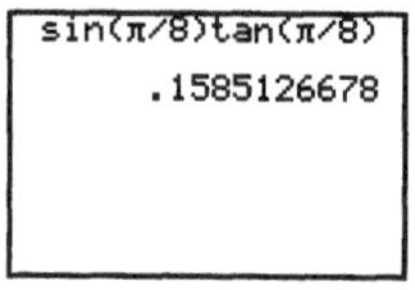

73. $62.5°\left(\frac{\pi \text{ rad}}{180°}\right) \Rightarrow 1.09083 = 1.091 \text{ rad}$

$\text{Area} = (5.92 \text{ in.})^2\left(\frac{1.091}{2}\right) \Rightarrow 19.117 = 19.1 \text{ in.}^2$

75. $\text{Area} = (38.4)^2 \arccos\frac{38.4 - 12.4}{38.4} - (38.4 - 12.4)\cdot\sqrt{2(38.4)(12.4) - (12.4)^2}$

$= 1474.56 \arccos(0.67708) - 26\sqrt{798.56}$

$= 1219.4741 - 734.728$

$= 484.7461 \Rightarrow 485 \text{ cm}^2$

77. $\text{Area} = \frac{1}{2}(64)^2(1.55 - \sin 1.55)$

$= 2048(1.55 - 0.999783)$

$= 2048(0.550217)$

$= 1126.844$

$= 1130 \text{ cm}^2$

Applications

79. $67.5°\left(\frac{\pi \text{ rad}}{180°}\right) \Rightarrow 1.178097 \text{ rad}$

Pad Area = pad area sector – pad sector hole area

$= \frac{(62.8)^2(1.178097)}{2} - \frac{(8.50)^2(1.178097)}{2}$

$= 2323.113036 - 42.55875$

$= 2280.554$

$= 2280 \text{ cm}^2$

81. $y = 4 \cos\left[25(2.00)\right]$

$= 4 \cos 50$

$= 4(0.964966)$

$= 3.8598$

$= 3.86 \text{ in.}$

Exercise 2 ◊ Arc Length

1. $r = 4.83$ in., $\theta = \frac{2\pi}{5}$: $\frac{2\pi}{5} = \frac{s}{4.83}$; $s = 4.83\left(\frac{2\pi}{5}\right) \Rightarrow 6.069557 = 6.07$ in.

3. $r = 284$ ft, $\theta = 46.4°$: $46.4°\left(\frac{\pi \text{ rad}}{180°}\right) = \frac{s}{284}$ $s = 46.4\left(\frac{\pi}{180}\right)(284) \Rightarrow 229.9925 = 230$ ft

5. $r = 64.8$ in,, $\theta = 38.5°$: $38.5°\left(\frac{\pi \text{ rad}}{180°}\right) = \frac{s}{64.8}$ $s = 38.5\left(\frac{\pi}{180}\right)(64.8) \Rightarrow 43.542 = 43.5$in.

7. $r = 263$ mm, $s = 582$ mm: $\theta = \frac{582}{263} \Rightarrow 2.2129 = 2.21$ rad

9. $r = 3.87$ m, $s = 15.8$ ft: $\theta = \left(\frac{15.8 \text{ ft}}{3.87 \text{ m}}\right)\left(\frac{\text{m}}{3.281 \text{ ft}}\right) \Rightarrow \frac{15.8}{(3.87)(3.281)} \Rightarrow 1.2443 = 1.24$ rad

11. $\theta = 77.2°$, $s = 1.11$ cm: $77.2\left(\frac{\pi \text{ rad}}{180}\right) = \frac{1.11}{r}$; $r = \frac{1.11}{77.2\left(\frac{\pi}{180}\right)} \Rightarrow 0.82381 = 0.824$ cm

13. $\theta = 12°55'$, $s = 28.2$ ft: $r = \frac{28.2}{\left(\frac{\pi}{180}\right)\left(12 + \frac{55}{60}\right)} \Rightarrow \frac{28.2}{0.225438} \Rightarrow 125.089 = 125$ ft

Applications

15. $\theta = 90° - 35.2° \Rightarrow 54.8°$

$s = \theta r \Rightarrow 54.8\left(\frac{\pi}{180}\right)(3960)$

$= 3787.504$

$= 3790 \text{ mi}$

17. $r = 3960 + 225 \Rightarrow 4185$

$\theta = 85.0\left(\frac{\pi}{180}\right) \Rightarrow 1.4835 \text{ rad}$

$1.4835 = \frac{s}{4185}; \quad s = 1.4835(4185) \Rightarrow 6208.44 = 6210 \text{ mi}$

19. $\frac{30°}{h}(10h) \Rightarrow \frac{30°(10h)}{h} \Rightarrow 30°(10) = 300°$

$s = 300\left(\frac{\pi}{180}\right)(85.5)$

$= 447.676$

$= 448 \text{ mm}$

21. $\theta = \frac{1}{4}(2\pi) \Rightarrow \frac{\pi}{2} \text{ rads}$

$\frac{\pi}{2} = \frac{s}{35}$

$s = 35\left(\frac{\pi}{2}\right) \Rightarrow 54.977 = 55 \text{ mm}$

23. $\frac{\pi}{12} = \frac{s}{155}; \quad s = 155\left(\frac{\pi}{12}\right) \Rightarrow 40.5789 \text{ mm}; \quad 40.5789(10 \text{ bits/mm}) = 405.789 = 406 \text{ bits}$

25. $s = 2(5.75°)\left(\frac{\pi \text{ rad}}{180°}\right)(1.25) \Rightarrow 0.25089 = 0.251 \text{ m}$

Why 2? The pendulum "swings 5.75° on *each* side of the vertical."

27. $\tan\beta = \frac{\frac{240-120}{2}}{350} \Rightarrow 0.171428$

$\tan\beta = \frac{120}{h}$

$0.171428 = \frac{120}{h}$

$h = \frac{120}{0.171428} \Rightarrow 700.002 = 700 \text{ mm}$

$R^2 = 700^2 + 120^2$

$R^2 = 504{,}400$

$R = 710 \text{ mm}$

$k = \sqrt{350^2 + 60^2}$

$= 355.105 \Rightarrow 355$

$r = R - k \Rightarrow 710 - 355 = 354.895 \text{ mm}$

$s = \text{circumference} \Rightarrow \pi d \Rightarrow \pi(240) = 753.982$

$\theta = \frac{s}{R} \Rightarrow \frac{753.982}{710} = 1.0616 \text{ rad} \Rightarrow 1.0616\left(\frac{180}{\pi}\right) = 60.845 \Rightarrow 60.8°$

29. $s = 28.3\left(\frac{\pi}{180}\right)(28.5)$
$= 14.0769 \Rightarrow 14.1 \text{ cm}$

31. $15°25'05''\left(\frac{\pi}{180}\right) = \frac{s}{325.500}$
$s = \left[\left(15 + \frac{25}{60} + \frac{5}{3600}\right)\left(\frac{\pi}{180}\right)\right](325.500)$
$= 87.59069 \Rightarrow 87.5907 \text{ ft}$

33. Given: $A = \frac{r^2\theta}{2}$ and $\theta = \frac{s}{r}$;

Show: area of a sector $= \frac{rs}{2}$

Since $A = \frac{r^2\theta}{2}$, then $A = \frac{r^2\left(\frac{s}{r}\right)}{2}$ $\left[\text{substitute } \frac{s}{r} \text{ for } \theta\right]$

Thus $A = \frac{r^2\left(\frac{s}{r}\right)}{2} \Rightarrow \frac{r^2}{2}\left(\frac{s}{r}\right) \Rightarrow \frac{\not{r}(r)(s)}{2\not{r}} \Rightarrow \frac{rs}{2}$

Exercise 3 ◊ Uniform Circular Motion
Angular Velocity

1. $1850\,\frac{\text{rev}}{\text{min}}\left(\frac{2\pi \text{ rad}}{\text{rev}}\right)\left(\frac{\text{min}}{60 \text{ s}}\right) \Rightarrow 193.73 = 194 \text{ rad/s}$

$1850\,\frac{\text{rev}}{\text{min}}\left(\frac{360°}{\text{rev}}\right)\left(\frac{\text{min}}{60 \text{ s}}\right) = 11{,}100 \text{ deg/s}$

3. $77.2\,\frac{\text{deg}}{\text{s}}\left(\frac{\text{rev}}{360°}\right)\left(\frac{60 \text{ s}}{\text{min}}\right) \Rightarrow 12.866 = 12.9 \text{ rev/min}$

$77.2\,\frac{\text{deg}}{\text{s}}\left(\frac{2\pi \text{ rad}}{360°}\right) \Rightarrow 1.3473 = 1.35 \text{ rad/s}$

5. $48.1\,\frac{\text{deg}}{\text{s}}\left(\frac{\text{rev}}{360°}\right)\left(\frac{60 \text{ s}}{\text{min}}\right) \Rightarrow 8.0166 = 8.01 \text{ rev/min}$

$48.1\,\frac{\text{deg}}{\text{s}}\left(\frac{2\pi \text{ rad}}{360°}\right) \Rightarrow 0.8395 = 0.840 \text{ rad/s}$

Linear Speed

7. $v = \omega r \Rightarrow 334\frac{\text{rev}}{\text{min}} \cdot \frac{2\pi \text{ rad}}{\text{rev}}\left(\frac{3.55}{12}\right) \Rightarrow 620.831 = 621 \text{ ft/min}$

9. $\omega = \frac{v}{r} \Rightarrow 56.5\,\frac{\text{ft}}{\text{min}} \div 1.14 \text{ ft} = 49.4\frac{\text{rad}}{\text{min}}$

$\Rightarrow 49.4\frac{\text{rad}}{\text{min}} \cdot \frac{1 \text{ rev}}{2\pi} \Rightarrow 7.8622 = 7.9\,\frac{\text{rev}}{\text{min}}$

Applications

11. $725\,\frac{\text{rev}}{\text{min}}(1 \text{ s})\left(\frac{\text{min}}{60 \text{ s}}\right)\left(\frac{360°}{\text{rev}}\right) = 4350°$

13. $\dfrac{2.00 \text{ rad}}{2550 \dfrac{\text{rev}}{\text{min}}\left(\dfrac{2\pi \text{ rad}}{\text{rev}}\right)\left(\dfrac{\text{min}}{60 \text{ s}}\right)} \Rightarrow 0.0074896 = 0.00749 \text{ s}$

15. $\omega = 3600$ rad/min; $v = 45.0$ m/min

$r = \dfrac{V}{\omega} \Rightarrow \dfrac{45.0\text{m}}{\text{min}}\left(\dfrac{\text{min}}{3600 \text{ rad}}\right)\left(\dfrac{1000 \text{ mm}}{\text{m}}\right) = 12.5$ mm; diameter is $2r \Rightarrow 2(12.5) = 25.0$ mm

17. $r = 3.25$ in.; $v = 55.0$ ft/min

$\omega = \dfrac{v}{r} \Rightarrow \dfrac{55.0 \dfrac{\text{ft}}{\text{min}}}{3.25 \text{ in.}}\left(\dfrac{12 \text{ in.}}{\text{ft}}\right) \Rightarrow 203.076 = 203 \text{ rad/min}$

$203 \dfrac{\text{rad}}{\text{min}}\left(\dfrac{1 \text{ rev}}{2\pi \text{ rad}}\right)\left(\dfrac{\text{min}}{60 \text{ s}}\right) \Rightarrow 0.53847 = 0.539 \text{ rev/s}$

In 10.0 seconds: $0.539(10) = 5.39$ rev

19. $v = \dfrac{2\pi \text{ rad}}{365 \text{ days}\left(\dfrac{24 \text{ h}}{\text{day}}\right)}(93.0 \times 10^6 \text{ mi}) \Rightarrow 66{,}705.049 = 66{,}700 \text{ mi/h}$

21. $v = \dfrac{1800 \dfrac{\text{rev}}{\text{min}}}{44}\left(\dfrac{2\pi \text{ rad}}{\text{rev}}\right)\left(\dfrac{21.7 \text{ ft}}{2}\right) \Rightarrow 2788.877 = 2790 \text{ ft/min}$

Why 44? $v = \omega r$, where $\omega = 1800$ rev/min and $r = 21.2$ ft. $v = \omega r$ is multiplied by the "gear ratio of 1:44."

23. $\omega = \dfrac{v}{r} \Rightarrow 110 \dfrac{\text{ft}}{\text{min}} \div \dfrac{2.45 \text{ in}}{12} = 85.749 = 85.7 \dfrac{\text{rev}}{\text{min}}$

CHAPTER 14 REVIEW PROBLEMS

1. $\dfrac{3\pi}{7} \Rightarrow \dfrac{3\pi}{7}\left(\dfrac{180°}{\pi}\right) \Rightarrow 77.142 = 77.1°$

3. $\dfrac{\pi}{9} \Rightarrow \dfrac{\pi}{9}\left(\dfrac{180°}{\pi}\right) = 20°$

5. $\dfrac{11\pi}{12} \Rightarrow \dfrac{11\pi}{12}\left(\dfrac{180°}{\pi}\right) = 165°$

7. $\dfrac{33.5 \text{ rev}}{1.45 \text{ min}}\left(\dfrac{2\pi \text{ rad}}{\text{rev}}\right)\left(\dfrac{\text{min}}{60 \text{ s}}\right) \Rightarrow 2.4193 = 2.42 \text{ rad/s}$

9. $300° \Rightarrow 300°\left(\dfrac{\pi}{180°}\right) \Rightarrow \dfrac{300\pi}{180} = \dfrac{5\pi}{3}$

11. $230° \Rightarrow 230°\left(\dfrac{\pi}{180°}\right) \Rightarrow \dfrac{230\pi}{180} = \dfrac{23\pi}{18}$

13. $\sin\dfrac{\pi}{9} \Rightarrow 0.34202 = 0.3420$

15. $\sin\left(\dfrac{2\pi}{5}\right)^2 \Rightarrow 0.999965 = 1.000$

17. $4\sin^2\dfrac{\pi}{9} - 4\sin\left(\dfrac{\pi}{9}\right)^2 \Rightarrow 4\left(\sin\dfrac{\pi}{9}\right)^2 - 4\sin\left(\dfrac{\pi}{9}\right)^2 \Rightarrow 0.467911 - 0.486182 \Rightarrow -0.018271 = -0.01827$

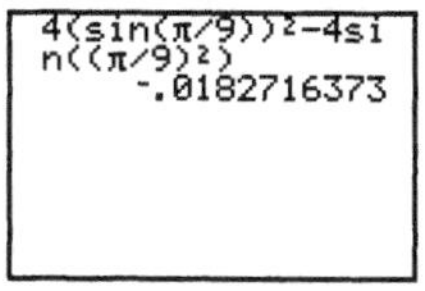

19. $r = 15$ cm; $v = 8.00$ ft/s

$$\omega = \frac{v}{r} \Rightarrow \left(\frac{8.00 \text{ ft/s}}{15.0 \text{ cm}}\right)\left(\frac{2.54 \text{ cm}}{\text{in.}}\right)\left(\frac{12 \text{ in.}}{\text{ft}}\right)\left(\frac{60 \text{ s}}{\text{min}}\right)\left(\frac{\text{rev}}{2\pi \text{ rad}}\right) = 155.233 = 155 \text{ rev/min}$$

21. $$v = \frac{2\pi \text{ rad}}{3(24) + 7 + \frac{35}{60}}(4250) \Rightarrow \frac{26{,}703.53756}{79.58333} \Rightarrow 335.5418 = 336 \text{ mi/h}$$

23. $\tan 0.837 \Rightarrow 1.10892 = 1.1089$

25. $\sin 4.22 \Rightarrow -0.881206 = -0.8812$

27. $\sec 3.38 \Rightarrow (\cos 3.38)^{-1} \Rightarrow -1.029107 = -1.0291$

29. $\sin (2.84)^2 \Rightarrow 0.97769 = 0.9777$

31. $\sin^2 (2.24)^2 \Rightarrow (\sin (2.24)^2)^2 \Rightarrow 0.909703 = 0.9097$

33. $\theta = \frac{s}{r} \Rightarrow 1.73 \text{ rad} = \frac{384 \text{ mm}}{s}$; $s = \frac{384}{1.73} \Rightarrow 221.965 = 222$ mm

35. $57.5°\left(\frac{\pi \text{ rad}}{180°}\right) \Rightarrow 1.00356 = 1.004$

$$V = \left[\frac{(287)^2(1.004)}{2} - \frac{(85)^2(1.004)}{2}\right]75.0 = (41{,}349.238 - 3626.95)75.0 = 2{,}829{,}171.6 \Rightarrow 2{,}830{,}000 \text{ mm}^3 \Rightarrow 2830 \text{ cm}^3$$

Chapter 15: Trigonometric, Parametric, and Polar Graphs

Exercise 1 ◊ Graphing the Sine Wave by Calculator

1. $y = 2 \sin x$

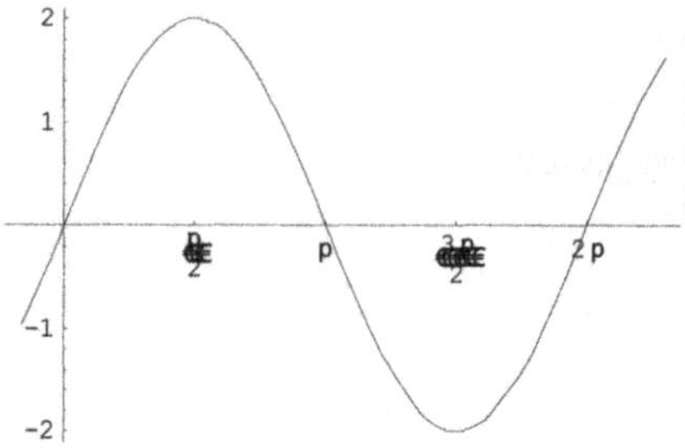

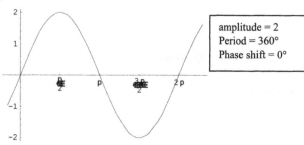

amplitude = 2
Period = 360°
Phase shift = 0°

3. $y = \sin 2x$

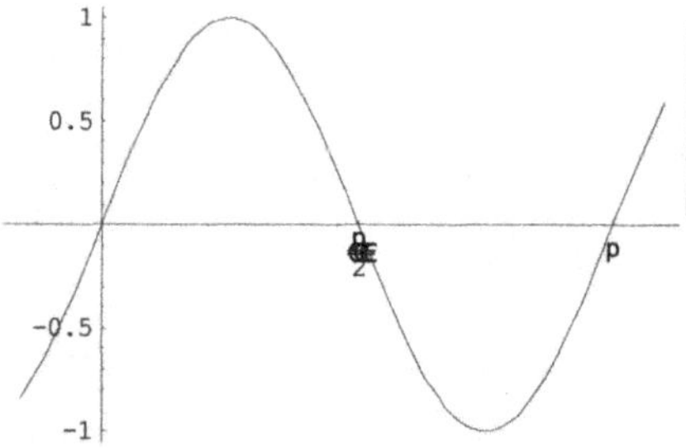

amplitude = 1
Period = 180°
Phase shift = 0°

5. $y = 3 \sin 2x$

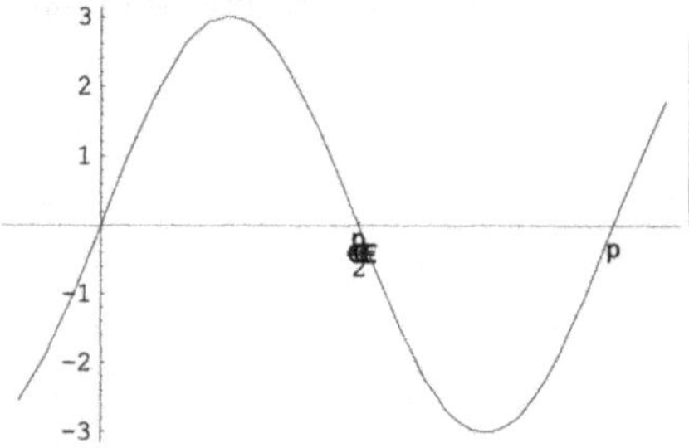

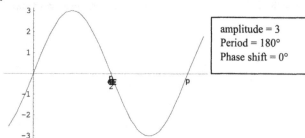

amplitude = 3
Period = 180°
Phase shift = 0°

7. $y = \sin (x + 15°)$

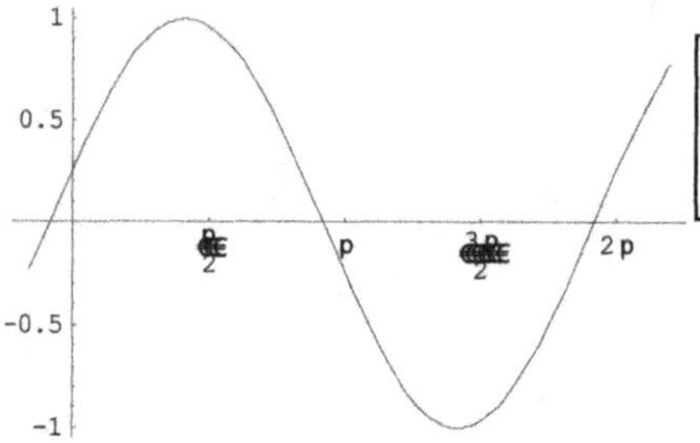

amplitude = 1
Period = 360°
Phase shift = −15°

9. $y = \sin\left(x - \frac{\pi}{2}\right)$

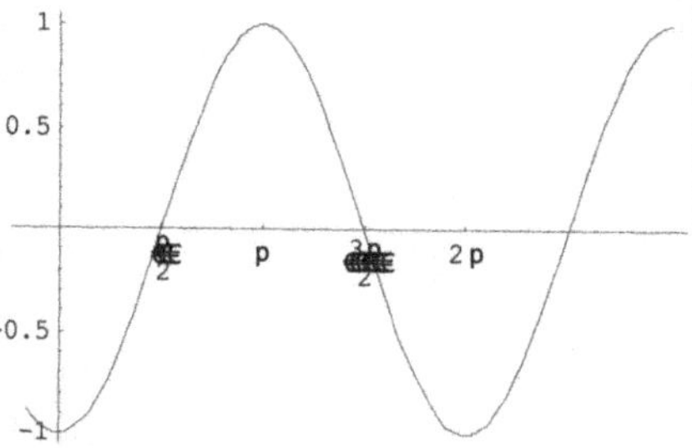

amplitude = 1
Period = 2π
Phase shift = $\dfrac{\pi}{2}$

11. $y = 3\sin(x + 45°)$

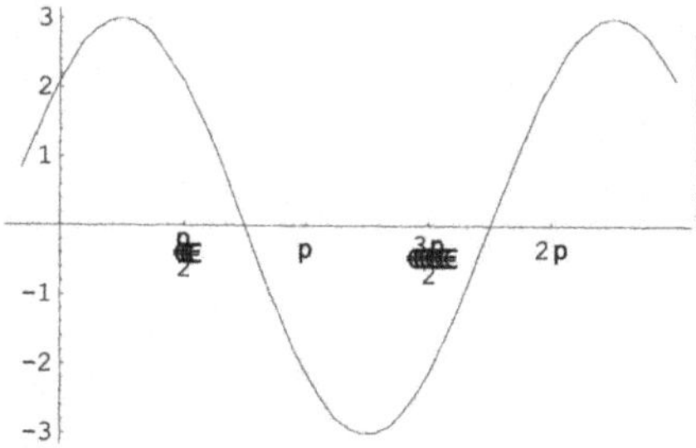

amplitude = 3
Period = 360°
Phase shift = −45°

13. $y = -4\sin\left(x - \dfrac{\pi}{4}\right)$

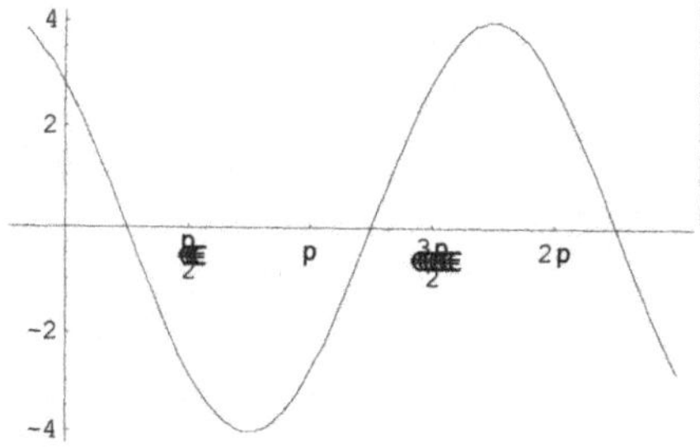

amplitude = 4
Period = 2π
Phase shift = $\dfrac{\pi}{4}$

15. $y = \sin(2x + 55°)$

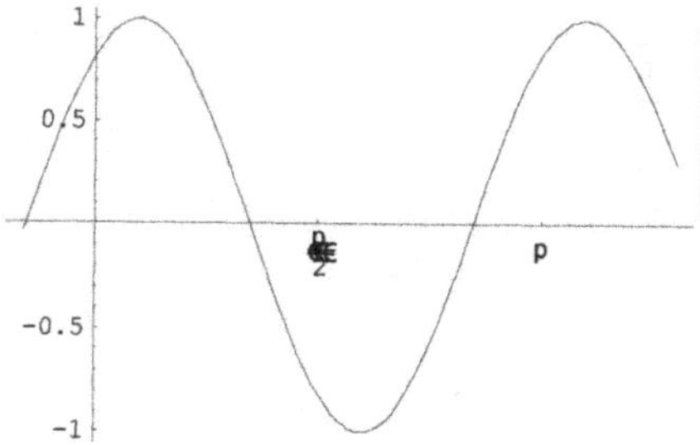

amplitude = 1
Period = 180°
Phase shift = −27.5°

17. $y = \sin\left(3x - \dfrac{\pi}{3}\right)$

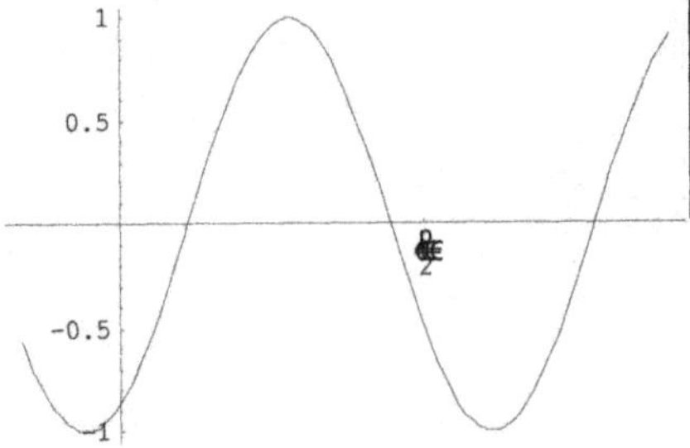

amplitude = 1

Period = $\dfrac{2\pi}{3}$

Phase shift = $\dfrac{\pi}{9}$

19. $y = 3\sin(2x + 55°)$

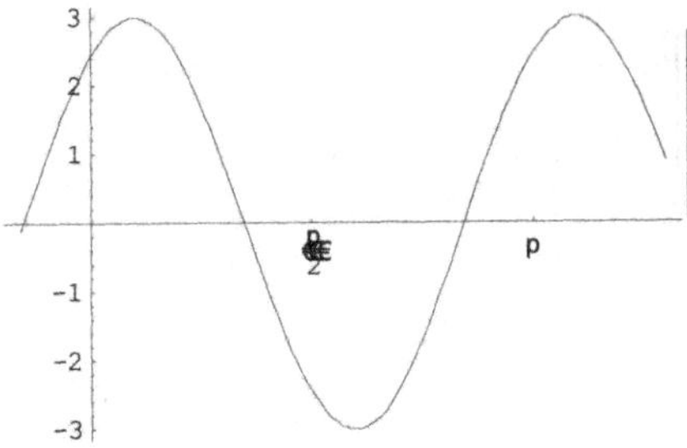

amplitude = 3
Period = 180°
Phase shift = −27.5°

21. $y = -2\sin\left(3x - \dfrac{\pi}{2}\right)$

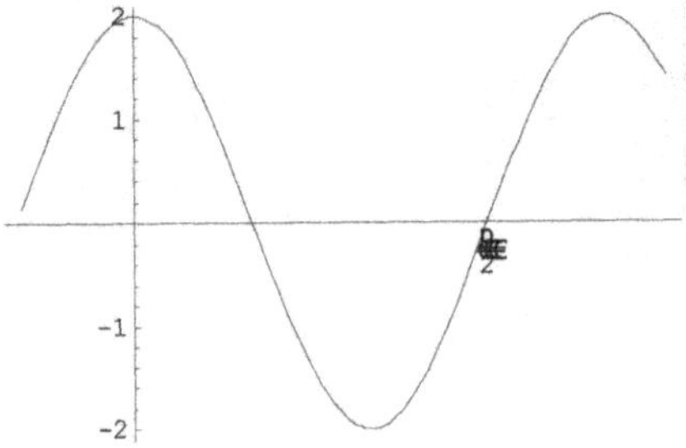

amplitude = 2

Period = $\dfrac{2\pi}{3}$

Phase shift = $\dfrac{\pi}{6}$

23. $y = 3.73\sin(4.32x + 55.4°)$

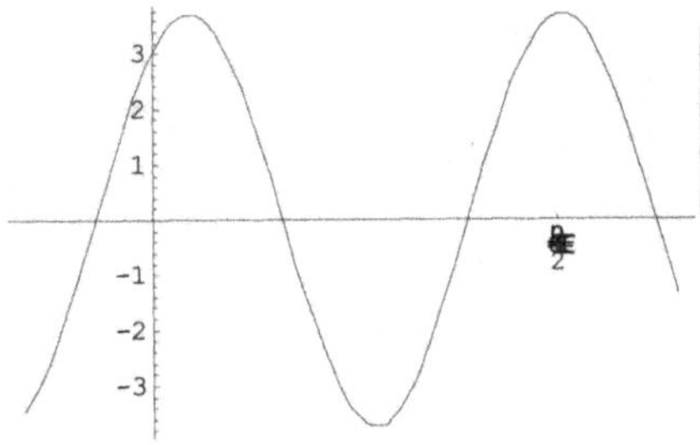

amplitude = 3.73
Period = 83.3°
Phase shift = −12.8°

Zeros and Instantaneous Value

25. $y = \sin(2x + 15°)$ $x = 15°$

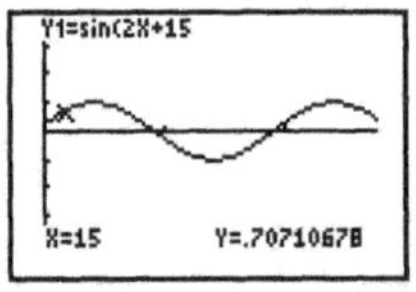

zeros at $x = 82.5°$ and $172.5°$; $y = 0.7071$ at $x = 15°$

27. $y = 2\sin(3x - 3)$ $x = 1$ radian

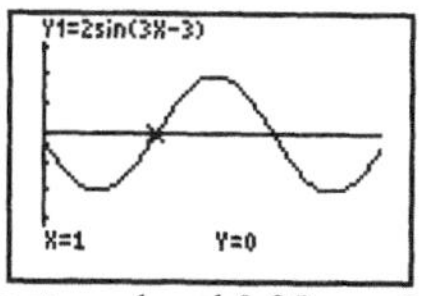

zeros at $x = 1$ and 2.05; $y = 0$ at $x = 1$ radian

Applications

29. $h = L\sin\theta$

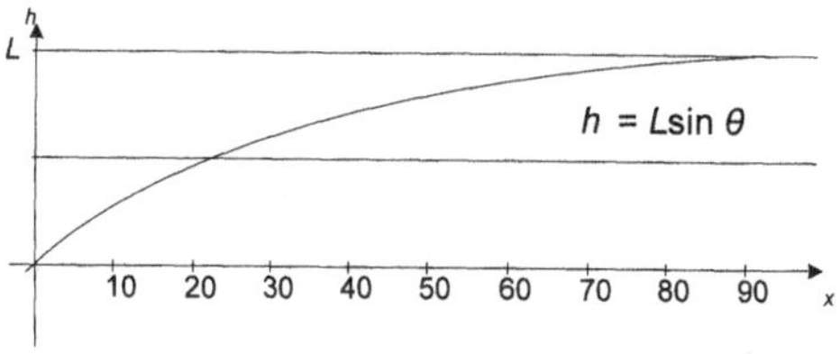

31. $x = r\sin\theta$

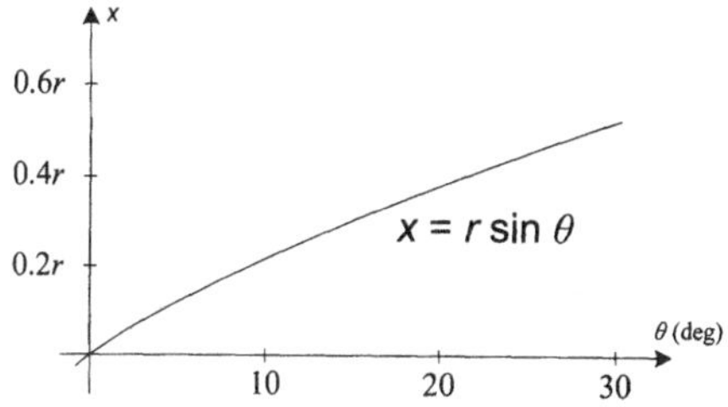

33. $x = r\sin(\theta + 38.6°)$

Exercise 2 ◊ Manual Graphing of the Sine Wave

1. See Solutions Exercise 1

Writing the Equation, Given the Amplitude, Period, and Phase Shift

3. $6\pi = \dfrac{2\pi}{b} \Rightarrow b = \dfrac{2\pi}{6\pi} = \dfrac{1}{3};$

$$-\frac{c}{b} = -\frac{\pi}{4} \Rightarrow -\frac{c}{\frac{1}{3}} = -\frac{\pi}{4} \Rightarrow \frac{\pi}{3} = 4c \Rightarrow c = \frac{\pi}{12}$$

Therefore, $y = -2\sin\left(\dfrac{x}{3} + \dfrac{\pi}{12}\right)$

Exercise 3 ◊ The Sine Wave as a Function of Time

1. 68 Hz. $P = \dfrac{2\pi}{\omega} \Rightarrow P = \dfrac{1}{f} \Rightarrow \dfrac{1}{68} \Rightarrow 0.014705 = 0.0147$ s

$\omega = 2\pi f \Rightarrow 2\pi(68) \Rightarrow 427.256 = 427$ rad/s

3. 5000 Hz. $P = \dfrac{1}{f} \Rightarrow \dfrac{1}{5000} = 0.0002$ s

$\omega = 2\pi f \Rightarrow 2\pi(5000) \Rightarrow 31{,}415.926 = 31{,}400$ rad/s

5. $\dfrac{1}{8}$ s. $f = \dfrac{1}{P} \Rightarrow \dfrac{1}{\frac{1}{8}} = 8$ Hz

$\omega = 2\pi f \Rightarrow 2\pi(8) \Rightarrow 50.265 = 50.3$ rad/s

7. $f = 60$ cycles/s; 200 cycles$\left(\dfrac{\text{s}}{60 \text{ cycles}}\right) \Rightarrow 3.3333 = 3.33$ s

9. $\omega = 455$ rad/s

$$P = \frac{2\pi}{\omega} \Rightarrow \frac{2\pi}{455} \Rightarrow 0.013809 = 0.0138 \text{ s}$$

$$f = \frac{1}{P} \Rightarrow \frac{1}{0.013809} \Rightarrow 72.415 = 72.4 \text{ Hz}$$

11. $\omega = 500$ rad/s

$$P = \frac{2\pi}{\omega} \Rightarrow \frac{2\pi}{500} \Rightarrow 0.012566 = 0.0126 \text{ s}$$

$$f = \frac{1}{P} \Rightarrow \frac{1}{0.012566} \Rightarrow 79.579 = 79.6 \text{ Hz}$$

13. $P = 400$ ms, amplitude = 10, $\varphi = 1.1$ rad $\left[\phi = \text{(phase shift) (angular velocity)} \Rightarrow -70\left(-\dfrac{2\pi}{400}\right) = 1.099 = 1.1 \text{ rad}\right]$

15. $y = a\sin(\omega t + \phi) \Rightarrow y = 5 \sin(750t + 15°)$

17. $y = 3 \sin 377t$

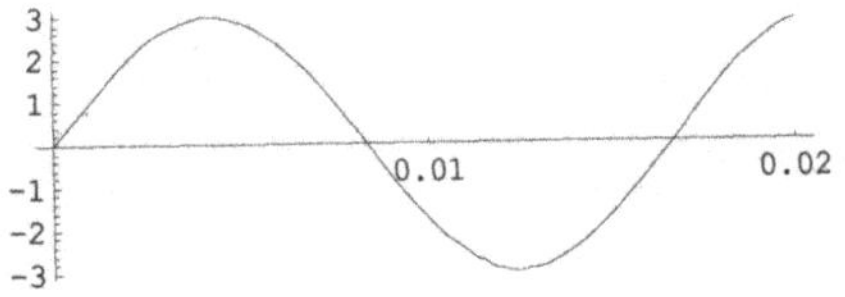

19. $y = 375\sin\left(55t + \dfrac{\pi}{4}\right)$

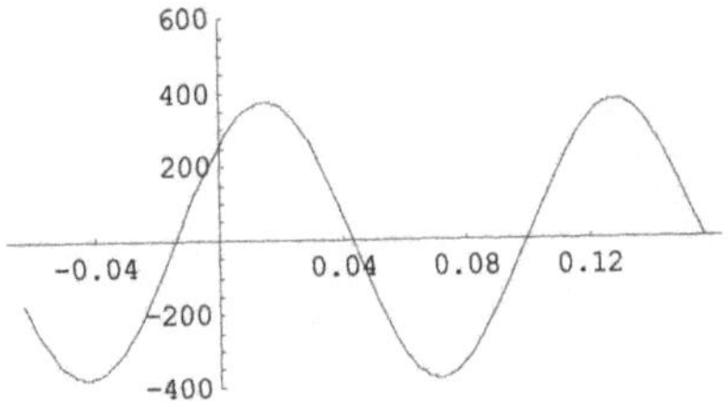

Mechanical Applications

21. $y = R \sin 16.0t$ $\left[16 \text{ from } 2\pi(2.55) = 16.022\right]$

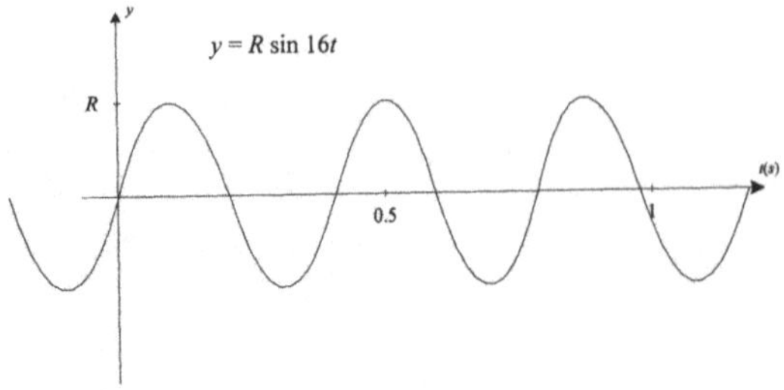

Alternating Current

23. From $v = V_m \sin(\omega t + \phi_1)$

$$V_{max} = 4.27 \text{ V}; P = \frac{2\pi}{\omega} \Rightarrow \frac{2\pi}{463} \Rightarrow 0.01357 \text{ s} \Rightarrow 13.6 \text{ ms}; \; f = \frac{1}{P} \Rightarrow \frac{1}{0.01357} \Rightarrow 73.691\text{Hz} = 73.7 \text{ Hz}; \phi = 27°$$

$$v(0.12) \Rightarrow 4.27\sin(463[0.12] + 27°) \Rightarrow 4.27\sin(55.56 + 0.471238) \Rightarrow -2.11214 = -2.11 \text{ V}$$

25. $f = \frac{1}{P} \Rightarrow \frac{1}{35} \Rightarrow 0.02857; \; \omega = \frac{2\pi}{P} \Rightarrow \frac{2\pi}{0.02857} \Rightarrow 219.92 = 220$

Therefore, $i = 49.2 \sin(220t + 63.2°)$ mA

Exercise 4 ◊ Graphs of Trigonometric Functions

The Cosine Curve

1. $y = 3 \cos x$

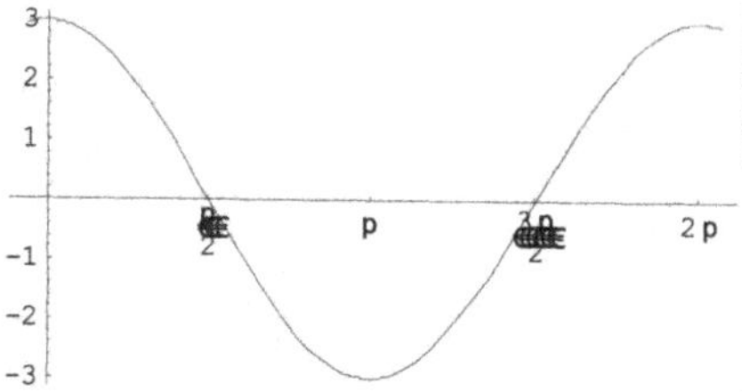

amplitude = 3
Period = 2π
Phase shift = 0

3. $y = \cos 3x$

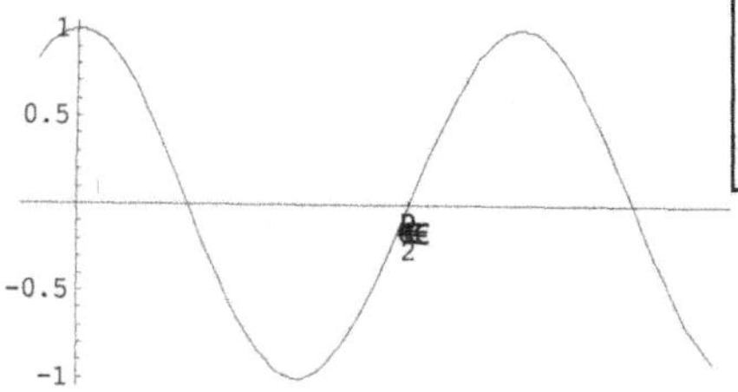

amplitude = 1
Period = $\frac{2\pi}{3}$
Phase shift = 0

5. $y = 2 \cos 3x$

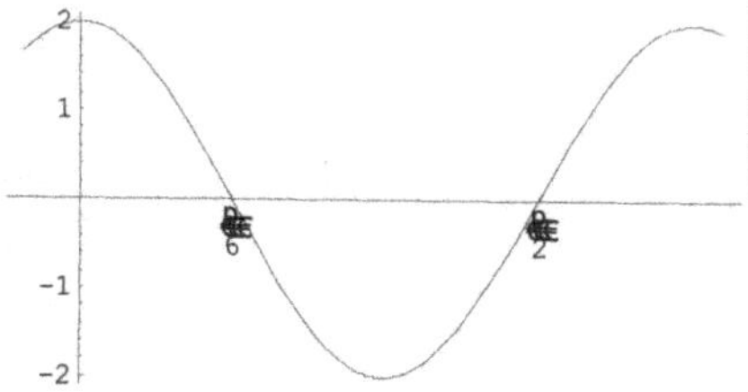

amplitude = 2
Period = $\frac{2\pi}{3}$
Phase shift = 0

7. $y = \cos(x - 1)$

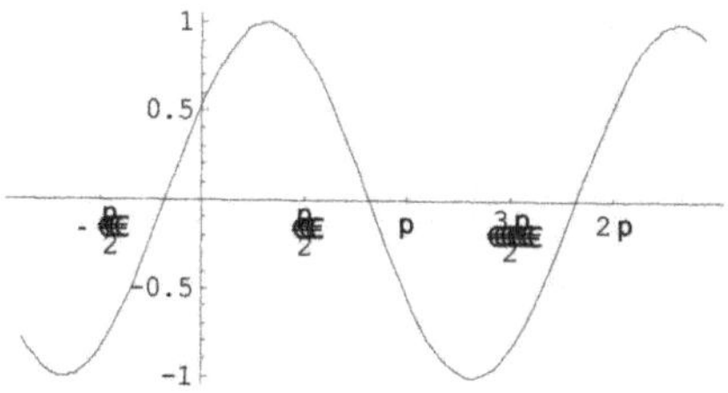

amplitude = 1
Period = 2π
Phase shift = 1

9. $y = 3\cos\left(x - \frac{\pi}{4}\right)$

amplitude = 3
Period = 2π
Phase shift = $\frac{\pi}{4}$

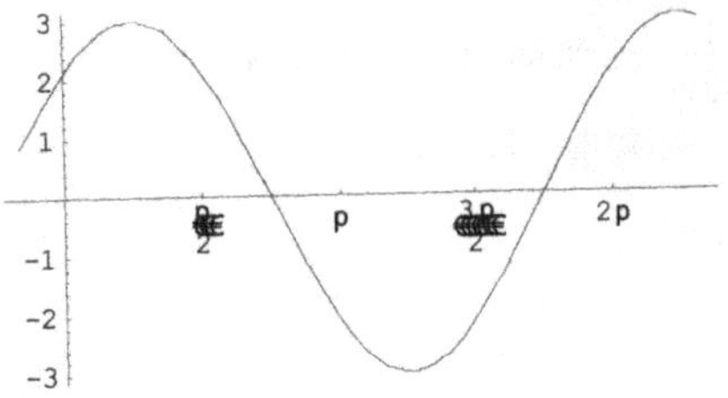

The Tangent Curve

11. $y = 2 \tan x$

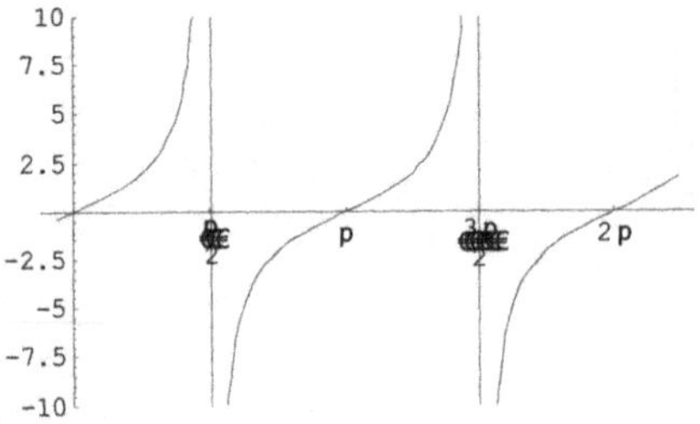

13. $y = 3 \tan 2x$

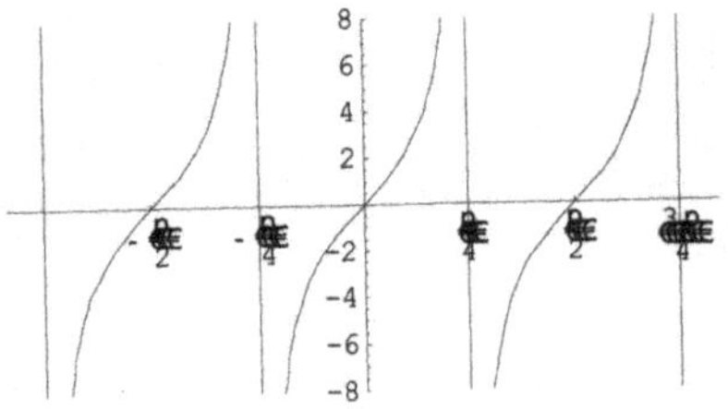

15. $y = 2 \tan (3x - 2)$

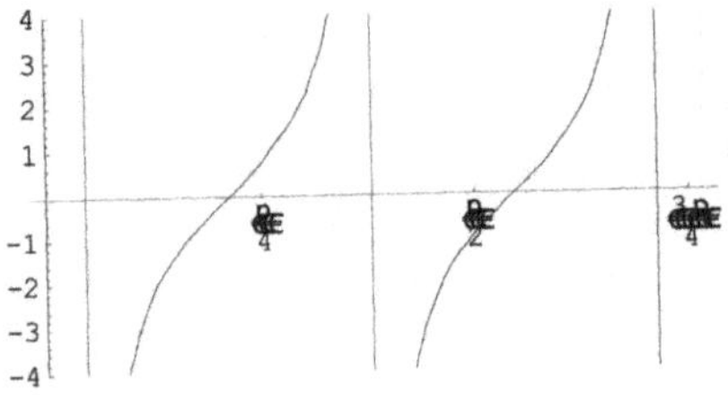

Cotangent, Secant, and Cosecant Curves

17. $y = 2 \cot 2x$

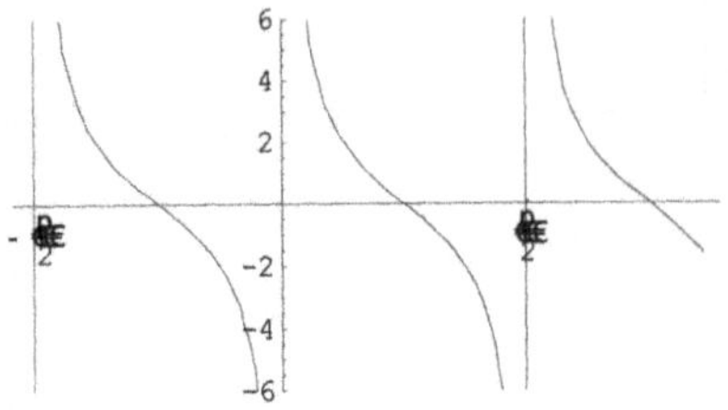

19. $y = 3 \csc 3x$

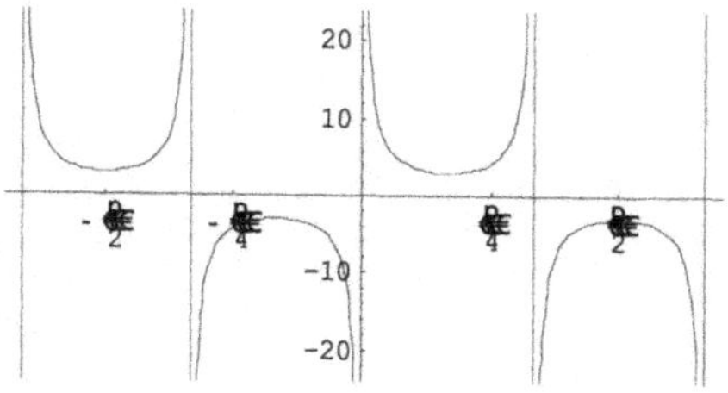

21. $y = 2 \sec (3x + 1)$

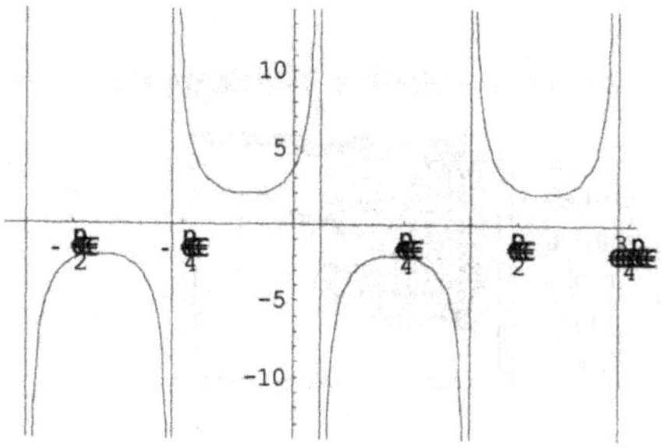

Applications

23. $y = r \cos \theta$

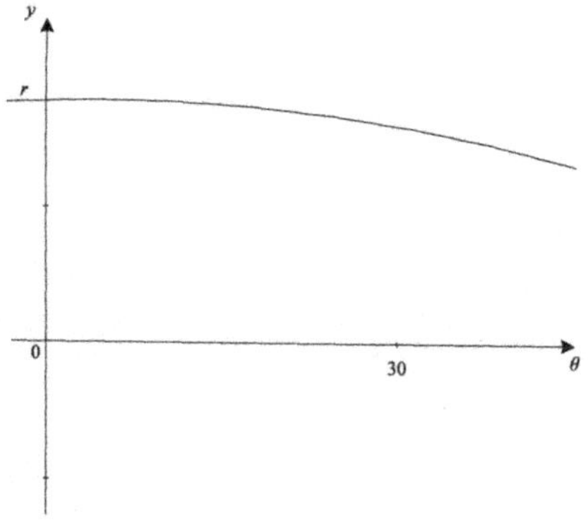

25. $y = 2.15 \tan \theta$ km

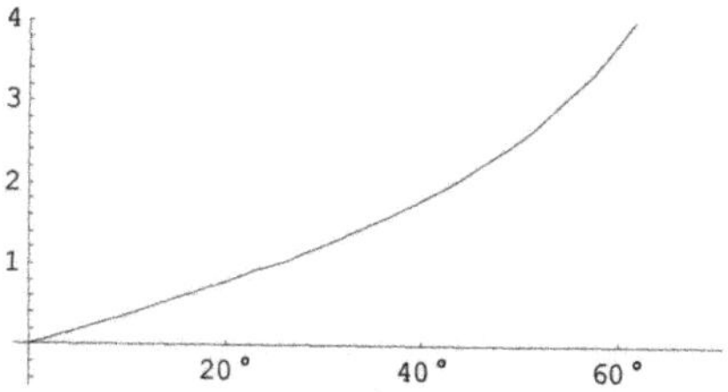

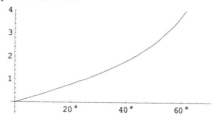

27. $F = \dfrac{fw}{f \sin \theta + \cos \theta} \Rightarrow \dfrac{0.55(5.35 \text{ kg})}{0.55 \sin \theta + \cos \theta}$

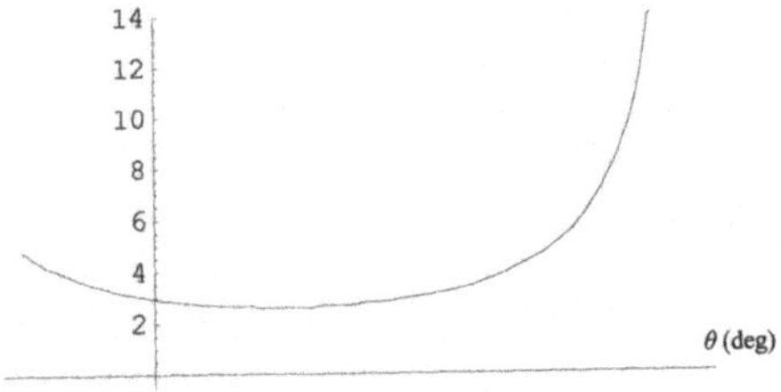

Exercise 5 ◊ Graphing a Parametric Equation

Graphing Parametric Equations by Calculator

To put the TI-83 in parametric mode, choose MODE, arrow down to the fourth line (**Func Par Pol Seq**), arrow over to **Par** and press **ENTER**.

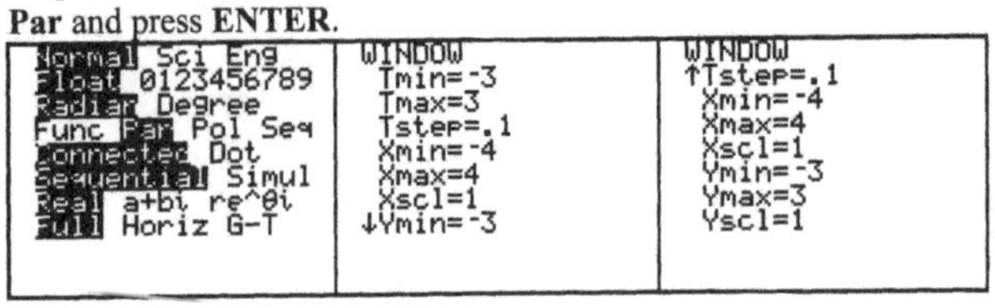

ZoomFit will also scale the graph to fit the window

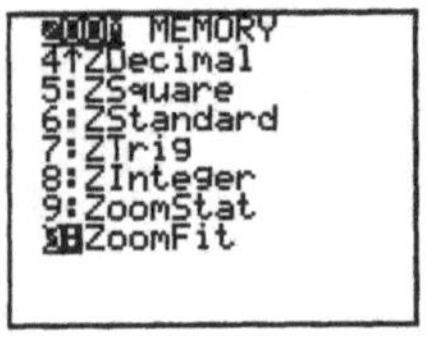

1. $x = t,\ y = t$

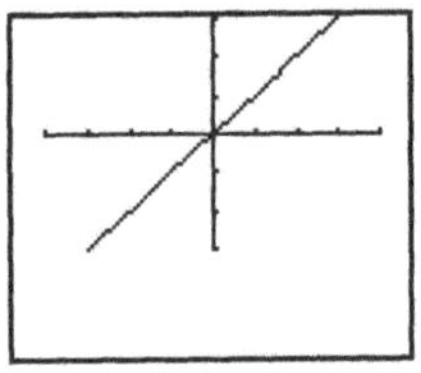

3. $x = -t,\ y = 2t^2$

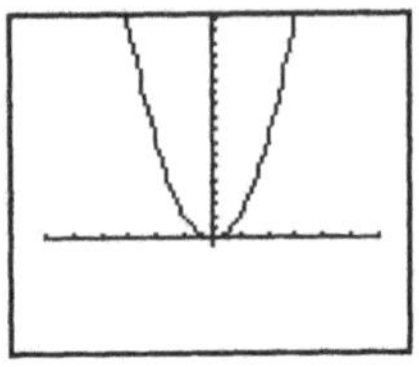

Graphing a Trigonometric Equation in Parametric Form

5. $x = \sin\theta;\ y = \sin\theta$

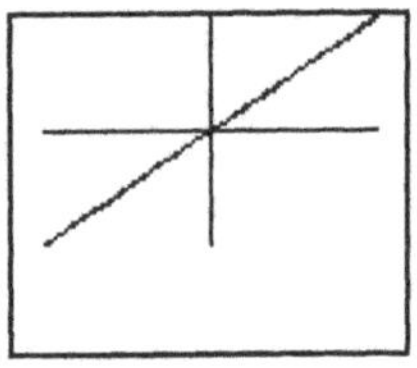

7. $x = \sin\theta;\ y = \sin 2\theta$

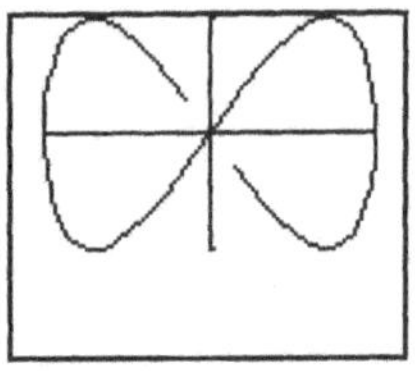

9. $x = \sin\theta;\ y = \sin\left(\theta + \frac{\pi}{4}\right)$

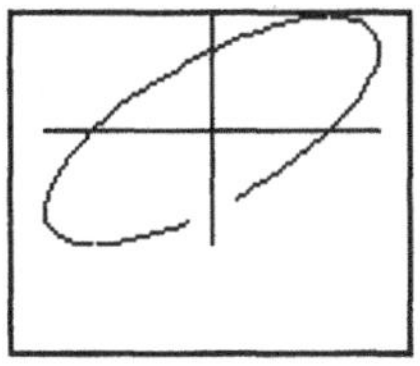

11. $x = \sin 2\theta;\ y = \sin 3\theta$

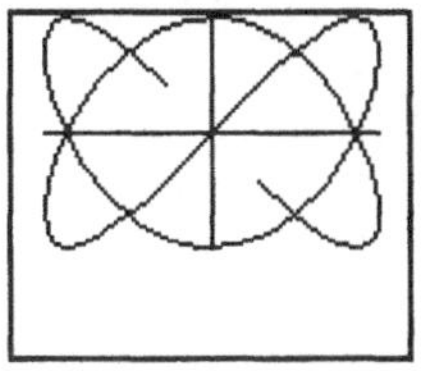

13. $x = 5.83 \sin 2\theta;\ y = 4.24 \sin \theta$

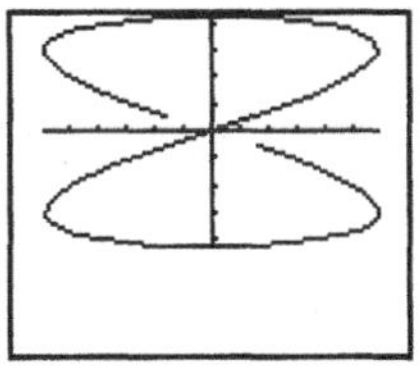

Applications

15. (a)

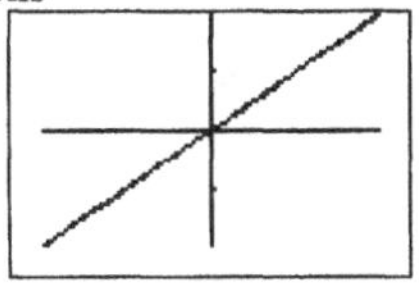

Multiplying the amplitude by a factor of n increases the slope

(b)

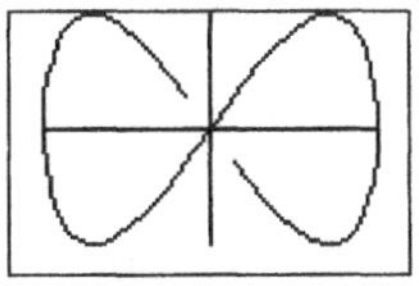

Multiplying the frequency by a factor of $2n$ closes the figure

(c)

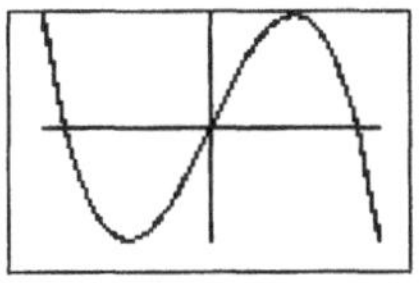

Multiplying the frequency by a factor of $2n - 1$ approaches the sine function

(d)

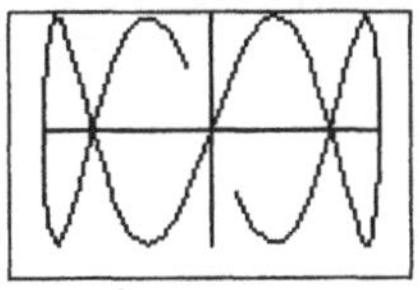

Multiplying the frequency by a factor of $2n$ closes the figure

(e)

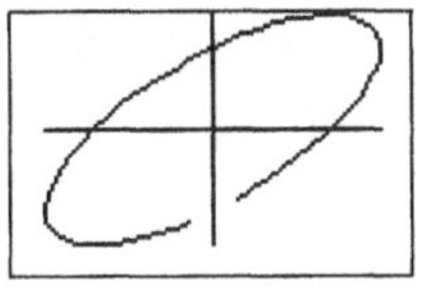

As the phase shift approaches 90°, the figure approaches the cosine function

(f) 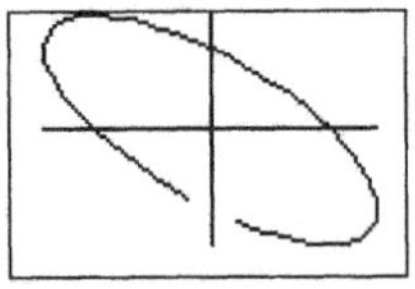

As the phase shift approaches 90°, the figure approaches the cosine function

(g) 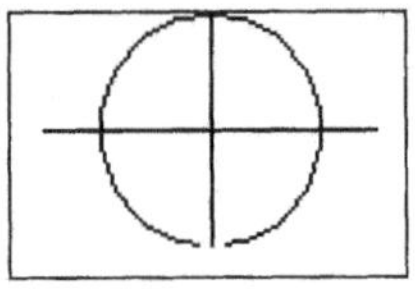

A phase shift of 90° is the cosine function

(h) 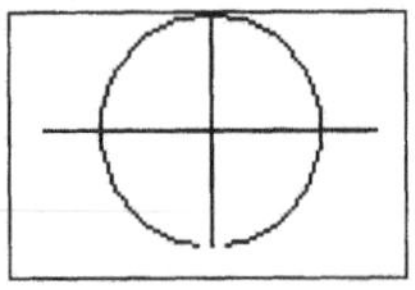

17. (a) 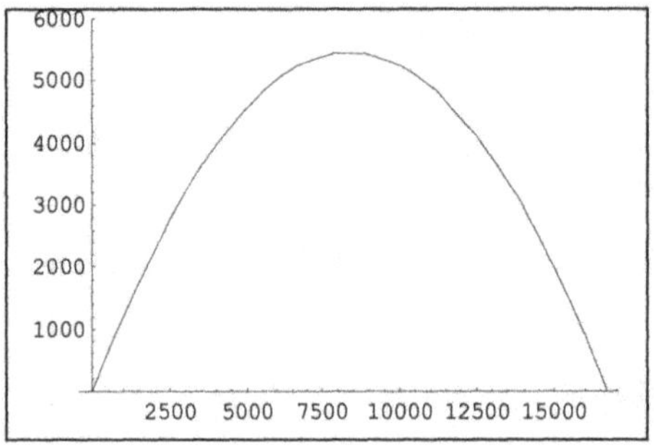

(b) 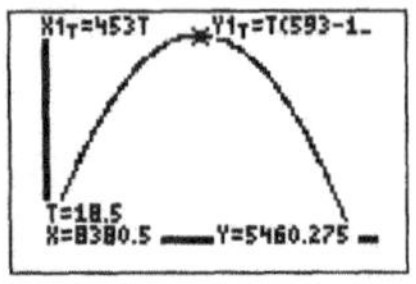

Maximum height: 5490 ft

(c) 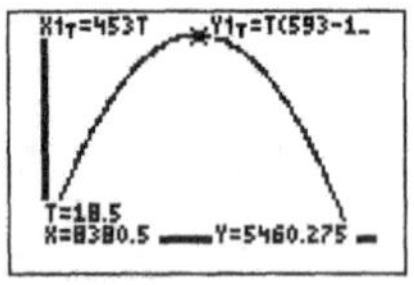

x value for which the height is at a maximum: 8380 ft

(d) 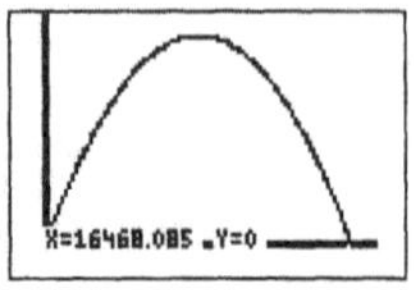

projectiles maximum distance, assuming the ground is level: 16,760 ft

(e) 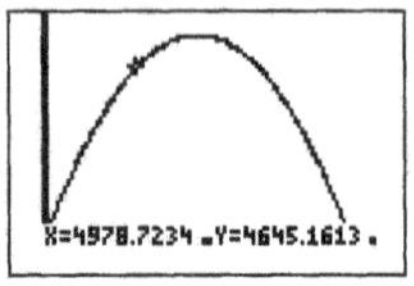

height when $x = 5000$ ft: 4600 ft

$5000 = 453t$; $t = 11.03752759$

$y = 593(11.03752759) - 16.1(11.03752759)^2$

$y = 4583.8389 \Rightarrow 4600$ ft

Exercise 6 ◊ Graphing in Polar Coordinates

1. $(4, 35°)$

3. $(2.5, 215°)$

5. $\left(2.7, \frac{\pi}{6}\right)$

7. $\left(-3, \frac{\pi}{2}\right)$

9. $(3.6, -20°)$

11. $\left(-1.8, -\frac{\pi}{6}\right)$

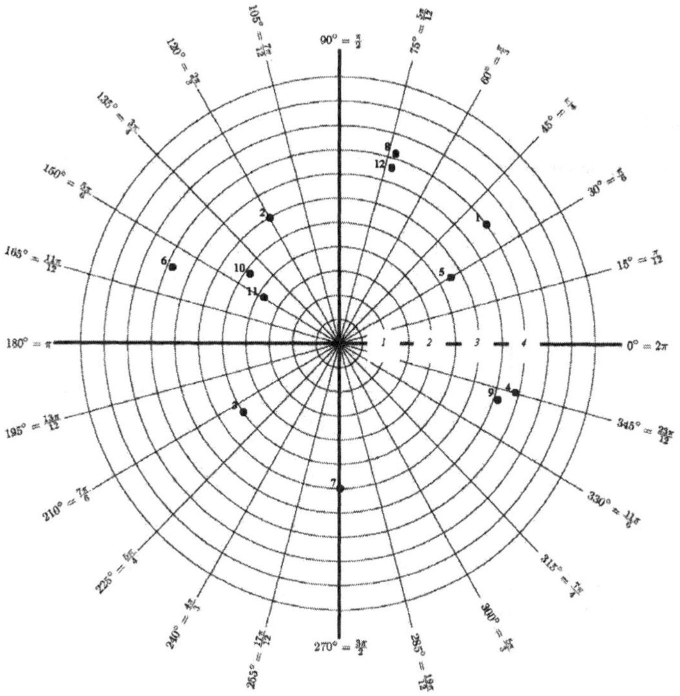

Graphing an Equation in Polar Coordinates

13. $r = 2\cos\theta$

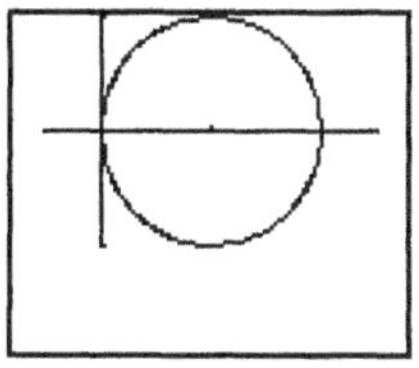

15. $r = 3\sin\theta + 3$

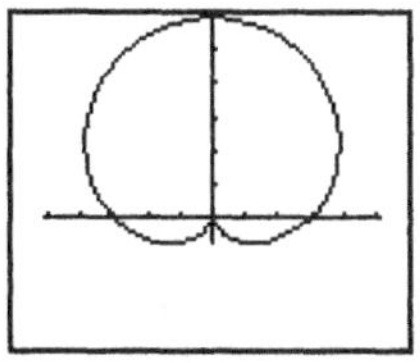

17. $r = 3\cos 2\theta$

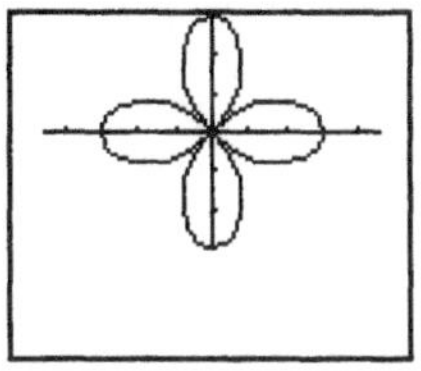

19. $r = \sin 2\theta - 1$

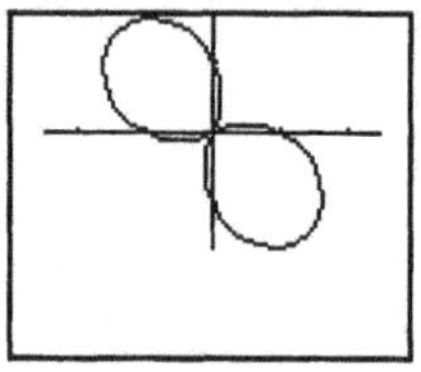

21. Lemniscate of Bernoulli

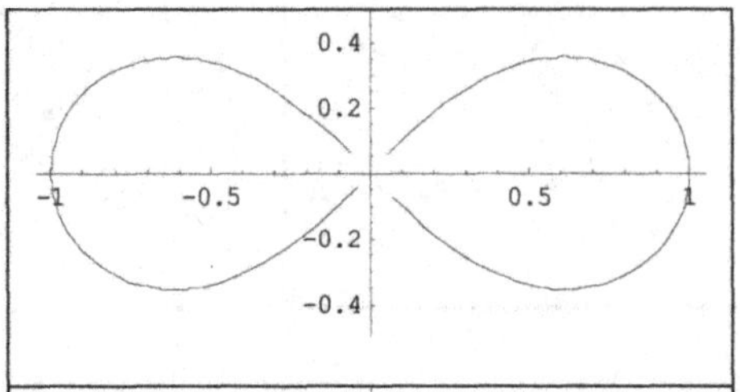

Bifolium

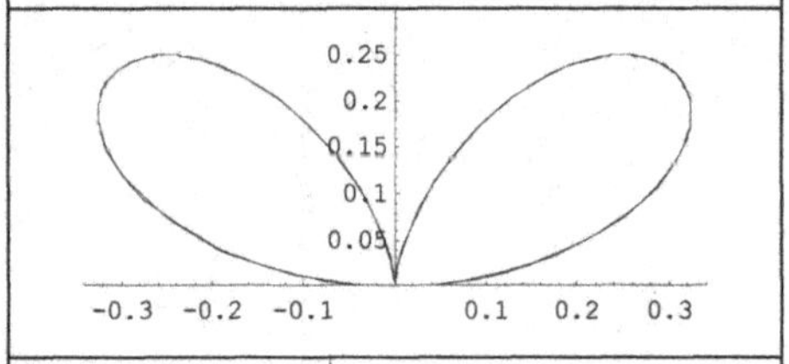

Three-leaved rose

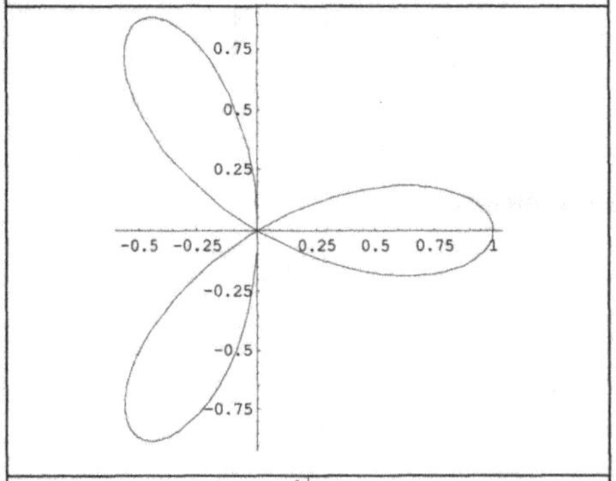

Four-leaved rose

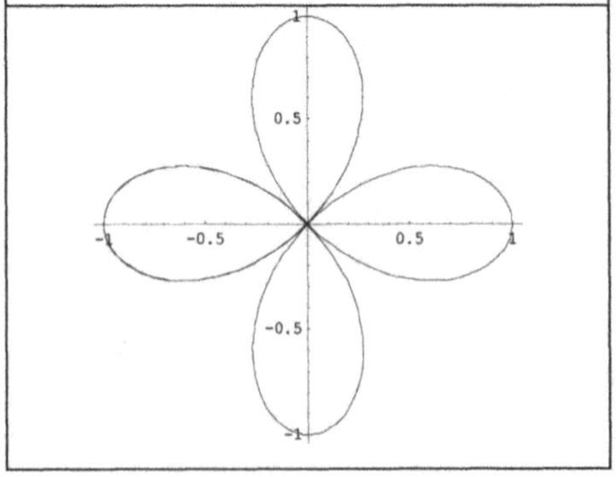

21. (continued) Four-leaved rose

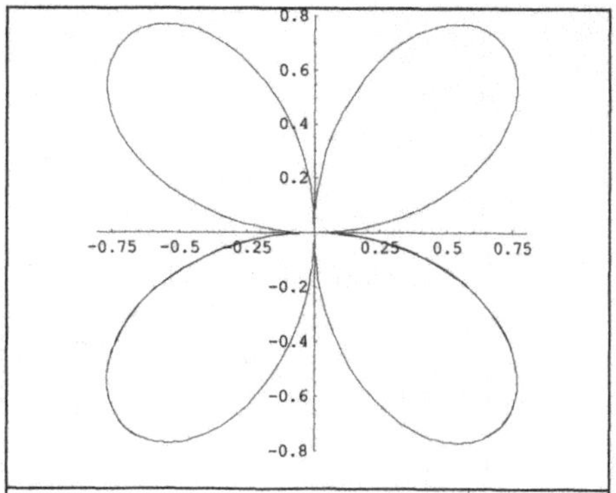

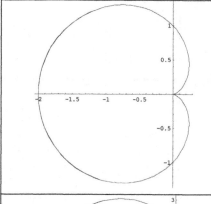

Cardioid

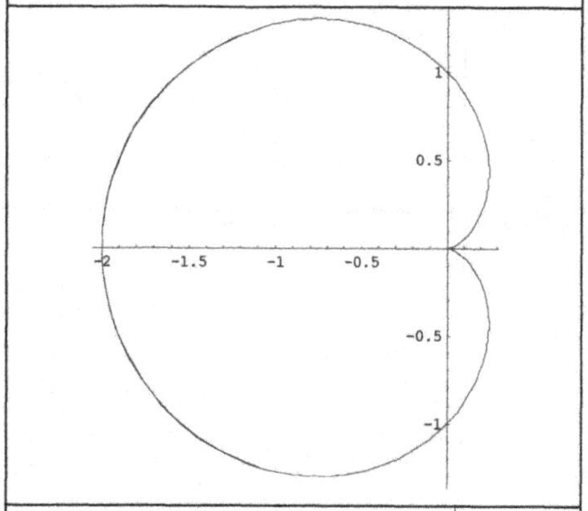

Limacon of Pascal

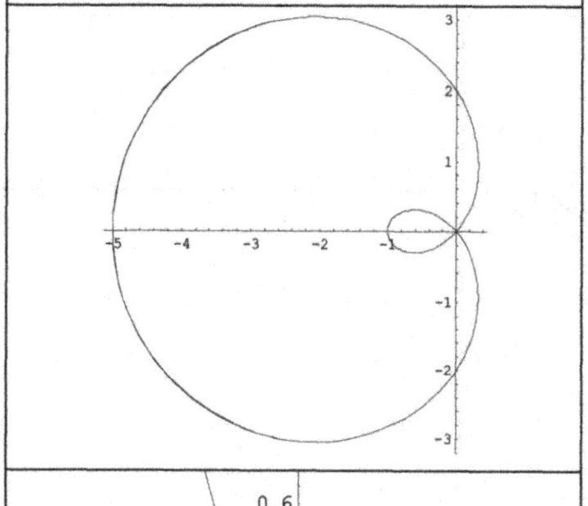

Strophoid

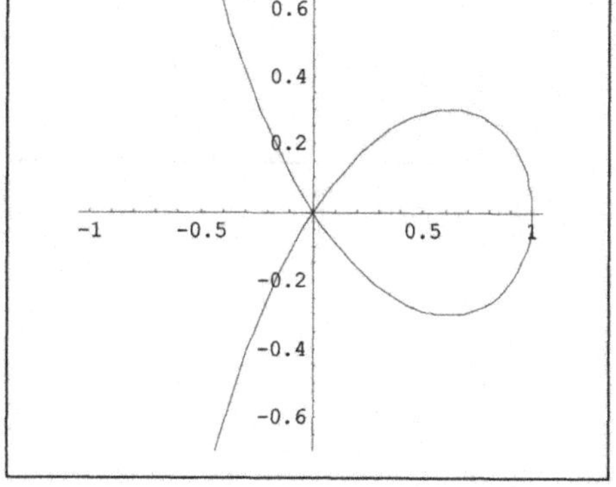

21. (continued) Cissoid of Diocles

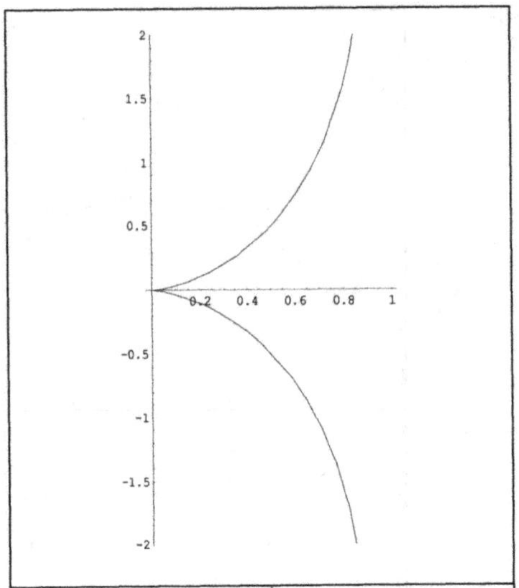

Conchoid of Nicodemus

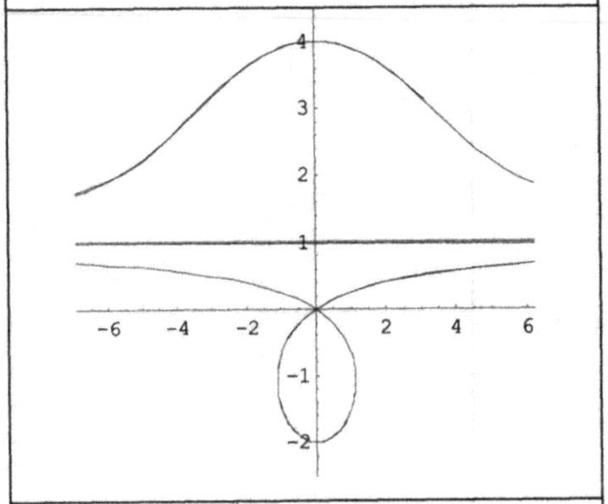

Spiral of Archimedes

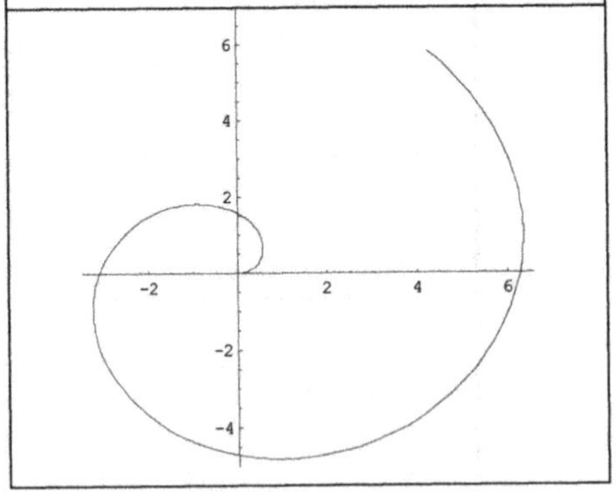

21. (continued) Hyperbolic spiral (with $a = 1$ and θ from 0 3.5π)

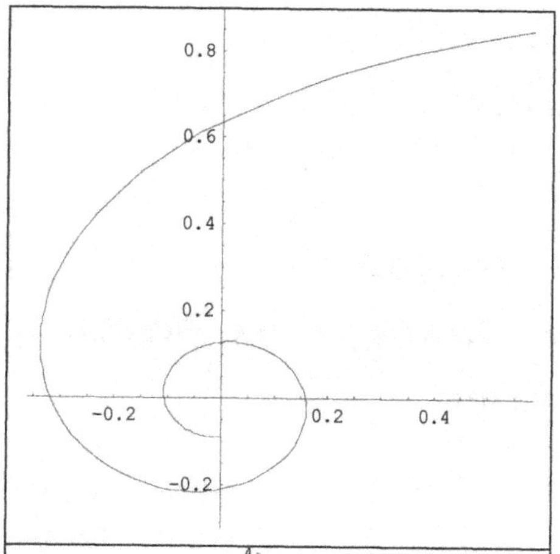

Parabolic spiral

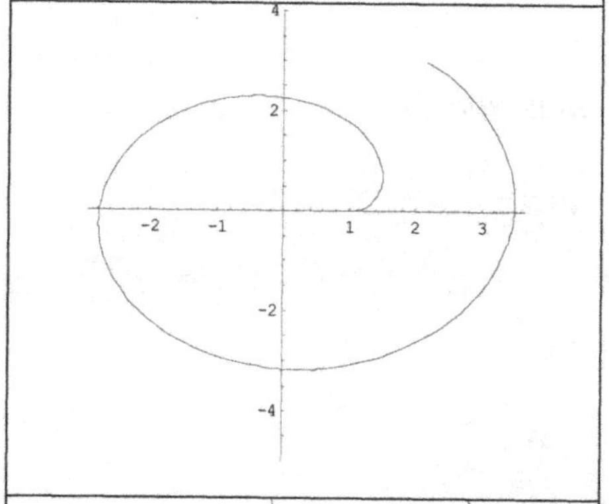

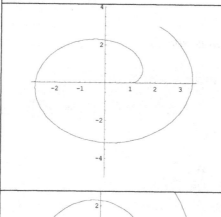

Logarithmic spiral

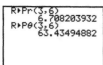

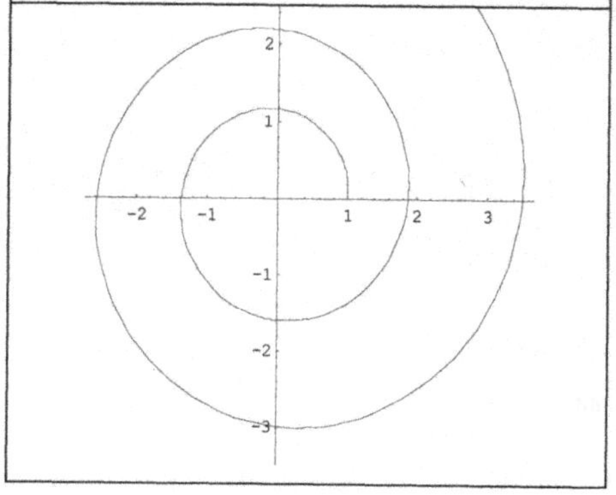

Transforming Between Rectangular and Polar Coordinates

23. (3.00, 6.00)

$$r = \sqrt{(3.00)^2 + (6.00)^2} \Rightarrow \sqrt{45} \Rightarrow 6.7082 = 6.71$$

$$\phi = \arctan\frac{6.00}{3.00} \Rightarrow 63.43494° = 63.4°$$

Thus, $(3.00, 6.00) \Rightarrow (6.71, 63.4°)$

```
R▸Pr(3,6)
        6.708203932
R▸Pθ(3,6)
        63.43494882
```

$r=\sqrt{(4.00)^2+(3.00)^2} \Rightarrow \sqrt{25}=5$

$\phi=\arctan\dfrac{3.00}{4.00} \Rightarrow 36.86989° = 36.9°$

Thus, $(4.00, 3.00) \Rightarrow (5, 36.9°)$

27. (–4.80, –5.90)

$r=\sqrt{(-4.80)^2+(-5.90)^2} \Rightarrow \sqrt{57.85} \Rightarrow 7.6059 = 7.61$

$\phi=\arctan\dfrac{-5.90}{-4.80} \Rightarrow 50.29008° = 51.0°$ in Quadrant III $\Rightarrow \phi = 180° + 51.0° = 231°$

Thus, $(-4.80, -5.90) \Rightarrow (7.61, 231°)$

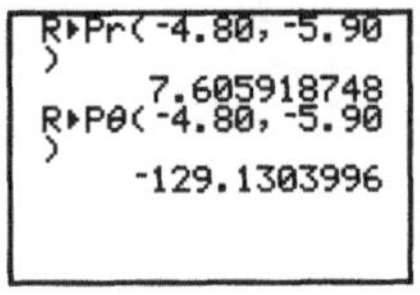

Now take $360 - 129.1303996 \Rightarrow 230.8696 = 231°$

29. (–312, –509)

$r=\sqrt{(-312)^2+(-509)^2} \Rightarrow \sqrt{356{,}425} \Rightarrow 597.0134 = 597$

$\phi=\arctan\dfrac{-509}{-312} \Rightarrow 58.49309° = 58.49°$ in Quadrant III $\Rightarrow \phi = 180° + 58.49° \Rightarrow 238.49° = 238°$

Thus, $(-312, -509) \Rightarrow (597, 238°)$

31. (5.00, 47.0°)

$x = 5.00 \cos 47.0° \Rightarrow 3.4099 = 3.41$

$y = 5.00 \sin 47.0° \Rightarrow 3.65676 = 3.66$

Thus, $(5.00, 47.0°) \Rightarrow (3.41, 3.66)$

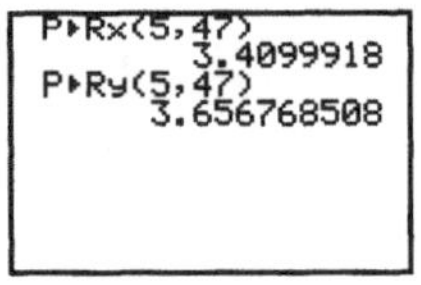

33. (445, 312°)

$x = 445 \cos 312° \Rightarrow 297.76311 = 298$

$y = 445 \sin 312° \Rightarrow -330.69944 = -331$

Thus, $(445, 312°) \Rightarrow (298, -331)$

35. $\left(-4.00, \dfrac{3\pi}{4}\right)$

$x = -4.00 \cos\dfrac{3\pi}{4} \Rightarrow 2.8282 = 2.83$

$y = -4.00 \sin\dfrac{3\pi}{4} \Rightarrow -2.82842 = -2.83$

Thus, $\left(-4.00, \dfrac{3\pi}{4}\right) \Rightarrow (2.83, -2.83)$

37. (15.0, –35.0°)

$x = 15.0 \cos -35° \Rightarrow 12.28728 = 12.3$

$y = 15.0 \sin -35° \Rightarrow -8.60364 = -8.60$

Thus, $(15.0, -35.0°) \Rightarrow (12.3, -8.60)$

39. $\left(-9.80, -\dfrac{\pi}{5}\right)$

$x = -9.80\cos -\dfrac{\pi}{5} \Rightarrow -7.92836 = -7.93$

$y = -9.80\sin -\dfrac{\pi}{5} \Rightarrow 5.76029 = 5.76$

Thus, $\left(-9.80, -\dfrac{\pi}{5}\right) \Rightarrow (-7.93, 5.76)$

Transforming an Equation

41. $r = 2 \sin\theta$

$r = 2\sin\theta \Rightarrow r^2 = 2r\sin\theta$ [multiply both sides by r]

$x^2 + y^2 = 2r\sin\theta$

$x^2 + y^2 = 2y$ [since $r\sin\theta = y$]

43. $r^2 = 1 - \tan\theta$

$x^2 + y^2 = 1 - \tan\theta$ [since $r^2 = x^2 + y^2$]

$x^2 + y^2 = 1 - \dfrac{y}{x}$

45. $r^2 = 4 - r\cos\theta$

$x^2 + y^2 = 4 - x$ [since $r^2 = x^2 + y^2$ and $r\cos\theta = x$]

47. $y = -3$

$r\sin\theta = -3$ [since $y = r\sin\theta$]

$r\sin\theta + 3 = 0$

$r\sin\theta = -3$

$r = \dfrac{-3}{\sin\theta}$

$r = \dfrac{-3}{\dfrac{1}{\csc\theta}}$ $\left[\text{since } \sin\theta = \dfrac{1}{\csc\theta}\right]$

$r = -3\csc\theta$

49. $x^2 + y^2 = 1$

$r^2 = 1$ [since $x^2 + y^2 = r^2$]

$r = 1$ [take square root of both sides]

49. $y = x^2$

$r\sin\theta = r^2 - y^2$ [since $x^2 = r^2 - y^2$]

$r\sin\theta = r^2 - r^2\sin^2\theta$ [since $y = r\sin\theta \Rightarrow y^2 = r^2\sin^2\theta$]

$r\sin\theta = r^2(1 - \sin^2\theta)$

$\sin\theta = r(1 - \sin^2\theta)$

$\sin\theta = r\cos^2\theta$ [since $\cos^2\theta + \sin^2 = 1 \Rightarrow \cos^2\theta = 1 - \sin^2\theta$]

Applications

53.

r (in.)	θ (degrees)	x (in.)	y (in.)
4.25	0	4.25	0.00
4.25	15	4.11	1.10
4.25	30	3.68	2.13
4.25	45	3.01	3.01
4.25	60	2.13	3.68

4.25	75	1.10	4.11
4.25	90	0.00	4.25

CHAPTER 15 REVIEW PROBLEMS

1. $y = 3\sin 2x$

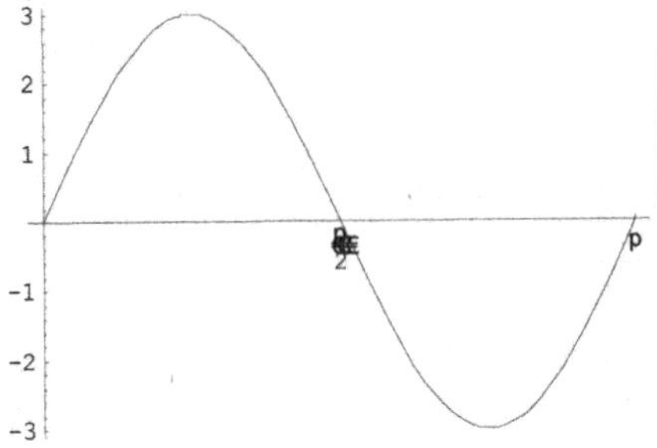

Period = π
amplitude = 3
Phase shift = 0

3. $y = 1.5\sin\left(3x + \frac{\pi}{2}\right)$

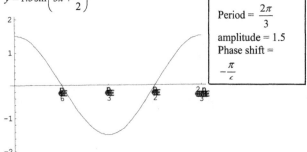

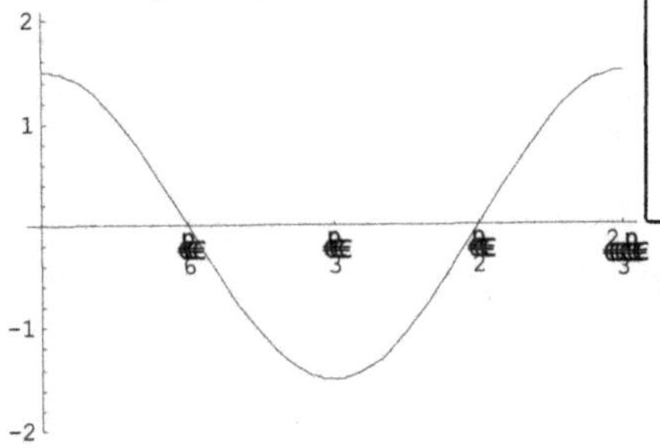

Period = $\frac{2\pi}{3}$
amplitude = 1.5
Phase shift =
$-\frac{\pi}{6}$

5. $y = 2.5\sin\left(4x + \frac{2\pi}{9}\right)$

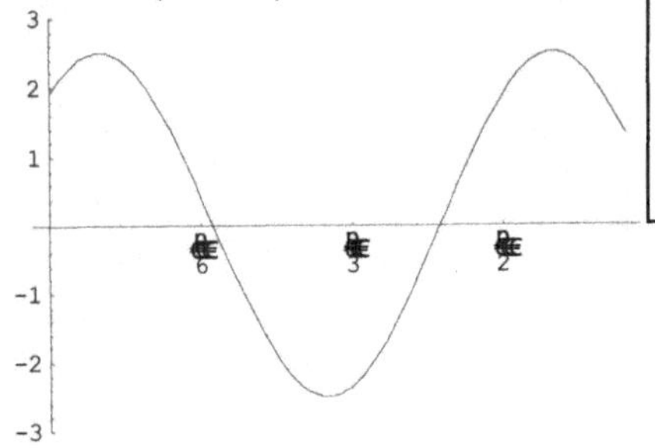

Period = $\frac{\pi}{2}$
amplitude = 2.5
Phase shift = $-\frac{\pi}{18}$

7. $y = 5\sin(bx + c)$

$$P = \frac{2\pi}{b} = 3\pi;\quad 3\pi b = 2\pi;\quad 3b = 2;\quad b = \frac{2}{3}$$

$$\text{phase shift} = -\frac{c}{b} = -\frac{c}{\frac{2}{3}} \Rightarrow -\frac{3c}{2} = -\frac{\pi}{6};\quad 2\pi = 18c;\quad c = \frac{2\pi}{18} = \frac{\pi}{9}$$

$$\text{Therefore, } y = 5\sin\left(\frac{2x}{3} + \frac{\pi}{9}\right)$$

9. (–2.2, 228°)

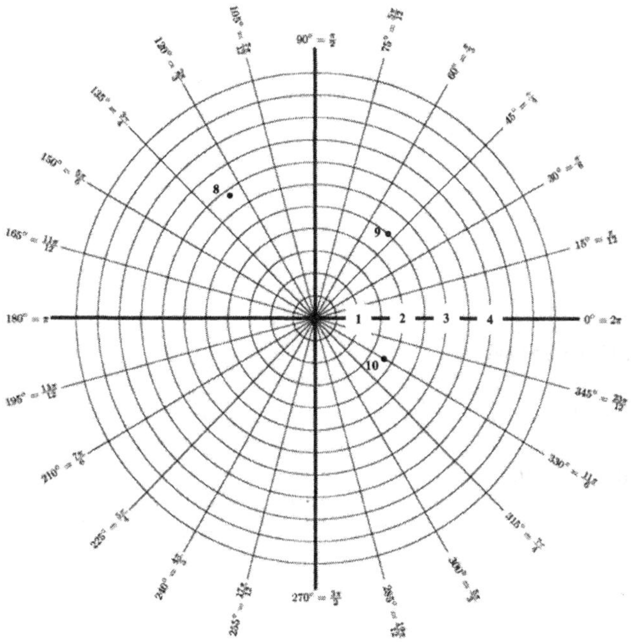

11. $r^2 = \cos 2\theta$

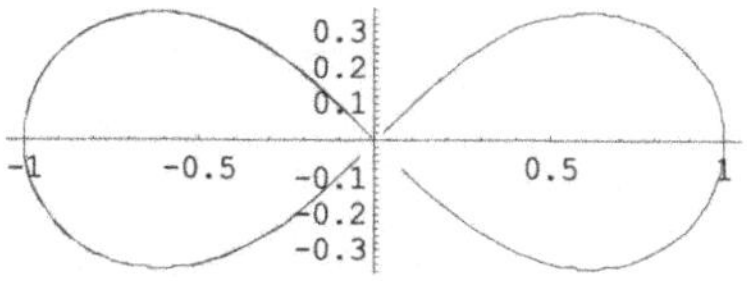

13. (7, 3)

$r = \sqrt{x^2 + y^2} \Rightarrow \sqrt{(7)^2 + (3)^2} \Rightarrow \sqrt{58} \Rightarrow 7.61577 = 7.62$

$\phi = \arctan \frac{y}{x} \Rightarrow \arctan \frac{3}{7} \Rightarrow 23.19859° = 23.2°$

Thus, $(7, 3) \Rightarrow (7.62, 23.2°)$

15. (–24, –52)

$\sqrt{(-24)^2 + (-52)^2} \Rightarrow \sqrt{3280} \Rightarrow 57.27128 = 57$

$\phi = \arctan \frac{-52}{-24} \Rightarrow 65.22845° \Rightarrow 65.22845° + 180° \Rightarrow 245.22465° = 245°$

Thus, $(-24, -52) \Rightarrow (57, 245°)$

17. $\left(-65, \frac{\pi}{9}\right)$

$x = -65 \cos \frac{\pi}{9} \Rightarrow -61.08002 = -61$

$y = -65 \sin \frac{\pi}{9} \Rightarrow -22.2313 = -22$

Thus, $\left(-65, \frac{\pi}{9}\right) \Rightarrow (-61, -22)$

19. $r(1 - \cos \theta) = 2$

$$r - r\cos\theta = 2$$
$$r - x = 2$$
$$\sqrt{x^2 + y^2} = x + 2$$
$$x^2 + y^2 = x^2 + 4x + 4$$
$$y^2 = 4x + 4$$

21. $x - 3y = 2$
$$r\cos\theta - 3(r\sin\theta) = 2$$
$$r(\cos\theta - 3\sin\theta) = 2$$

23. $f = \frac{1}{P} \Rightarrow \frac{1}{2.5} = 0.4 \text{ Hz}$
$$\omega = 2\pi f \Rightarrow 2\pi(0.4) \Rightarrow 2.51327 = 2.51 \text{ rad/s}$$

25. $P = \frac{2\pi}{\omega} \Rightarrow \frac{2\pi}{44.8} \Rightarrow 0.14024 = 0.14 \text{ s}$
$$f = \frac{1}{P} \Rightarrow \frac{1}{0.14024} \Rightarrow 7.13063 = 7.13 \text{ Hz}$$

27. $f = \frac{500 \text{ cycles}}{2 \text{ s}} = 250 \text{ Hz}$

29. $f = 82 \text{ Hz} = \frac{\omega}{2\pi}$
$$\omega = 82(2\pi) \Rightarrow 515.22119 = 515 \text{ rad/s}$$
Thus, $i = 92.6\sin(515t + 28.3°)$ mA

31. $v_{max} = 27.4 \text{ V}$
$$P = \frac{2\pi}{\omega} \Rightarrow \frac{2\pi}{736} \Rightarrow 0.0085369 \text{ s} = 0.00854 \text{ s} = 8.54 \text{ ms}$$
$$f = \frac{1}{P} \Rightarrow \frac{1}{8.54 \text{ ms}} \Rightarrow 0.117096 = 117 \text{ Hz}$$
$$\phi = 37° \Rightarrow 37\left(\frac{2\pi \text{ rad}}{360°}\right) \Rightarrow 0.6457717 = 0.646 \text{ rad}$$
$$v(0.25) = 27.4\sin[(736)(0.25) + 0.646] \Rightarrow 27.4\sin 184.646 = 17.817 \text{ V} \Rightarrow 17.8 \text{ V}$$

Chapter 16: Trigonometric Identities and Equations

Exercise 1 ◊ Fundamental Identities

1. $\tan x - \sec x \Rightarrow \dfrac{\sin x}{\cos x} - \dfrac{1}{\cos x} = \dfrac{\sin x - 1}{\cos x}$

3. $\tan\theta \sec\theta \Rightarrow \dfrac{\sin\theta}{\cos\theta}\left(\dfrac{1}{\sin\theta}\right) \Rightarrow \dfrac{\sin\theta}{\cos\theta\sin\theta} = \dfrac{1}{\cos\theta}$

5. $\dfrac{\tan\theta}{\csc\theta} + \dfrac{\sin\theta}{\tan\theta} \Rightarrow \dfrac{\frac{\sin\theta}{\cos\theta}}{\frac{1}{\sin\theta}} + \dfrac{\sin\theta}{\frac{\sin\theta}{\cos\theta}} \Rightarrow \dfrac{\sin^2\theta}{\cos\theta} + \cos\theta \Rightarrow \dfrac{\sin^2\theta}{\cos\theta} + \dfrac{\cos^2\theta}{\cos\theta} \Rightarrow \dfrac{\sin^2\theta + \cos^2\theta}{\cos\theta} = \dfrac{1}{\cos\theta}$

7. $1 - \sec^2 x \Rightarrow 1 - \left(1 + \tan^2 x\right) = -\tan^2 x$

9. $\dfrac{\cos\theta}{\cot\theta} \Rightarrow \dfrac{\cos\theta}{\frac{\cos\theta}{\sin\theta}} \Rightarrow \dfrac{\cos\theta\sin\theta}{\cos\theta} = \sin\theta$

11. $\tan\theta\csc\theta \Rightarrow \dfrac{\sin\theta}{\cos\theta}\left(\dfrac{1}{\sin\theta}\right) \Rightarrow \dfrac{\sin\theta}{\cos\theta\sin\theta} \Rightarrow \dfrac{1}{\cos\theta} = \sec\theta$

13. $\sec x \sin x \Rightarrow \left(\dfrac{1}{\cos x}\right)\sin x \Rightarrow \dfrac{\sin x}{\cos x} = \tan x$

15. $\csc\theta\tan\theta - \tan\theta\sin\theta \Rightarrow \tan\theta\left(\csc\theta - \sin\theta\right) \Rightarrow \dfrac{\sin\theta}{\cos\theta}\left(\dfrac{1}{\sin\theta} - \sin\theta\right) \Rightarrow \dfrac{\sin\theta}{\cos\theta\sin\theta} - \dfrac{\sin\theta\sin\theta}{\cos\theta}$

$\Rightarrow \dfrac{1 - \sin^2\theta}{\cos\theta} \Rightarrow \dfrac{\cos^2\theta}{\cos\theta} = \cos\theta$

17. $\cot\theta\tan^2\theta\cos\theta \Rightarrow \left(\dfrac{\cos\theta}{\sin\theta}\right)\left(\dfrac{\sin\theta}{\cos\theta}\right)^2\cos\theta \Rightarrow \dfrac{\cos^2\theta\sin^2\theta}{\cos^2\theta\sin\theta} \Rightarrow \dfrac{\sin^2\theta}{\sin\theta} = \sin\theta$

19. $\dfrac{\sin\theta}{\cos\theta\tan\theta} \Rightarrow \dfrac{\sin\theta}{\cos\theta\left(\frac{\sin\theta}{\cos\theta}\right)} \Rightarrow \dfrac{\sin\theta}{\sin\theta} = 1$

21. $\dfrac{1}{\sec^2 x} + \dfrac{1}{\csc^2 x} \Rightarrow \cos^2 x + \sin^2 x = 1$

23. $\csc x - \cot x\cos x \Rightarrow \dfrac{1}{\sin x} - \left(\dfrac{\cos x}{\sin x}\right)\cos x \Rightarrow \dfrac{1}{\sin x} - \dfrac{\cos^2 x}{\sin x} \Rightarrow \dfrac{1 - \cos^2 x}{\sin x} \Rightarrow \dfrac{\sin^2 x}{\sin x} = \sin x$

25. $\dfrac{\sec x - \csc x}{1 - \cot x} \Rightarrow \dfrac{\frac{1}{\cos x} - \frac{1}{\sin x}}{1 - \cot x} \Rightarrow \dfrac{\frac{\sin x}{\cos x\sin x} - \frac{\cos x}{\sin x\cos x}}{1 - \frac{\cos x}{\sin x}} \Rightarrow \dfrac{\frac{\sin x - \cos x}{\sin x\cos x}}{\frac{\sin x - \cos x}{\sin x}} \Rightarrow \dfrac{\sin x}{\sin x\cos x} \Rightarrow \dfrac{1}{\cos x} = \sec x$

27. $\sec^2 x\left(1 - \cos^2 x\right) \Rightarrow \dfrac{1}{\cos^2 x}\left(\sin^2 x\right) \Rightarrow \dfrac{\sin^2 x}{\cos^2 x} = \tan^2 x$

29. $\cos\theta\sec\theta - \dfrac{\sec\theta}{\cos\theta} \Rightarrow \cos\theta\left(\dfrac{1}{\cos\theta}\right) - \dfrac{\frac{1}{\cos\theta}}{\cos\theta} \Rightarrow 1 - \dfrac{1}{\cos^2\theta} \Rightarrow 1 - \sec^2\theta = -\tan^2\theta$

31. $\tan x\cos x = \sin x$

$$\left(\frac{\sin x}{\cos x}\right)\cos x = \sin x$$

$$\frac{\sin x\cos x}{\cos x} = \sin x$$

$$\sin x = \sin x$$

33. $$\frac{\sin x}{\csc x} + \frac{\cos x}{\sec x} = 1$$

$$\frac{\sin x}{\frac{1}{\sin x}} + \frac{\cos x}{\frac{1}{\cos x}} = 1$$

$$\sin x\left(\frac{\sin x}{1}\right) + \cos x\left(\frac{\cos x}{1}\right) = 1$$

$$\sin^2 x + \cos^2 x = 1$$

$$1 = 1$$

35. $$(\cos^2\theta + \sin^2\theta)^2 = 1$$

$$(1)^2 = 1$$

$$1 = 1$$

37. $$\frac{\csc\theta}{\sec\theta} = \cot\theta$$

$$\frac{\frac{1}{\sin\theta}}{\frac{1}{\cos\theta}} = \cot\theta$$

$$\left(\frac{1}{\sin\theta}\right)\left(\frac{\cos\theta}{1}\right) = \cot\theta$$

$$\frac{\cos\theta}{\sin\theta} = \cot\theta$$

$$\cot\theta = \cot\theta$$

39. $$\cos x + 1 = \frac{\sin^2 x}{1 - \cos x}$$

$$\cos x + 1 = \frac{1 - \cos^2 x}{1 - \cos x}$$

$$\cos x + 1 = \frac{(1 - \cos x)(1 + \cos x)}{1 - \cos x}$$

$$\cos x + 1 = 1 + \cos x$$

41. $$\cot^2 x - \cos^2 x = \cos^2 x\cot^2 x$$

$$\left(\frac{\cos x}{\sin x}\right)^2 - \cos^2 x = \cos^2 x\cot^2 x$$

$$\frac{\cos^2 x}{\sin^2 x} - \frac{\cos^2 x\sin^2 x}{\sin^2 x} = \cos^2 x\cot^2 x$$

$$\frac{\cos^2 x - \cos^2 x\sin^2 x}{\sin^2 x} = \cos^2 x\cot^2 x$$

$$\frac{\cos^2 x(1 - \sin^2 x)}{\sin^2 x} = \cos^2 x\cot^2 x$$

$$\frac{\cos^2 x(\cos^2 x)}{\sin^2 x} = \cos^2 x\cot^2 x$$

$$\left(\frac{\cos^2 x}{\sin^2 x}\right)\cos^2 x = \cos^2 x\cot^2 x$$

$$\cot^2 x\cos^2 x = \cos^2 x\cot^2 x$$

43. $1 = (\csc x - \cot x)(\csc x + \cot x)$

$$1 = \csc^2 x - \cot^2 x$$

$$1 = 1 + \cot^2 x - \cot^2 x$$

$$1 = 1$$

45. $\dfrac{\tan x + 1}{1 - \tan x} = \dfrac{\sin x + \cos x}{\cos x - \sin x}$

$$\frac{\frac{\sin x}{\cos x} + 1}{1 - \frac{\sin x}{\cos x}} = \frac{\sin x + \cos x}{\cos x - \sin x}$$

$$\frac{\frac{\sin x}{\cos x} + \frac{\cos x}{\cos x}}{\frac{\cos x}{\cos x} - \frac{\sin x}{\cos x}} = \frac{\sin x + \cos x}{\cos x - \sin x}$$

$$\frac{\frac{\sin x + \cos x}{\cos x}}{\frac{\cos x - \sin x}{\cos x}} = \frac{\sin x + \cos x}{\cos x - \sin x}$$

$$\frac{\sin x + \cos x}{\cos x - \sin x} = \frac{\sin x + \cos x}{\cos x - \sin x}$$

47. $\dfrac{\sin\theta + 1}{1 - \sin\theta} = (\tan\theta + \sec\theta)^2$

$$\frac{\sin\theta + 1}{1 - \sin\theta} = \left(\frac{\sin\theta}{\cos\theta} + \frac{1}{\cos\theta}\right)^2$$

$$\frac{\sin\theta + 1}{1 - \sin\theta} = \left(\frac{\sin\theta + 1}{\cos\theta}\right)^2$$

$$\frac{\sin\theta + 1}{1 - \sin\theta} = \frac{(\sin\theta + 1)(\sin\theta + 1)}{\cos^2\theta}$$

$$\frac{\sin\theta + 1}{1 - \sin\theta} = \frac{(\sin\theta + 1)(\sin\theta + 1)}{1 - \sin^2\theta}$$

$$\frac{\sin\theta + 1}{1 - \sin\theta} = \frac{(\sin\theta + 1)(\sin\theta + 1)}{(1 - \sin\theta)(1 + \sin\theta)}$$

$$\frac{\sin\theta + 1}{1 - \sin\theta} = \frac{\sin\theta + 1}{1 - \sin\theta}$$

49. $(\sec\theta - \tan\theta)(\tan\theta + \sec\theta) = 1$

$$\sec^2\theta - \tan^2\theta = 1$$

$$1 + \tan^2\theta - \tan^2\theta = 1$$

$$1 = 1$$

Applications

51. (a) Let T = the tension in the rope. Then $T\cos\theta = W$ and $T\sin\theta = W$

(b) $\dfrac{W}{F}=\dfrac{T\cos\theta}{T\sin\theta}=\cot\theta$ and $F=\cot\theta$ or $F=W\tan\theta$

53. $\dfrac{(a\cos\alpha)^2}{a^2}+\dfrac{(b\sin\alpha)^2}{b^2}=1$

$\dfrac{a^2\cos^2\alpha}{a^2}+\dfrac{b^2\sin^2\alpha}{b^2}=1$

$\cos^2\alpha+\sin^2\alpha=1$

$1=1$

55. (a) $(Z\cos\theta)^2+(Z\sin\theta)^2 \Rightarrow Z^2\cos^2\theta+Z^2\sin^2\theta \Rightarrow Z^2(\cos^2\theta+\sin^2) \Rightarrow Z^2(1)=Z^2$

(b) $\dfrac{Z\sin\theta}{Z\cos\theta} \Rightarrow \dfrac{\sin\theta}{\cos\theta}=\tan\theta$

Exercise 2 ◊ Sum or Difference of Two Angles

1. $\sin(\theta+30°) \Rightarrow \sin\theta\cos 30°+\cos\theta\sin 30° \Rightarrow \dfrac{\sqrt{3}}{2}\sin\theta+\dfrac{1}{2}\cos\theta=\dfrac{1}{2}\left(\sqrt{3}\sin\theta+\cos\theta\right)$

3. $\sin(x+60°) \Rightarrow \sin x\cos 60°+\cos x\sin 60° \Rightarrow \dfrac{1}{2}\sin x+\dfrac{\sqrt{3}}{2}\cos x=\dfrac{1}{2}\left(\sin x+\sqrt{3}\cos x\right)$

5. $\cos\left(x+\dfrac{\pi}{2}\right)=\cos x\cos\dfrac{\pi}{2}-\sin x\sin\dfrac{\pi}{2}=-\sin x\left[\text{since } \cos\dfrac{\pi}{2}=0 \text{ and } \sin\dfrac{\pi}{2}=1\right]$

7. $\sin(\theta+2\varphi)=\sin\theta\cos 2\varphi+\cos\theta\sin 2\varphi$

9. $\cos 2x\cos 9x+\sin 2x\sin 9x \Rightarrow \cos(2x-9x) \Rightarrow \cos(-7x)=\cos 7x$

11. $\sin 3\theta\cos 2\theta-\cos 3\theta\sin 2\theta \Rightarrow \sin(3\theta-2\theta)=\sin\theta$

13. $\sin(\alpha+\beta)+\sin(\alpha-\beta)=2\sin\alpha\cos\beta$

$\sin\alpha\cos\beta+\cos\alpha\sin\beta+\sin\alpha\cos\beta-\cos\alpha\sin\beta=2\sin\alpha\cos\beta$

$\sin\alpha\cos\beta+\sin\alpha\cos\beta=2\sin\alpha\cos\beta$

$2\sin\alpha\cos\beta=2\sin\alpha\cos\beta$

15. $\sin\left(x+\dfrac{\pi}{6}\right)-\sin\left(x-\dfrac{\pi}{6}\right)=\cos x$

$\sin x\cos\dfrac{\pi}{6}+\cos x\sin\dfrac{\pi}{6}-\left(\sin x\cos\dfrac{\pi}{6}-\cos x\sin\dfrac{\pi}{6}\right)=\cos x$

$\sin x\cos\dfrac{\pi}{6}+\cos x\sin\dfrac{\pi}{6}-\sin x\cos\dfrac{\pi}{6}+\cos x\sin\dfrac{\pi}{6}=\cos x$

$\cos x\sin\dfrac{\pi}{6}+\cos x\sin\dfrac{\pi}{6}=\cos x$

$\dfrac{\cos x}{2}+\dfrac{\cos x}{2}=\cos x$

$\dfrac{2\cos x}{2}=\cos x$

$\cos x=\cos x$

17. $\cos(x+60°)+\cos(60°-x)=\cos x$

$\cos 60°\cos x-\sin x\sin 60°+\cos 60°\cos x+\sin 60°\sin x=\cos x$

$\cos 60°\cos x+\cos 60°\cos x=\cos x$

$$\frac{\cos x}{2}+\frac{\cos x}{2}=\cos x$$

$$\frac{2\cos x}{2}=\cos x$$

$$\cos x=\cos x$$

19. $\dfrac{1+\tan x}{1-\tan x}=\tan\left(\dfrac{\pi}{4}+x\right)$

$$\frac{1+\tan x}{1-\tan x}=\frac{\tan\frac{\pi}{4}+\tan x}{1-\tan\frac{\pi}{4}\tan x}$$

$$\frac{1+\tan x}{1-\tan x}=\frac{1+\tan x}{1-(1)\tan x}$$

$$\frac{1+\tan x}{1-\tan x}=\frac{1+\tan x}{1-\tan x}$$

Applications

21. (b) $P\cos\theta-W\sin\theta=f=N\tan\varphi=[P\sin\theta+W\cos\theta]\tan\varphi$

$$P(\cos\theta-\sin\theta\tan\phi)=W\sin\theta+W\cos\theta\tan\phi$$

$$P=\frac{W(\sin\theta+\cos\theta\tan\phi)}{\cos\theta-\sin\theta\tan\phi}$$

$$P=W\left[\frac{\dfrac{\sin\theta+\cos\theta\tan\phi}{\cos\theta}}{\dfrac{\cos\theta-\sin\theta\tan\phi}{\cos\theta}}\right]$$

$$P=W\left[\frac{\tan\theta+\tan\phi}{1-\tan\theta\tan\phi}\right]$$

$$P=W\tan(\theta+\phi)$$

Exercise 3 ◊ Functions of Double Angles and Half-Angles

Double Angles

1. $2\sin^2 x+\cos 2x\Rightarrow 2\sin^2 x+(\cos^2 x-\sin^2 x)\Rightarrow\cos^2 x+\sin^2 x=1$

3. $\dfrac{2\tan x}{1+\tan^2 x}\Rightarrow\dfrac{2\tan x}{\sec^2 x}\Rightarrow\dfrac{2\left(\dfrac{\sin x}{\cos x}\right)}{\dfrac{1}{\cos^2 x}}\Rightarrow\left(\dfrac{2\sin x}{\cos x}\right)\left(\dfrac{\cos^2 x}{1}\right)\Rightarrow 2\sin x\cos x=\sin 2x$

5. $\dfrac{2\tan\theta}{1-\tan^2\theta}=\tan 2\theta$

$$\frac{\tan\theta+\tan\theta}{1-\tan\theta\tan\theta}=\tan 2\theta$$

$$\tan(\theta+\theta)=\tan 2\theta$$

$$\tan 2\theta=\tan 2\theta$$

7. $\tan\theta+\cot\theta=2\csc 2\theta$

$$\frac{\sin\theta}{\cos\theta}+\frac{\cos\theta}{\sin\theta}=2\csc 2\theta$$

$$\frac{\sin^2\theta}{\cos\theta\sin\theta}+\frac{\cos^2\theta}{\sin\theta\cos\theta}=2\csc 2\theta$$

$$\frac{\sin^2\theta+\cos^2\theta}{\cos\theta\sin\theta}=2\csc 2\theta$$

$$\frac{2}{2\cos\theta\sin\theta}=2\csc 2\theta$$

$$\frac{2}{\sin 2\theta}=2\csc 2\theta$$

$$2\csc 2\theta=2\csc 2\theta$$

9. $$\frac{1+\cot^2 x}{\cot^2 x-1}=\sec 2x$$

$$\frac{1+\left(\frac{\cos x}{\sin x}\right)^2}{\left(\frac{\cos x}{\sin x}\right)^2-1}=\sec 2x$$

$$\frac{\frac{\sin^2 x}{\sin^2 x}+\frac{\cos^2 x}{\sin^2 x}}{\frac{\cos^2 x}{\sin^2 x}-\frac{\sin^2 x}{\sin^2 x}}=\sec 2x$$

$$\frac{\frac{\sin^2 x+\cos^2 x}{\sin^2 x}}{\frac{\cos^2 x-\sin^2 x}{\sin^2 x}}=\sec 2x$$

$$\frac{\sin^2 x+\cos^2 x}{\cos^2 x-\sin^2 x}=\sec 2x$$

$$\frac{1}{\cos^2 x-\sin^2 x}=\sec 2x$$

$$\frac{1}{\cos 2x}=\sec 2x$$

$$\sec 2x=\sec 2x$$

11. $$\frac{2\cos 2x}{\sin 2x-2\sin^2 x}=1+\cot x$$

$$\frac{2\left(\cos^2 x-\sin^2 x\right)}{2\sin x\cos x-2\sin^2 x}=1+\cot x$$

$$\frac{2(\cos x-\sin x)(\cos x+\sin x)}{2\sin x(\cos x-\sin x)}=1+\cot x$$

$$\frac{(\cos x+\sin x)}{\sin x}=1+\cot x$$

$$\frac{\cos x}{\sin x}+\frac{\sin x}{\sin x}=1+\cot x$$

$$\cot x+1=1+\cot x$$

13. $$\frac{\cot^2 x-1}{2\cot x}=\cot 2x$$

$$\frac{\frac{1}{\tan^2 x}-1}{\frac{2}{\tan x}}=\cot 2x$$

$$\frac{\frac{1}{\tan^2 x} - \frac{\tan^2 x}{\tan^2 x}}{\frac{2}{\tan x}} = \cot 2x$$

$$\frac{\frac{1-\tan^2 x}{\tan^2 x}}{\frac{2}{\tan x}} = \cot 2x$$

$$\frac{1-\tan^2 x}{2\tan x} = \cot 2x$$

$$\frac{1}{\tan 2x} = \cot 2x$$

$$\cot 2x = \cot 2x$$

Half-Angles

15. $4\cos^2\frac{x}{2}\sin^2\frac{x}{2} = 1-\cos^2 x$

$$4\left(\frac{1+\cos x}{2}\right)\left(\frac{1-\cos x}{2}\right) = 1-\cos^2 x$$

$$4\left(\frac{1-\cos^2 x}{4}\right) = 1-\cos^2 x$$

$$1-\cos^2 x = 1-\cos^2 x$$

17. $\dfrac{\cos^2\frac{\theta}{2} - \cos\theta}{\sin^2\frac{\theta}{2}} = 1$

$$\frac{\frac{1+\cos\theta}{2} - \cos\theta}{\frac{1-\cos\theta}{2}} = 1$$

$$\frac{\frac{1+\cos\theta - 2\cos\theta}{2}}{\frac{1-\cos\theta}{2}} = 1$$

$$\frac{\frac{1-\cos\theta}{2}}{\frac{1-\cos\theta}{2}} = 1$$

$$1 = 1$$

19. $\left(\cos\frac{x}{2} + \sin\frac{x}{2}\right)^2 = \sin x + 1$

$$\cos^2\frac{x}{2} + 2\sin\frac{x}{2}\cos\frac{x}{2} + \sin^2\frac{x}{2} = \sin x + 1$$

$$1 + 2\sin\frac{x}{2}\cos\frac{x}{2} = \sin x + 1$$

$$1 + \sin 2\left(\frac{x}{2}\right) = \sin x + 1$$

$$1 + \sin x = \sin x + 1$$

21. $$\frac{1-\tan^2\frac{\theta}{2}}{1+\tan^2\frac{\theta}{2}}=\cos\theta$$

$$\frac{1-\frac{1-\cos\theta}{1+\cos\theta}}{1+\frac{1-\cos\theta}{1+\cos\theta}}=\cos\theta$$ [this follows by squaring both sides of Formula 136c, Tnagent of Half an Angle]

$$\frac{\frac{1+\cos\theta}{1+\cos\theta}-\frac{1-\cos\theta}{1+\cos\theta}}{\frac{1+\cos\theta}{1+\cos\theta}+\frac{1-\cos\theta}{1+\cos\theta}}=\cos\theta$$

$$\frac{\frac{1+\cos\theta-(1-\cos\theta)}{1+\cos\theta}}{\frac{1+\cos\theta+(1-\cos\theta)}{1+\cos\theta}}=\cos\theta$$

$$\frac{2\cos\theta}{2}=\cos\theta$$

$$\cos\theta=\cos\theta$$

Applications

23. (a) Set $y = 0$ and solve $t\left[v_0\sin\theta-\frac{1}{2}gt\right]=0$. This gives us $t = 0$ and $t=\frac{2v_0\sin\theta}{g}$.

(b) $x=(v_0\cos\theta)\left(\frac{2v_0\sin\theta}{g}\right)\Rightarrow\frac{v_0^2(2\sin\theta\cos\theta)}{g}=\frac{v_0^2\sin 2\theta}{g}$

Exercise 4 ◊ Evaluating a Trigonometric Expression

1. 5.27 sin 45.8° – 1.73 ⇒ 2.04811 = 2.05

3. 3.72(sin 28.3° + cos 72.3°) ⇒ 2.89461 = 2.89

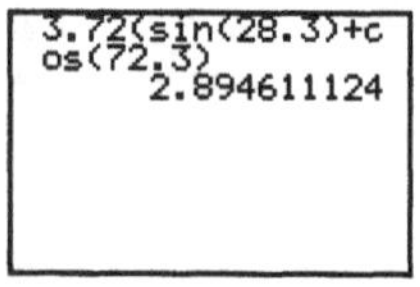

5. 2.84(5.28 cos 2 – 2.82) + 3.35 ⇒ –10.899 = –10.9

7. sin 35° + cos 35° ⇒ 1.392728 = 1.39

9. cos 270° cos 150° + sin 270° sin 150° = –0.500

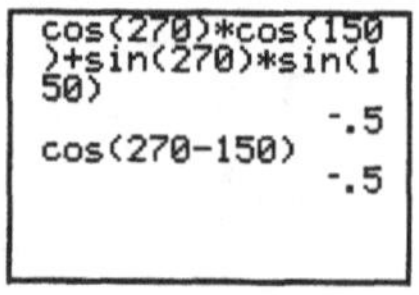

11. $\sin^2 75°$ ⇒ 0.933012 = 0.933

13. $(\cos^2 206° + \sin 206°)^2$ ⇒ 0.13650 = 0.137

Applications

15. $y=125\tan 35.5°-\frac{16.1(125)^2}{376^2}\left(\sec^2 35.5°\right)$

$y = 89.16163 - 1.77938(1.50878)$

$y = 86.4769 \Rightarrow 86.5 \text{ ft}$

17. $T \cong 2\pi\sqrt{\frac{1.25}{32.2}\left(1+\frac{1}{4}\sin^2\frac{7.83°}{2}-\frac{9}{64}\sin^4\frac{7.83°}{2}\right)}$

$T \cong 2\pi\sqrt{0.03881(1+0.00116542-0.0000030559)}$

$T \cong 2\pi\sqrt{0.038855111}$

$T \cong 2\pi(0.197117)$

$T \cong 1.2385$

$T \cong 1.24 \text{ s}$

$x = 2.84056 + \sqrt{76.5625 - 2.6896}$

19. Miter angle: $\mu = \tan^{-1}\left(\cos\theta\,\tan\frac{\gamma}{2}\right)$; bevel angle: $\beta = \sin^{-1}\left(\sin\theta\sin\frac{\gamma}{2}\right)$

(a) for $\theta = 45°$ and $\gamma = 90°$

$$\mu = \tan^{-1}\left(\cos 45°\,\tan\frac{90°}{2}\right) \Rightarrow 35.264 = 35.3°$$

$$\beta = \sin^{-1}\left(\sin 45°\sin\frac{90°}{2}\right) \Rightarrow 30°$$

(b) for $\theta = 45°$ and $\gamma = 125°$

$$\mu = \tan^{-1}\left(\cos 45°\,\tan\frac{125°}{2}\right) \Rightarrow 53.640 = 53.6°$$

$$\beta = \sin^{-1}\left(\sin 45°\sin\frac{125°}{2}\right) \Rightarrow 38.845 = 38.8°$$

21. Since a hexagon has six sides, $\gamma = \frac{360}{6} = 60°$ with $\theta = 18°$

Miter $\Rightarrow \mu = \tan^{-1}\left(\cos\theta\,\tan\frac{\gamma}{2}\right) \Rightarrow \tan^{-1}\left(\cos 18°\,\tan\frac{60°}{2}\right) \Rightarrow 28.771 = 28.8°$

Bevel $\Rightarrow \beta = \sin^{-1}\left(\sin\theta\sin\frac{\gamma}{2}\right) \Rightarrow \sin^{-1}\left(\sin 18°\sin\frac{60°}{2}\right) \Rightarrow 8.888 = 8.9°$

Exercise 5 ◊ Solving a Trigonometric Equation

1. $\sin x = \frac{1}{2}$

$x = \arcsin\frac{1}{2}$

$x = 30°, 150°$

3. $1 - \tan x = 0$

$\tan x = 1$

$x = \arctan 1$

$x = 45°, 225°$

5. $4\sin^2 x = 3$

$\sin^2 x = \frac{3}{4}$

$x = \arcsin \pm \sqrt{\frac{3}{4}}$

$x = \arcsin \pm \frac{\sqrt{3}}{2}$

For $\arcsin \frac{\sqrt{3}}{2}$, $x = 60°, 120°$; for $\arcsin -\frac{\sqrt{3}}{2}$, $x = 240°, 300°$

7. $3 \sin x - 1 = 2 \sin x$

$3\sin x - 2\sin x = 1$

$\sin x = 1$

$x = \arcsin 1$

$x = 90°$

9. $2 \cos^2 x = 1 + 2 \sin^2 x$

$2\cos^2 x - 2\sin^2 x = 1$

$2(\cos^2 x - \sin^2 x) = 1$

$\cos 2x = \frac{1}{2}$

$2x = \arccos \frac{1}{2}$

$2x = 60°, 300°, 420°, 660°$

$x = 30°, 150°, 210°, 330°$

11. $4 \sin^4 x = 1$

$\sin^4 x = \frac{1}{4}$

$\sin x = \pm\sqrt[4]{\frac{1}{4}}$

$\sin x = \pm\sqrt{\frac{1}{2}}$

$\sin x = \pm 0.7071$

$x = \arcsin(\pm 0.7071)$

For $\arcsin(0.7071); x = 45°, 135°$; for $\arcsin(-0.7071)$, $x = 225°, 315°$

13. $1 + \tan x = \sec^2 x$

$1 + \tan x = 1 + \tan^2 x$

$\tan x = \tan^2 x$

$\tan x - \tan^2 x = 0$

$\tan x(1 - \tan x) = 0$

$\tan x = 0;\ \tan x = 1$

$x = \arctan 0 \Rightarrow 0°$

$x = \arctan 1 \Rightarrow 45°$

$x = 0°, 45°, 180°, 225°$

15. $3 \cot x = \tan x$

$\frac{3}{\tan x} = \tan x$

$\tan^2 x = 3$

$\tan x = \pm\sqrt{3}$

$x = \arctan \pm \sqrt{3}$

For $\arctan\sqrt{3}$, $x = 60°, 300°$; for $\arctan -\sqrt{3}$, $x = 120°, 300°$

17. $3\sin\frac{x}{2} - 1 = 2\sin^2\frac{x}{2}$

$2\sin^2\frac{x}{2} - 3\sin^2\frac{x}{2} + 1 = 0$

$\left(2\sin\frac{x}{2} - 1\right)\left(\sin\frac{x}{2} - 1\right) = 0$

$2\sin\frac{x}{2} - 1 = 0;\ 2\sin\frac{x}{2} = 1 \qquad \sin\frac{x}{2} - 1 = 0;\ \sin\frac{x}{2} = 1$

$2\sin\frac{x}{2} = 1;\ \sin\frac{x}{2} = \frac{1}{2};\ x = 2\arcsin\frac{1}{2}$

$\sin\frac{x}{2} = 1;\ x = 2\arcsin 1$

$x = 60°, 180°, 300°$

19. $4\cos^2 x + 4\cos x = -1$

$4\cos^2 x + 4\cos x + 1 = 0$

$\cos x = \frac{-4 \pm \sqrt{(4)^2 - 4(4)(1)}}{2(4)}$

$\cos x = -\frac{1}{2}$

$x = \arccos\left(-\frac{1}{2}\right)$

$x = 120°, 240°$

21. $3\tan x = 4\sin^2 x \tan x$

$3\tan x - 4\sin^2 x \tan x = 0$

$\tan x\left(3 - 4\sin^2 x\right) = 0$

$\tan x = 0 \qquad 3 - 4\sin^2 x = 0;\ 4\sin^2 x = 3$

$x = \arctan 0$

$x = \arcsin \pm\frac{\sqrt{3}}{2}$

$x = 0°, 60°, 120°, 180°, 240°, 300°$

23. $\sec x = -\csc x$

$\frac{1}{\cos x} = -\frac{1}{\sin x}$

$\frac{\sin x}{\cos x} = -1$

$\tan x = -1$

$x = \arctan(-1)$

$x = 135°, 315°$

25. $\sin x = 2\sin x \cos x$

$$\sin x - 2\sin x\cos x = 0$$
$$\sin x(1 - 2\cos x) = 0$$
$$\sin x = 0;\quad 1 - 2\cos x = 0;\quad \cos x = \frac{1}{2}$$
$$x = \arcsin 0$$
$$x = \arccos\left(\frac{1}{2}\right)$$
$$x = 0°, 60°, 180°, 300°$$

Applications

27. $i = 274\sin(144t + 35°)$ mA

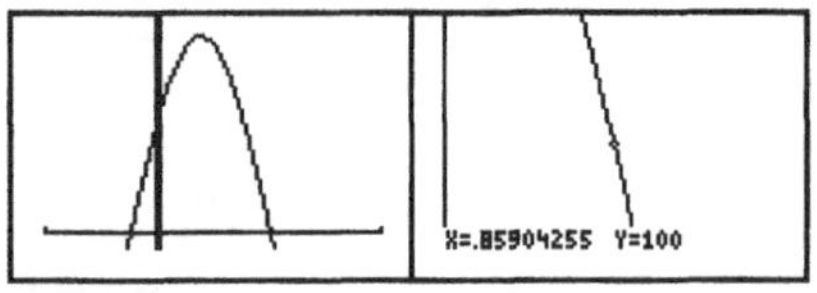

$t = 0.858$

29. $0 = 224\tan\theta - \dfrac{16.1(224)^2}{(125)^2}\sec^2\theta$

$$0 = 224\tan\theta - 51.70135\sec^2\theta$$
$$0 = 224\left(\frac{\sin\theta}{\cos\theta}\right) - 51.70135\left(\frac{1}{\cos^2\theta}\right)$$
$$0 = 224\left(\frac{\sin\theta\cos\theta}{\cos^2\theta}\right) - \frac{51.70135}{\cos^2\theta}$$
$$0 = \frac{224\sin\theta\cos\theta - 51.70135}{\cos^2\theta}$$
$$0 = 224\sin\theta\cos\theta - 51.70135$$
$$51.70135 = 224\sin\theta\cos\theta$$
$$\frac{51.70135}{112} = 2\sin\theta\cos\theta$$
$$\frac{51.70135}{112} = 2\sin\theta$$
$$2\theta = \arcsin 0.461619 \Rightarrow 2\theta = 27.4916° \Rightarrow \theta = 13.7458° = 13.7°$$
$$2\theta = 180° - 27.4916° \Rightarrow 2\theta = 152.5084° \Rightarrow \theta = 76.2542° = 76.3°$$

CHAPTER 16 REVIEW PROBLEMS

1. $\dfrac{\cot\alpha\cot\beta + 1}{\cot\beta - \cot\alpha} = \cot(\alpha - \beta)$

$$\frac{\left(\frac{1}{\tan\alpha}\right)\left(\frac{1}{\tan\beta}\right) + 1}{\frac{1}{\tan\beta} - \frac{1}{\tan\alpha}} = \frac{1}{\tan(\alpha - \beta)}$$
$$\frac{1 + \tan\alpha\tan\beta}{\tan\alpha - \tan\beta} = \frac{1 + \tan\alpha\tan\beta}{\tan\alpha - \tan\beta}$$

3. $\sin\theta\cot\dfrac{\theta}{2} = \cos\theta + 1$

$$\sin\theta\left(\frac{1}{\tan\frac{\theta}{2}}\right)=\cos\theta+1$$

$$\sin\theta\left(\frac{1+\cos\theta}{\sin\theta}\right)=\cos\theta+1$$

$$1+\cos\theta==\cos\theta+1$$

5. $$\frac{\sin\theta\sec\theta}{\tan\theta}=1$$

$$\frac{\sin\theta\left(\frac{1}{\cos\theta}\right)}{\frac{\sin\theta}{\cos\theta}}=1$$

$$\frac{\frac{\sin\theta}{\cos\theta}}{\frac{\sin\theta}{\cos\theta}}=1$$

$$1=1$$

7. $$\frac{\sec\theta}{\sec\theta-1}+\frac{\sec\theta}{\sec\theta+1}=2\csc^2\theta$$

$$\frac{\sec\theta(\sec\theta+1)}{(\sec\theta-1)(\sec\theta+1)}+\frac{\sec\theta(\sec\theta-1)}{(\sec\theta+1)(\sec\theta-1)}=2\csc^2\theta$$

$$\frac{\sec\theta(\sec\theta+1)+\sec\theta(\sec\theta-1)}{(\sec\theta-1)(\sec\theta+1)}=2\csc^2\theta$$

$$\frac{\sec^2\theta+\sec\theta+\sec^2\theta-\sec\theta}{\sec^2\theta-1}=2\csc^2\theta$$

$$\frac{2\sec^2\theta}{\tan^2\theta}=2\csc^2\theta$$

$$\frac{2\cos^2\theta}{\cos^2\theta\sin^2\theta}=2\csc^2\theta$$

$$2\csc^2\theta=2\csc^2\theta$$

9. $$(\sec\theta-1)(\sec\theta+1)=\tan^2\theta$$

$$\sec^2\theta-1=\tan^2\theta$$

$$\tan^2\theta=\tan^2\theta$$

11. $$\frac{\cos\theta}{1-\tan\theta}+\frac{\sin\theta}{1-\cot\theta}=\sin\theta+\cos\theta$$

$$\frac{\cos\theta}{1-\frac{\sin\theta}{\cos\theta}}+\frac{\sin\theta}{1-\frac{\cos\theta}{\sin\theta}}=\sin\theta+\cos\theta$$

$$\frac{\cos\theta}{\frac{\cos\theta}{\cos\theta}-\frac{\sin\theta}{\cos\theta}}+\frac{\sin\theta}{\frac{\sin\theta}{\sin\theta}-\frac{\cos\theta}{\sin\theta}}=\sin\theta+\cos\theta$$

$$\frac{\cos\theta}{\frac{\cos\theta-\sin\theta}{\cos\theta}}+\frac{\sin\theta}{\frac{\sin\theta-\cos\theta}{\sin\theta}}=\sin\theta+\cos\theta$$

$$\frac{\cos^2\theta - \sin^2\theta}{\cos\theta - \sin\theta} = \sin\theta + \cos\theta$$
$$\frac{(\cos\theta - \sin\theta)(\cos\theta + \sin\theta)}{\cos\theta - \sin\theta} = \sin\theta + \cos\theta$$
$$\cos\theta + \sin\theta = \sin\theta + \cos\theta$$

13. $\tan^2\theta(1+\cot^2\theta) = \sec^2\theta$

$$\tan^2\theta\csc^2\theta = \sec^2\theta$$
$$\frac{\sin^2\theta}{\cos^2\theta}\left(\frac{1}{\sin^2\theta}\right) = \sec^2\theta$$
$$\frac{\sin^2\theta}{\cos^2\theta\sin^2\theta} = \sec^2\theta$$
$$\frac{1}{\cos^2\theta} = \sec^2\theta$$
$$\sec^2\theta = \sec^2\theta$$

15. $1+\sin^2\theta = 3\sin\theta$

$$\sin\theta(2\sin\theta-1)-(2\sin\theta-1)=0$$
$$(\sin\theta-1)(2\sin\theta-1)=0$$

$\sin\theta - 1 = 0;\ \sin\theta = 1 \qquad 2\sin\theta - 1 = 0;\ \sin\theta = \frac{1}{2}$

$$\theta = \arcsin 1$$
$$\theta = \arcsin\frac{1}{2}$$
$$\theta = 30°, 90°, 150°$$

17. $\cos\theta - 2\cos^3\theta = 0$

$$\cos\theta(1-2\cos^2\theta) = 0$$

$\cos\theta = 0 \qquad 1-2\cos^2\theta = 0;\ \cos\theta = \pm\sqrt{\frac{1}{2}}$

$$\theta = \arccos 0$$

$\theta = \arccos\left(\sqrt{\frac{1}{2}}\right)$ and $\theta = \arccos\left(-\sqrt{\frac{1}{2}}\right)$

For $\arccos\left(\sqrt{\frac{1}{2}}\right)$; $\theta = 45°, 90°, 315°$; for $\arccos\left(-\sqrt{\frac{1}{2}}\right)$, $\theta = 135°, 225°, 270°$

19. $\sin\theta = 1 - 3\cos\theta$

$$\sin^2\theta = 1 - 6\cos\theta + 9\cos^2\theta$$
$$1-\cos^2\theta = 1 - 6\cos\theta + 9\cos^2\theta$$
$$10\cos^2\theta - 6\cos\theta = 0$$
$$5\cos^2\theta - 3\cos\theta = 0$$
$$\cos\theta(5\cos\theta - 3) = 0$$

$\cos\theta = 0 \qquad 5\cos\theta - 3 = 0;\ \cos\theta = \frac{3}{5}$

$\theta = \arccos 0$

$\theta = \arccos\left(\frac{3}{5}\right)$

$\theta = 90°, 306.9°$

21. $16\cos^4\frac{\theta}{2} = 9$

$\cos^4\frac{\theta}{2} = \frac{9}{16}$

$\cos\frac{\theta}{2} = \pm\sqrt[4]{\frac{9}{16}}$

$\cos\frac{\theta}{2} = \pm 0.8660$

$\frac{\theta}{2} = \arccos(\pm 0.8660)$

$\frac{\theta}{2} = 30°, 150°$

$\theta = 60°, 300°$

23. $3.85(\cos 52.5° + \sin 22.6°) \Rightarrow 3.82326 = 3.82$

25. $\sin^2 3.53 + \cos^2 1.77 \Rightarrow 0.182584 = 0.183$

Chapter 17: Ratio, Proportion, and Variation

Exercise 17.1 • Ratio and Proportion

Find the value of x. $a:b=c:d \Rightarrow \frac{a}{b}=\frac{c}{d} \Rightarrow ad=bc$

1. $3:x=4:6 \Rightarrow 4x=18 \Rightarrow x=\frac{9}{2}$

3. $4:6=x:4 \Rightarrow 6x=16 \Rightarrow x=\frac{8}{3}$

5. $x:(14-x)=4:3 \Rightarrow 3x=4(14-x) \Rightarrow 3x=56-4x \Rightarrow 7x=56 \Rightarrow x=8$

7. $x:6=(x+6):10\frac{1}{2} \Rightarrow \frac{21}{2}x=6(x+6) \Rightarrow 21x=12x+72 \Rightarrow 9x=72 \Rightarrow x=8$

Insert the missing quantity. Let a variable (like z) represent the missing quantity, and solve for the variable.

9. $\frac{x}{3}=\frac{z}{9} \Rightarrow 3z=9x \Rightarrow z=\frac{9x}{3}=3x$

11. $\frac{5a}{7b}=\frac{z}{-7b} \Rightarrow z(7b)=5a(-7b) \Rightarrow z=\frac{-35ab}{7b}=-5a$

13. $\frac{x+2}{5x}=\frac{z}{5} \Rightarrow z(5x)=5(x+2) \Rightarrow z=\frac{5(x+2)}{5x} \Rightarrow z=\frac{x+2}{x}$

Find the mean proportional between the following. $a:b=b:c \Rightarrow b^2=ac \Rightarrow b=\pm\sqrt{ac}$.

15. 3 and 48 $\Rightarrow b=\pm\sqrt{3(48)}=\pm\sqrt{144}=\pm 12$

17. 5 and 45 $\Rightarrow b=\pm\sqrt{5(45)}=\pm\sqrt{225}=\pm 15$

Applications

19. turn ratio $=\frac{N_2}{N_1}=15 \Rightarrow 15N_1=N_2 \Rightarrow N_1=\frac{N_2}{15}=\frac{4500}{15}=300$ turns

21. Actual mechanical advantage $R_a=\frac{\text{output force}}{\text{input force}}=\frac{W}{F}=\frac{279\text{ lb}}{106\text{ lb}}=2.63$

23. a) $R_a = \dfrac{\text{output force}}{\text{input force}} = \dfrac{W}{F} = \dfrac{74.8 \text{ lb}}{34.8 \text{ lb}} = 2.15$

b) $R_i = \dfrac{\text{distance traveled by the force}}{\text{distance traveled by the weight}} = \dfrac{d}{s} = \dfrac{28.7 \text{ in.}}{12.8 \text{ in.}} = 2.24$

c) $E = \dfrac{R_a}{R_i} = \dfrac{2.15}{2.24} = .960$ or 96.0%

Exercise 17.2 • Similar Figures

1. Let $h =$ height of larger container.

$$\left(\frac{h}{0.755}\right)^3 = \frac{40}{20} \Rightarrow h^3 = \frac{40(0.755)^3}{20} = 0.861 \Rightarrow h = \sqrt[3]{0.861} = 0.951 \text{ m}$$

3. Let $W =$ weight of new firebox.

$$\frac{W}{327} = (1.25)^3 \Rightarrow W = 1.25^3(327) = 639 \text{ lb}$$

5. Let $s =$ side of the original square.

$$\frac{s^2}{s^2+2450} = \left(\frac{s}{s+15}\right)^2 \Rightarrow s^2(s+15)^2 = s^2\left(s^2+2450\right) \Rightarrow (s+15)(s+15) = s^2+2450$$

$$\Rightarrow s^2+30s+225 = s^2+2450 \Rightarrow 30s = 2225 \Rightarrow s = 74.2 \text{ mm}$$

7. Let $A =$ area of the window.

$$\frac{A}{18.2 \text{ in}^2} = \left(\frac{4}{1}\right)^2 \Rightarrow A = (4)^2 18.2 \text{ in}^2 = 291.2 \text{ in}^2\left(\frac{1 \text{ ft}}{12 \text{ in}}\right)^2 = 2.02 \text{ ft}^2$$

9. Let $d_1 =$ the diameter of the new fence, $d_2 =$ the diameter of the old fence, and $A =$ the area enclosed by the old fence.

$$\frac{\frac{1}{5}A}{A} = \left(\frac{d_1}{d_2}\right)^2 \Rightarrow \frac{1}{\sqrt{5}} = \frac{d_1}{d_2}$$

Set up a ratio with circumference ($C = \pi d$) and cost.

$$\frac{\pi d_1}{\pi d_2} = \frac{x}{756} \Rightarrow \frac{1}{\sqrt{5}} = \frac{x}{756} \Rightarrow x = \frac{756}{\sqrt{5}} = \$338$$

11. Let $A =$ area of the industrial park.

$$\frac{A}{2.75 \text{ in}^2} = \left(\frac{10000}{1}\right)^2 \Rightarrow A = (10000)^2 2.75 \text{ in}^2 \Rightarrow A = 2.75\times10^8 \text{ in}^2\left(\frac{1 \text{ ft}^2}{144 \text{ in}^2}\right)\left(\frac{1 \text{ acre}}{43560 \text{ ft}^2}\right) = 43.8 \text{ acres}$$

13. Let $A =$ area of the woodland.

$$\frac{A}{1.75 \text{ in}^2} = \left(\frac{24000}{1}\right)^2 \Rightarrow A = (24000)^2 1.75 \text{ in}^2 \Rightarrow A = 1.01\times10^9 \text{ in}^2\left(\frac{1 \text{ ft}^2}{144 \text{ in}^2}\right)\left(\frac{1 \text{ acre}}{43560 \text{ ft}^2}\right) = 161 \text{ acres}$$

15. Let $V =$ volume of the roof.

$$\frac{V}{837\text{ in}^3}=\left(\frac{25}{1}\right)^3 \Rightarrow V=(25)^3 837\text{ in}^3=1.308\times10^7\text{ in}^3\left(\frac{1\text{ ft}^3}{1728\text{ in}^3}\right)=7570\text{ ft}^3$$

Exercise 17.3 • Direct Variation

1. $\frac{y}{x}=\frac{56}{21} \Rightarrow \frac{y}{74}=\frac{56}{21} \Rightarrow y=\frac{56}{21}(74)=197$

3. $\frac{q}{p}=\frac{135}{846} \Rightarrow \frac{q}{448}=\frac{135}{846} \Rightarrow q=\frac{135}{846}(448)=71.5$

5. $k=\frac{y}{x}=\frac{45}{9}=5 \Rightarrow y=5x$

$x=11 \Rightarrow y=5(11)=55$

$y=75 \Rightarrow 75=5x \Rightarrow x=\frac{75}{5}=15$

7. $k=\frac{y}{x}=\frac{187}{154}=1.214 \Rightarrow y=1.214x$

$x=115 \Rightarrow y=1.214(115)=140$

$x=125 \Rightarrow y=1.214(125)=152$

$y=167 \Rightarrow 167=1.214x \Rightarrow x=\frac{167}{1.214}=138$

Applications

9. $\frac{n}{10.0\text{ kg}}=\frac{2500\text{ balls}}{3.65\text{ kg}} \Rightarrow n=\frac{2500}{3.65}(10.0)=6850\text{ balls}$

11. $\frac{d}{35.0\text{ mi/gal}}=\frac{250\text{ mi}}{21.0\text{ mi/gal}} \Rightarrow d=\frac{250}{21.0}(35.0)=417\text{ mi}$

13. $\frac{R}{75.0\text{ mi}}=\frac{155\ \Omega}{2.60\text{ mi}} \Rightarrow R=\frac{155}{2.60}(75.0)=4470\ \Omega$

15. $\frac{p}{7.5\text{ h}}=\frac{1850\text{ parts}}{55\text{ min}} \Rightarrow p=\frac{1850}{55}(7.5)\left(\frac{60\text{ min}}{1\text{ h}}\right)=15{,}136\text{ parts}$

17. $\frac{P}{5000\text{ gal/min}}=\frac{41.2\text{ MW}}{5625\text{ gal/min}} \Rightarrow P=\frac{41.2}{5625}(5000)=36.6\text{ MW}$

Exercise 17.4 • The Power Function

1. $y=kx^2 \Rightarrow 726=k(163)^2 \Rightarrow k=\frac{726}{(163)^2}=0.02733 \Rightarrow y=0.02733(274)^2=2050$

3. $y=kx^3 \Rightarrow 4.83=k(1.33)^3 \Rightarrow k=\frac{4.83}{(1.33)^3}=2.053 \Rightarrow y=2.053(3.38)^3=79.3$

5. $y = kx^2 \Rightarrow 285.0 = k(112.0)^2 \Rightarrow k = \dfrac{285.0}{(112.0)^2} = 0.022720 \Rightarrow y = 0.022720(351.0)^2 = 2799$

7. $y = kx^3 \Rightarrow 638 = k(145)^3 \Rightarrow k = \dfrac{638}{(145)^3} = 0.00020927 \Rightarrow y = 0.00020927(68.3)^3 = 66.7$

9. $k = \dfrac{y}{x^4} = \dfrac{29.7}{18.2^4} = 0.0002707 \Rightarrow y = 0.0002707x^4$

$x = 75.6 \Rightarrow y = 0.0002707(75.6)^4 = 8840$

$y = 154 \Rightarrow 154 = 0.0002707x^4 \Rightarrow x = \left(\dfrac{154}{0.0002707}\right)^{1/4} = 27.5$

11. $k = \dfrac{y}{\sqrt[3]{x}} = \dfrac{275}{\sqrt[3]{782}} = 29.85 \Rightarrow y = 29.85\sqrt[3]{x}$

$x = 315 \Rightarrow y = 29.85\sqrt[3]{315} = 203$

$y = 148 \Rightarrow 148 = 29.85\sqrt[3]{x} \Rightarrow x = \left(\dfrac{148}{29.85}\right)^3 = 122$

13. $y = 0.553x^5$

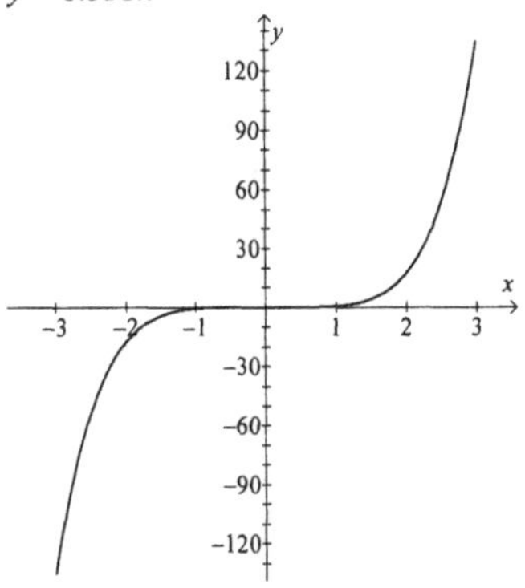

Applications

In 15-18 the initial velocity is 0, so the distance fallen by the object is directly proportional to the time squared ($d = at^2$).

15. $a = \dfrac{d}{t^2} = \dfrac{176 \text{ m}}{(6.00 \text{ s})^2} = 4.889 \Rightarrow d = 4.889(9.00 \text{ s})^2 = 396 \text{ m}$

17. $a = \dfrac{d}{t^2} = \dfrac{129 \text{ ft}}{(2.00 \text{ s})^2} = 32.25 \Rightarrow 525 \text{ ft} = 32.25t^2 \Rightarrow t = \sqrt{\dfrac{525}{32.25}} = 4.03 \text{ s}$

In 19-21 the power dissipated in a resistor is directly proportional to the square of the current in the resistor ($P = RI^2$).

19. $R = \dfrac{P}{I^2} = \dfrac{486\ W}{(2.75 \text{ A})^2} = 64.26 \Rightarrow P = 64.26(3.45 \text{ A})^2 = 765 \text{ W}$

21. $\frac{3P_1}{P_1} = \left(\frac{I_2}{I_1}\right)^2 \Rightarrow \frac{I_2}{I_1} = \sqrt{3} = 1.73$ The current needs to be increased by a factor of 1.73.

In 22-25 the exposure time for a photograph is directly proportional to the square of the f stop ($t = kf^2$).

23. $k = \frac{t}{f^2} = \frac{1/100 \text{ s}}{(5.6)^2} = 0.0003189 \Rightarrow \frac{1}{50}\text{ s} = 0.0003189(f)^2 \Rightarrow f = \sqrt{\frac{1/50}{0.0003189}} = 7.92 \approx \text{f}8$

25. $k = \frac{t}{f^2} = \frac{24 \text{ s}}{(22)^2} = 0.0496 \Rightarrow 8 \text{ s} = 0.0496(f)^2 \Rightarrow f = \sqrt{\frac{8}{0.0496}} = 12.7 \approx \text{f}13$

Exercise 17.5 • Inverse Variation

1. $y = \frac{k}{x} \Rightarrow 385 = \frac{k}{832} \Rightarrow k = 385(832) = 320320 \Rightarrow y = \frac{320320}{226} = 1420$

3. $y = k\frac{1}{x} \Rightarrow y' = k\frac{1}{(2x)} = \frac{1}{2}\left(k\frac{1}{x}\right) = \frac{1}{2}y$ y is halved.

5. $k = xy = 306(125) = 38250 \Rightarrow y = \frac{38250}{x}$

$x = 622 \Rightarrow y = \frac{38250}{622} = 61.5$

$y = 418 \Rightarrow 418 = \frac{38250}{x} \Rightarrow x = \frac{38250}{418} = 91.5$

7. $k = \sqrt{x}y = \sqrt{3567}(1828) = 109176 \Rightarrow y = \frac{109176}{\sqrt{x}}$

$y = 1136 \Rightarrow 1136 = \frac{109176}{\sqrt{x}} \Rightarrow x = \left(\frac{109176}{1136}\right)^2 = 9236$

$x = 5725 \Rightarrow y = \frac{109176}{\sqrt{5725}} = 1443$

9. $y = k\frac{1}{\sqrt{x}} \Rightarrow y' = k\frac{1}{\sqrt{0.5x}} = \frac{1}{\sqrt{0.5}}\left(k\frac{1}{\sqrt{x}}\right) = \frac{1}{\sqrt{0.5}}y$ $\frac{1}{\sqrt{0.5}} \approx 1.414$ so y increases by 41.4%.

11. $y = \frac{1}{x}$

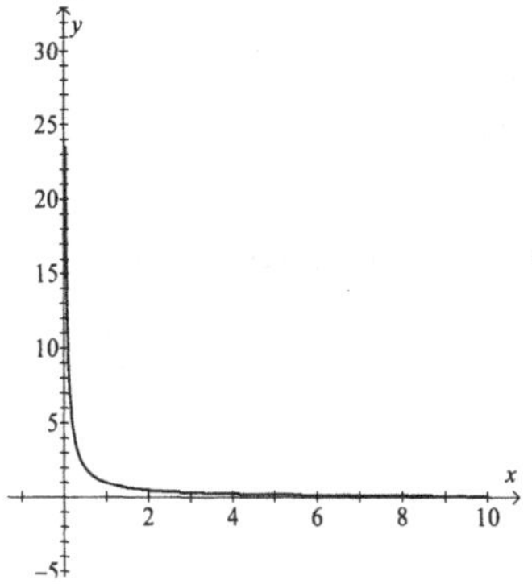

Applications

In 12-14, assuming a constant temperature, the pressure is inversely proportional to the volume ($P = \frac{k}{V}$).

13. $k = PV = \left(14.7 \text{ lb/in.}^2\right)\left(175 \text{ in.}^3\right) = 2572.5 \Rightarrow P = \frac{2572.5}{25.0 \text{ in.}^3} = 103 \text{ lb/in.}^2$

In 15-17 two bodies attract each other with a force that is inversely proportional to the square of the distance between them ($F = \frac{k}{d^2}$).

15. $k = Fd^2 = \left(3.75 \times 10^{-5} \text{ dyne}\right)(18.0 \text{ cm})^2 = 0.01215 \Rightarrow F = \frac{0.01215}{(52.0 \text{ cm})^2} = 4.49 \times 10^{-6} \text{ dyne}$

17. $k = Wd^2 = (150 \text{ lb})(3960 \text{ mi})^2 = 2.352 \times 10^9 \Rightarrow W = \frac{2.352 \times 10^9}{([3960 + 1500] \text{ mi})^2} = 79 \text{ lb}$

In 19-22 the intensity of illumination on the surface is inversely proportional to the square of the distance between the source of light and the surface ($I = \frac{k}{d^2}$).

19. $k = I_1 d_1^2 = I_2 d_2^2 \Rightarrow (800 \text{ lux}) d_1^2 = I_2 (2d_1)^2 \Rightarrow I_2 = \frac{800}{2^2} = 200 \text{ lux}$

21. $k = Id^2 = (426 \text{ lux})(7.50 \text{ m})^2 = 23962.5 \Rightarrow d = \sqrt{\frac{23962.5}{850 \text{ lux}}} = 5.31 \text{ m}$

Electrical

23. $I = \frac{k}{R} \Rightarrow I' = \frac{k}{(1.1R)} = \frac{1}{1.1}\left(\frac{k}{R}\right) = 0.909I$ The current will change by a factor of 0.909.

25. $X_c = \frac{k}{C} \Rightarrow X_c' = \frac{k}{(0.75C)} = \frac{1}{0.75}\left(\frac{k}{C}\right) = 1.333X_c$ The capacitive reactance increases by 33.3%.

Exercise 17.6 • Functions of More Than One Variable

Joint Variation

1. $y = kwx \Rightarrow k = \dfrac{y}{wx} = \dfrac{483}{(383)(742)} = 0.001700 \Rightarrow y = 0.001700(756)(274) = 352$

3. $y = kwx \Rightarrow y' = k(1.12w)(0.930x) = (1.12)(0.930)(kwx) = 1.0416y$ y will increase by 4.2%.

5. $k = \dfrac{y}{wx} = \dfrac{127}{(46.2)(18.3)} = 0.1502 \Rightarrow y = 0.1502wx$

$w = 19.5$ and $x = 41.2 \Rightarrow y = 0.1502(19.5)(41.2) = 121$

$x = 8.86$ and $y = 155 \Rightarrow w = \dfrac{155}{0.1502(8.86)} = 116$

$w = 12.2$ and $y = 79.8 \Rightarrow x = \dfrac{79.8}{0.1502(12.2)} = 43.5$

Combined Variation

7. $y = k\dfrac{\sqrt{w}}{x^3} \Rightarrow y' = k\dfrac{\sqrt{3w}}{\left(\frac{1}{2}x\right)^3} = \dfrac{\sqrt{3}}{1/8}\left(k\dfrac{\sqrt{w}}{x^3}\right) = 8\sqrt{3}y = 13.9y$ y is changed by a factor of 13.9.

9. $y = k\dfrac{x^{3/2}}{w} \Rightarrow k = \dfrac{yw}{x^{3/2}} = \dfrac{284(361)}{(858)^{3/2}} = 4.08 \Rightarrow y = 4.08\dfrac{x^{3/2}}{w}$

Geometry

11. $A = kbh \Rightarrow A' = k(1.15b)(0.75h) = (1.15)(0.75)kbh = .8625A$ The area will decrease 13.75%.

Electrical

13. $R = k\dfrac{L}{A} = k\dfrac{L}{\pi(d/2)^2} = k\dfrac{L}{\pi d^2/4} = k\dfrac{4L}{\pi d^2} \Rightarrow R' = k\dfrac{4(3L)}{\pi(3d)^2} = \dfrac{3}{9}\left(k\dfrac{4L}{\pi d^2}\right) = \dfrac{1}{3}R$

The resistance will change by a factor of $\dfrac{1}{3}$.

Gravitation

15. $F = k\dfrac{m_1m_2}{d^2} \Rightarrow F' = k\dfrac{(3m_1)(3m_2)}{(2d)^2} = \dfrac{9}{4}\left(\dfrac{m_1m_2}{d^2}\right) = \dfrac{9}{4}F$ The force will change by a factor of $\dfrac{9}{4}$.

Illumination

17. $I = k\dfrac{L}{d^2} \Rightarrow k\dfrac{(75.0\text{ W})}{(8.00\text{ ft})^2} = k\dfrac{L}{(12.00\text{ ft})^2} \Rightarrow L = \dfrac{75.0(144.0)}{64.0} = 169\text{ W}$

Gas Laws

19. $V = k\frac{t}{p} \Rightarrow k = \frac{Vp}{t} \Rightarrow k = \frac{(4.45\text{ m}^3)(225\text{ kPa})}{(305\text{ K})} = 3.283 \Rightarrow V = 3.283\frac{(354\text{ K})}{(325\text{ kPa})} = 3.58\text{ m}^3$

Work

21. $P = knt \Rightarrow k = \frac{P}{nt} \Rightarrow \frac{\$5123.73}{(5\text{ workers})(3.0\text{ weeks})} = \frac{\$6148.48}{(6\text{ workers})t} \Rightarrow t = \frac{6148.48(5)(3.0)}{5123.73(6)} = 3.0\text{ weeks}$

Strength of Materials

23. $S = k\frac{wd^2}{l} \Rightarrow k = \frac{Sl}{wd^2} = \frac{(15000\text{ lb})(16.0\text{ ft})}{(8.00\text{ in.})(11.5\text{ in.})^2} = 226.8 \Rightarrow S = 226.8\frac{(6.50\text{ in.})(13.4\text{ in.})^2}{21.0\text{ ft}} = 12{,}600\text{ lb}$

Mechanics

25. $\omega = k\frac{\sqrt{t}}{l} \Rightarrow k = \frac{\omega l}{\sqrt{t}} = \frac{(325\text{ vib/s})(1.00\text{ m})}{\sqrt{115\ N}} = 30.31 \Rightarrow \omega = 30.31\frac{\sqrt{95.0\ N}}{0.750\text{ m}} = 394\text{ times/s}$

Fluid Flow

27. $t = kr^2\sqrt{h} \Rightarrow t' = k(1.25r)^2\sqrt{2h} = (1.25)^2\sqrt{2}\left(kr^2\sqrt{h}\right) = 2.21t$ The time will increase by a factor of 2.21.

Chapter 17 • Review Problems

1. $y = k\frac{1}{x} \Rightarrow k = xy \Rightarrow 822(736) = 583y \Rightarrow y = \frac{822(736)}{583} = 1040$

3. $y = kxz \Rightarrow y' = k(1.15x)(0.96z) = (1.15)(0.96)(kxz) = 1.104y$ y will increase by 10.4%.

5. $f = k\sqrt{d} \Rightarrow k = \frac{f}{\sqrt{d}} \Rightarrow \frac{225\text{ L/min}}{\sqrt{3.46\text{ m}}} = \frac{f}{\sqrt{1.00\text{ m}}} \Rightarrow f = \frac{225\sqrt{1.00}}{\sqrt{3.46}} = 121\text{ L/min}$

7. $S = kd^2 \Rightarrow S' = 3kd^2 = k\left(\sqrt{3}d\right)^2$ The diameter needs to be increased by a factor of $\sqrt{3} \approx 1.73$.

9. $t = kd^{3/2}$

   Earth: $(1\text{ yr}) = kd^{3/2}$

   Saturn: $t = k\left(\frac{19}{2}d\right)^{3/2} = \left(\frac{19}{2}\right)^{3/2}\left(kd^{3/2}\right) = \left(\frac{19}{2}\right)^{3/2}(1) = 29\text{ yr}$

11. $N = k\frac{1}{L} \Rightarrow k = NL$

$N_1 = 24(60) = 1440$ min/day and $N_2 = 1425$ min/day

$N_1L_1 = N_2L_2 \Rightarrow 1440L_1 = 1425(30.00\text{ in.}) \Rightarrow L_1 = \frac{1425(30.00)}{1440} = 29.6875\text{ in.}$

Shorten pendulum $30.00 - 29.6875 = 0.3125$ in.

13. $F = kAw^2 \Rightarrow kA'(12\text{ mi/h})^2 = kA(35\text{ mi/h})^2 \Rightarrow A' = \frac{35^2}{12^2}A = 8.5A$

The area needs to be increased by a factor of 8.5.

15. $t = k\frac{1}{v}$

current: $26.3\text{ h} = k\frac{1}{v}$

new: $t = k\frac{1}{(1.15v)} = \frac{1}{1.15}\left(k\frac{1}{v}\right) = \frac{1}{1.15}(26.3) = 22.87\text{ h} = 22\text{ h }52\text{ min}$

17. $F = k\frac{m_1m_2}{d^2}$ and take 239,000 = 2.39

Moon: $F = k\frac{m_{\text{moon}}m_{\text{ship}}}{d^2}$ Earth: $F = k\frac{(82.0m_{\text{moon}})m_{\text{ship}}}{(2.39-d)^2}$

$$k\frac{m_{\text{moon}}m_{\text{ship}}}{(2.39-d)^2} = k\frac{(82.0m_{\text{moon}})m_{\text{ship}}}{d^2}$$

$(2.39-d)^2\,82.0 = d^2 \Rightarrow (2.39-d)\sqrt{82.0} = \pm d \Rightarrow 21.65 - 9.06d = \pm d \Rightarrow (9.06 \pm 1)d = 21.65$

$d = \frac{21.65}{9.06+1} = 2.15$ or 215,000 mi

$\left(\text{exclude } \frac{21.65}{9.06-1} = 2.68\right.$ or 268,000 mi, which is larger than the 239,000 mi between the earth and the moon.$\left.\right)$

19. $F = k\frac{L}{r} \Rightarrow k = \frac{Fr}{L} \Rightarrow \frac{F(4.5\text{ cm})}{18\text{ m}} = \frac{(35\text{ kg})(3.6\text{ cm})}{12\text{ m}} \Rightarrow F = \frac{35(3.6)(18)}{12(4.5)} = 42\text{ kg}$

21. $I = kd^{3/2} \Rightarrow I' = k(2d)^{3/2} = 2^{3/2}\left(kd^{3/2}\right) = 2.8$ The current will be increased by a factor of 2.8.

23. $F = kv^2\sin\frac{\theta}{2} \Rightarrow \frac{F_2}{F_1} = \frac{k(1.40v)^2\sin(40^\circ/2)}{kv^2\sin(55^\circ/2)} = 1.40^2\frac{\sin(40^\circ/2)}{\sin(55^\circ/2)} = 1.45$

There is a 45% increase in the force.

25. $C = kv^3 \Rightarrow k = \frac{C}{v^3} \Rightarrow \frac{C}{(10\text{ knots})^3} = \frac{1584\text{ gal}}{(15\text{ knots})^3} \Rightarrow C = \frac{1584(10)^3}{15^3} = 469\text{ gal}$

27. $\frac{A}{52.0\text{ m}^2} = \left(\frac{1.40}{1}\right)^2 \Rightarrow A = 1.40^2(52.0) = 102\text{ m}^2$

29. $t = k\frac{1}{d^2} \Rightarrow k = td^2 \Rightarrow (9.0\text{ h})d^2 = (5.0\text{ h})(1.5\text{ in.})^2 \Rightarrow d = \sqrt{\frac{(5.0)1.5^2}{9.0}} = 1.1\text{ in.}$

31. $\frac{A}{1.10\text{ m}^2} = \left(\frac{1}{0.25}\right)^2 \Rightarrow A = (4)^2 1.10 = 17.6\text{ m}^2$

33. $\frac{W}{5.80\text{ tons}} = \left(\frac{1.30}{1}\right)^3 \Rightarrow W = (1.30)^3(5.80) = 12.7\text{ tons}$

35. $\frac{W}{18.0\text{ oz}} = \left(\frac{9.00\text{ in.}}{4.50\text{ in.}}\right)^3 \Rightarrow W = 2^3(18.0) = 144\text{ oz}$

37. $S = kC \Rightarrow k = \frac{S}{C} \Rightarrow \frac{S}{12.7} = \frac{\$900}{9.40} \Rightarrow S = \frac{900}{9.40}(12.7) = \$1216\text{ per week}$

39. $C = \frac{5.25}{3.5} = 1.5 \Rightarrow C = \frac{7.875}{5.25} = 1.5$

41. $\frac{A}{746\text{ cm}^2} = \left(\frac{1.01}{1}\right)^2 \Rightarrow A = 1.01^2(746) = 761\text{ cm}^2$

Chapter 18: Exponential and Logarithmic Functions

Exercise 18.1 • The Exponential Function

1. $y = 0.2(3.2)^x$

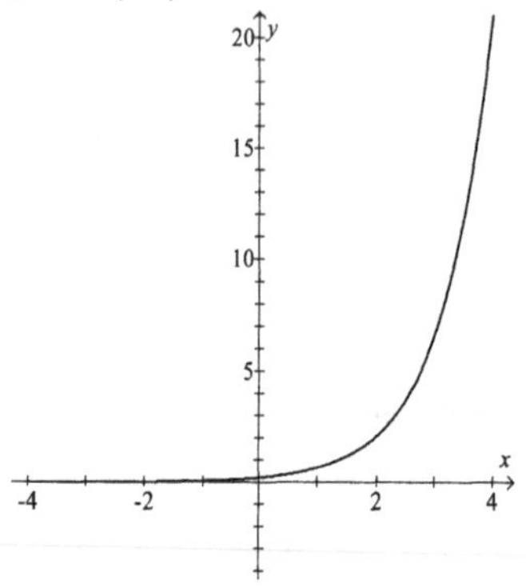

3. $y = 5\left(1 - e^{-x}\right)$

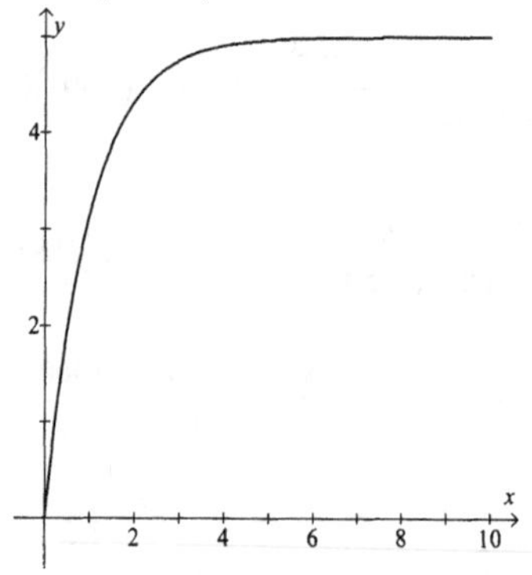

5. $y = \frac{4}{2}\left(e^{x/4} + e^{-x/4}\right) = 2\left(e^{x/4} + e^{-x/4}\right)$

   a) $y(-5) = y(5) = 7.55$ m

   b) $7.55 - y(0) = 7.55 - 4 = 3.55$ m

   c) $h = 3.55 + 3 = 6.55$ m

Compound Interest

7. $a = \frac{y}{(1+n)^t} = \frac{10000}{(1+0.075)^{10}} = \$4852$

9. $y = a\left(1 + \frac{n}{m}\right)^{mt}$

   a) $y = 1\left(1 + \frac{0.10}{1}\right)^{20\times1} = \$6.73$ b) $y = 1\left(1 + \frac{0.10}{12}\right)^{20\times12} = \$7.33$ c) $y = 1\left(1 + \frac{0.10}{365}\right)^{20\times365} = \$7.39$

11. $R = \frac{ny}{(1+n)^t - 1} = \frac{0.06(100000)}{(1+0.06)^{25} - 1} = \$1823$

Exponential Growth

13. $y = ae^{nt} = 201e^{0.0500(7)} = 285$ units

15. $y = ae^{nt} = 11.4e^{0.0163(5)} = 12.4$ million

17. $y = ae^{nt} = 12.0e^{0.0830(5)} = 18.2$ barrels

Exponential Growth to an Upper Limit

19. $y = a\left(1 - e^{-nt}\right) = 801\left(1 - e^{-0.0500(20)}\right) = 506^\circ$ C

21. $i=\frac{E}{R}\left(1-e^{-Rt/L}\right)=\frac{249}{6.25}\left(1-e^{-6.25(0.0750)/186}\right)=0.100\text{ A}$

Exponential Decay

23. $y=ae^{-nt}=1495e^{-0.0200(60)}=450°\text{ F}$ The temperature will drop $1495-450=1045°\text{ F}$

25. $P=29.92e^{-h/5}=29.92e^{-\left(\frac{30100}{5280}\right)/5}=9.568\text{ in. Hg}$

27. $y=\frac{E}{R}e^{-t/RC}=\frac{220}{2700}e^{-0.045/\left(2700\times130\times10^{-6}\right)}=0.072\text{ A}=72\text{ mA}$

29. $y=ae^{-nt}=ae^{-0.24(5)}=0.30a$ 30% of the original moisture will be left.

Exercise 18.2 • Logarithms

Converting between Logarithmic and Exponential Forms

1. $3^4=81 \Rightarrow \log_3 81=4$

3. $4^6=4096 \Rightarrow \log_4 4096=6$

5. $x^5=995 \Rightarrow \log_x 995=5$

7. $\log_{10}100=2 \Rightarrow 10^2=100$

9. $\log_5 125=3 \Rightarrow 5^3=125$

11. $\log_3 x=57 \Rightarrow 3^{57}=x$

Common Logarithms

13. $\log 27.6=1.441$

15. $\log 5.93=0.773$

17. $\log 48.3=1.684$

19. $\log 836=2.922$

21. $\log 27.4=1.438$

Antilogarithms for Common Logarithms

23. $\log x=2.957 \Rightarrow x=10^{2.957}=906$

25. $\log x=0.886 \Rightarrow x=10^{0.886}=7.69$

27. $\log x=0.366 \Rightarrow x=10^{0.366}=2.32$

29. $\log x=4.974 \Rightarrow x=10^{4.974}=94,200$

Natural Logarithms

31. $\ln 48.3=3.8774$

33. $\ln 2365=7.7685$

35. $\ln 1.845=0.6125$

37. $\ln 1.374=0.3177$

39. $\ln 1.364=0.3104$

Antilogarithms for Natural Logarithms

41. $\ln x=4.263 \Rightarrow x=e^{4.263}=71.0$

43. $\ln x=1.845 \Rightarrow x=e^{1.845}=6.33$

45. $\ln x = 5.937 \Rightarrow x = e^{5.937} = 379$

47. $\ln x = 4.9715 \Rightarrow x = e^{4.9715} = 144.2$

Applications

49. $t = \dfrac{\log y - \log a}{\log(1+n)} = \dfrac{\log 30000 - \log 10000}{\log(1+0.0800)} = 14.3 \text{ yr}$

51. $t = \dfrac{\log\left(\dfrac{an}{R-an}+1\right)}{\log(1+n)} = \dfrac{\log\left(\dfrac{200000(0.12)}{30000-200000(0.12)}+1\right)}{\log(1+0.12)} = \dfrac{\log 5}{\log 1.12} = 14 \text{ yr}$

53. $h = 60470 \log \dfrac{B_2}{B_1} = 60470 \log\left(\dfrac{29.14 \text{ in. Hg}}{26.22 \text{ in. Hg}}\right) = 2773 \text{ ft}$

55. $\text{EET} = \dfrac{1}{n}\ln\left(\dfrac{nR}{r}+1\right) = \dfrac{1}{0.0700}\ln\left(\dfrac{0.0700(207\times10^9)}{6.00\times10^9}+1\right) = 17.5 \text{ yr}$

Exercise 18.3 • Properties of Logarithms

1. $\log\dfrac{2}{3} = \log 2 - \log 3$

3. $\log ab = \log a + \log b$

5. $\log xyz = \log x + \log y + \log z$

7. $\log\dfrac{3x}{4} = \log 3x - \log 4 = \log 3 + \log x - \log 4$

9. $\log\dfrac{1}{2x} = \log 1 - \log 2x = 0 - (\log 2 + \log x) = -\log 2 - \log x$

11. $\log\dfrac{abc}{d} = \log abc - \log d = \log a + \log b + \log c - \log d$

13. $\log 3 + \log 4 = \log(3\cdot 4) = \log 12$

15. $\log 2 + \log 3 - \log 4 = \log\left(\dfrac{2\cdot 3}{4}\right) = \log\dfrac{3}{2}$

17. $4\log 2 + 3\log 3 - 2\log 4 = \log 2^4 + \log 3^3 - \log 4^2 = \log\left(\dfrac{16\cdot 27}{16}\right) = \log 27$

19. $3\log a - 2\log b + 4\log c = \log a^3 - \log b^2 + \log c^4 = \log\left(\dfrac{a^3c^4}{b^2}\right)$

21. $\log\frac{x}{a}+2\log\frac{y}{b}+3\log\frac{z}{c}=\log\frac{x}{a}+\log\left(\frac{y}{b}\right)^2+\log\left(\frac{z}{c}\right)^3=\log\left(\frac{xy^2z^3}{ab^2c^3}\right)$

23. $\log_2 x+2\log_2 y=x$

$\log_2 x+\log_2 y^2=x$

$\log_2 xy^2=x$

$xy^2=2^x$

25. $\log\left(p^2-q^2\right)-\log(p+q)=2$

$\log\frac{p^2-q^2}{p+q}=2$

$\log\frac{(p-q)(p+q)}{p+q}=2$

$\log(p-q)=2$

$p-q=10^2$

$p-q=100$

27. $\log_e e=1$

29. $\log_3 3^2=2$

31. $\log_{10}10^x=x$

33. $2^{\log_2 3y}=3y$

$\log N=\frac{\ln N}{\ln 10}$

35. $\log N=\frac{8.36}{\ln 10}=3.63$

37. $\log N=\frac{3.775}{\ln 10}=1.639$

39. $\log N=\frac{5.26}{\ln 10}=2.28$

$\ln N=\ln 10(\log N)$

41. $\ln N=\ln 10(84.9)=195$

43. $\ln N=\ln 10(-3.82)=-8.80$

45. $\ln N=\ln 10(2.37)=5.46$

Exercise 18.4 • Exponential Equations

1. $2^x=7$

$\log 2^x=\log 7$

$x\log 2=\log 7$

$x=\frac{\log 7}{\log 2}=2.81$

3. $(1.15)^{x+2}=12.5$

$\log(1.15)^{x+2}=\log 12.5$

$(x+2)\log 1.15=\log 12.5$

$x+2=\frac{\log 12.5}{\log 1.15}$

$x=\frac{\log 12.5}{\log 1.15}-2=16.1$

5. $(15.4)^{\sqrt{x}} = 72.8$

$$\log(15.4)^{\sqrt{x}} = \log 72.8$$
$$\sqrt{x}\log 15.4 = \log 72.8$$
$$\sqrt{x} = \frac{\log 72.8}{\log 15.4}$$
$$x = \left(\frac{\log 72.8}{\log 15.4}\right)^2 = 2.46$$

7. $5.62e^{3x} = 188$

$$e^{3x} = \frac{188}{5.62}$$
$$\ln\left(e^{3x}\right) = \ln\left(\frac{188}{5.62}\right)$$
$$3x = \ln\left(\frac{188}{5.62}\right)$$
$$x = \frac{\ln\left(\frac{188}{5.62}\right)}{3} = 1.17$$

9. $e^{2x-1} = 3e^{x+3}$

$$\frac{e^{2x-1}}{e^{x+3}} = 3$$
$$e^{2x-1-(x+3)} = 3$$
$$e^{x-4} = 3$$
$$x - 4 = \ln 3$$
$$x = \ln 3 + 4 = 5.10$$

11. $5^{2x} = 7^{3x-2}$

$$\log\left(5^{2x}\right) = \log\left(7^{3x-2}\right)$$
$$2x\log 5 = (3x-2)\log 7$$
$$2x\log 5 = 3x\log 7 - 2\log 7$$
$$2x\log 5 - 3x\log 7 = -2\log 7$$
$$x = \frac{-2\log 7}{2\log 5 - 3\log 7} = 1.49$$

13. $10^{3x} = 3\left(10^x\right)$

$$\frac{10^{3x}}{10^x} = 3$$
$$10^{3x-x} = 3$$
$$10^{2x} = 3$$
$$\log\left(10^{2x}\right) = \log(3)$$
$$2x = \log 3$$
$$x = \frac{\log 3}{2} = 0.239$$

15. $2^{3x+1} = 3^{2x+1}$

$$\log\left(2^{3x+1}\right) = \log\left(3^{2x+1}\right)$$
$$(3x+1)\log 2 = (2x+1)\log 3$$
$$3x\log 2 + \log 2 = 2x\log 3 + \log 3$$
$$3x\log 2 - 2x\log 3 = \log 3 - \log 2$$
$$x = \frac{\log 3 - \log 2}{3\log 2 - 2\log 3} = -3.44$$

17. $7e^{1.5x} = 2e^{2.4x}$

$$\frac{7}{2} = \frac{e^{2.4x}}{e^{1.5x}}$$
$$3.5 = e^{0.9x}$$
$$\ln\left(e^{0.9x}\right) = \ln 3.5$$
$$0.9x = \ln 3.5$$
$$x = \frac{\ln 3.5}{0.9} = 1.39$$

Applications

19.
$$i = \frac{E}{R}e^{-t/RC}$$
$$0.0165 = \frac{325}{1.35}e^{-t/\left[1.35\left(3210\times10^{-6}\right)\right]}$$
$$e^{-t/\left[1.35\left(3210\times10^{-6}\right)\right]} = \frac{0.0165}{325/1.35}$$
$$\ln\left(e^{-t/0.004334}\right) = \ln\left(6.854\times10^{-5}\right)$$
$$\frac{-t}{0.004334} = \ln\left(6.854\times10^{-5}\right)$$
$$t = (-0.004334)\ln\left(6.854\times10^{-5}\right) = 0.0416 \text{ s}$$

21.
$$T = 2005e^{-0.0620t}$$
$$500 = 2005e^{-0.0620t}$$
$$e^{-0.0620t} = \frac{500}{2005}$$
$$-0.0620t = \ln\left(\frac{500}{2005}\right)$$
$$t = \frac{\ln(500/2005)}{-0.0620} = 22.4 \text{ s}$$

23.
$$P = 9000e^{0.02t}$$
$$27000 = 9000e^{0.02t}$$
$$e^{0.02t} = \frac{27000}{9000}$$
$$0.02t = \ln(3)$$
$$t = \frac{\ln 3}{0.02} = 55 \text{ yr}$$

25.
$$d = 64.0e^{0.00676h}$$
$$64.5 = 64.0e^{0.00676h}$$
$$e^{0.00676h} = \frac{64.5}{64.0}$$
$$0.00676h = \ln(64.5/64.0)$$
$$h = \frac{\ln(64.5/64.0)}{0.00676} = 1.15 \text{ mi}$$

27.
$$y = a(1+n)^t$$
$$3a = a(1+0.12)^t$$
$$1.12^t = \frac{3a}{a}$$
$$\log\left(1.12^t\right) = \log(3)$$
$$t\log 1.12 = \log 3$$
$$t = \frac{\log 3}{\log 1.12} = 9.7 \text{ yr}$$

29. $t = \frac{\ln 2}{n} = \frac{\ln 2}{0.0350} = 19.8 \text{ yr}$

31. $t = \frac{\ln 2}{n} = \frac{\ln 2}{0.0164} = 42.3 \text{ yr}$

Exercise 18.5 • Solving Logarithmic Equations

1.
$$x = \log_3 9$$
$$3^x = 9$$
$$3^x = 3^2$$
$$x = 2$$

3.
$$x = \log_8 2$$
$$8^x = 2$$
$$\left(2^3\right)^x = 2$$
$$3x = 1$$
$$x = \frac{1}{3}$$

5.
$$x = \log_{27} 9$$
$$27^x = 9$$
$$\left(3^3\right)^x = 3^2$$
$$3x = 2$$
$$x = \frac{2}{3}$$

7. $x = \log_8 4$

$$8^x = 4$$
$$\left(2^3\right)^x = 2^2$$
$$3x = 2$$
$$x = \frac{2}{3}$$

9. $\log_x 8 = 3$

$$x^3 = 8$$
$$x^3 = 2^3$$
$$x = 2$$

11. $\log_x 27 = 3$

$$x^3 = 27$$
$$x^3 = 3^3$$
$$x = 3$$

13. $\log_5 x = 2$

$$x = 5^2$$
$$x = 25$$

15. $\log_{36} x = \frac{1}{2}$

$$x = 36^{\frac{1}{2}}$$
$$x = \sqrt{36} = 6$$

17. $x = \log_{25} 125$

$$25^x = 125$$
$$\left(5^2\right)^x = 5^3$$
$$2x = 3$$
$$x = \frac{3}{2}$$

19. $\log(2x+5) = 2$

$$2x + 5 = 10^2$$
$$2x + 5 = 100$$
$$2x = 95$$
$$x = 47.5$$

21. $\log\left(2x + x^2\right) = 2$

$$2x + x^2 = 10^2$$
$$x^2 + 2x = 100$$
$$x^2 + 2x - 100 = 0$$
$$x = \frac{-2 \pm \sqrt{4 - 4(-100)}}{2}$$
$$x = -1 \pm \sqrt{101}$$
$$x = -11.0 \text{ or } 9.05$$

23. $\ln 6 + \ln(x-2) = \ln(3x-2)$

$$\ln\left[6(x-2)\right] = \ln(3x-2)$$
$$6x - 12 = 3x - 2$$
$$3x = 10$$
$$x = \frac{10}{3}$$

25. $\ln(5x+2) - \ln(x+6) = \ln 4$

$$\ln\frac{5x+2}{x+6} = \ln 4$$
$$\frac{5x+2}{x+6} = 4$$
$$5x + 2 = 4(x+6)$$
$$5x + 2 = 4x + 24$$
$$x = 22$$

27. $\ln x + \ln(x+2) = 1$

$$\ln\left[x(x+2)\right] = 1$$
$$x^2 + 2x = e$$
$$x^2 + 2x - e = 0$$
$$x = \frac{-2 \pm \sqrt{4 + 4e}}{2} = -1 \pm \sqrt{1+e}$$
$$x = -2.928 \text{ or } 0.928$$

$x = -2.928$ cannot be a solution since a logarithm cannot be taken of a negative number, so $x = 0.928$ is the only solution.

29. $2\log x - \log(1-x) = 1$

$$\log\left(\frac{x^2}{1-x}\right) = 1$$

$$\frac{x^2}{1-x} = 10$$

$$x^2 = 10(1-x)$$

$$x^2 + 10x - 10 = 0$$

$$x = \frac{-10 \pm \sqrt{100+40}}{2}$$

$$x = \frac{-10 \pm \sqrt{140}}{2}$$

$x = -10.916$ or 0.916

$x = 0.916$ is the only solution that works since we cannot take the logarithm of a negative number.

31. $\log(x^2-4) - 1 = \log(x+2)$

$$\log(x^2-4) - \log(x+2) = 1$$

$$\log\left(\frac{x^2-4}{x+2}\right) = 1$$

$$\frac{(x-2)(x+2)}{x+2} = 10$$

$$x - 2 = 10$$

$$x = 12$$

33. $\log(x^2-1) - 2 = \log(x+1)$

$$\log(x^2-1) - \log(x+1) = 2$$

$$\log\left(\frac{x^2-1}{x+1}\right) = 2$$

$$\frac{(x-1)(x+1)}{x+1} = 10^2 = 100$$

$$x - 1 = 100$$

$$x = 101$$

Applications

35. $h = 60470\log\frac{B_2}{B_1}$

$$3909 = 60470\log\left(\frac{29.66}{B_1}\right)$$

$$\log\left(\frac{29.66}{B_1}\right) = \frac{3909}{60470}$$

$$\frac{29.66}{B_1} = 10^{0.06464}$$

$$B_1 = \frac{29.66}{10^{0.06464}} = 25.56 \text{ in. Hg}$$

37. $G = 10\log\frac{P_2}{P_1}$

$$-3.25 = 10\log\left(\frac{P_2}{2750}\right)$$

$$\log\left(\frac{P_2}{2750}\right) = \frac{-3.25}{10}$$

$$\frac{P_2}{2750} = 10^{-0.325}$$

$$P_2 = 2750\left(10^{-0.325}\right) = 1300 \text{ kW}$$

39.
$$q=\frac{2\pi k(t_1-t_2)}{\ln(r_2/r_1)}$$
$$215=\frac{2\pi(0.044)528}{\ln(r_2/(11.4/2))}$$
$$\ln(r_2/5.7)=\frac{2\pi(0.044)528}{215}$$
$$\ln\left(\frac{r_2}{5.7}\right)=0.679$$
$$\frac{r_2}{5.7}=e^{0.679}$$
$$r_2=5.7e^{0.679}=11.2\text{ in.}$$

41.
$$pH=-\log_{10}C$$
$$-\log_{10}C=7.0$$
$$C=10^{-7.0}$$

43.
$$R=\log\frac{a}{a_0}=\log a-\log a_0$$
$$6.2=\log a_2-\log a_0$$
$$-(4.6=\log a_1-\log a_0)$$
$$1.6=\log a_2-\log a_1$$
$$1.6=\log\frac{a_2}{a_1}$$
$$\frac{a_2}{a_1}=10^{1.6}=40$$
The stronger is 40 times greater than the weaker.

45. $G=10\log\frac{P_2}{P_1}=10\log\left(\frac{0.5P_1}{P_1}\right)=-3.01\text{ dB}$

47. $G=10\log\frac{P_2}{P_1}=10\log\left(\frac{400\text{ W}}{20\text{ W}}\right)=13\text{ dB}$

49. $G_V=20\log\frac{V_2}{V_1}=20\log\left(\frac{84\text{ V}}{1.0\text{ V}}\right)=38\text{ dB}$

51. $G_V=20\log\frac{V_2}{V_1}=20\log\left(\frac{0.9V_1}{V_1}\right)=-0.915\text{ dB}$

53. $G=20\log\frac{d_f}{d_n}=20\log\left(\frac{2d_n}{d_n}\right)=6.02\text{ dB}$

Chapter 18 • Review Problems

1.
$$x^{5.2}=352$$
$$\log_x 352=5.2$$

3.
$$24^{1.4}=x$$
$$\log_{24}x=1.4$$

5.
$$\log_x 5.2=124$$
$$x^{124}=5.2$$

7.
$$\log_{81}27=x$$
$$81^x=27$$
$$\left(3^4\right)^x=3^3$$
$$4x=3$$
$$x=\frac{3}{4}$$

9.
$$\log_x 32=-\frac{5}{7}$$
$$x^{-5/7}=32$$
$$\left(x^{-5/7}\right)^7=\left(2^5\right)^7$$
$$x^{-5}=2^{35}$$
$$\left(x^{-5}\right)^{-1/5}=\left(2^{35}\right)^{-1/5}$$
$$x=2^{-7}$$
$$x=\frac{1}{128}$$

11. $\log\frac{3x}{z}=\log 3+\log x-\log z$

13. $\log 5+\log 2=\log(5\cdot 2)=\log 10$

15. $\frac{1}{2}\log p - \frac{1}{4}\log q = \log p^{1/2} - \log q^{1/4} = \log \frac{\sqrt{p}}{\sqrt[4]{q}}$

17. $\log 364 = 2.5611$

19. $\log 18.6 = 1.2695$

21. $\log x = 2.846 \Rightarrow x = 10^{2.846} = 701.5$

23. $\log x = -0.473 \Rightarrow x = 10^{-0.473} = 0.337$

25. $\ln 84.72 = 4.4394$

27. $\ln 0.00873 = -4.7410$

29. $\ln x = 1.473 \Rightarrow x = e^{1.473} = 4.362$

31. $\log x = \frac{4.837}{\ln 10} = 2.101$

33. $\ln x = 5.837 \ln 10 = 13.44$

35.
$$4.88^x = 152$$
$$x \log 4.88 = \log 152$$
$$x = \frac{\log 152}{\log 4.88} = 3.17$$

37.
$$3\log x - 3 = 2\log x^2$$
$$\log x^3 - \log x^4 = 3$$
$$\log \frac{x^3}{x^4} = 3$$
$$\frac{1}{x} = 10^3$$
$$x = \frac{1}{10^3} = 0.001$$

39. $y = a\left(1 + \frac{n}{m}\right)^{mt} = 1500\left(1 + \frac{.065}{4}\right)^{4\times 5} = \$2071$

41. $y = ae^{-nt} = 2250e^{-0.0500\times 20.0} = 828$ rev/min

43. $t = \frac{\ln 2}{n} = \frac{\ln 2}{0.030} = 23$ yr

47.
$$G_V = 20\log \frac{V_2}{V_1}$$
$$20\log\left(\frac{2500 \text{ V}}{V_1}\right) = -0.50 \text{ dB}$$
$$\log\left(\frac{2500}{V_1}\right) = -0.025$$
$$\frac{2500}{V_1} = 10^{-0.025}$$
$$V_1 = \frac{2500}{10^{-0.025}} = 2600 \text{ V}$$

Chapter 19: Complex Numbers

Exercise 19.1 • Complex Numbers in Rectangular Form

Addition and Subtraction of Complex Numbers

1. $(3-2i)+(-4+3i)=(3-4)+(-2+3)i=-1+i$

3. $(a-3i)+(a+5i)=(a+a)+(-3+5)i=2a+2i$

5. $\left(\frac{1}{2}+\frac{i}{3}\right)+\left(\frac{1}{4}-\frac{i}{6}\right)=\left(\frac{1}{2}+\frac{1}{4}\right)+\left(\frac{1}{3}-\frac{1}{6}\right)i=\frac{3}{4}+\frac{i}{6}$

7. $(2.28-1.46i)+(1.75+2.66i)=(2.28+1.75)+(-1.46+2.66)i=4.03+1.20i$

Powers of i

9. $i^5=i^4i=\left(i^2\right)^2 i=(-1)^2 i=i$

11. $i^{21}=i^{20}i=\left(i^2\right)^{10} i=(-1)^{10} i=i$

Multiplication of Complex Numbers

13. $7\times 2i=14i$

15. $3i\times 5i=15i^2=15(-1)=-15$

17. $4\times 2i\times 3i\times 4i=96i^3=96(-i)=-96i$

19. $(5i)^2=25i^2=25(-1)=-25$

21. $2(3-4i)=6-8i$

23. $4i(5-2i)=20i-8i^2=20i-(-1)8=8+20i$

25. $(3-5i)(2+6i)=6+18i-10i-30i^2=6+30+8i=36+8i$

27. $(6+3i)(3-8i)=18-48i+9i-24i^2=18+24-39i=42-39i$

29. $(5-2i)^2=(5-2i)(5-2i)=25-10i-10i+4i^2=25-4-20i=21-20i$

Graphing Complex Numbers

31. $2+5i$
33. $3-2i$
35. $5i$

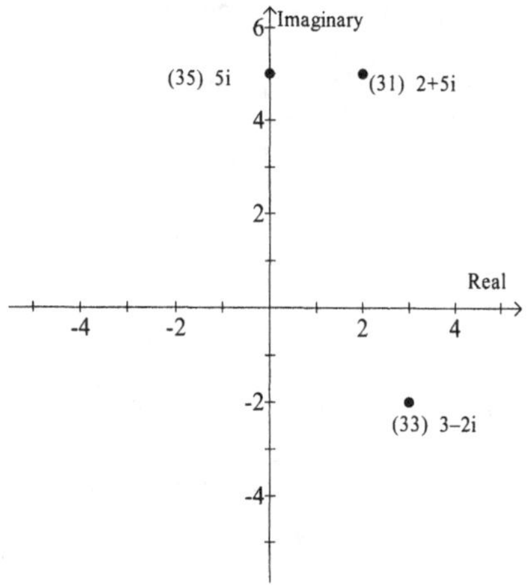

37. The conjugate of $2-3i$ is $2+3i$.

39. The conjugate of $p+qi$ is $p-qi$.

41. The conjugate of $-mi+n=n-mi$ is $n+mi$.

Division of Complex Numbers

43. $8i \div 4 = \dfrac{8i}{4} = 2i$

45 $12i \div 6i = \dfrac{12i}{6i} = 2$

47. $(4+2i) \div 2 = \dfrac{4+2i}{2} = \dfrac{2(2+i)}{2} = 2+i$

49. $(-2+3i) \div (1-i) = \dfrac{-2+3i}{1-i} \cdot \dfrac{1+i}{1+i} = \dfrac{-2-2i+3i+3i^2}{1-i+i-i^2} = \dfrac{-2-3+i}{1+1} = -\dfrac{5}{2} + \dfrac{i}{2}$

51. $(7i+2) \div (3i-5) = \dfrac{2+7i}{-5+3i} \cdot \dfrac{-5-3i}{-5-3i} = \dfrac{-10-6i-35i-21i^2}{25+15i-15i-9i^2} = \dfrac{-10+21-41i}{25+9} = \dfrac{11}{34} - \dfrac{41}{34}i$

Exercise 19.2 • Complex Numbers in Polar Form

Converting from Rectangular to Polar Form

1. $5+4i$

$r = \sqrt{5^2+4^2} = \sqrt{41} = 6.40$

$\tan\theta = \dfrac{4}{5} \Rightarrow \theta = \arctan\left(\dfrac{4}{5}\right) = 38.7^\circ$

$5+4i = 6.40\angle 38.7^\circ$

3. $4-3i$

$r = \sqrt{(4)^2+(-3)^2} = \sqrt{25} = 5$

$\tan\theta_{\text{Ref}} = \left|\dfrac{-3}{4}\right| \Rightarrow \theta_{\text{Ref}} = \arctan\left|\dfrac{-3}{4}\right| = 37^\circ$

In Quadrant IV, $\theta = 360^\circ - 37^\circ = 323^\circ$

$4-3i = 5\angle 323^\circ$

5. $-5-2i$

$r=\sqrt{(-5)^2+(-2)^2}=\sqrt{29}=5.39$

$\tan\theta_{\text{Ref}}=\left|\frac{-2}{-5}\right| \Rightarrow \theta_{\text{Ref}}=\arctan\left|\frac{-2}{-5}\right|=22^\circ$

In Quadrant III, $\theta=180^\circ+22^\circ=202^\circ$

$-5-2i=5.39\angle 202^\circ$

7. $-9-5i$

$r=\sqrt{(-9)^2+(-5)^2}=\sqrt{106}=10.3$

$\tan\theta_{\text{Ref}}=\left|\frac{-5}{-9}\right| \Rightarrow \theta_{\text{Ref}}=\arctan\left|\frac{-5}{-9}\right|=29^\circ$

In Quadrant III, $\theta=180^\circ+29^\circ=209^\circ$

$-9-5i=10.3\angle 209^\circ$

9. $-4-7i$

$r=\sqrt{(-4)^2+(-7)^2}=\sqrt{65}=8.06$

$\tan\theta_{\text{Ref}}=\left|\frac{-7}{-4}\right| \Rightarrow \theta_{\text{Ref}}=\arctan\left|\frac{-7}{-4}\right|=60^\circ$

In Quadrant III, $\theta=180^\circ+60^\circ=240^\circ$

$-4-7i=8.06\angle 240^\circ$

Converting from Rectangular to Polar Form

11. $9\angle 59^\circ$

$a=9\cos 59^\circ=4.64$

$b=9\sin 59^\circ=7.71$

$4.64+7.71i$

13. $7\angle -53^\circ$

$a=7\cos\left(-53^\circ\right)=4.21$

$b=7\sin\left(-53^\circ\right)=-5.59$

$4.21-5.59i$

Multiplying Complex Numbers in Polar Form

15. $\left(5.82\angle 44.8^\circ\right)\left(2.77\angle 10.1^\circ\right)=(5.82\times 2.77)\angle\left(44.8^\circ+10.1^\circ\right)=16.1\angle 54.9^\circ$

17. $\left(8\angle 45^\circ\right)\left(7\angle 15^\circ\right)=(8\times 7)\angle\left(45^\circ+15^\circ\right)=56\angle 60^\circ$

Dividing Complex Numbers in Polar Form

19. $\frac{58.3\angle 77.4^\circ}{12.4\angle 27.2^\circ}=\frac{58.3}{12.4}\angle\left(77.4^\circ-27.2^\circ\right)=4.70\angle 50.2^\circ$

21. $\frac{50\angle 72^\circ}{5\angle 12^\circ}=\frac{50}{5}\angle\left(72^\circ-12^\circ\right)=10\angle 60^\circ$

Exercise 19.3 • Complex Numbers on the Calculator

See Examples for commands for specific calculator. Problems are answered in previous sections without calculators.

Exercise 19.4 • Vector Operations Using Complex Numbers

Components of a Vector

The following are done on a TI-83 Plus with the calculator in radian and rectangular modes.

1. 385∠83.5°

```
385e^(83.5°i
    43.58+382.53i
```

The horizontal component is 43.6, and the vertical component is 383.

The following are done on a TI-89 with the calculator in degree and rectangular modes.

3. 28.4∠73.0°

```
F1▾   F2▾     F3▾  F4▾   F5     F6▾
Tools Algebra Calc Other PrgmIO Clean Up

■( 28.4 ∠ 73.)
              8.30336 + 27.1591·i
(28.4 ∠ 73.)
MAIN        DEG AUTO     FUNC      1/30
```

The horizontal component is 8.30, and the vertical component is 27.2.

Resultants of Several Vectors

The following are done on a TI-83 Plus with the calculator in radian and rectangular modes. The angle in the resultant will need to be converted back to degree mode.

5. 284∠33.6° and 184∠11.5°

```
284e^(33.6°i)+18
4e^(11.5°i)
459.723e^(.435i)
.435*180/π
           24.924
```

460∠24.9°

The following are done on a TI-89 with the calculator in degree and rectangular modes.

7. 16.8∠22.5°, 38.5∠73.4°, and 6.48∠125.5°

```
F1▾   F2▾     F3▾  F4▾   F5     F6▾
Tools Algebra Calc Other PrgmIO Clean Up

■( 16.8 ∠ 22.5) + ( 38.5 ∠ 73.▶
           ( 53.6642 ∠ 64.9084)
...5 ∠ 73.4)+(6.48 ∠ 125.5...
MAIN        DEG AUTO     FUNC      1/30
```

53.7∠64.9°

Applications

9. 58.4 lb∠38.5°

```
58.4e^(38.5°i
    45.704+36.355i
```

The horizontal component is 45.7 lb, and the vertical component is 36.4 lb.

11. 753 lb∠48.3° and 824 lb∠0°

```
F1▾   F2▾     F3▾  F4▾   F5     F6▾
Tools Algebra Calc Other PrgmIO Clean Up

■(753 ∠ 48.3)+(824 ∠ 0)
            (1439.27 ∠ 22.9936)
(753 ∠ 48.3)+(824 ∠ 0)
MAIN      DEG AUTO      FUNC      1/30
```

1440 lb∠23.0°

Exercise 19.5 • Alternating Current Applications

1. $I_{eff} = \dfrac{I_m}{\sqrt{2}} = \dfrac{69.3}{\sqrt{2}} = 49.0 \text{ A}$

3. $V_m = V_{eff}\sqrt{2} = 634\sqrt{2} = 897 \text{ V}$

Alternating Current and Voltage in Complex Form

5. $i = 250\sin\left(\omega t + 25^\circ\right)$

$\text{I} = I_{eff}\angle 33^\circ = \dfrac{250}{\sqrt{2}}\angle 33^\circ = 177\angle 33^\circ$

7. $v = 57\sin\left(\omega t - 90^\circ\right)$

$\text{V} = v_{eff}\angle -90^\circ = \dfrac{57}{\sqrt{2}}\angle -90^\circ = 40\angle -90^\circ$

9. $v = 144\sin(\omega t)$

$\text{V} = v_{eff}\angle 0^\circ = \dfrac{144}{\sqrt{2}}\angle 0^\circ = 102\angle 0^\circ$

11. $\text{V} = 150\angle 0^\circ$

$v = 150\sqrt{2}\sin(\omega t) = 212\sin(\omega t)$

13. $\text{V} = 300\angle -90^\circ$

$v = 300\sqrt{2}\sin\left(\omega t - 90^\circ\right) = 424\sin\left(\omega t - 90^\circ\right)$

15. $\text{I} = 7.5\angle 0^\circ$

$i = 7.5\sqrt{2}\sin(\omega t) = 11\sin(\omega t)$

Complex Impedance

17. In rectangular form $\mathbf{Z} = R + Xi = 155 + 0i$.

$\phi = \arctan\dfrac{X}{R} = \arctan\dfrac{0}{155} = 0^\circ$

$Z = \sqrt{155^2} = 155$

In polar form $\mathbf{Z} = 155\angle 0^\circ$.

19. $X = X_L - X_C = 0 - 18 = -18$

In rectangular form $\mathbf{Z} = R + Xi = 0 - 18i$.

In polar form $\mathbf{Z} = 18\angle 270^\circ$.

21. $X = X_L - X_C = 0 - 42 = -42$

In rectangular form $\mathbf{Z} = R + Xi = 72 - 42i$.

$\phi = \arctan\dfrac{X}{R} = \arctan\dfrac{-42}{72} = -30.3^\circ$

$Z = \sqrt{(-42)^2 + 72^2} = 83.4$

In polar form $\mathbf{Z} = 83.4\angle -30.3^\circ$.

23. $X = X_L - X_C = 148 - 720 = -572$

In rectangular form $\mathbf{Z} = R + Xi = 552 - 572i$.

$\phi = \arctan\dfrac{X}{R} = \arctan\dfrac{-572}{552} = -46.0^\circ$

$Z = \sqrt{(-572)^2 + 552^2} = 795$

In polar form $\mathbf{Z} = 795\angle -46.0^\circ$.

25. a) $\mathbf{I} = \frac{1.7}{\sqrt{2}}\angle 25^\circ = 1.202\angle 25^\circ$

$\mathbf{Z} = R + (X_L - X_C)i = 176 + 308i$

$\phi = \arctan\frac{X}{R} = \arctan\frac{308}{176} = 60.26^\circ$

$Z = \sqrt{308^2 + 176^2} = 354.74$

$\mathbf{Z} = 354.74\angle 60.26^\circ$

$\mathbf{V} = \mathbf{IZ} = \left(1.202\angle 25^\circ\right)\left(354.74\angle 60.26^\circ\right)$

$= 426.4\angle 85.3^\circ$

$v = 426.4\sqrt{2}\sin\left(\omega t + 85.3^\circ\right)$

$= 603\sin\left(\omega t + 85.3^\circ\right)$

b) $\mathbf{I} = \frac{43}{\sqrt{2}}\angle -30^\circ = 30.4\angle -30^\circ$

$\mathbf{Z} = R + (X_L - X_C)i$

$= 4.78 + (2.35 - 7.21)i = 4.78 - 4.86i$

$\phi = \arctan\frac{X}{R} = \arctan\frac{-4.86}{4.78} = -45.48^\circ$

$Z = \sqrt{(-4.86)^2 + 4.78^2} = 6.817$

$\mathbf{Z} = 6.817\angle -45.48^\circ$

$\mathbf{V} = \mathbf{IZ} = \left(30.4\angle -30^\circ\right)\left(6.817\angle -45.48^\circ\right)$

$= 207.2\angle -75.5^\circ$

$v = 207.2\sqrt{2}\sin\left(\omega t - 75.5^\circ\right)$

$= 293\sin\left(\omega t - 75.5^\circ\right)$

Chapter 19 • Review Problems

1. $i^{17} = i^{16}i = \left(i^2\right)^8 i = (-1)^8 i = i$

3. $(7 - 3i) + (2 + 5i) = 7 + 2 + (-3 + 5)i = 9 + 2i$

5. $52\cos 50^\circ = 33.42 \quad 52\sin 50^\circ = 39.83$

$28\cos 12^\circ = 27.39 \quad 28\sin 12^\circ = 5.82$

$52\angle 50^\circ + 28\angle 12^\circ = (33.42 + 39.83i) + (27.39 + 5.82i) = 60.8 + 45.7i$

7. $(2 - i)(3 + 5i) = 6 + 10i - 3i - 5i^2 = 6 + 5 + 7i = 11 + 7i$

9. $\left(2\angle 20^\circ\right)\left(6\angle 18^\circ\right) = (2 \times 6)\angle\left(20^\circ + 18^\circ\right) = 12\angle 38^\circ$

11. $(9 - 3i) \div (4 + i) = \frac{9 - 3i}{4 + i} \cdot \frac{4 - i}{4 - i} = \frac{36 - 9i - 12i + 3i^2}{16 - 4i + 4i - i^2} = \frac{36 - 3 - 21i}{16 + 1} = \frac{33}{17} - \frac{21}{17}i$

13. $\left(16\angle 85^\circ\right) \div \left(8\angle 40^\circ\right) = \frac{16\angle 85^\circ}{8\angle 40^\circ} = \frac{16}{8}\angle\left(85^\circ - 40^\circ\right) = 2\angle 45^\circ$

15. $7+4i$

17. $6\angle 135^\circ = -3\sqrt{2}+3\sqrt{2}i$

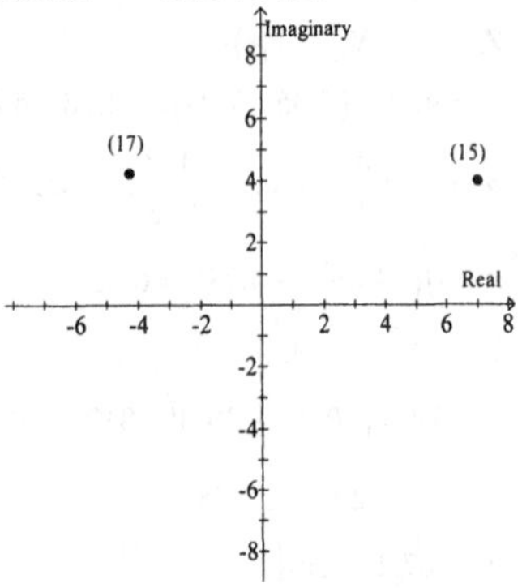

19. $(-4+3i)^2 = (-4+3i)(-4+3i) = 16-12i-12i+9i^2 = 16-9-24i = 7-24i$

21. $\left(5\angle 10^\circ\right)^3 = 5^3\angle 3\left(10^\circ\right) = 125\angle 30^\circ$

23. $i = 45\sin\left(\omega t + 32^\circ\right) \Rightarrow \mathbf{I} = \frac{45}{\sqrt{2}}\angle 32^\circ = 32\angle 32^\circ$

25. $\text{Period} = 0.4\text{ s} = \frac{2\pi}{\omega} \Rightarrow \omega = \frac{2\pi}{0.4} = 15.7\text{ rad/s}$

$\phi = (0.05\text{ s})(15.7\text{ rad/s}) = 0.785\text{ rad} = 45^\circ$

$i = 1.25\sin\left(15.7t - 45^\circ\right) \Rightarrow \mathbf{I} = \frac{1.25}{\sqrt{2}}\angle -45^\circ = 0.884\angle -45^\circ$

27. a) $x^2+9 = (x+3i)(x-3i)$

b) $b^2+25 = (b+5i)(b-5i)$

c) $4y^2+z^2 = (2y+zi)(2y-zi)$

d) $25a^2+9b^2 = (5a+3bi)(5a-3bi)$

Chapter 20: Sequences, Series, and the Binomial Theorem

Exercise 20.1 • Sequences and Series

Take $n=1,2,3,4,$ and 5

1. $u_n = 3n;$ $\quad 3(1)+3(2)+3(3)+3(4)+3(5)+\cdots+3n+\cdots$
$= 3+6+9+12+15+\cdots+3n+\cdots$

3. $u_n = \dfrac{n+1}{n^2};$ $\quad \dfrac{(1)+1}{(1)^2}+\dfrac{(2)+1}{(2)^2}+\dfrac{(3)+1}{(3)^2}+\dfrac{(4)+1}{(4)^2}+\dfrac{(5)+1}{(5)^2}+\cdots+\dfrac{n+1}{n^2}+\cdots$
$= 2+\dfrac{3}{4}+\dfrac{4}{9}+\dfrac{5}{16}+\dfrac{6}{25}+\cdots+\dfrac{n+1}{n^2}+\cdots$

5. $2+4+6+\cdots$ Each term is even. The general term is $u_n = 2n$. The next two terms are 8 and 10.

7. $\dfrac{2}{4}+\dfrac{4}{5}+\dfrac{8}{6}+\dfrac{16}{7}+\cdots$ The numerators look to be increasing by powers of 2. The denominators are increasing by 1 after starting at 4. The general term is $u_n = \dfrac{2^n}{n+3}$. The next two terms are $\dfrac{32}{8}$ and $\dfrac{64}{9}$.

Recursion Relations

9. $1+5+9+\cdots$ Each term is four more than the previous term. The general term is $u_n = u_{n-1}+4$. The next two terms are 13 and 27.

11. $3+9+81+\cdots$ Each term is the square of the previous term. The general term is $u_n = (u_{n-1})^2$. The next two terms are 6561 and 43,046,721.

Exercise 20.2 • Arithmetic and Harmonic Progressions

General Term Use $a_n = a+(n-1)d$.

1. Find the fifteenth term of an AP with first term 4 and common difference 3.
$a=4,\ n=15,\ d=3$ $\quad a_{15} = 4+(15-1)3 = 46$

3. Find the twelfth term of an AP with first term -1 and common difference 4.
$a=-1,\ n=12,\ d=4$ $\quad a_{12} = -1+(12-1)4 = 43$

5. Find the eleventh term of the AP: $9, 13, 17, \cdots$.
$a=9,\ n=11,\ d=4$ $\quad a_{11} = 9+(11-1)4 = 49$

7. Find the ninth term of the AP: $x, x+3y, x+6y, \cdots$.
$a=x,\ n=9,\ d=3y$ $\quad a_9 = x+(9-1)3y = x+24y$

9. First term is 3 and thirteenth term is 55.

$a=3$

$55=3+(13-1)d \Rightarrow 12d=52 \Rightarrow d=\frac{52}{12}=\frac{13}{3}=4\frac{1}{3}$

The first five terms are $3, 7\frac{1}{3}, 11\frac{2}{3}, 16, 20\frac{1}{3}, \cdots$.

11. Seventh term is 41 and fifteenth term is 89.

$89=a+14d$ Then $41=a+6(6)$

$\underline{41=a+\ 6d}$ $a=5$

$48= \quad 8d \Rightarrow d=6$

The first five terms are $5, 11, 17, 23, 29, \cdots$.

13. Find the first term of an AP whose common difference is 3 and whose seventh term is 11.

$d=3$

$11=a+(7-1)3$

$11=a+(6)3 \Rightarrow a=-7$

Sum of an Arithmetic Progression Use $s_n=\frac{n}{2}[2a+(n-1)d]$

15. Find the sum of the first 12 terms of the AP: 3, 6, 9, 12, ...

$a=3,\ d=3,$ and $n=12 \Rightarrow s_{12}=\frac{12}{2}[2(3)+(12-1)3]=6[6+33]=234$

17. Find the sum of the first 9 terms of the AP: 5, 10, 15, 20, ...

$a=5,\ d=5,$ and $n=9 \Rightarrow s_9=\frac{9}{2}[2(5)+(9-1)5]=\frac{9}{2}[10+40]=225$

19. How many terms of the AP 4, 7, 10, ... will give a sum of 375?

$a=4,\ d=3,$ and $s_n=375 \Rightarrow 375=\frac{n}{2}(2(4)+(n-1)3)$

$$750=8n+3n^2-3n$$

$$3n^2+5n-750=0$$

$$(3n+50)(n-15)=0$$

$$n=-\frac{50}{3} \text{ and } 15$$

n cannot be negative, so 15 terms is the answer.

Arithmetic Means

21. Insert two arithmetic means between 5 and 20.

We want an AP with four terms, with the first term of 5 and the fourth term of 20.

$20=5+3d \Rightarrow 3d=15 \Rightarrow d=5$

The arithmetic means are 10 and 15.

23. Insert four arithmetic means between –6 and –9.
We want an AP with six terms, with the first term of –6 and the sixth term of –9.

$-9=-6+5d \Rightarrow 5d=-3 \Rightarrow d=-\frac{3}{5}$

The arithmetic means are $-6\frac{3}{5}, -7\frac{1}{5}, -7\frac{4}{5}$, and $-8\frac{2}{5}$.

Harmonic Progressions

25. Find the fourth term of the harmonic progression: $\frac{3}{5}, \frac{3}{8}, \frac{3}{11}, \cdots$.

Taking the reciprocals gives the AP $\frac{5}{3}, \frac{8}{3}, \frac{11}{3}, \cdots$ with $a=\frac{5}{3}$ and $d=1$

$a_4=\frac{5}{3}+(4-1)1=\frac{5}{3}+3=\frac{14}{3}$

The fourth term in the harmonic progression is the reciprocal of $\frac{14}{3}$, or $\frac{3}{14}$.

Harmonic Means

27. The harmonic mean between two numbers a and b will be the reciprocal of the arithmetic mean between $\frac{1}{a}$ and $\frac{1}{b}$.

$\frac{1}{b}=\frac{1}{a}+2d \Rightarrow 2d=\frac{1}{b}-\frac{1}{a} \Rightarrow 2d=\frac{a-b}{ab} \Rightarrow d=\frac{a-b}{2ab}$

The arithmetic mean between $\frac{1}{a}$ and $\frac{1}{b}$ will be $\frac{1}{a}+\frac{a-b}{2ab}=\frac{2b+a-b}{2ab}=\frac{a+b}{2ab}$.

Hence, the harmonic mean between a and b will be the reciprocal of $\frac{a+b}{2ab}$, or $\frac{2ab}{a+b}$.

29. Insert three harmonic means between $\frac{6}{21}$ and $\frac{6}{5}$.

Taking reciprocals, our AP has five terms with the first term of $\frac{21}{6}$ and fifth term of $\frac{5}{6}$.

$\frac{5}{6}=\frac{21}{6}+4d \Rightarrow 4d=-\frac{16}{6} \Rightarrow d=-\frac{4}{6}$

The arithmetic means are $\frac{17}{6}, \frac{13}{6}$, and $\frac{9}{6}$.

Taking reciprocals gives the harmonic means of $\frac{6}{17}, \frac{6}{13}$, and $\frac{6}{9}$.

Applications

30. Initial loan is for \$10,000 with an annual payment of \$1000 plus 8% of the the unpaid balance.
a) Interest each year: 1st: $10000\times0.08=800$
2nd: $9000\times0.08=720$
3rd: $8000\times0.08=640$, etc.
This forms an AP with $a=800$ and $d=-80$. The interest for any given year is $a_n=800+(n-1)(-80)$.
b) The total interest will be $s_{10}=\frac{10}{2}[2(800)+9(-80)]=5[1600-720]=\$4400$.

31. A person puts in \$50 at the beginning of each month for 36 months. The person withdraws the interest which has a rate of 1% per month at the beginning of each month.

a) Interest each month: 1st: $50\times 0.01=0.50$

2nd: $100\times 0.01=1.00$

3rd: $150\times 0.01=1.50$, etc.

This forms an AP with $a=0.50$ and $d=0.50$. The interest for any given month is $a_n=0.5+(n-1)(0.5)$.

AP is \$0.50, \$1.00, \$1.50, ...

b) The total interest earned over the 36 months following the first deposit will be.

$$s_{36}=\frac{36}{2}\left[2(0.5)+35(0.5)\right]=18[1+17.5]=\$333.$$

33. Beginning salary is \$40,000 with an increase of \$2500 at the end of each year.

$a=40000$ and $d=2500$

Total amount over 10 years is $s_{10}=\frac{10}{2}\left[2(40000)+9(2500)\right]=5[80000+22500]=\$512,500$

35. Total distance fallen during the first t s is

$$s_t=\frac{t}{2}\left[2\frac{g}{2}+(t-1)g\right]=\frac{t}{2}[g+tg-g]=\frac{t}{2}[tg]=\frac{1}{2}gt^2$$

Exercise 20.3 • Geometric Progressions

General Term of a Geometric Progression Use $a_n=ar^{n-1}$

1. Find the fifth term of a GP with $a=5$ and $r=2$.

$$a_5=5\left(2^{5-1}\right)=5(16)=80$$

3. Find the sixth term of a GP with $a=-3$ and $r=5$.

$$a_6=-3\left(5^{6-1}\right)=-3(3125)=-9375$$

Sum of first n Terms of a Geometric Progression Use $s_n=\frac{a\left(1-r^n\right)}{1-r}=\frac{a-ra_n}{1-r}$

5. Find the sum of the first ten terms of the GP in problem 1.

$$s_{10}=\frac{5\left(1-2^{10}\right)}{1-2}=\frac{5(-1023)}{-1}=5115$$

7. Find the sum of the first eight terms of the GP in problem 3.

$$s_8=\frac{-3\left(1-5^8\right)}{1-5}=\frac{-3(-309,624)}{-4}=-292,968$$

Geometric Means Use $b=\pm\sqrt{ac}$

9. Insert a geometric mean between 5 and 45.

$b=\pm\sqrt{5\cdot 45}=\pm\sqrt{225}=\pm 15$

The GP is either $5, 15, 45$ or $5, -15, 45$

11. Insert a geometric mean between –10 and –90.

$b=\pm\sqrt{(-10)(-90)}=\pm\sqrt{900}=\pm 30$

The GP is either

$-10, 30, -90$ or $-10, -30, -90$

13. Insert two geometric means between 8 and 216.
$a = 8, a_4 = 216$, and $n = 4$

$$a_4 = ar^{4-1} \Rightarrow 216 = 8r^3 \Rightarrow r^3 = \frac{216}{8} = 27 \Rightarrow r = 3$$

The geometric means are 24 and 72.

15. Insert three geometric means between 5 and 1280.
$a = 5, a_5 = 1280$, and $n = 5$

$$a_5 = ar^{5-1} \Rightarrow 1280 = 5r^4 \Rightarrow r^4 = \frac{1280}{5} = 256 \Rightarrow r = \pm 4$$

The geometric means are either 20, 80, and 320 or -20, 80, and -320.

17. Using $y = e^{0.5t}$

t	0	1	2	3	4	5	6	7	8	9	10
y	1	1.65	2.72	4.48	7.39	12.18	20.09	33.12	54.60	90.02	148.41

$$r = \frac{a_{t+1}}{a_t} = \frac{e^{(t+1)/2}}{e^{t/2}} = e^{\frac{t+1}{2} - \frac{t}{2}} = e^{\frac{1}{2}} \approx 1.649$$

19. A casting starts at 1800° F and the temperature cools by 10% each minute.
$a = 1800, r = 0.9$, and $n = 60 \Rightarrow a_{60} = 1800(0.9)^{60} = 3.2°$ F

21. Radioactive material decays at a rate of 8% per year.
$r = 0.92$ and $a_n = 0.5a$

$$0.5a = a(0.92)^n$$

$$\log 0.5 = \log 0.92^n$$

$$n = \frac{\log 0.5}{\log 0.92} = 8.3 \text{ yr}$$

23. A ball dropped from a height of 10.0 ft rebounds to half the previous height on each bounce.
The total distance of a bounce is the distance up to the top and down to the ground. If the ball were thrown up to 10 ft and then let fall the distance traveled the first time the ball hit the ground would be 20 ft. Then 20(.5) = 10 ft for the second, 10(.5) = 5 ft for the third, etc. The total distance traveled when the ball hits the ground can be found by subtracting 10 ft (Since the first bounce only travels down and not both up and down) from the following.

$$a = 20.0, r = 0.50, \text{ and } n = 5 \Rightarrow s_5 = \frac{20\left(1 - 0.50^5\right)}{1 - 0.50} = 38.75 \text{ ft}$$

The total distance traveled will be 38.75 - 10 = 28.8 ft.

25. A person has two biological parents, who each have 2 biological parents, etc.

$$a = 2, r = 2, \text{ and } n = 5 \Rightarrow s_5 = \frac{2\left(1 - 2^5\right)}{1 - 2} = \frac{2(-31)}{-1} = 62 \text{ ancestors}$$

27. A reaction proceeds 15% faster for each 10° C increase in temperature.
$r = 1.15$ and $n = 5 \Rightarrow a_5 = a(1.15)^5 = 2.01a$ The speed of the reaction has increased by a factor of 2.

29. Energy consumption increase by 7.00% each year.
$r = 1.07$ and $n = 10.0 \Rightarrow a_{10} = a(1.07)^{10} = 1.97a$ The energy consumption has increased by a factor of 1.97.

31. A person deposited \$10,000 into a account that pays 6% interest compounded annually.
$a = 10000,\ r = 1.06$ and $n = 50 \Rightarrow a_{50} = 10000(1.06)^{50} = \$184{,}202$

33. A machine having a initial book value of \$100,000 depreciates by 40% each year.
$a = 100000,\ r = 0.60$ and $n = 5 \Rightarrow a_5 = 100000(0.60)^5 = \$7776$

Exercise 20.4 • Infinite Geometric Progressions

Limits

1. $\lim\limits_{b\to 0} b - c + 5 = 0 - c + 5 = 5 - c$

3. $\lim\limits_{b\to 0} \dfrac{3+b}{c+4} = \dfrac{3+0}{c+4} = \dfrac{3}{c+4}$

Sum of an Infinite, Decreasing Geometric Progression If $|r| < 1$, use $S = \lim\limits_{n\to\infty} s_n = \lim\limits_{n\to\infty} \dfrac{a - ra_n}{1-r} = \dfrac{a}{1-r}$.

5. 144, 72, 36, 18, ⋯
$a = 144$ and $r = 0.5$
$$S = \frac{144}{1-0.5} = 288$$

7. 10, 2, 0.4, 0.08, ⋯
$a = 10$ and $r = 0.2$
$$S = \frac{10}{1-0.2} = 12.5$$

Applications

9. A pendulum first swing is 10 in. Every swing after is 78% of the length of the previous swing.
$a = 10$ and $r = 0.78 \Rightarrow S = \dfrac{10}{1-0.78} = 45.5$ in. The pendulum travels 45.5 in. before it comes to rest.

Exercise 20.5 • The Binomial Theorem

Factorial Notation

1. $6! = 6\cdot 5\cdot 4\cdot 3\cdot 2\cdot 1 = 720$

3. $\dfrac{7!}{5!} = \dfrac{7\cdot 6\cdot 5\cdot 4\cdot 3\cdot 2\cdot 1}{5\cdot 4\cdot 3\cdot 2\cdot 1} = 7\cdot 6 = 42$

5. $\dfrac{7!}{3!\,4!} = \dfrac{7\cdot 6\cdot 5\cdot 4\cdot 3\cdot 2\cdot 1}{(3\cdot 2\cdot 1)(4\cdot 3\cdot 2\cdot 1)} = 7\cdot 5 = 35$

Binomials Raised to an Integral Power

The coefficients for the following were taken from Pascal's Triangle.

7. $(x+y)^7 = x^7 + 7x^6y + 21x^5y^2 + 35x^4y^3 + 35x^3y^4 + 21x^2y^5 + 7xy^6 + y^7$

9. $(3a-2b)^4 = (3a)^4 + 4(3a)^3(-2b) + 6(3a)^2(-2b)^2 + 4(3a)^1(-2b)^3 + (-2b)^4$
$= 81a^4 - 216a^3b + 216a^2b^2 - 96ab^3 + 16b^4$

11. $\left(x^{1/2}+y^{2/3}\right)^5=\left(x^{1/2}\right)^5+5\left(x^{1/2}\right)^4y^{2/3}+10\left(x^{1/2}\right)^3\left(y^{2/3}\right)^2+10\left(x^{1/2}\right)^2\left(y^{2/3}\right)^3+5\left(x^{1/2}\right)^1\left(y^{2/3}\right)^4+\left(y^{2/3}\right)^5$
$=x^{5/2}+5x^2y^{2/3}+10x^{3/2}y^{4/3}+10xy^2+5x^{1/2}y^{8/3}+y^{10/3}$

13. $\left(\frac{a}{b}-\frac{b}{a}\right)^6=\left(\frac{a}{b}\right)^6+6\left(\frac{a}{b}\right)^5\left(-\frac{b}{a}\right)+15\left(\frac{a}{b}\right)^4\left(-\frac{b}{a}\right)^2+20\left(\frac{a}{b}\right)^3\left(-\frac{b}{a}\right)^3+15\left(\frac{a}{b}\right)^2\left(-\frac{b}{a}\right)^4+6\left(\frac{a}{b}\right)\left(-\frac{b}{a}\right)^5+\left(-\frac{b}{a}\right)^6$
$=\left(\frac{a}{b}\right)^6-6\left(\frac{a}{b}\right)^4+15\left(\frac{a}{b}\right)^2-20+15\left(\frac{b}{a}\right)^2-6\left(\frac{b}{a}\right)^4+\left(\frac{b}{a}\right)^6$

15. $\left(2a^2+\sqrt{b}\right)^5=\left(2a^2\right)^5+5\left(2a^2\right)^4b^{1/2}+10\left(2a^2\right)^3\left(b^{1/2}\right)^2+10\left(2a^2\right)^2\left(b^{1/2}\right)^3+5\left(2a^2\right)\left(b^{1/2}\right)^4+\left(b^{1/2}\right)^5$
$=32a^{10}+80a^8b^{1/2}+80a^6b+40a^4b^{3/2}+10a^2b^2+b^{5/2}$

Verify the first four terms of each expansion

17. $\left(x^2+y^3\right)^8=\left(x^2\right)^8+8\left(x^2\right)^7\left(y^3\right)+\frac{8\cdot7}{2}\left(x^2\right)^6\left(y^3\right)^2+\frac{8\cdot7\cdot6}{3\cdot2}\left(x^2\right)^5\left(y^3\right)^3+\cdots$
$=x^{16}+8x^{14}y^3+28x^{12}y^6+56x^{10}y^9+\cdots$

19. $\left(a-b^4\right)^9=(a)^9+9(a)^8\left(-b^4\right)+\frac{9\cdot8}{2}(a)^7\left(-b^4\right)^2+\frac{9\cdot8\cdot7}{3\cdot2}(a)^6\left(-b^4\right)^3+\cdots$
$=a^9-9a^8b^4+36a^7b^8-84a^6b^{12}+\cdots$

General Term For the expansion of $(a+b)^n$ the rth term $=\frac{n!}{(r-1)!(n-r+1)!}a^{n-r+1}b^{r-1}$.

21. 7th term of $\left(a^2-2b^3\right)^{12}$

$n=12$ and $r=7$ gives $n-r+1=12-7+1=6$ and $\frac{n!}{(r-1)!(n-r+1)!}=\frac{12!}{6!6!}=924$

7th term $=924\left(a^2\right)^6\left(-2b^3\right)^6=59{,}136a^{12}b^{18}$

23. 4th term of $(2a-3b)^7$

$n=7$ and $r=4$ gives $n-r+1=7-4+1=4$ and $\frac{n!}{(r-1)!(n-r+1)!}=\frac{7!}{3!4!}=35$

4th term $=35(2a)^4(-3b)^3=-15{,}120a^4b^3$

25. 5th term of $\left(x-2\sqrt{y}\right)^{25}$

$n=25$ and $r=5$ gives $n-r+1=25-5+1=21$ and $\frac{n!}{(r-1)!(n-r+1)!}=\frac{25!}{4!21!}=12650$

5th term $=12650(x)^{21}\left(-2y^{1/2}\right)^4=202{,}400x^{21}y^2$

Binomials Raised to a Fractional or Negative Power
Write the first four terms of each infinite binomial series.

27. $$(1-a)^{2/3}=(1)^{2/3}+\frac{2}{3}(1)^{-1/3}(-a)+\frac{\left(\frac{2}{3}\right)\left(-\frac{1}{3}\right)}{2}(1)^{-4/3}(-a)^2+\frac{\left(\frac{2}{3}\right)\left(-\frac{1}{3}\right)\left(-\frac{4}{3}\right)}{3\cdot 2}(1)^{-7/3}(-a)^3\ldots$$
$$=1-\frac{2a}{3}-\frac{a^2}{9}-\frac{4a^3}{81}\ldots$$

29. $$(1+5a)^{-5}=(1)^{-5}+(-5)(1)^{-6}(5a)+\frac{(-5)(-6)}{2}(1)^{-7}(5a)^2+\frac{(-5)(-6)(-7)}{3\cdot 2}(1)^{-8}(5a)^3\ldots$$
$$=1-25a+375a^2-4375a^3\ldots$$

31. $$\frac{1}{\sqrt[6]{1-a}}=(1-a)^{-1/6}$$
$$=(1)^{-1/6}+\left(-\frac{1}{6}\right)(1)^{-7/6}(-a)+\frac{\left(-\frac{1}{6}\right)\left(-\frac{7}{6}\right)}{2}(1)^{-13/6}(-a)^2+\frac{\left(-\frac{1}{6}\right)\left(-\frac{7}{6}\right)\left(-\frac{13}{6}\right)}{3\cdot 2}(1)^{-19/6}(-a)^3\ldots$$
$$=1+\frac{a}{6}+\frac{7a^2}{72}+\frac{91a^3}{1296}\ldots$$

Chapter 20 • Review Problems

1. Find the sum of the first 7 terms of the AP: –4, –1, 2 ...
$a=-4,\ d=3,$ and $n=7\ \Rightarrow\ s_7=\frac{7}{2}[2(-4)+(7-1)3]=\frac{7}{2}[(-8)+18]=35$

3. Insert four arithmetic means between 3 and 18.
We want an AP with six terms, with the first term of 3 and the sixth term of 18.
$18=3+5d\ \Rightarrow\ 5d=15\ \Rightarrow\ d=3$
The arithmetic means are 6, 9, 12, and 15.

5. Insert four harmonic means between 2 and 12.
Taking reciprocals, our AP has six terms with the first term of $\frac{1}{2}$ and sixth term of $\frac{1}{12}$.
$\frac{1}{12}=\frac{1}{2}+5d\ \Rightarrow\ 5d=-\frac{5}{12}\ \Rightarrow\ d=-\frac{1}{12}$
The arithmetic means are $\frac{5}{12}, \frac{4}{12}, \frac{3}{12},$ and $\frac{2}{12}$.
Taking reciprocals gives the harmonic means of $2\frac{2}{5}$, 3, 4, and 6.

7. Find the sum of the first 10 terms of the AP: $1, 2\frac{2}{3}, 4\frac{1}{3}, \ldots$.
$a=1,\ d=\frac{5}{3},$ and $n=10\ \Rightarrow\ s_{10}=\frac{10}{2}\left[2(1)+(10-1)\frac{5}{3}\right]=5[2+15]=85$

9. $(a-2)^5 = a^5 + 5a^4(-2) + 10a^3(-2)^2 + 10a^2(-2)^3 + 5a(-2)^4 + (-2)^5$

$$= a^5 - 10a^4 + 40a^3 - 80a^2 + 80a - 32$$

11. Find the fifth term of a GP with $a = 3$ and $r = 2$.

$a_5 = 3\left(2^{5-1}\right) = 3(16) = 48$

13. $\lim_{b\to 0}(2b - x + 9) = 0 - x + 9 = 9 - x$

15. 4, 2, 1, $\cdots$

$a = 4$ and $r = 0.5 \Rightarrow S = \dfrac{4}{1-0.5} = 8$

17. $1, -\dfrac{2}{5}, \dfrac{4}{25}, \cdots$

$a = 1$ and $r = -\dfrac{2}{5} \Rightarrow S = \dfrac{1}{1+2/5} = \dfrac{1}{7/5} = \dfrac{5}{7}$

19. 150, 30, 6, $\cdots$

$a = 150$ and $r = \dfrac{1}{5} \Rightarrow S = \dfrac{150}{1-1/5} = \dfrac{150}{4/5} = \dfrac{375}{2}$ or $187\dfrac{1}{2}$

21. Find the sixth term of a GP with $a = 5$ and $r = 3$.

$a_6 = 5\left(3^{6-1}\right) = 5(243) = 1215$

23. Find the fifth term of a GP with $a = -4$ and $r = 4$.

$a_5 = -4\left(4^{5-1}\right) = -4(256) = -1024$

25. Insert two geometric means between 8 and 125.

$a = 8$, $a_4 = 125$, and $n = 4$

$a_4 = ar^{4-1} \Rightarrow 125 = 8r^3 \Rightarrow r^3 = \dfrac{125}{8} = \Rightarrow r = \dfrac{5}{2}$

The geometric means are 20 and 50.

27. $\dfrac{4!\,5!}{2!} = \dfrac{(4\cdot 3\cdot 2\cdot 1)(5\cdot 4\cdot 3\cdot 2\cdot 1)}{(2\cdot 1)} = 1440$

29. $\dfrac{7!\,3!}{5!} = \dfrac{(7\cdot 6\cdot 5\cdot 4\cdot 3\cdot 2\cdot 1)(3\cdot 2\cdot 1)}{5\cdot 4\cdot 3\cdot 2\cdot 1} = 42\cdot 6 = 252$

31. $\left(\dfrac{2x}{y^2} - y\sqrt{x}\right)^7 = \left(\dfrac{2x}{y^2}\right)^7 + 7\left(\dfrac{2x}{y^2}\right)^6\left(-yx^{1/2}\right) + \dfrac{7\cdot 6}{2}\left(\dfrac{2x}{y^2}\right)^5\left(-yx^{1/2}\right)^2 + \dfrac{7\cdot 6\cdot 5}{3\cdot 2}\left(\dfrac{2x}{y^2}\right)^4\left(-yx^{1/2}\right)^3 + \cdots$

$$= 128\frac{x^7}{y^{14}} - 448\frac{x^{13/2}}{y^{11}} + 672\frac{x^6}{y^8} - 560\frac{x^{11/2}}{y^5} + \cdots$$

33. $\frac{1}{\sqrt{1+a}}=(1+a)^{-1/2}$

$$=(1)^{-1/2}+\left(-\frac{1}{2}\right)(1)^{-3/2}(a)+\frac{\left(-\frac{1}{2}\right)\left(-\frac{3}{2}\right)}{2}(1)^{-5/2}(a)^2+\frac{\left(-\frac{1}{2}\right)\left(-\frac{3}{2}\right)\left(-\frac{5}{2}\right)}{3\cdot 2}(1)^{-7/2}(a)^3\ldots$$

$$=1-\frac{a}{2}+\frac{3a^2}{8}-\frac{5a^3}{16}\ldots$$

35. 8th term of $(3a-b)^{11}$

$n=11$ and $r=8$ gives $n-r+1=11-8+1=4$ and $\frac{n!}{(r-1)!(n-r+1)!}=\frac{11!}{7!4!}=330$

8th term $=330(3a)^4(-b)^7=-26,730a^4b^7$

Chapter 21: Introduction to Statistics and Probability

Exercise 1 ◊ Definitions and Terminology

1. Discrete since it is obtained by counting
3. Categorical since it belongs to one of several categories
5. Categorical since it belongs to one of several categories
7. 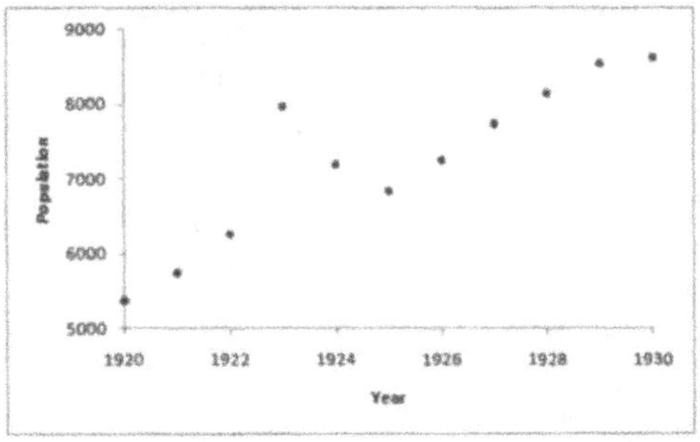

9. 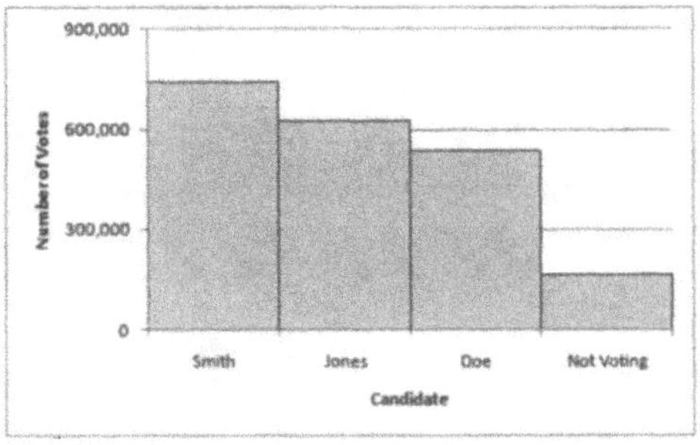

Exercise 2 ◊ Frequency Distributions

1. (a) Range = 172 – 111 = 61

(b, c)

Class Limits		Class Midpoints	Absolute Frequency	Relative Frequency
110.5	115.5	113	3	7.5%
115.5	120.5	118	4	10.0%
120.5	125.5	123	1	2.5%
125.5	130.5	128	3	7.5%
130.5	135.5	133	0	0.0%
135.5	140.5	138	2	5.0%
140.5	145.5	143	2	5.0%
145.5	150.5	148	6	15.0%
150.5	155.5	153	7	17.5%
155.5	160.5	158	2	5.0%
160.5	165.5	163	5	12.5%
165.5	170.5	168	3	7.5%
170.5	175.5	173	2	5.0%

3. (a) Range = 972 – 584 = 388

(b, c)

Class Limits		Class Midpoints	Absolute Frequency	Relative Frequency
500.1	550	525.05	0	0.0%
550.1	600	575.05	1	3.3%
600.1	650	625.05	2	6.7%
650.1	700	675.05	4	13.3%
700.1	750	725.05	3	10.0%
750.1	800	775.05	7	23.3%
800.1	850	825.05	1	3.3%
850.1	900	875.05	7	23.3%
900.1	950	925.05	4	13.3%
950.1	1000	975.05	1	3.3%

5.

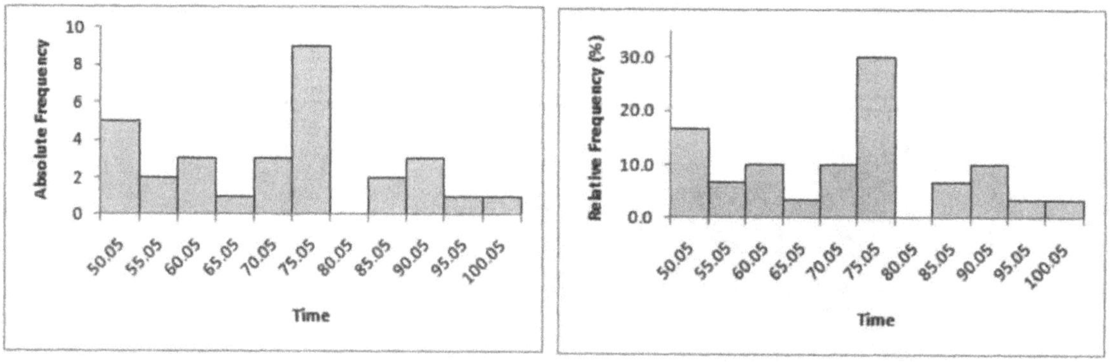

7.

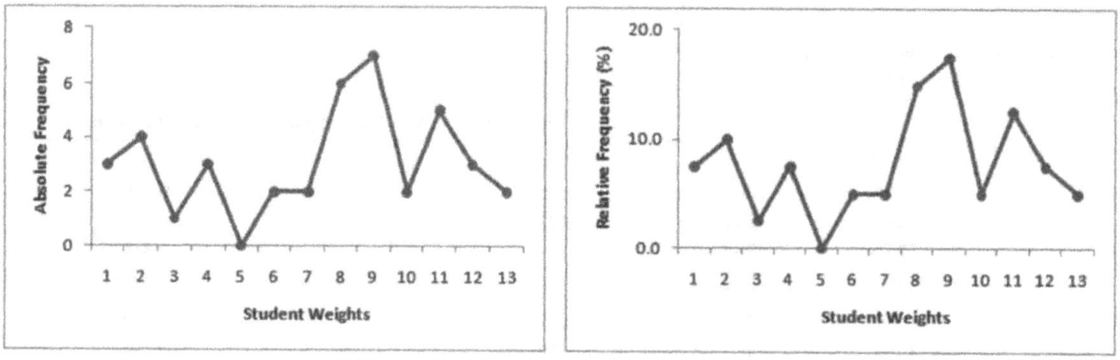

9.

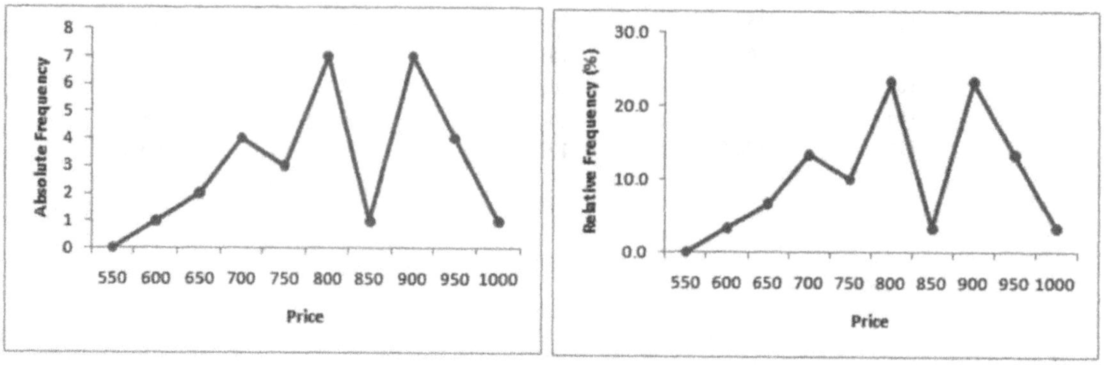

11.

Cumulative Frequency	
Absolute	*Relative*
5	16.7
7	23.3
10	33.3
11	36.6
14	46.7
23	76.7
23	76.7
25	83.3
28	93.3
29	96.7
30	100.0

13.

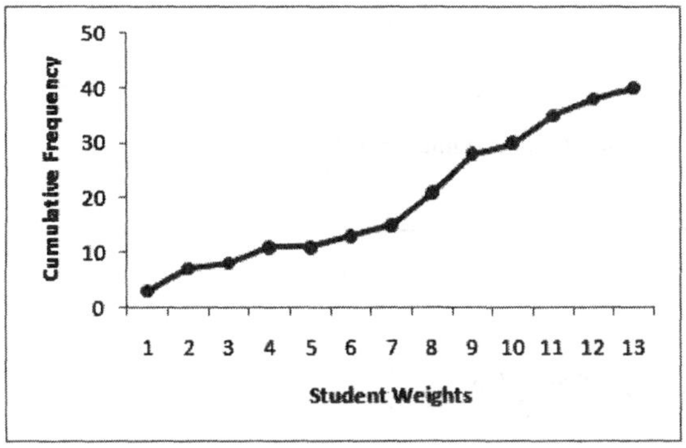

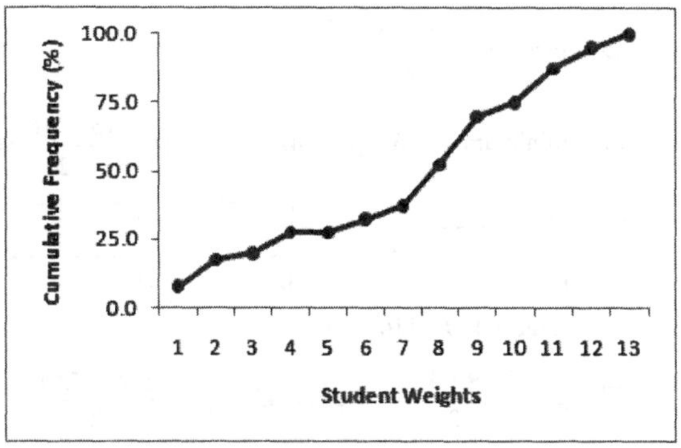

15.

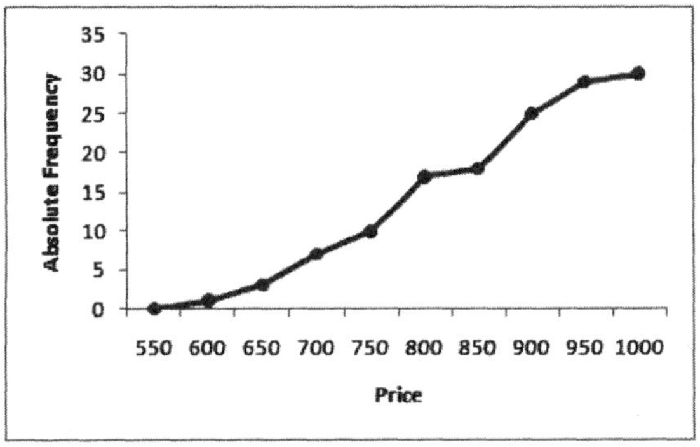

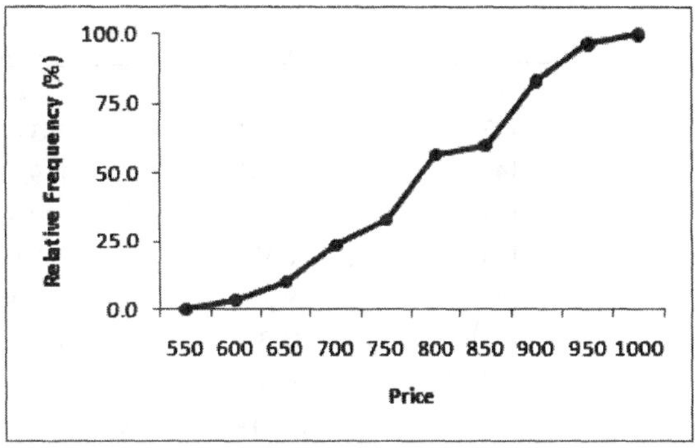

17.

4	3.5	8.9	9.6	8.4	9.4			
5	4.8	9.3	9.3	9.3	0.3			
6	1.4	9.3	6.3	9.3				
7	2.5	1.2	1.4	4.5	3.6	1.4	4.9	2.7	2.8
8	8.2	4.6	9.4	5.7	3.8			
9	9.2	2.4						

Exercise 3 ◊ Numerical Descriptions of Data

1. $\text{mean} = \dfrac{85+74+69+59+60+96+84+48+89+76+96+68+98+79+76}{15} = 77.13 = 77$

3. $\dfrac{\sum \text{student weights}}{40} = \dfrac{5810}{40} = 145.25 = 145 \text{ lb}$

5. $\dfrac{\sum \text{prices}}{30} = \dfrac{23873}{30} = 795.6667 = \$796$

7. $86(0.15) + 92(0.15) + 68(0.15) + 75(0.15) + 82(0.30) + 88(0.10) = 81.55 = 81.6$

9. $\text{midrange} = \dfrac{116 + 199}{2} = 157.5$

11. none

13. mode = 59.3

15. 48, 59, 60, 68, 69, 74, 76, **76**, 79, 84, 85, 89, 96, 96, 98: median = 76

17. Since there are an even number of data points, the median $= \dfrac{149 + 149}{2} = 149 \text{ lb}$

19. $\text{median} = \dfrac{789 + 795}{2} = \$792$

21. minimum $= 116$; $Q_1 = 142 = 68.5$; $Q_2 = \dfrac{153 + 163}{2} = 158$; $Q_3 = 164$; maximum $= 199$

23.

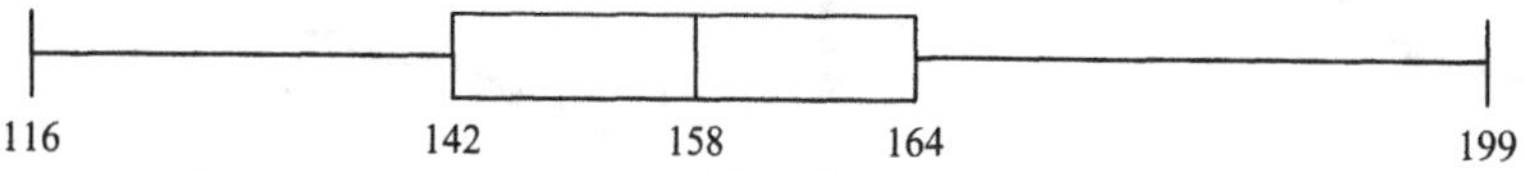

25. range = 199 – 116 = 83

27. $Q_1 = \dfrac{3.59 + 4.96}{2} = 4.275 = 4.28$ $Q_2 = \dfrac{6.89 + 7.91}{2} = 7.40$ $Q_3 = \dfrac{9.44 + 10.6}{2} = 10.02$

quartile range = 10.02 – 4.28 = 5.74

29.

x	$x - \mu$	$(x - \mu)^2$
116	-38.63	1492.277
127	-27.63	763.4169
142	-12.63	159.5169
153	-1.63	2.6569
163	8.37	70.0569
164	9.37	87.7969
173	18.37	337.4569
199	44.37	1968.697
	$\Sigma(x - \mu) = 0$	$\Sigma(x - \mu)^2 = 4881.875$

Since this is a sample, $s^2 = \dfrac{\sum(x-\mu)^2}{n-1} \Rightarrow \dfrac{4881.875}{8-1} = 697.411 = 697.4$

$$s = \sqrt{\frac{\sum(x-\mu)^2}{n-1}} \Rightarrow \sqrt{\frac{4881.875}{8-1}} = \sqrt{697.411} = 26.409 = 26.4$$

31.

x	$x - \mu$	$(x - \mu)^2$
48.4	-20.68	427.6624
48.5	-20.58	423.5364
48.9	-20.18	407.2324
49.4	-19.68	387.3024
49.6	-19.48	379.4704
50.3	-18.78	352.6884
54.8	-14.28	203.9184
59.3	-9.78	95.6484
59.3	-9.78	95.6484
59.3	-9.78	95.6484
61.4	-7.68	58.9824
66.3	-2.78	7.7284
69.3	0.22	0.0484
69.3	0.22	0.0484
71.2	2.12	4.4944
71.4	2.32	5.3824
71.4	2.32	5.3824
72.5	3.42	11.6964
72.7	3.62	13.1044
72.8	3.72	13.8384
73.6	4.52	20.4304
74.5	5.42	29.3764
74.9	5.82	33.8724
83.8	14.72	216.6784
84.6	15.52	240.8704
85.7	16.62	276.2244
88.2	19.12	365.5744
89.4	20.32	412.9024
92.4	23.32	543.8224
99.2	30.12	907.2144
	$\Sigma(x - \mu) = 0$	$\Sigma(x - \mu)^2 = 6036.428$

Since this is a population, $\sigma^2 = \dfrac{\sum(x-\mu)^2}{n} \Rightarrow \dfrac{6036.428}{30} = 201.214 = 201$

$$\sigma = \sqrt{\frac{\sum(x-\mu)^2}{n}} \Rightarrow \sqrt{\frac{6036.428}{30}} = \sqrt{201.214} = 14.185 = 14.2$$

Exercise 4 ◊ Introduction to Probability

1. $\dfrac{\text{number of ways to draw green}}{\text{total number of outcomes}} \Rightarrow \dfrac{8}{8+7} = \dfrac{8}{15}$

3. $\dfrac{\text{number of ways to get 2 heads and 2 tails}}{\text{total number of outcomes}} \Rightarrow \dfrac{6}{16} = \dfrac{3}{8}$

5. $\dfrac{\text{number of ways to get sum of 9}}{\text{total number of outcomes}} \Rightarrow \dfrac{4}{36} = \dfrac{1}{9}$

7. $\frac{\text{number of ways to get sum of 6}}{\text{total number of outcomes}} \cdot \frac{\text{number of ways to get sum of 6}}{\text{total number of outcomes}} \Rightarrow \frac{1}{6} \cdot \frac{1}{6} = \frac{1}{36}$

9. $P(\text{brown hair}) + P(\text{blue eyes}) - P(\text{brown hair and blue eyes}) \Rightarrow 0.55 + 0.15 - 0.07 = 0.63$

11. $P(7) + P(9) \Rightarrow \frac{1}{6} + \frac{1}{9} = \frac{5}{18}$

13. From Figure 26-3: $\frac{4}{30} = \frac{2}{15}$ or 0.133

15. $P(5) = \frac{5!}{(5-3)!3!}(0.4)^3(0.6)^{5-3} = 0.2304 = 0.230$

17. $P(7) = \frac{7!}{(7-6)!6!}(0.8)^6(0.2)^{7-6} = 0.36700 = 0.367$

19.

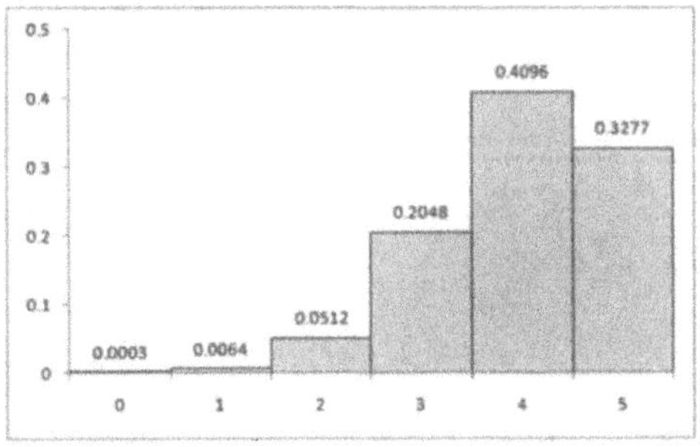

21. $P(7) = \frac{10!}{(10-7)!7!}(0.5)^7(0.5)^{10-7} = 0.1172 = 0.117$

23. $P(10) = \frac{15!}{(15-10)!10!}(0.15)^{10}(0.85)^{15-10} = 7.6836 \times 10^{-6} = 7.68 \times 10^{-6}$

Exercise 5 ◊ The Normal Curve

1. From Table 7, the area between the mean and 1.5 standard deviations is 0.4332.

3. The area between 0 and 0.8 standard deviations is 0.2881. Since the total area to the right of the mean is 0.5, the area to the left of $z = 0.8$ is $1 - (0.2881 + 0.5)$ or 0.2119.

5. $z(130) = \frac{130-163}{18} = -1.8 \Rightarrow \text{area} = 0.4641;\ z(170) = \frac{170-163}{18} = 0.4 \Rightarrow \text{area} = 0.1554$

   Area between 130 lb and 170 lb $\Rightarrow 0.4641 + 0.1554 = 0.6195$

   $\Rightarrow 0.6195(1000) = 619.5$ or 620 students

7. $z(195) = \frac{195-163}{18} = 1.8 \Rightarrow \text{area} = 0.5 - 0.4641 = 0.0359$

   $\Rightarrow 0.0359(1000) = 35.9$ or 36 students

9. $z(60) = \frac{60-82.6}{7.4} = -3.1 \Rightarrow \text{area} = 0.5 - 0.4990 = 0.001$

   $\Rightarrow 0.001(1000) = 0.5$ or 1 student

Exercise 6 ◊ Standard Errors

1. $\bar{x} \pm SE_x \Rightarrow \bar{x} \pm \frac{\sigma}{\sqrt{n}} \Rightarrow 69.47 \pm \frac{2.35}{\sqrt{49}} = 69.47 \pm 0.3357 \Rightarrow 69.47 \pm 0.34$ or 69.13 to 69.81

3. $s \pm SE_s = s \pm \frac{\sigma}{\sqrt{2n}} \Rightarrow 2.35 \pm \frac{2.35}{\sqrt{2(49)}} = 2.35 \pm 0.2374 = 2.35 \pm 0.24$ or 2.11 to 2.59

5. $164.0 \pm \frac{16.31}{\sqrt{32}} = 164.0 \pm 2.8832 = 164.0 \pm 2.88$ or 161.12 to 166.88

7. $16.31 \pm \frac{16.31}{\sqrt{2(32)}} = 16.31 \pm 2.0388 = 16.31 \pm 2.04$ or 14.27 to 18.35

9. $p \pm SE_p \Rightarrow p \pm \sqrt{\frac{p(1-p)}{n}}$; where $p = \frac{13}{52} \Rightarrow 0.25,\ n = 200$

$\Rightarrow 0.25 \pm \sqrt{\frac{0.25(1-0.25)}{200}} = 0.25 \pm 0.0306 = 0.25 \pm 0.031$ or 0.219 to 0.281

Exercise 7 ◊ Process Control

1. Central line $\Rightarrow \overline{p} = \frac{554}{20(1000)} = 0.0277$

UCL $\Rightarrow \overline{p} + 3SE_p = \overline{p} + 3\left(\sqrt{\frac{\overline{p}(1-\overline{p})}{n}}\right) \Rightarrow 0.0277 + 3\left(\sqrt{\frac{0.0277(1-0.0277)}{1000}}\right) = 0.0277 + 0.0156 = 0.0433$

LCL $\Rightarrow \overline{p} - 3SE_p = \overline{p} - 3\left(\sqrt{\frac{\overline{p}(1-\overline{p})}{n}}\right) \Rightarrow 0.0277 - 3\left(\sqrt{\frac{0.0277(1-0.0277)}{1000}}\right) = 0.0277 - 0.0156 = 0.0121$

3. Central line $\Rightarrow \overline{p} = \frac{2990}{20(500)} = 0.299$

Upper control limit $\Rightarrow \overline{p} + 3SE_p = \overline{p} + 3\left(\sqrt{\frac{\overline{p}(1-\overline{p})}{n}}\right) \Rightarrow 0.299 + 3\left(\sqrt{\frac{0.299(1-0.299)}{500}}\right) = 0.299 + 0.0614 = 0.3604$

Lower control limit $\Rightarrow \overline{p} - 3SE_p = \overline{p} - 3\left(\sqrt{\frac{\overline{p}(1-\overline{p})}{n}}\right) \Rightarrow 0.299 - 3\left(\sqrt{\frac{0.299(1-0.299)}{500}}\right) = 0.299 - 0.0614 = 0.2376$

5. $\overline{\overline{X}} = \frac{\sum \overline{X}}{N} \Rightarrow \frac{4177.3}{21} = 198.919 = 198.9$

UCL $= \overline{\overline{X}} + A_2\overline{R}$ (where $A_2 = 0.577$ by Table 11) $\Rightarrow 198.9 + 0.577(24.8) = 198.9 + 14.31 = 213.2$

LCL $= \overline{\overline{X}} - A_2\overline{R}$ (where $A_2 = 0.577$ by Table 11) $\Rightarrow 198.9 - 0.577(24.8) = 198.9 - 14.31 = 184.6$

7.

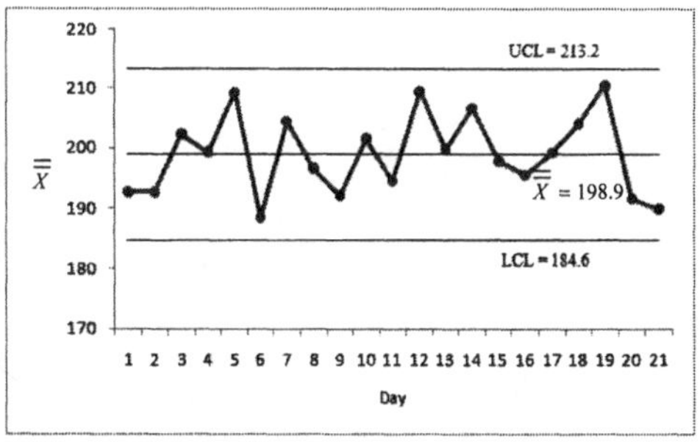

9. $\overline{\overline{X}} = \frac{\sum \overline{X}}{N} \Rightarrow \frac{313.92}{21} = 14.948 = 14.9$

$$\text{UCL} = \bar{\bar{X}} + A_2\bar{R}\ \left(\text{where } A_2 = 0.577 \text{ by Table 11}\right) \Rightarrow 14.9 + 0.577(5.9) = 14.9 + 3.4 = 18.3$$

$$\text{LCL} = \bar{\bar{X}} - A_2\bar{R}\ \left(\text{where } A_2 = 0.577 \text{ by Table 11}\right) \Rightarrow 14.9 - 0.577(5.9) = 14.9 - 3.427 = 11.5$$

11.

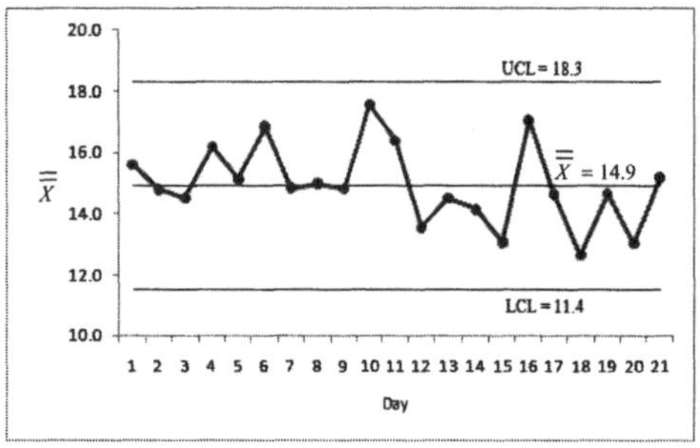

Exercise 8 ◊ Regression

1.

x	y	xy	x^2	y^2
–8	–6.238	49.904	64	38.91264
–6.66	–3.709	24.70194	44.3556	13.75668
–5.33	–0.712	3.79496	28.4089	0.506944
–4	1.887	–7.548	16	3.560769
–2.66	4.628	–12.3105	7.0756	21.41838
–1.33	7.416	–9.86328	1.7689	54.99706
0	10.2	0	0	104.04
1.33	12.93	17.1969	1.7689	167.1849
2.66	15.7	41.762	7.0756	246.49
4	18.47	73.88	16	341.1409
5.33	21.32	113.6356	28.4089	454.5424
6.66	23.94	159.4404	44.3556	573.1236
8	26.7	213.6	64	712.89
9.33	29.61	276.2613	87.0489	876.7521
10.6	32.35	342.91	112.36	1046.523
12	35.22	422.64	144	1240.448
31.93	229.712	1710.005	666.6269	5896.287

$$r = \frac{n\sum xy - \sum x \sum y}{\sqrt{n\sum x^2 - \left(\sum x\right)^2}\sqrt{n\sum y^2 - (y)^2}} \Rightarrow \frac{16(1710.005) - (31.93)(229.712)}{\sqrt{16(666.6269) - 31.93^2}\sqrt{16(5896.287) - 229.712^2}} = \frac{20025.37584}{20025.835} = 1.00$$

3.

x	y	xy	x^2	y^2
−11	−65.3	718.3	121	4264.09
−9.33	−56.78	529.7574	87.0489	3223.968
−7.66	−47.26	362.0116	58.6756	2233.508
−6	−37.21	223.26	36	1384.584
−4.33	−27.9	120.807	18.7489	778.41
−2.66	−18.39	48.9174	7.0756	338.1921
−1	−9.277	9.277	1	86.06273
0.66	0.081	0.05346	0.4356	0.006561
2.33	9.404	21.91132	5.4289	88.43522
4	18.93	75.72	16	358.3449
5.66	27.86	157.6876	32.0356	776.1796
7.33	37.78	276.9274	53.7289	1427.328
9	46.64	419.76	81	2175.29
10.6	56.69	600.914	112.36	3213.756
12.3	64.74	796.302	151.29	4191.268
14	75.84	1061.76	196	5751.706
23.9	75.848	5423.366	977.828	30291.13

$$r = \frac{n\sum xy - \sum x \sum y}{\sqrt{n\sum x^2 - \left(\sum x\right)^2}\sqrt{n\sum y^2 - (y)^2}} \Rightarrow \frac{16(5423.366) - (23.9)(75.848)}{\sqrt{16(977.828) - (23.9)^2}\sqrt{16(30291.13) - 75.848^2}} = \frac{84961.0888}{84964.90213} = 0.99995 = 1.00$$

5.

x	y	xy	x^2
−20	82.29	−1645.8	400
−18.5	73.15	−1353.28	342.25
−17	68.11	−1157.87	289
−15.6	59.31	−925.236	243.36
−14.1	53.65	−756.465	198.81
−12.6	45.9	−578.34	158.76
−11.2	38.69	−433.328	125.44
−9.73	32.62	−317.393	94.6729
−8.26	24.69	−203.939	68.2276
−6.8	18.03	−122.604	46.24
−5.33	11.31	−60.2823	28.4089
−3.86	3.981	−15.3667	14.8996
−2.4	−2.968	7.1232	5.76
−0.93	−9.986	9.28698	0.8649
0.53	−16.92	−8.9676	0.2809
2	−23.86	−47.72	4
−143.78	457.997	−7610.18	2020.975

$$\text{slope} = \frac{n\sum xy - \sum x \sum y}{n\sum x^2 - \left(\sum x\right)^2} \Rightarrow \frac{16(-7610.18) - (-143.78)(457.997)}{16(2020.975) - (-143.78)^2} = \frac{-55912.07134}{11662.9116} = -4.794 = -4.79$$

$$y\text{-intercept} = \frac{\sum x^2 \sum y - \sum x \sum xy}{n\sum x^2 - \left(\sum x\right)^2} \Rightarrow \frac{(2020.975)(457.997) - (-143.78)(-7610.18)}{16(2020.975) - (-143.78)^2} = \frac{-168591.1933}{11662.9116} = -14.455 = -14.5$$

Therefore the equation of the least squares line is $y = -4.79x - 14.5$

CHAPTER 21 REVIEW PROBLEMS

1. Continuous

3. Categorical

5. 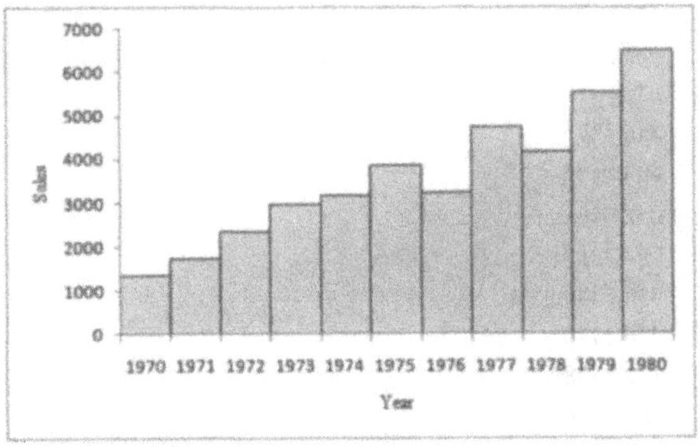

7. $\dfrac{\text{number of ways to get sum of 8}}{\text{total number of outcomes}} = \dfrac{5}{36}$

9. P(brown hair + let-handed) – P(brown hair and left-handed) ⇒ 0.72 + 0.06 – 0.03 = 0.75

11. From Table 7, the area between the mean and 0.5 standard deviations is 0.1915

13. The area between 0 and 0.4 standard deviations is 0.1554. But since the total area to the right of the mean (0) is 0.5, the area to the right of $z = 0.4$ is 0.5 – 0.1554 or 0.3446.

15. range = 193 – 112 = 81

17. 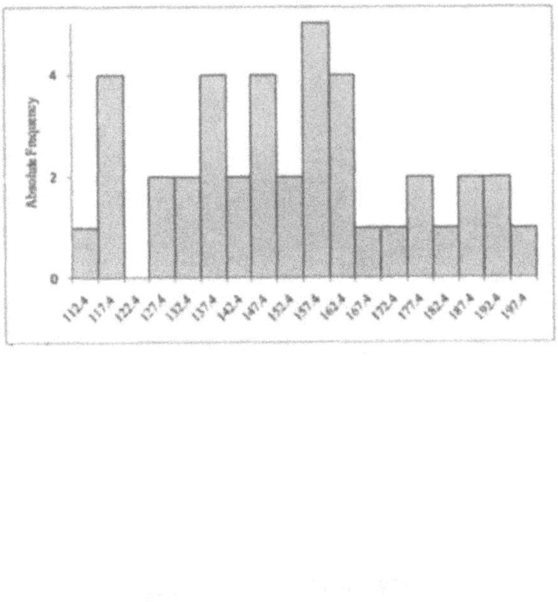

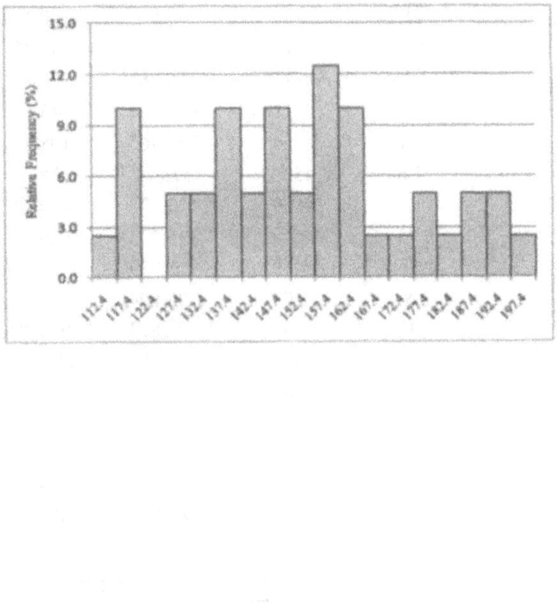

19.

Class Limits		Cumulative Absolute Frequency	Cumulative Relative Frequency
107.5	112.4	1	2.5%
112.5	117.4	5	12.5%
117.5	122.4	5	12.5%
122.5	127.4	7	17.5%
127.5	132.4	9	22.5%
132.5	137.4	13	32.5%
137.5	142.4	15	37.5%
142.5	147.4	19	47.5%
147.5	152.4	21	52.5%
152.5	157.4	26	65.0%
157.5	162.4	30	75.0%
162.5	167.4	31	77.5%
167.5	172.4	32	80.0%
172.5	177.4	34	85.0%
177.5	182.4	35	87.5%
182.5	187.4	37	92.5%
187.5	192.4	39	97.5%
192.5	197.4	40	100.0%

21. $\text{mean} = \dfrac{\sum \text{data}}{40} = 150.15 = 150$

23. mode = 137 and 153

25.

x	$x - \mu$	$(x - \mu)^2$
112	–38.15	1455.423
116	–34.15	1166.223
116	–34.15	1166.223
117	–33.15	1098.923
117	–33.15	1098.923
125	–25.15	632.5225
127	–23.15	535.9225
129	–21.15	447.3225
131	–19.15	366.7225
133	–17.15	294.1225
137	–13.15	172.9225
137	–13.15	172.9225
137	–13.15	172.9225
138	–12.15	147.6225
141	–9.15	83.7225
144	–6.15	37.8225
144	–6.15	37.8225
145	–5.15	26.5225
146	–4.15	17.2225
148	–2.15	4.6225
152	1.85	3.4225
153	2.85	8.1225
153	2.85	8.1225
153	2.85	8.1225
154	3.85	14.8225
154	3.85	14.8225
159	8.85	78.3225
161	10.85	117.7225
162	11.85	140.4225
162	11.85	140.4225
167	16.85	283.9225
168	17.85	318.6225
173	22.85	522.1225
174	23.85	568.8225
182	31.85	1014.423
183	32.85	1079.123
183	32.85	1079.123
188	37.85	1432.623
192	41.85	1751.423
193	42.85	1836.123
	$\Sigma(x - \mu) = 0$	$\Sigma(x - \mu)^2 = 19557.1$

Since this is a population, $\sigma = \sqrt{\frac{\sum(x-\mu)^2}{n}} \Rightarrow \sqrt{\frac{19557.1}{400}} = \sqrt{488.928} = 22.11 = 22.1$

27. $\bar{x} \pm 2SE_x \Rightarrow \bar{x} \pm 2\left(\frac{\sigma}{\sqrt{n}}\right) \Rightarrow 150 \pm 2\left(\frac{22.1}{\sqrt{40}}\right) = 150 \pm 6.77 \Rightarrow 150 \pm 7$ or 143 to 157

29. $s \pm 2SE_s = s \pm 2\left(\frac{\sigma}{\sqrt{2n}}\right) \Rightarrow 22.1 \pm 2\left(\frac{22.1}{\sqrt{2(40)}}\right) = 22.1 \pm 4.942 = 22.1 \pm 4.94$ or 17.16 to 27.04

31. $z(60) = \frac{60 - 79.3}{11.6} = -1.7 \Rightarrow \text{area} = 0.5 - 0.4554 = 0.0446 \Rightarrow 0.0446(300) = 13.38$ or 13 students

33. $Q_1 = \frac{327 + 486}{2} = 406.5 = 407 \quad Q_2 = \frac{639 + 797}{2} = 718 \quad Q_3 = \frac{974 + 1136}{2} = 1055$

quartile range = 1055 – 407 = 648

35. $\bar{\bar{X}} = \frac{\sum \bar{X}}{N} \Rightarrow \frac{15575.2}{21} = 741.67 = 742$

$\text{UCL} = \bar{\bar{X}} + A_2\bar{R}$ (where $A_2 = 0.577$ by Table 11) $\Rightarrow 742 + 0.577(456) = 742 + 263 = 1005$

$\text{LCL} = \bar{\bar{X}} - A_2\bar{R}$ (where $A_2 = 0.577$ by Table 11) $\Rightarrow 742 - 0.577(456) = 742 - 263 = 479$

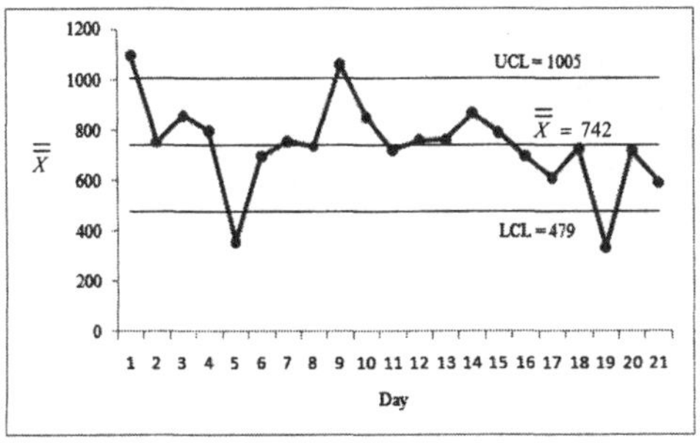

37.

x	y	xy	x^2	y^2
5.0	6.882	34.41	25.0	47.362
11.2	–7.623	–85.378	125.44	58.11
17.4	–22.45	–390.63	302.76	504.003
23.6	–36.09	–851.724	556.96	1302.488
29.8	–51.13	–1523.674	888.04	2614.277
36.0	–64.24	–2312.64	1296.0	4126.778
42.2	–79.44	–3352.368	1780.84	6310.714
48.2	–94.04	–4532.728	2323.24	8843.522
54.6	–107.8	–5885.88	2981.16	11620.84
60.8	–122.8	–7466.24	3696.64	15079.84
67.0	–138.6	–9286.2	4489.0	19209.96
73.2	–151.0	–11053.2	5358.24	22801.0
79.4	–165.3	–13124.82	6304.36	27324.09
85.6	–177.6	–15202.56	7327.36	31541.76
91.8	–193.9	–17800.02	8427.24	37597.21
98.0	–208.9	–20472.2	9604.0	43639.21
823.8	–1614.031	–113305.852	55486.28	232621.162

$$r = \frac{n\sum xy - \sum x \sum y}{\sqrt{n\sum x^2 - \left(\sum x\right)^2}\sqrt{n\sum y^2 - (y)^2}} \Rightarrow \frac{16(-113305.852) - (823.8)(-1614.031)}{\sqrt{16(55486.852) - 823.8^2}\sqrt{16(232621.162) - (-1614.031)^2}} = \frac{-483254.8942}{483301.1589} = -0.9999 = -1.00$$

$$\text{slope} = \frac{n\sum xy - \sum x \sum y}{n\sum x^2 - \left(\sum x\right)^2} \Rightarrow \frac{16(-113305.852)-(823.8)(-1614.031)}{16(55486.28)-823.8^2} = \frac{-483254.8942}{209134.04} = -2.311 = -2.31$$

$$y\text{-intercept} = \frac{\sum x^2 \sum y - \sum x \sum xy}{n\sum x^2 - \left(\sum x\right)^2} \Rightarrow \frac{(55486.28)(-1614.031)-(823.8)(-113305.852)}{16(55486.28)-823.8^2} = \frac{3784784.883}{209134.04} = 18.09 = 18.1$$

Therefore the equation of the least squares line is $y = -2.31x + 18.1$

39. $\frac{4}{52} \pm \sqrt{\frac{\frac{4}{52}\left(1-\frac{4}{52}\right)}{200}} = 0.0769 \pm 0.0188$ or 0.0581 to 0.0957

Chapter 22: Analytic Geometry

Exercise 22.1 • The Straight Line

Directed Distance AB

1. $A(3,0);\ B(5,0)$
   $AB = 5 - 3 = 2$

3. $B(-8,-2);\ A(-8,-5)$
   $AB = -2 - (-5) = -2 + 5 = 3$

5. $A(3.95,-2.07);\ B(-3.95,-2.07)$
   $AB = -3.95 - 3.95 = -7.90$

Increments

7. $A(2,4);\ B(5,7)$
   $\Delta x = x_2 - x_1 = 5 - 2 = 3$
   $\Delta y = y_2 - y_1 = 7 - 4 = 3$

9. $A(-4,4);\ B(5,-8)$
   $\Delta x = x_2 - x_1 = 5 - (-4) = 5 + 4 = 9$
   $\Delta y = y_2 - y_1 = -8 - 4 = -12$

Length of a Line Segment

11. $(5,0)$ and $(2,0)$
    $d = \sqrt{(2-5)^2 + (0-0)^2} = \sqrt{9} = 3$

13. $(-2,0)$ and $(7,0)$
    $d = \sqrt{(7-(-2))^2 + (0-0)^2} = \sqrt{81} = 9$

15. $(0,-2.74)$ and $(0,3.86)$
    $d = \sqrt{(0-0)^2 + (3.86-(-2.74))^2}$
    $= \sqrt{43.56} = 6.60$

17. $(5.59,3.25)$ and $(8.93,3.25)$
    $d = \sqrt{(8.93-5.59)^2 + (3.25-3.25)^2}$
    $= \sqrt{11.156} = 3.34$

19. $(8.38,-3.95)$ and $(2.25,-4.99)$ $\Rightarrow$ $d = \sqrt{(2.25-8.38)^2 + (-4.99-(-3.95))^2} = \sqrt{38.66} = 6.22$

Slope

21. $m = \frac{\text{rise}}{\text{run}} = \frac{6}{4} = \frac{3}{2}$

23. $m = \frac{\text{rise}}{\text{run}} = \frac{-9}{-3} = 3$

25. $(5,2)$ and $(3,6)$
    $m = \frac{y_2 - y_1}{x_2 - x_1} = \frac{6-2}{3-5} = \frac{4}{-2} = -2$

27. $(3,-3)$ and $(-6,2)$
    $m = \frac{y_2 - y_1}{x_2 - x_1} = \frac{2-(-3)}{-6-3} = \frac{5}{-9} = -\frac{5}{9}$

Angle of Inclination

29. $m = \tan 77.9^\circ = 4.66$

31. $m = \tan 58^\circ 14' = \tan\left(58 + \frac{14}{60}\right)^\circ = 1.615$

33. $m = \tan 132.8^\circ = -1.080$

35. $\theta = \arctan 1.84 = 61.5^\circ$

37. $\arctan(-2.75) = -70^\circ$

$\theta = 180^\circ - 70^\circ = 110^\circ$

39. $\arctan(-15) = -86.2^\circ$

$\theta = 180^\circ - 86.2^\circ = 93.8^\circ$

41. $(-2.5,-3.1)$ and $(5.8,4.2)$

$$m = \frac{y_2 - y_1}{x_2 - x_1} = \frac{4.2-(-3.1)}{5.8-(-2.5)} = \frac{7.3}{8.3} = 0.8795$$

$\theta = \arctan(0.8795) = 41.3^\circ$

43. $(x,3)$ and $(x+5,8)$

$$m = \frac{y_2 - y_1}{x_2 - x_1} = \frac{8-3}{(x+5)-x} = \frac{5}{5} = 1$$

$\theta = \arctan(1) = 45.0^\circ$

Slopes of Parallel and Perpendicular Lines

45. $m = 2 \Rightarrow m_{\text{parallel}} = 2$ and $m_{\text{perpendicular}} = -\frac{1}{2}$

47. $m = -1.85 \Rightarrow m_{\text{parallel}} = -1.85$ and $m_{\text{perpendicular}} = -\frac{1}{(-1.85)} = 0.541$

49. $m = -5.372 \Rightarrow m_{\text{parallel}} = -5.372$ and $m_{\text{perpendicular}} = -\frac{1}{(-5.372)} = 0.1862$

Angle Between Two Lines

51. $m_1 = 3$ and $m_2 = -2$

$$\tan\phi = \frac{m_2 - m_1}{1 + m_1 m_2} = \frac{-2-3}{1+3(-2)} = \frac{-5}{-5} = 1 \Rightarrow \phi = \arctan(1) = 45^\circ$$

53. $\phi = 86^\circ - 22^\circ = 64^\circ$

Applications

55. $y_2 - y_1 = 65.2 - 22.3 = 42.9$ and $x_2 - x_1 = 92.7 - 26.8 = 65.9$

$$d = \sqrt{(x_2 - x_1)^2 + (y_2 - y_1)^2} = \sqrt{(65.9)^2 + (42.9)^2} = \sqrt{6183.22} = 78.6 \text{ mm}$$

57. $\frac{d}{2055} = \cos 12.3^\circ \Rightarrow d = 2055 \cos 12.3^\circ = 2008 \text{ ft}$

59. $\frac{1250}{d} = \cos 7^\circ \Rightarrow d = \frac{1250}{\cos 7^\circ} = 1260 \text{ m}$

61. $\sin\phi = \frac{12}{755} \Rightarrow \phi = \arcsin\left(\frac{12}{755}\right) = 0.911^\circ$

63. $\tan\phi = \frac{2}{3} \Rightarrow \phi = \arctan\left(\frac{2}{3}\right) = 33.7^\circ$

Exercise 22.2 • Equation of a Straight Line

Slope-Intercept Form of a Straight Line

1. $y = 4x - 3$

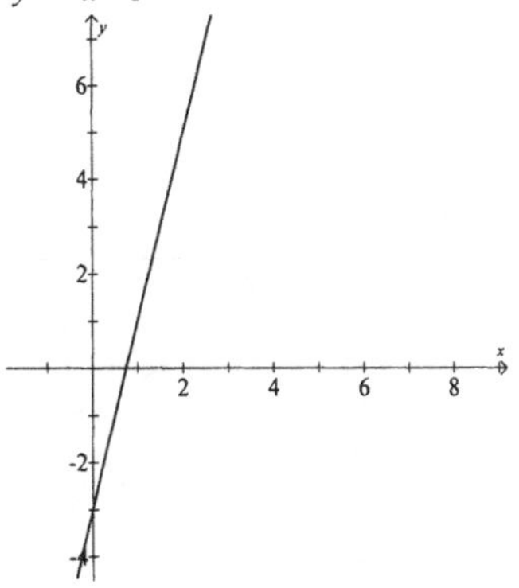

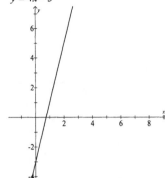

3. $y = 3x - 1$

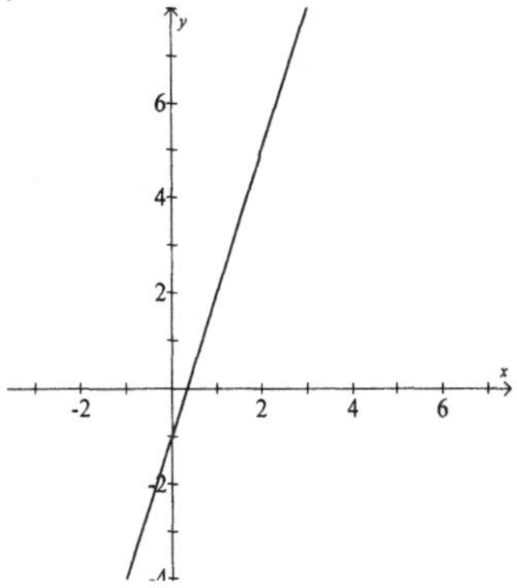

5. $y = 2.3x - 1.5$

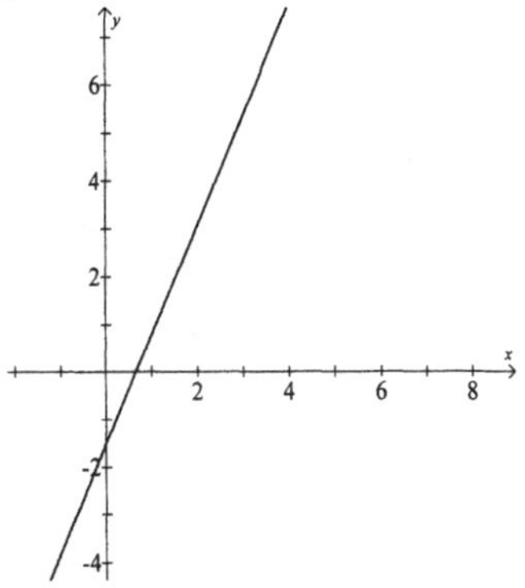

7. $m = \dfrac{1}{2}$ and $b = 0$

$y = \dfrac{1}{2}x + 0 \Rightarrow 2y = x \Rightarrow x - 2y = 0$

9. $x = 2 \Rightarrow x - 2 = 0$

11. $y = 3x - 5$

$m = 3$ and $b = -5$

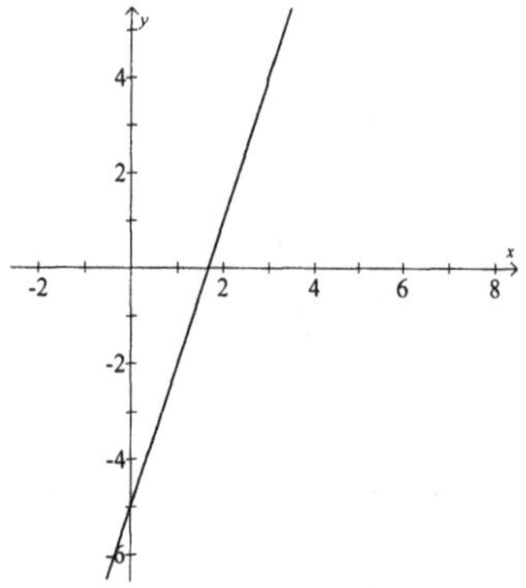

13. $y = -\frac{1}{2}x - \frac{1}{4}$

$m = -\frac{1}{2}$ and $b = -\frac{1}{4}$

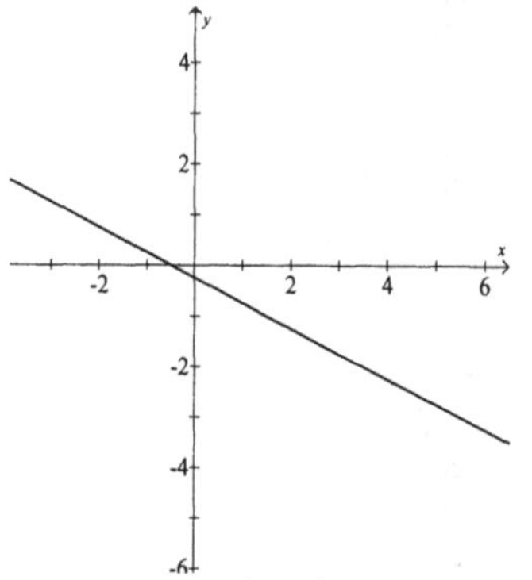

15. $m = 2$ through $(3,4)$

$\frac{y-4}{x-3} = 2$

$y - 4 = 2(x-3)$

$y - 4 = 2x - 6$

$y = 2x - 2$

17. $m = -4$ through $(-2,5)$

$\frac{y-5}{x-(-2)} = -4$

$y - 5 = -4(x+2)$

$y - 5 = -4x - 8$

$4x + y + 3 = 0$

19. Slope of the given line is 5, and a parallel line will also have slope of 5.

$m = 5$ and $b = 3$

$y = 5x + 3$

$5x - y + 3 = 0$

21. Slope of the given line is 3, so a perpendicular line will have slope of $-\frac{1}{3}$.

$m = -\frac{1}{3}$ and $b = -5$

$y = -\frac{1}{3}x - 5$

$3y = -x - 15$

$x + 3y + 15 = 0$

23. Slope of the given line is 5, and a parallel line will also have slope of 5.

$m = 5$ through $(-2,5)$

$\frac{y-5}{x-(-2)} = 5$

$y - 5 = 5(x+2)$

$y - 5 = 5x + 10$

$5x - y + 15 = 0$

25. Slope of the given line is 5, so a perpendicular line will have slope of $-\frac{1}{5}$.

$$m=-\frac{1}{5} \text{ through } (-4,2)$$
$$\frac{y-2}{x-(-4)}=-\frac{1}{5}$$
$$5(y-2)=-1(x+4)$$
$$5y-10=-x-4$$
$$x+5y-6=0$$

27. A line parallel to the x-axis will be a horizontal line with all points on the line having the same y-coordinate. The horizontal line that passes through the point $(5,2)$ is $y=2$ or in general form $y-2=0$.

29. Through $(3,5)$ and $(-1,2)$

$$\frac{y-5}{x-3}=\frac{2-5}{-1-3}$$
$$\frac{y-5}{x-3}=\frac{3}{4}$$
$$4(y-5)=3(x-3)$$
$$4y-20=3x-9$$
$$3x-4y+11=0$$

31. Through $(5,0)$ and $(0,-3)$

$$\frac{y-0}{x-5}=\frac{-3-0}{0-5}$$
$$\frac{y}{x-5}=\frac{3}{5}$$
$$5y=3(x-5)$$
$$5y=3x-15$$
$$3x-5y-15=0$$

33. $y=3x+2$

$$3x-y+2=0$$

35. $y=-2x+6$

$$2x+y-6=0$$

37. $x+2y+5=0$

$$2y=-x-5$$
$$y=-\frac{1}{2}x-\frac{5}{2}$$

39. $-7x+3y+8=0$

$$3y=7x-8$$
$$y=\frac{7}{3}x-\frac{8}{3}$$

Applications
Spring Constant

41. $F=kx$

$$\frac{F-0}{L-L_0}=k \Rightarrow F=kL-kL_0$$

Velocity of Uniformly Accelerated Body

43. $v=v_0+at$

a) $21.8=v_0+2.15(5.25) \Rightarrow v_0=21.8-2.15(5.25)=10.5$ m/s

b) $v=10.5+2.15(25.0)=64.3$ m/s

Resistance Change with Temperature

45. $\alpha = \frac{1}{234.5t_1}$, $t_1 = 20.0$, $R_1 = 148.4$, $t = 75.0$

$$R = R_1\left[1+\alpha\left(t-t_1\right)\right]$$
$$= 148.4\left[1+\frac{1}{234.5(20.0)}(75.0-20.0)\right]$$
$$= 150.1\ \Omega$$

Thermal Expansion

47. $\frac{L-L_0}{t-t_0} = L_0\alpha$

$$L - L_0 = L_0\alpha t - L_0\alpha t_0$$
$$L = L_0 + L_0\alpha t - L_0\alpha t_0$$
$$L = L_0\left[1+\alpha\left(t-t_0\right)\right]$$
$$L = L_0\left[1+\alpha\Delta t\right] \quad \text{where } \Delta t = t - t_0$$

Fluid Pressure

49. $P = 0.432x + 20.6$

$$P = 30.0$$
$$0.432x + 20.6 = 30.0$$
$$0.432x = 9.4$$
$$x = \frac{9.4}{0.432} = 21.8 \text{ ft}$$

Temperature Gradient

51. $m = \frac{t_0 - t_i}{38.0 - 0} = \frac{-5.0 - 25.0}{38.0} = -\frac{30.0}{38.0} = -0.789$

$$t = -0.789x + 25.0$$
$$-0.789x + 25.0 = 0$$
$$x = \frac{25.0}{0.789} = 31.7 \text{ cm}$$
$$m = -0.789$$

Straight-Line Depreciation

53. $m = \frac{S-P}{L}$

$$y = \left(\frac{S-P}{L}\right)t + P$$
$$P = 15428,\ S = 2264,\ L = 20,\ t = 15$$
$$y = \left(\frac{2264-15428}{20}\right)15 + 15428 = \$5555$$

Exercise 22.3 • The Circle

Standard Equation of a Circle

1. center at $(0,0)$; radius = 7

$x^2 + y^2 = 7^2$

$x^2 + y^2 = 49$

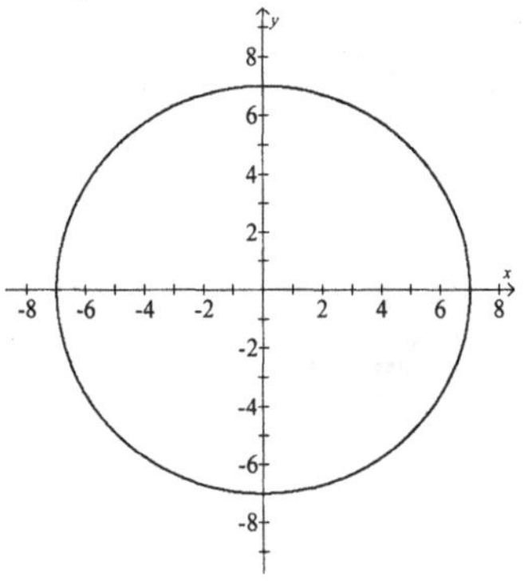

3. center at $(2,3)$; radius = 5

$h = 2$, $k = 3$, and $r = 5$

$(x-2)^2 + (y-3)^2 = 5^2$

$(x-2)^2 + (y-3)^2 = 25$

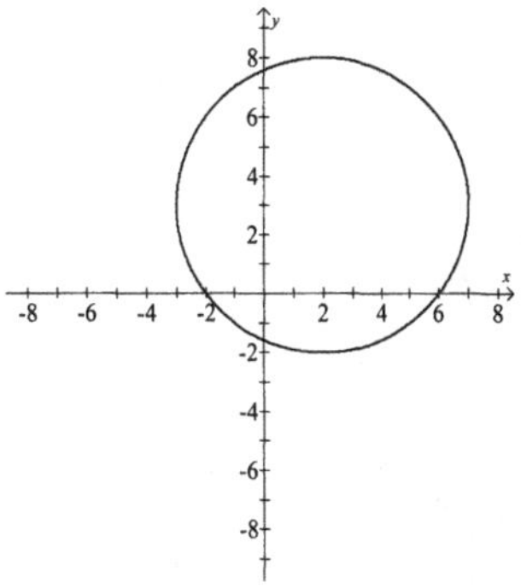

5. center at $(5,-3)$; radius = 4

$h = 5$, $k = -3$, and $r = 4$

$(x-5)^2 + (y-(-3))^2 = 4^2$

$(x-5)^2 + (y+3)^2 = 16$

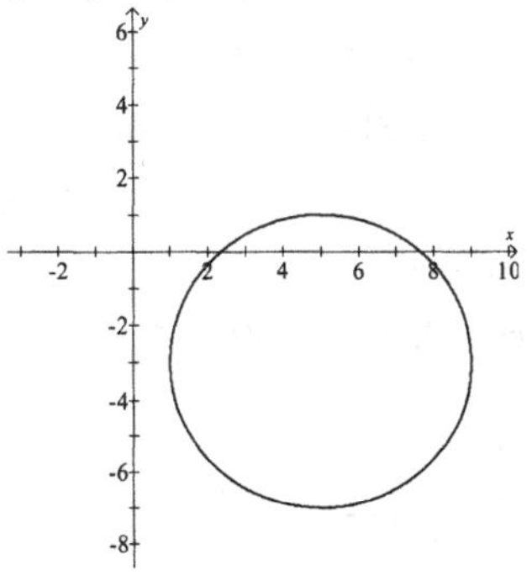

7. $x^2 + y^2 = 49$

center at $(0,0)$ and $r = \sqrt{49} = 7$

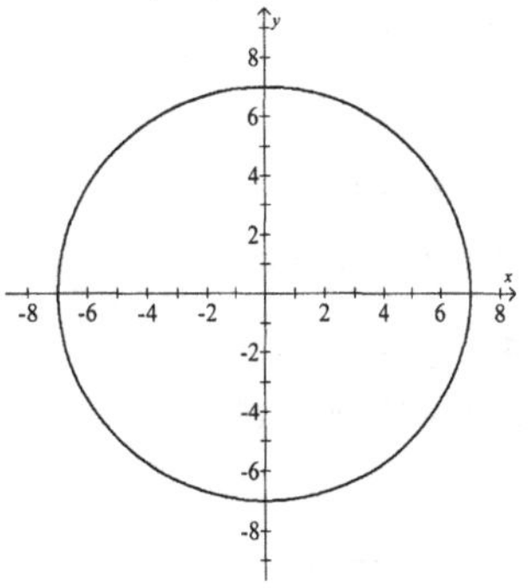

9. $(x-2)^2+(y+4)^2=16$

$x-h=x-2 \qquad y-k=y+4$

$h=2 \qquad k=-4$

center at $(2,-4)$ and $r=\sqrt{16}=4$

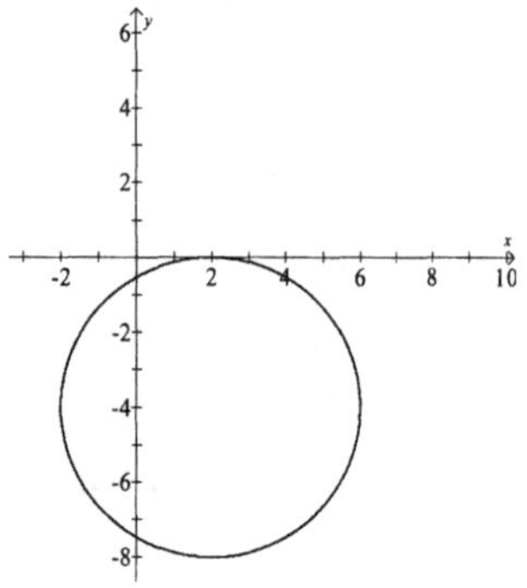

11. $(y+5)^2+(x-3)^2=36$

$x-h=x-3 \qquad y-k=y+5$

$h=3 \qquad k=-5$

center at $(3,-5)$ and $r=\sqrt{36}=6$

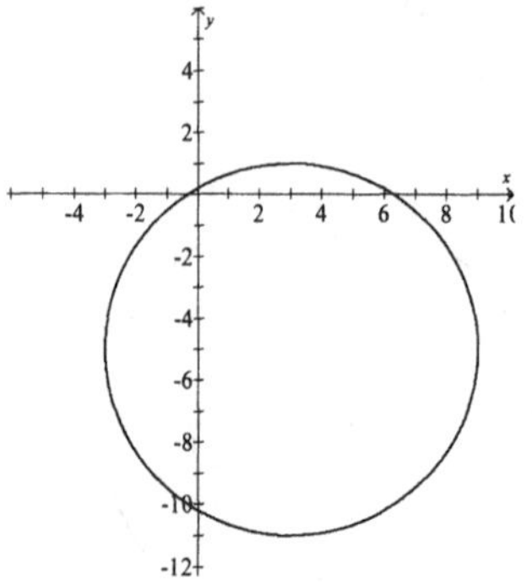

General Equation of a Circle

13. $(x+2)^2+(y+3)^2=16$

$x^2+4x+4+y^2+6y+9=16$

$x^2+y^2+4x+6y-3=0$

15. $(x-5)^2+(y+4)^2=42$

$x^2-10x+25+y^2+8y+16=42$

$x^2+y^2-10x+8y-1=0$

17. $x^2+y^2-8x=0$

$\left(x^2-8x+16\right)+y^2=0+16$

$(x-4)^2+y^2=16$

center at $(4,0)$, $r=\sqrt{16}=4$

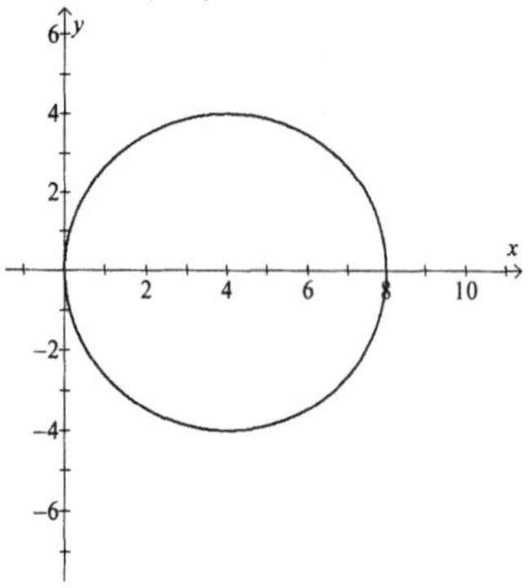

19. $x^2+y^2-10x+12y+25=0$

$\left(x^2-10x+25\right)+\left(y^2+12y+36\right)$

$=-25+25+36$

$(x-5)^2+(y+6)^2=36$

center at $(5,-6)$, $r=\sqrt{36}=6$

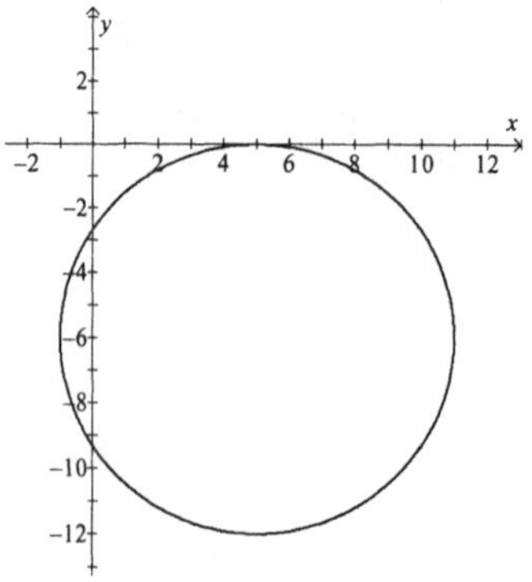

21. $x^2+y^2+6x-2y=15$

$\left(x^2+6x+9\right)+\left(y^2-2y+1\right)=15+9+1$

$(x+3)^2+(y-1)^2=25$

center at $(-3,1)$, $r=\sqrt{25}=5$

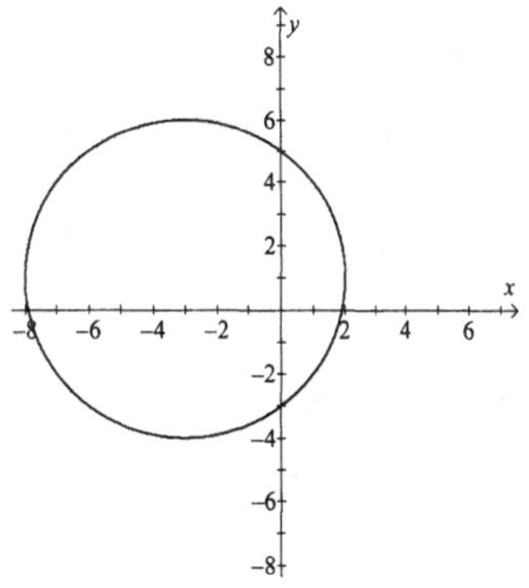

Applications

23. Using the figure below.

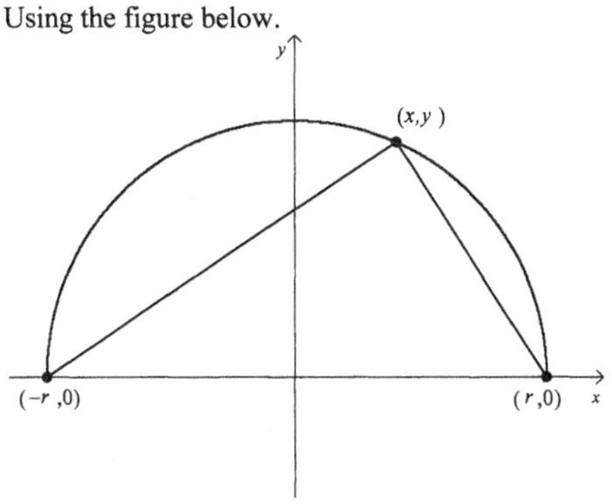

Equation of the circle: $x^2+y^2=r^2$ or $y^2=r^2-x^2$

Slope of line 1: $m_1=\dfrac{y-0}{x-(-r)}=\dfrac{y}{x+r}$

Slope of line 2: $m_2=\dfrac{y-0}{x-r}=\dfrac{y}{x-r}$

$$-\frac{1}{m_1}=-\frac{1}{\left(\dfrac{y}{x+r}\right)}=-\frac{x+r}{y}\cdot\frac{x-r}{x-r}$$

$$=-\frac{x^2-r^2}{y(x-r)}=\frac{y^2}{y(x-r)}=\frac{y}{x-r}=m_2$$

The lines are perpendicular since $m_2=-\dfrac{1}{m_1}$, so they intersect at a right angle.

25. Left Arch: $h_1=0,\ k_1=5.80,\ r=3.60$

$x^2+(y-5.80)^2=12.96$

$x^2+y^2-11.6y+20.68=0$

Right Arch: $h_2=7.20,\ k_2=4.20,\ r=5.80$

$(x-7.20)^2+(y-4.20)^2=33.64$

$x^2-14.40x+y^2-8.40y+35.84=0$

Take Right - Left = $-14.40x+3.20y=-15.16$

$x=0.22y+1.05$

Plug into the equation for the Left Arch and solve for y:

$(0.22y+1.05)^2+y^2-11.6y+20.68=0$

$1.05y^2-11.14y+21.78=0$

$y=2.58$ ft or 8.02 ft

Drop the 2.58 value since we need the higher value. The height of the column is 8.02 ft.

27. $h = 6.00, k = 0, r = 10.0 \Rightarrow (x-6.00)^2 + y^2 = 100$

when $y = 3.00$: $(x-6.00)^2 + 3.00^2 = 100 \Rightarrow (x-6.00)^2 = 91.0 \Rightarrow x-6 = \pm 9.54$

$x = -3.54$ ft or 15.5 ft

Drop the 15.5 value since we need the lower value. The width of the arch is 2(3.54) = 7.08 ft.

Exercise 22.4 • The Parabola

Standard Equation of a Parabola: Vertex at the Origin

1. $y^2 = 8x$

$4p = 8 \Rightarrow p = 2$

$F(2,0)$ and $L = |4(2)| = 8$

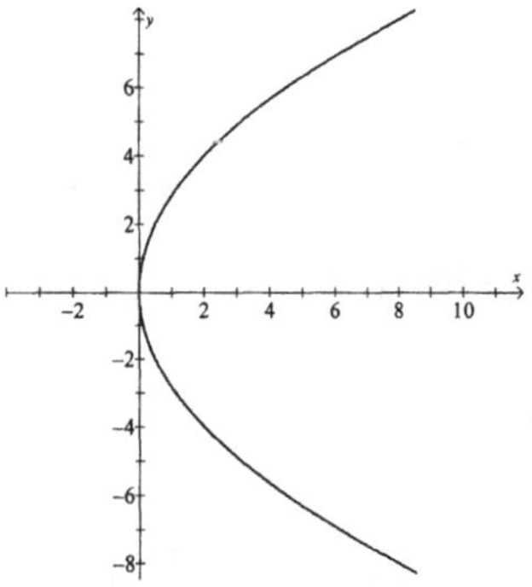

3. $7x^2 + 12y = 0 \Rightarrow x^2 = -\frac{12}{7}y$

$4p = -\frac{12}{7} \Rightarrow p = -\frac{3}{7}$

$F\left(0, -\frac{3}{7}\right)$ and $L = \left|4\left(-\frac{3}{7}\right)\right| = \frac{12}{7}$

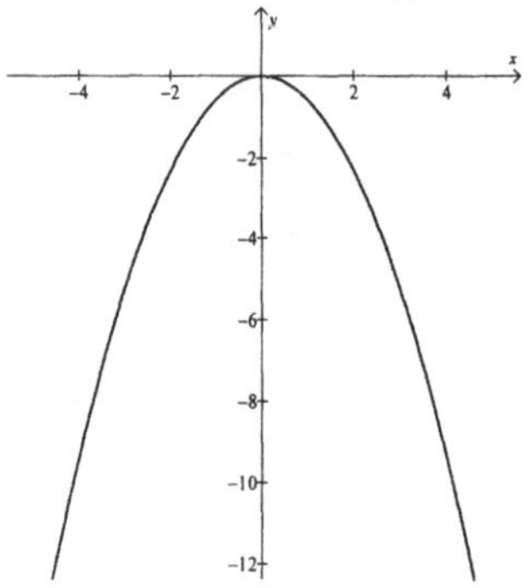

5. $x^2 = 4py$

$6^2 = 4p4 \Rightarrow 4p = \frac{36}{4} = 9 \Rightarrow p = \frac{9}{4}$

$x^2 = 9y$ and $F\left(0, \frac{9}{4}\right)$

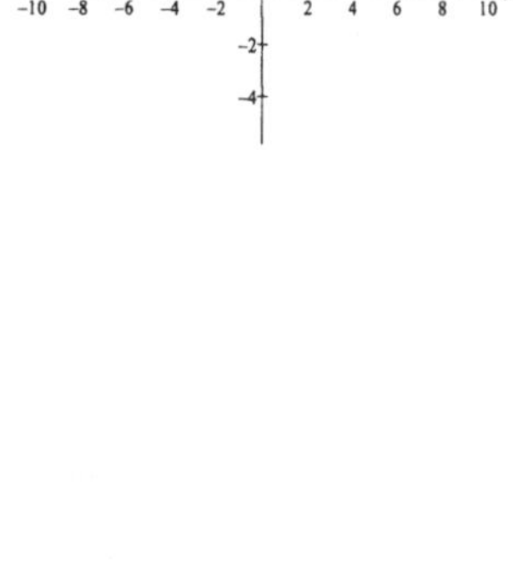

7. $y^2 = 4px$

$2^2 = 4p3 \Rightarrow 4p = \frac{4}{3} \Rightarrow p = \frac{1}{3}$

$y^2 = \frac{4}{3}x$ or $3y^2 = 4x$ and $F\left(\frac{1}{3}, 0\right)$

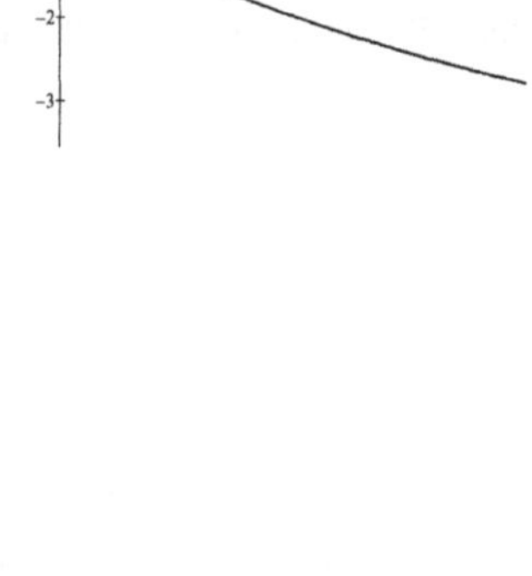

Standard Equation of a Parabola: Vertex Not at the Origin

9. $(y-5)^2 = 12(x-3)$

This is a horizontal parabola opening to the right with $h=3$ and $k=5$, so the vertex is at $V(3,5)$.

Also $4p=12$ gives $p=3$. The focus is 3 units to the right of the vertex, at $F(6,5)$.

The focal width is $L=|4p|=12$.

The axis is the horizontal line that crosses through the vertex, so it is $y=5$.

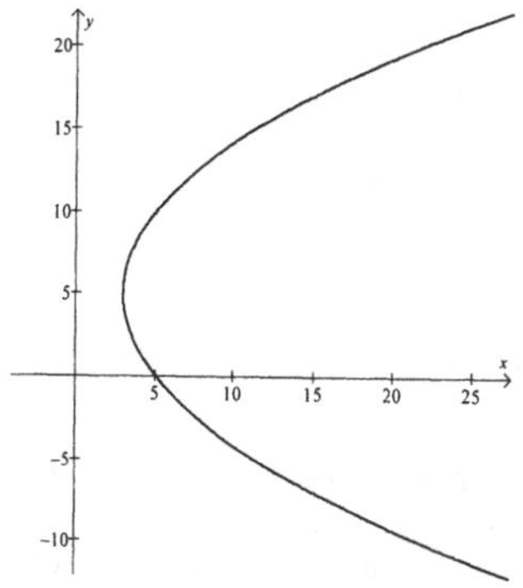

11. $(x-3)^2 = 24(y+1)$

This is a vertical parabola opening up with $h=3$ and $k=-1$, so the vertex is at $V(3,-1)$.

Also $4p=24$ gives $p=6$. The focus is 6 units above the vertex, at $F(3,5)$.

The focal width is $L=|4p|=24$.

The axis is the vertical line that crosses through the vertex, so it is $x=3$.

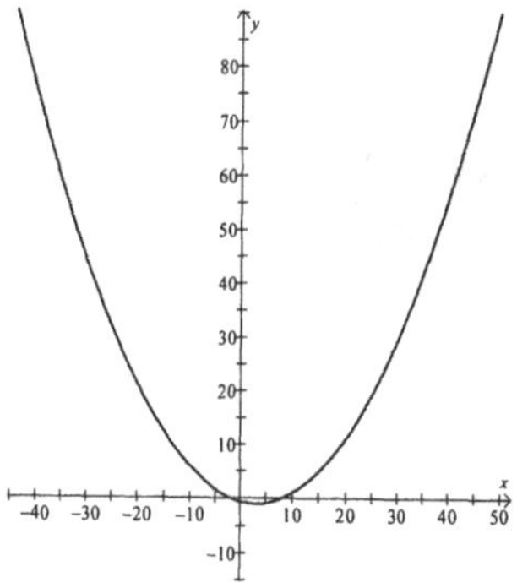

13. $3x+2y^2+4y-4=0$

$2y^2+4y=-3x+4$

$y^2+2y+1=-\frac{3}{2}x+2+1$

$(y+1)^2=-\frac{3}{2}(x-2)$

This is a parabola opening to the left with vertex at $V(2,-1)$.

Also $4p=-\frac{3}{2} \Rightarrow p=-\frac{3}{8}$ The focus is $\frac{3}{8}$ of a unit to the left of the vertex, at $F\left(\frac{13}{8},-1\right)$

The focal width is $L=|4p|=\frac{3}{2}$. The axis is $y=-1$.

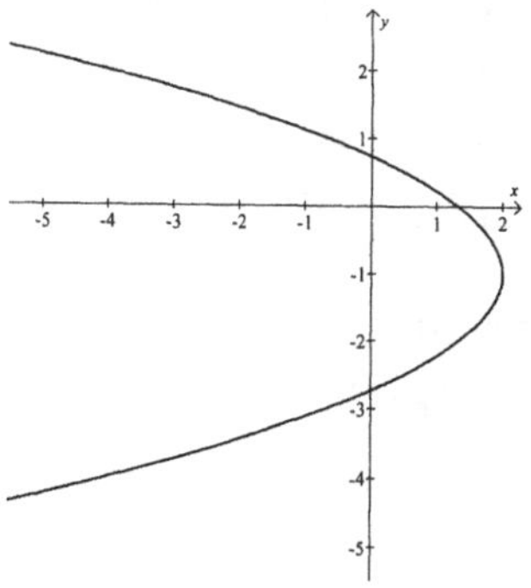

15. $y - 3x + x^2 + 1 = 0$

$$x^2 - 3x + \frac{9}{4} = -y - 1 + \frac{9}{4}$$

$$\left(x - \frac{3}{2}\right)^2 = -\left(y - \frac{5}{4}\right)$$

This is a parabola opening down with vertex at $V\left(\frac{3}{2}, \frac{5}{4}\right)$.

Also $4p = -1 \Rightarrow p = -\frac{1}{4}$, so the focus is $\frac{1}{4}$ of a unit below the vertex, at $F\left(\frac{3}{2}, 1\right)$.

The focal width is $L = |4p| = 1$. The axis is $x = \frac{3}{2}$.

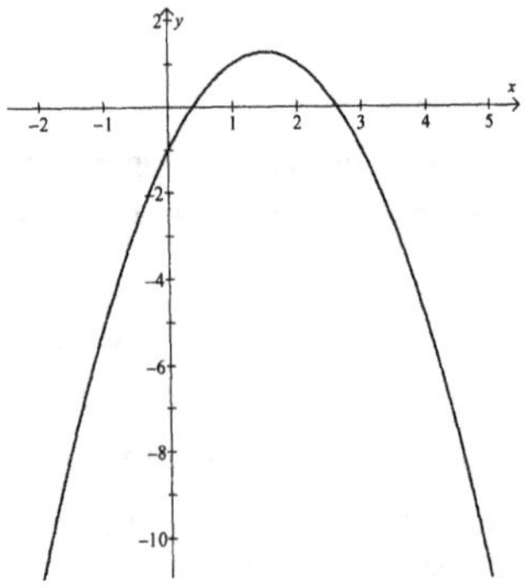

17. vertex at $(1,2)$; $L = 8$; axis is $y = 2$

opens to the right

$L = 8 \Rightarrow 4p = 8 \Rightarrow p = 2$

$F(3,2)$

$(y-2)^2 = 8(x-1)$

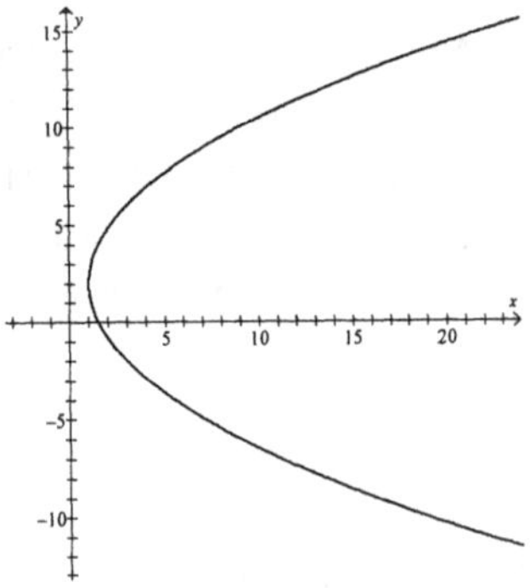

19. vertex at $(0,2)$; axis is $x = 0$; passes through $(-4,-2)$

axis is a vertical line so the parabola is vertical

$(x)^2 = 4p(y-2)$

at $(-4,-2)$: $(-4)^2 = 4p(-2-2) \Rightarrow p = \dfrac{16}{4(-4)} = -1$

$L = |4(-1)| = 4$; $F(0,1)$

$x^2 = -4(y-2)$

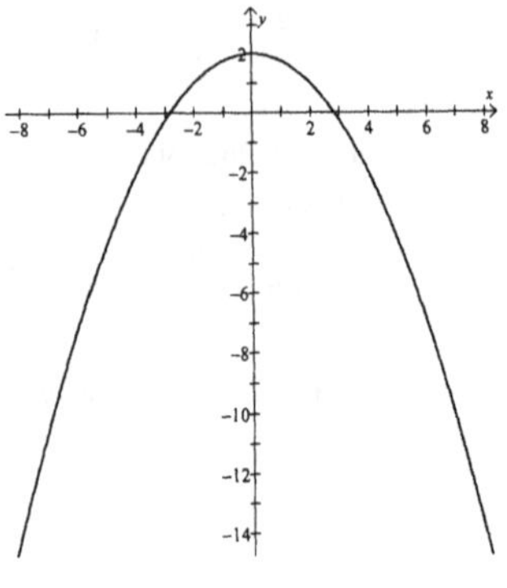

General Equation of a Parabola

21. $(x+6)^2 = 4(y-5)$

$x^2 + 12x + 36 = 4y - 20$

$x^2 + 12x - 4y + 56 = 0$

23. $(y-5)^2 = 9(x+8)$

$y^2 - 10y + 25 = 9x + 72$

$y^2 - 9x - 10y - 47 = 0$

25. $x^2 - 3x + 7y + 4 = 0$

$x^2 - 3x + \frac{9}{4} = -7y - 4 + \frac{9}{4}$

$\left(x - \frac{3}{2}\right)^2 = -7y - \frac{7}{4} \Rightarrow \left(x - \frac{3}{2}\right)^2 = -7\left(y + \frac{1}{4}\right)$

vertex at $V\left(\frac{3}{2}, -\frac{1}{4}\right)$

$4p = -7 \Rightarrow p = -\frac{7}{4}$

focus at $F\left(\frac{3}{2}, -\frac{1}{4} - \frac{7}{4}\right) = F\left(\frac{3}{2}, -2\right)$

$L = |4p| = 7$

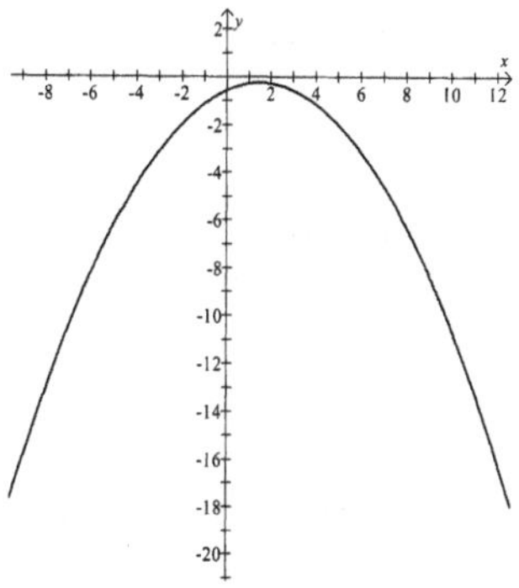

Trajectories

27. $h = 70.0$ and $k = 85.0$

$(x - 70.0)^2 = 4p(y - 85.0)$

at $(0,0)$: $(-70.0)^2 = 4p(-85.0) \Rightarrow 4p = \frac{70.0^2}{-85.0} = -57.6$

$(x - 70.0)^2 = -57.6(y - 85.0)$

when $x = 95$, $(95.0 - 70.0)^2 = -57.6(y - 85.0)$

$$y = \frac{625 - 57.6(85.0)}{-57.6} = 74.1 \text{ ft}$$

29. $h = 0$ and $k = 3520$

$x^2 = 4p(y - 3520)$

at $(2150,0)$: $(2150)^2 = 4p(-3520) \Rightarrow 4p = \frac{2150^2}{-3520} = -1310$

$x^2 = -1310(y - 3520)$

when $x = 1000$, $1000^2 = -1310(y - 3520)$

$$y = \frac{1000^2 - 1310(3520)}{-1310} = 2760 \text{ m}$$

Parabolic Arch

31. Take the bottom of the sag in the cable as the origin.
The point at the top of the right tower is $(500, 200)$.

$$x^2 = 4py \Rightarrow 500^2 = 4p(200) \Rightarrow 4p = \frac{500^2}{200} = 1250$$

$$x^2 = 1250y$$

Parabolic Reflector

33. $h = 0, k = 0 \Rightarrow x^2 = 4py$

at $(125, 32)$: $5^2 = 4p(3) \Rightarrow p = \frac{5^2}{12} = 2.08$ ft

35. $x = 0, k = 75$

$$x^2 = 4p(y - 75)$$

at $(300, 0)$: $(300)^2 = 4p(-75) \Rightarrow 4p = \frac{300^2}{-75} = -1200$

$$x^2 = -1200(y - 75)$$

37. at $x = 10.0$: $(10 - 16)^2 = -3070\left(y - \frac{1}{12}\right)$

$$y = \frac{36 - 3070(1/12)}{-3070} = 0.0716 \text{ ft}$$

Exercise 22.5 • The Ellipse

Standard Equation of Ellipse, Center at Origin

1. $\frac{x^2}{25} + \frac{y^2}{16} = 1$

$a = \sqrt{25} = 5,\ b = \sqrt{16} = 4$

$c = \sqrt{a^2 - b^2} = \sqrt{5^2 - 4^2} = \sqrt{9} = 3$

$V(\pm 5, 0)$ and $F(\pm 3, 0)$

3. $3x^2 + 4y^2 = 12$

$\frac{x^2}{4} + \frac{y^2}{3} = 1$

$a = \sqrt{4} = 2,\ b = \sqrt{3}$

$c = \sqrt{a^2 - b^2} = \sqrt{2^2 - (\sqrt{3})^2} = \sqrt{1} = 1$

$V(\pm 2, 0)$ and $F(\pm 1, 0)$

5. $4x^2 + 3y^2 = 48$

$\frac{x^2}{12} + \frac{y^2}{16} = 1$

$a = \sqrt{16} = 4,\ b = \sqrt{12}$

$c = \sqrt{a^2 - b^2} = \sqrt{4^2 - (\sqrt{12})^2} = \sqrt{4} = 2$

$V(0, \pm 4)$ and $F(0, \pm 2)$

7. $V(\pm 5, 0),\ F(\pm 4, 0)$

$a = 5,\ c = 4$

$b = \sqrt{a^2 - c^2} = \sqrt{5^2 - 4^2} = \sqrt{9} = 3$

vertices on horizontal line,
so ellipse is horizontal

$\frac{x^2}{25} + \frac{y^2}{9} = 1$

9. $a = \frac{12}{2} = 6$, through $\left(3, \sqrt{3}\right)$

$$\frac{x^2}{36} + \frac{y^2}{b^2} = 1$$

at $\left(3, \sqrt{3}\right)$: $\frac{9}{36} + \frac{3}{b^2} = 1$

$$\frac{1}{4} + \frac{3}{b^2} = 1$$

$$b^2 + 12 = 4b^2$$

$$3b^2 = 12$$

$$b^2 = 4 \Rightarrow b = 2$$

horizontal major axis means horizontal ellipse

$$\frac{x^2}{36} + \frac{y^2}{4} = 1$$

11. $a = \frac{26}{2} = 13,\ c = \frac{24}{2} = 12$

$$b = \sqrt{a^2 - c^2} = \sqrt{13^2 - 12^2} = \sqrt{25} = 5$$

horizontal major axis means horizontal ellipse

$$\frac{x^2}{169} + \frac{y^2}{25} = 1$$

13. through $(1,4)$ and $(-6,1)$

at $(1,4)$: $\frac{1}{a^2} + \frac{16}{b^2} = 1 \qquad (1)$

at $(-6,1)$: $\frac{36}{a^2} + \frac{1}{b^2} = 1 \qquad (2)$

Multiply (2) by 16, $\frac{576}{a^2} + \frac{16}{b^2} = 16 \qquad (3)$

Taking $(3) - (1)$, $\frac{576}{a^2} - \frac{1}{a^2} = 15 \Rightarrow a^2 = \frac{576 - 1}{15} = \frac{115}{3}$

Plug into (1), $\frac{3}{115} + \frac{16}{b^2} = 1 \Rightarrow \frac{16}{b^2} = 1 - \frac{3}{115} = \frac{112}{115} \Rightarrow b^2 = 16\left(\frac{115}{112}\right) = \frac{115}{7}$

$$\frac{3x^2}{115} + \frac{7y^2}{115} = 1$$

Standard Equation of an Ellipse, Center Not At Origin

15. $\frac{(x-2)^2}{16} + \frac{(y+2)^2}{9} = 1$

$h = 2,\ k = -2,\ a = 4,\ b = 3$

$c = \sqrt{16 - 9} = \sqrt{7} = 2.65$

$C(2,-2)$

$V(2 \pm a, -2) = (6,-2)$ and $(-2,-2)$

$F(2 \pm c, -2) = (4.65,-2)$ and $(-0.65,-2)$

17. $5x^2+20x+9y^2-54y+56=0$

$5\left(x^2+4x+4\right)+9\left(y^2-6y+9\right)=-56+20+81$

$5(x+2)^2+9(y-3)^2=45$

$\dfrac{(x+2)^2}{9}+\dfrac{(y-3)^2}{5}=1$

$h=-2,\ k=3,\ a=3,\ b=\sqrt{5}$

$c=\sqrt{9-5}=\sqrt{4}=2$

$C(-2,3)$

$V(-2\pm a,3)=(-5,3)$ and $(1,3)$

$F(-2\pm c,3)=(-4,3)$ and $(0,3)$

19. $7x^2-14x+16y^2+32y=89$

$7\left(x^2-2x+1\right)+16\left(y^2+2y+1\right)=89+7+16$

$7(x-1)^2+16(y+1)^2=112$

$\dfrac{(x-1)^2}{16}+\dfrac{(y+1)^2}{7}=1$

$h=1,\ k=-1,\ a=4,\ b=\sqrt{7}$

$c=\sqrt{16-7}=\sqrt{9}=3$

$C(1,-1)$

$V(1\pm a,-1)=(-3,-1)$ and $(5,-1)$

$F(1\pm c,-1)=(-2,-1)$ and $(4,-1)$

21. $25x^2+150x+9y^2-36y+36=0$

$25\left(x^2+6x+9\right)+9\left(y^2-4y+4\right)=-36+225+36$

$25(x+3)^2+9(y-2)^2=225$

$\dfrac{(x+3)^2}{9}+\dfrac{(y-2)^2}{25}=1$

$h=-3,\ k=2,\ a=5,\ b=3$

$c=\sqrt{25-9}=\sqrt{16}=4$

$C(-3,2)$

$V(-3,2\pm a)=(-3,-3)$ and $(-3,7)$

$F(-3,2\pm c)=(-3,-2)$ and $(-3,6)$

23. $C(0,3),\ a=\dfrac{12}{2}=6,\ b=\dfrac{6}{2}=3$

major axis is vertical, so ellipse is vertical

$\dfrac{x^2}{9}+\dfrac{(y-3)^2}{36}=1$

25. $C(-2,-3),\ V(-2,1)$

center and vertex on vertical line, so ellipse is vertical

$a=1-(-3)=4 \qquad c=\dfrac{4}{2}=2$

$b=\sqrt{16-4}=\sqrt{12}$

$\dfrac{(x+2)^2}{12}+\dfrac{(y+3)^2}{16}=1$

General Equation

27. $\dfrac{x^2}{16}+\dfrac{y^2}{36}=4$

Multiply through by LCD of 144.

$9x^2+4y^2=576$

$9x^2+4y^2-576=0$

29. $\dfrac{(x+5)^2}{9}+\dfrac{(y+7)^2}{4}=22$

Multiply through by the LCD of 36.

$$4\left(x^2+10x+25\right)+9\left(y^2+14y+49\right)=22(36)$$

$$4x^2+40x+100+9y^2+126y+441=792$$

$$4x^2+9y^2+40x+126y-251=0$$

31. $x^2+9y^2-4x+18y+4=0$

$$\left(x^2-4x+4\right)+9\left(y^2+2y+1\right)=-4+4+9$$

$$(x-2)^2+9(y+1)^2=9$$

$$\frac{(x-2)^2}{9}+\frac{(y+1)^2}{1}=1$$

$h=2,\ k=-1,\ a=3,\ b=1$

$c=\sqrt{9-1}=\sqrt{8}=2.83$

$C(2,-1)$

$V(2\pm a,-1)=(-1,-1)$ and $(5,-1)$

$F(2\pm c,-1)=(-0.83,-1)$ and $(4.83,-1)$

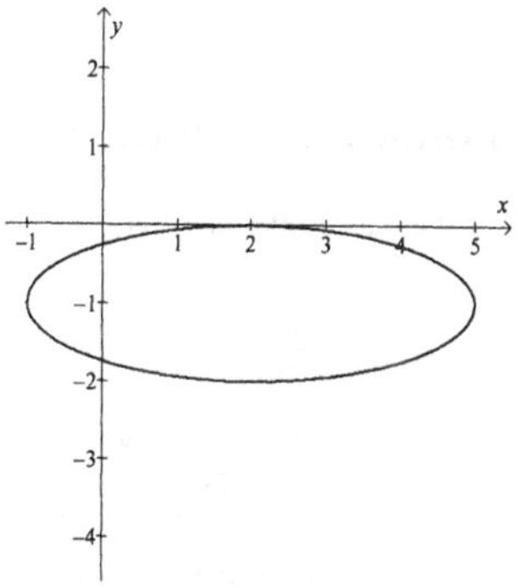

33. $49x^2+81y^2+294x+810y-1503=0$

$$49\left(x^2+6x+9\right)+81\left(y^2+10y+25\right)=1503+441+2025$$

$$49(x+3)^2+81(y+5)^2=3969$$

$$\frac{(x+3)^2}{81}+\frac{(y+5)^2}{49}=1$$

$h=-3,\ k=-5,\ a=9,\ b=7$

$c=\sqrt{81-49}=\sqrt{32}=5.66$

$C(-3,-5)$

$V(-3\pm a,-5)=(-12,-5)$ and $(6,-5)$

$F(-3\pm c,-5)=(-8.66,-5)$ and $(2.66,-5)$

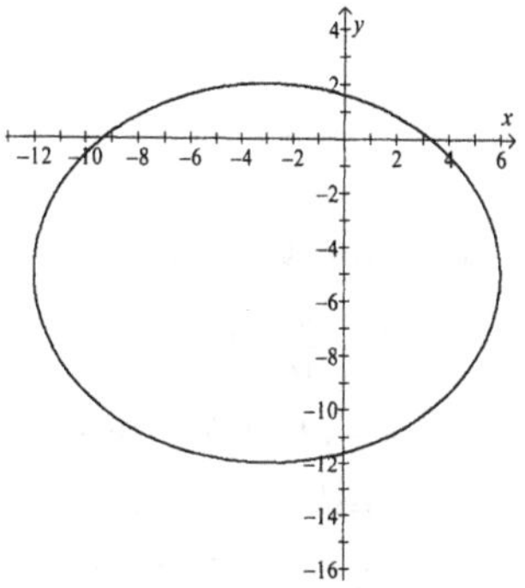

Focal Width

35. $a=\dfrac{44}{2}=22,\ b=\dfrac{22}{2}=11\ \Rightarrow\ L=\dfrac{2b^2}{a}=\dfrac{2(11)^2}{22}=11$

Applications

37. Take the center of the ellipse to be the origin.

$a=\dfrac{25}{2}=12.5,\ c=\dfrac{15}{2}=7.5\ \Rightarrow\ b=\sqrt{12.5^2-7.5^2}=\sqrt{100}=10$

The width of the chamber will be 2(10) = 20 cm.

39. Let the center be the origin.

$$a = \frac{8}{2} = 4,\ b = \frac{4}{2} = 2 \ \Rightarrow\ \frac{x^2}{16} + \frac{y^2}{4} = 1$$

Water level of 1 ft will be equivalent to $y = -1$.

$$\text{at } y = -1:\ \frac{x^2}{16} + \frac{(-1)^2}{4} = 1 \ \Rightarrow\ \frac{x^2}{16} = 1 - \frac{1}{4} = \frac{3}{4} \ \Rightarrow\ x^2 = 12 \ \Rightarrow\ x = \sqrt{12} \text{ (Drop the negative)}$$

The width of the stream is $w = 2x = 2\sqrt{12} = 6.9$ ft.

Exercise 22.6 • The Hyperbola

Standard Equation of Hyperbola with Center at Origin

1. $\frac{x^2}{16} - \frac{y^2}{25} = 1$

$a = \sqrt{16} = 4,\ b = \sqrt{25} = 5$

$c = \sqrt{a^2 + b^2} = \sqrt{16 + 25} = \sqrt{41}$

$V(\pm 4, 0)$ and $F(\pm\sqrt{41}, 0)$

transverse axis is horizontal

slope of asymptotes $= \pm\frac{b}{a} = \pm\frac{5}{4}$

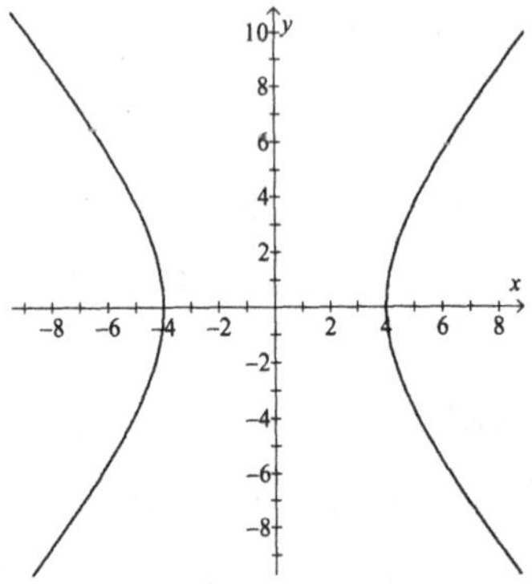

3. $16x^2 - 9y^2 = 144$

$\frac{x^2}{9} - \frac{y^2}{16} = 1$

$a = \sqrt{9} = 3,\ b = \sqrt{16} = 4$

$c = \sqrt{a^2 + b^2} = \sqrt{9 + 16} = \sqrt{25} = 5$

$V(\pm 3, 0)$ and $F(\pm 5, 0)$

transverse axis is horizontal

slope of asymptotes $= \pm\frac{b}{a} = \pm\frac{4}{3}$

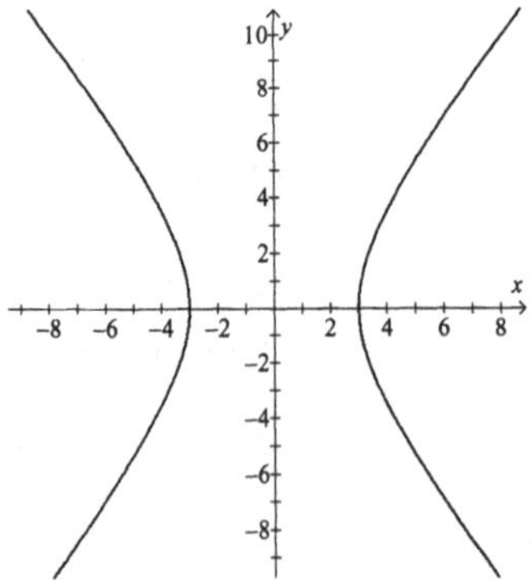

5. $x^2 - 4y^2 = 16$

$\frac{x^2}{16} - \frac{y^2}{4} = 1$

$a = \sqrt{16} = 4,\ b = \sqrt{4} = 2$

$c = \sqrt{a^2 + b^2} = \sqrt{16+4} = \sqrt{20} = 2\sqrt{5}$

$V(\pm 4, 0)$ and $F(\pm 2\sqrt{5}, 0)$

transverse axis is horizontal

slope of asymptotes $= \pm\frac{b}{a} = \pm\frac{2}{4} = \pm\frac{1}{2}$

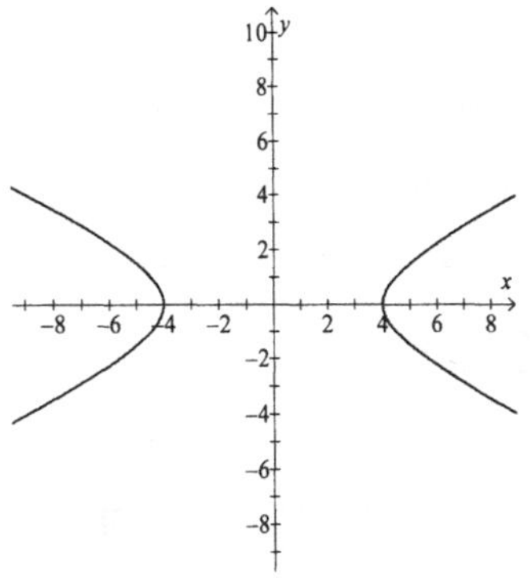

7. $V(\pm 5, 0) \Rightarrow a = 5$

$F(\pm 13, 0) \Rightarrow c = 13$

vertices and foci on horizontal line

so transverse axis is horizontal

$b = \sqrt{c^2 - a^2} = \sqrt{169-25} = \sqrt{144} = 12$

$\frac{x^2}{25} - \frac{y^2}{144} = 1$

9. distance between foci $= 8 \Rightarrow c = \frac{8}{2} = 4$

transverse axis $= 6 \Rightarrow a = \frac{6}{2} = 3$

transverse axis is horizontal

$b = \sqrt{c^2 - a^2} = \sqrt{16-9} = \sqrt{7}$

$\frac{x^2}{9} - \frac{y^2}{7} = 1$

11. conjugate axis = transverse axis $\Rightarrow b = a$

transverse axis is vertical

$\frac{y^2}{a^2} - \frac{x^2}{a^2} = 1$

at $(3,5)$: $\frac{25}{a^2} - \frac{9}{a^2} = 1$

$\frac{16}{a^2} = 1 \Rightarrow a^2 = b^2 = 16$

$\frac{y^2}{16} - \frac{x^2}{16} = 1$

13. transverse axis $= 10 \Rightarrow a = \frac{10}{2} = 5$

transverse axis is vertical

$\frac{y^2}{25} - \frac{x^2}{b^2} = 1$

at $(8,10)$: $\frac{100}{25} - \frac{64}{b^2} = 1$

$4b^2 - 64 = b^2$

$3b^2 = 64 \Rightarrow b^2 = \frac{64}{3}$

$\frac{y^2}{25} - \frac{3x^2}{64} = 1$

Standard Equation of Hyperbola with Center Not at Origin

15. $\frac{(x-2)^2}{25} - \frac{(y+1)^2}{16} = 1 \Rightarrow C(2,-1)$

$a = \sqrt{25} = 5,\ b = \sqrt{16} = 4$

$c = \sqrt{a^2 + b^2} = \sqrt{25+16} = \sqrt{41} = 6.4$

$V(2 \pm 5, -1) = (-3,-1)$ and $(7,-1)$

$F(2 \pm \sqrt{41}, -1) = (-4.4,-1)$ and $(8.4,-1)$

transverse axis is horizontal

slope of asymptotes $= \pm\frac{b}{a} = \pm\frac{4}{5}$

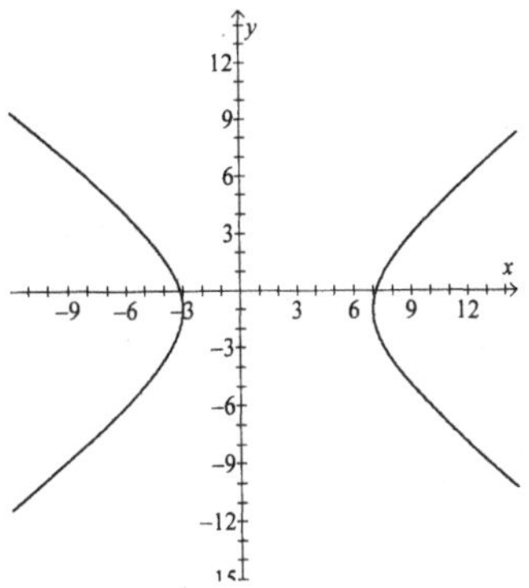

17. $C(3,2)$

transverse axis $= 8 \Rightarrow a = \frac{8}{2} = 4$

transverse axis is vertical

conjugate axis $= 4 \Rightarrow b = \frac{4}{2} = 2$

$$\frac{(y-2)^2}{16} - \frac{(x-3)^2}{4} = 1$$

19. foci at $(5,-2)$ and $(-3,-2) \Rightarrow c = \frac{8}{2} = 4$

$C(1,-2)$

foci on horizontal line so transverse axis is horizontal

vertex halfway between center and focus so $a = \frac{4}{2} = 2$

$b = \sqrt{c^2 - a^2} = \sqrt{16-4} = \sqrt{12}$

$$\frac{(x-1)^2}{4} - \frac{(y+2)^2}{12} = 1$$

Write each standard equation in general form.

21. $\frac{x^2}{25} - \frac{y^2}{9} = 1$

multiply both sides by the LCD of 225

$9x^2 - 25y^2 = 225 \Rightarrow 9x^2 - 25y^2 - 225 = 0$

23. $\frac{(x-3)^2}{16} - \frac{(y-4)^2}{25} = 12$

multiply both sides by the LCD of 400

$25(x^2 - 6x + 9) - 16(y^2 - 8y + 16) = 4800$

$25x^2 - 150x + 225 - 16y^2 + 128y - 256 = 4800$

$25x^2 - 16y^2 - 150x + 128y - 4831 = 0$

25. $16x^2 - 64x - 9y^2 - 54y = 161$

$16\left(x^2 - 4x + 4\right) - 9\left(y^2 + 6y + 9\right) = 161 + 64 - 81$

$16(x-2)^2 - 9(y+3)^2 = 144$

$\dfrac{(x-2)^2}{9} - \dfrac{(y+3)^2}{16} = 1$

$C(2,-3)$

$a = \sqrt{9} = 3,\ b = \sqrt{16} = 4$

$c = \sqrt{a^2 + b^2} = \sqrt{9+16} = \sqrt{25} = 5$

$V(2 \pm 3, -3) = (-1,-3)$ and $(5,-3)$

$F(2 \pm 5, -3) = (-3,-3)$ and $(7,-3)$

transverse axis is horizontal

slope of asymptotes $= \pm\dfrac{b}{a} = \pm\dfrac{4}{3}$

27. $4x^2 + 8x - 5y^2 - 10y + 19 = 0$

$4\left(x^2 + 2x + 1\right) - 5\left(y^2 + 2y + 1\right) = -19 + 4 - 5$

$4(x+1)^2 - 5(y+1)^2 = -20$

$\dfrac{(y+1)^2}{4} - \dfrac{(x+1)^2}{5} = 1$

$C(-1,-1)$

$a = \sqrt{4} = 2,\ b = \sqrt{5}$

$c = \sqrt{a^2 + b^2} = \sqrt{4+5} = \sqrt{9} = 3$

$V(-1, -1 \pm 2) = (-1,-3)$ and $(-1,1)$

$F(-1, -1 \pm 3) = (-1,-4)$ and $(-1,2)$

transverse axis is vertical

slope of asymptotes $= \pm\dfrac{a}{b} = \pm\dfrac{2}{\sqrt{5}}$

Hyperbola Whose Asymptotes are Coordinate Axes

29. Hyperbola with center at the origin and asymptotes on the coordinate axes will be of the form $xy = k$. Find k by using the given vertex of $(6,6)$, $k = 6(6) = 36$. The equation is $xy = 36$.

Applications

31. $a = 18.0, c = 26.0 \Rightarrow b^2 = c^2 - a^2 = 676 - 324 = 352 \Rightarrow \dfrac{x^2}{324} - \dfrac{y^2}{352} = 1$

33. $pv = c$

$c = 25.0(1000) = 25{,}000$

$pv = 25{,}000$

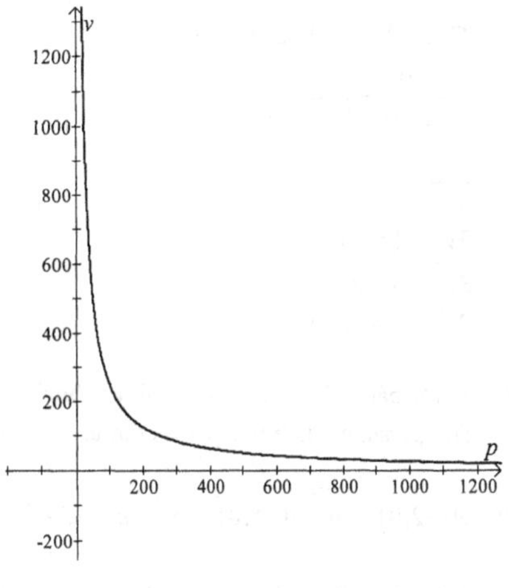

Chapter 22 • Review Problems

1. $(3,0)$ and $(7,0)$ $\Rightarrow$ $d=\sqrt{(3-7)^2+(0-0)^2}=\sqrt{16}=4$

3. $m=\tan 34.8^\circ=0.695$ $\Rightarrow$ $m_{\text{perpendicular}}=-\dfrac{1}{(-0.695)}=-1.44$

5. $m_{\text{perpendicular}}=-\dfrac{1}{1.55}=-0.645$

   $\arctan(-0.645)=-32.8^\circ$ $\Rightarrow$ $\theta=180^\circ-32.8^\circ=147^\circ$

7. $m=\dfrac{a}{2b}$ $\Rightarrow$ $m_{\text{perpendicular}}=-\dfrac{1}{a/2b}=-\dfrac{2b}{a}$

9. $2y-5=3(x-4)$

   $2y-5=3x-12$

   $y=\dfrac{3}{2}x-\dfrac{7}{2}$ $\Rightarrow$ $m=\dfrac{3}{2}$ and $b=-\dfrac{7}{2}$

11. Through $(-5,-1)$ and $(-2,6)$

    $\dfrac{y-(-1)}{x-(-5)}=\dfrac{6-(-1)}{-2-(-5)}$

    $\dfrac{y+1}{x+5}=\dfrac{7}{3}$

    $3(y+1)=7(x+5)$

    $3y+3=7x+35$

    $7x-3y+32=0$

13. $m=5$ through $(-4,7)$

    $\dfrac{y-7}{x-(-4)}=5$

    $y-7=5(x+4)$

    $y-7=5x+20$

    $5x-y+27=0$

15. Through $(-3,0)$ and $(0,7)$

    $\dfrac{y-0}{x-(-3)}=\dfrac{7-0}{0-(-3)}$

    $\dfrac{y}{x+3}=\dfrac{7}{3}$

    $3y=7(x+3)$

    $3y=7x+21$

    $7x-3y+21=0$

17. A line parallel to the x-axis will be a horizontal line with all points on the line having the same y-coordinate. The horizontal line that passes through the point $(2,5)$ is $y=5$ or in general form $y-5=0$.

19. $A(-2,0)$ and $B(-5,0)$ $\Rightarrow$ $AB=-5-(-2)=-3$

21. A line parallel to the y-axis will be a vertical line with all points on the line having the same x-coordinate. The vertical line that passes through the point $(-3,6)$ is $x=-3$ or in general form $x+3=0$.

23. Vertices of the triangle are $(6,4)$, $(5,-2)$, and $(-3,-4)$.
The lengths of the three sides are
$a=\sqrt{(5-6)^2+(-2-4)^2}=\sqrt{1+36}=\sqrt{37}=6.08$
$b=\sqrt{(-3-6)^2+(-4-4)^2}=\sqrt{81+64}=\sqrt{145}=12.0$
$c=\sqrt{(-3-5)^2+(-4-(-2))^2}=\sqrt{64+4}=\sqrt{68}=8.25$
Then by Hero's Formula,
$s=\dfrac{a+b+c}{2}=\dfrac{6.08+12.0+8.25}{2}=13.17$
$A=\sqrt{s(s-a)(s-b)(s-c)}=\sqrt{13.17(13.17-6.08)(13.17-12.0)(13.17-8.25)}=23.2$

25. $x^2+6x+4y=3$
$x^2+6x+9=-4y+3+9$
$(x+3)^2=-4y+12$
$(x+3)^2=-4(y-3)$ Parabola
vertex at $V(-3,3)$
$4p=-4 \Rightarrow p=-1$
focus at $F(-3,3-1)=F(-3,2)$
$L=|4p|=4$

27. $25x^2-200x+9y^2-90y=275$
$25(x^2-8x+16)+9(y^2-10y+25)=275+400+225$
$25(x-4)^2+9(y-5)^2=900$
$\dfrac{(x-4)^2}{36}+\dfrac{(y-5)^2}{100}=1$ Ellipse
$h=4,\ k=5,\ a=\sqrt{100}=10,\ b=\sqrt{36}=6,$ and $c=\sqrt{100-36}=\sqrt{64}=8$
$C(4,5)$
$V(4,5\pm10)=(4,-5)$ and $(4,15)$
$F(4,5\pm8)=(4,-3)$ and $(4,13)$

29. $x^2+y^2=9 \Rightarrow$ center at $(0,0)$ and $r=\sqrt{9}=3$

30. The tangent at point (x_1,y_1) with slope m can be found using point-slope form of a line: $y-y_1=m(x-x_1)$.

31. The normal at point (x_1,y_1) will have then have slope $-\dfrac{1}{m}$.

$y-y_1=-\dfrac{1}{m}(x-x_1) \Rightarrow m(y-y_1)=-x+x_1 \Rightarrow x-x_1=m(y-y_1)=0$.

32. The x-intercept A of the tangent

$y=0 \Rightarrow 0-y_1=m(x-x_1) \Rightarrow -y_1=mx-mx_1 \Rightarrow mx=mx_1-y_1 \Rightarrow x=x_1-\dfrac{y_1}{m}$ or $\left(x_1-\dfrac{y_1}{m},0\right)$

33. The y-intercept B of the tangent

$x=0 \Rightarrow y-y_1=m(0-x_1) \Rightarrow y-y_1=-mx_1 \Rightarrow y=y_1-mx_1$ or $(0, y_1-mx_1)$

34. The distance between (x_1, y_1) and $\left(x_1-\frac{y_1}{m}, 0\right)$ is

$$d=\sqrt{\left(x_1-x_1+\frac{y_1}{m}\right)^2+(y_1-0)^2}=\sqrt{\frac{y_1^2}{m^2}+y_1^2}=\sqrt{\frac{y_1^2}{m^2}+\frac{m^2 y_1^2}{m^2}}=\sqrt{\frac{y_1^2}{m^2}(1+m^2)}=\frac{y_1}{m}\sqrt{1+m^2}$$

35. The distance between (x_1, y_1) and $(0, y_1-mx_1)$ is

$$d=\sqrt{(x_1-0)^2+(y_1-y_1+mx_1)^2}=\sqrt{x_1^2+m^2x_1^2}=\sqrt{x_1^2(1+m^2)}=x_1\sqrt{1+m^2}$$

36. The x-intercept C of the normal

$y=0 \Rightarrow x-x_1+m(0-y_1)=0 \Rightarrow x=x_1+my_1$ or $(x_1+my_1, 0)$

37. The y-intercept D of the normal

$x=0 \Rightarrow 0-x_1+m(y-y_1)=0 \Rightarrow my-my_1=x_1 \Rightarrow my=my_1+x_1 \Rightarrow y=y_1+\frac{x_1}{m}$ or $\left(0, y_1+\frac{x_1}{m}\right)$

38. The distance between (x_1, y_1) and $(x_1+my_1, 0)$ is

$$d=\sqrt{(x_1-x_1-my_1)^2+(y_1-0)^2}=\sqrt{m^2y_1^2+y_1^2}=\sqrt{y_1^2(m^2+1)}=y_1\sqrt{(m^2+1)}$$

39. The distance between (x_1, y_1) and $\left(0, y_1+\frac{x_1}{m}\right)$ is

$$d=\sqrt{(x_1-0)^2+\left(y_1-y_1-\frac{x_1}{m}\right)^2}=\sqrt{x_1^2+\frac{x_1^2}{m^2}}=\sqrt{\frac{m^2x_1^2}{m^2}+\frac{x_1^2}{m^2}}=\sqrt{\frac{x_1^2}{m^2}(m^2+1)}=\frac{x_1}{m}\sqrt{m^2+1}$$

41. $C(0,0)$, $a=\frac{20}{2}=10$, $b=c$

major axis is horizontal

$a^2=c^2+b^2=b^2+b^2=2b^2$

$b^2=\frac{a^2}{2}=\frac{100}{2}=50$

$\frac{x^2}{100}+\frac{y^2}{50}=1 \Rightarrow x^2+2y^2=100$

43. vertex at $(0,0)$; focus is $(-4.25, 0)$

$p=-4.25 \Rightarrow 4p=-17$

axis is horizontal line so parabola is horizontal

$y^2=-17x$

45. Center at $(-5,0)$ and $r=5 \Rightarrow (x+5)^2+y^2=25$

47. $F(2,1)$ and $(-6,1)$

$c=\dfrac{2-(-6)}{2}=4 \Rightarrow C(-2,1)$

$2a=10 \Rightarrow a=5$

foci on horizontal line, so ellipse is horizontal

$b=\sqrt{a^2-c^2}=\sqrt{25-16}=\sqrt{9}=3$

$\dfrac{(x+2)^2}{25}+\dfrac{(y-1)^2}{9}=1$

49. $y^2+4x-6y=16$

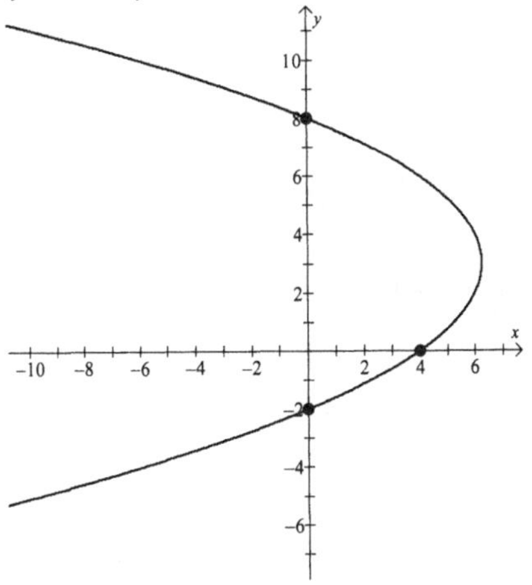

x-int: $(4,0)$

y-int: $(0,8)$ and $(0,-2)$

51. $a=\dfrac{15}{2}=7.50,\ b=5 \Rightarrow \dfrac{x^2}{56.25}+\dfrac{y^2}{25}=1$

at $x=6.0$: $\dfrac{36}{56.25}+\dfrac{y^2}{25}=1 \Rightarrow \dfrac{y^2}{25}=1-\dfrac{36}{56.25}=0.36 \Rightarrow y=\sqrt{0.36(25)}=3.0$ m

53. Take the lower left point of the arch to be the origin.

$h=3$ and $k=5 \Rightarrow (x-3)^2=4p(y-5)$

at $(0,0)$: $(-3)^2=4p(-5) \Rightarrow 4p=\dfrac{9}{-5}$

$(x-3)^2=-\dfrac{9}{5}(y-5)$

when $y=2$, $(x-3)^2=-\dfrac{9}{5}(2-5) \Rightarrow 5(x^2-6x+9)=27 \Rightarrow 5x^2-30x+18=0$

$x=\dfrac{30\pm\sqrt{900-4(5)(18)}}{10}=0.676$ m and 5.324 m

Width of the arch will be 5.324 - 0.676 = 4.65 m.

Chapter 23: Derivatives of Algebraic Functions

Exercise 23.1 • Limits

Simple Limits

1. $\lim_{x\to 2}\left(x^2+2x-7\right)=2^2+2(2)-7=1$

3. $\lim_{x\to 2}\frac{x^2-x-1}{x+3}=\frac{2^2-2-1}{2+3}=\frac{1}{5}$

5. $\lim_{x\to 5}\frac{x^2-25}{x-5}=\lim_{x\to 5}\frac{(x-5)(x+5)}{x-5}=\lim_{x\to 5}(x+5)=5+5=10$

7. $\lim_{x\to 1}\frac{x^2+2x-3}{x-1}=\lim_{x\to 1}\frac{(x-1)(x+3)}{x-1}=\lim_{x\to 1}(x+3)=1+3=4$

9. $\lim_{x\to 0}\frac{\sin x}{\tan x}=\lim_{x\to 0}\frac{\sin x}{\sin x/\cos x}=\lim_{x\to 0}\sin x\frac{\cos x}{\sin x}=\lim_{x\to 0}\cos x=1$

11. $\lim_{x\to -3}\frac{x^2+2x-3}{x+3}=\lim_{x\to -3}\frac{(x+3)(x-1)}{x+3}=\lim_{x\to -3}(x-1)=-3-1=-4$

Limits Involving Zero or Infinity

13. $\lim_{x\to 0}\left(4x^2-5x-8\right)=4(0)^2-5(0)-8=-8$

15. $\lim_{x\to 0}\frac{\sqrt{x}-4}{\sqrt[3]{x}+5}=\frac{\sqrt{0}-4}{\sqrt[3]{0}+5}=-\frac{4}{5}$

17. $\lim_{x\to 0}\left(\frac{1}{2+x}-\frac{1}{2}\right)\cdot\frac{1}{x}=\lim_{x\to 0}\left(\frac{2-(2+x)}{(2+x)2}\right)\cdot\frac{1}{x}=\lim_{x\to 0}\left(\frac{2-2-x}{(2+x)2}\right)\cdot\frac{1}{x}=\lim_{x\to 0}\left(\frac{-1}{(2+x)2}\right)=\frac{-1}{(2+0)2}=-\frac{1}{4}$

19. $\lim_{x\to\infty}\frac{2x+5}{x-4}=\lim_{x\to\infty}\frac{2+\frac{5}{x}}{1-\frac{4}{x}}=\frac{2+0}{1-0}=2$

21. $\lim_{x\to\infty}\frac{x^2+x-3}{5x^2+10}=\lim_{x\to\infty}\frac{1+\frac{1}{x}-\frac{3}{x^2}}{5+\frac{10}{x^2}}=\frac{1+0-0}{5+0}=\frac{1}{5}$

Limits Depending on Direction of Approach

23. $\lim\limits_{x \to 0^+} \frac{7}{x} = +\infty$

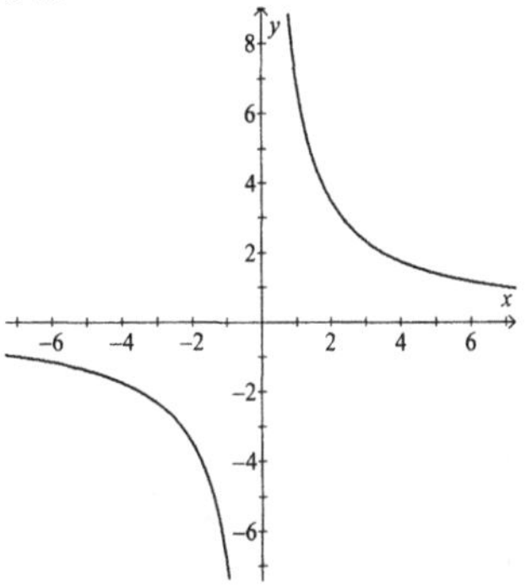

25. $\lim\limits_{x \to 0^+} \frac{x+1}{x} = +\infty$

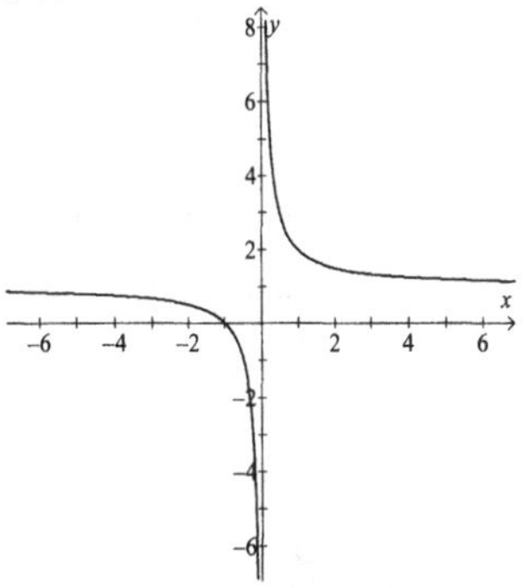

27. $\lim\limits_{x \to 2^-} \frac{5+x}{x-2} = -\infty$

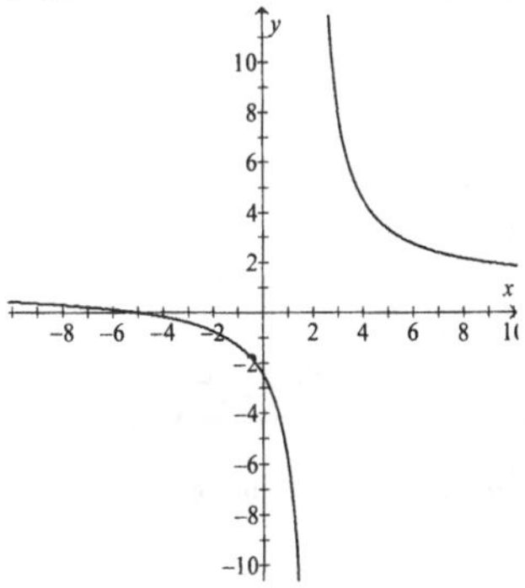

When the Limit is an Expression

29. $\lim\limits_{d \to 0} \left(x^2 + 2d + d^2\right) = x^2 + 2(0) + 0^2 = x^2$

31. $\lim\limits_{d \to 0} \frac{(x+d)^2 - x^2}{x^2(x+d)} = \frac{(x+0)^2 - x^2}{x^2(x+0)} = \frac{0}{x^3} = 0$

33. $\lim\limits_{d \to 0} \frac{3(x+d) - 3x}{d} = \lim\limits_{d \to 0} \frac{3x + 3d - 3x}{d} = \lim\limits_{d \to 0} \frac{3d}{d} = 3$

35. $\lim\limits_{d \to 0} \left[3x + d - \frac{1}{(x+d+2)(x-2)}\right] = 3x + 0 - \frac{1}{(x+0+2)(x-2)} = 3x - \frac{1}{(x+2)(x-2)} = 3x - \frac{1}{x^2 - 4}$

37. $\lim\limits_{d \to 0} \frac{(x+d)^3 - x^3}{d} = \lim\limits_{d \to 0} \frac{x^3 + 3x^2 d + 3xd^2 + d^3 - x^3}{d} = \lim\limits_{d \to 0} \frac{d\left(3x^2 + 3xd + d^2\right)}{d} = \lim\limits_{d \to 0} \left(3x^2 + 3xd + d^2\right) = 3x^2$

39. $$\lim_{d\to 0}\frac{(x+d)^2-2(x+d)-x^2+2x}{d}=\lim_{d\to 0}\frac{x^2+2dx+d^2-2x-2d-x^2+2x}{d}=\lim_{d\to 0}\frac{2dx+d^2-2d}{d}$$
$$=\lim_{d\to 0}\frac{d(2x+d-2)}{d}=\lim_{d\to 0}(2x+d-2)=2x-2$$

Exercise 23.2 • Rate of Change and the Tangent

A TI-83+ calculator was used for the following problems.

1. $y=x^2$

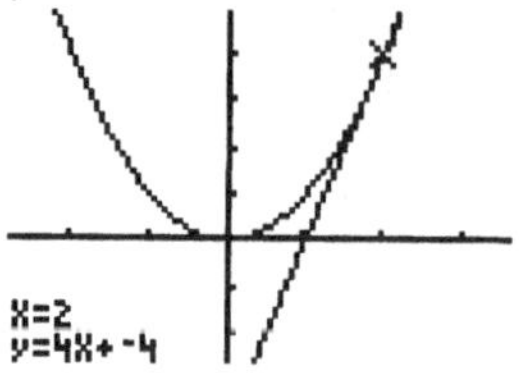

$m=4$

3. $y=\sqrt{x}$

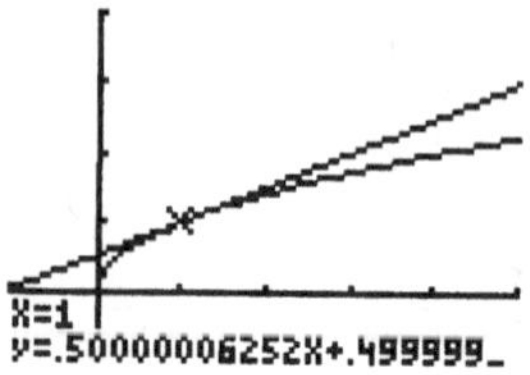

$m=0.5$

5. $y=x-x^2$

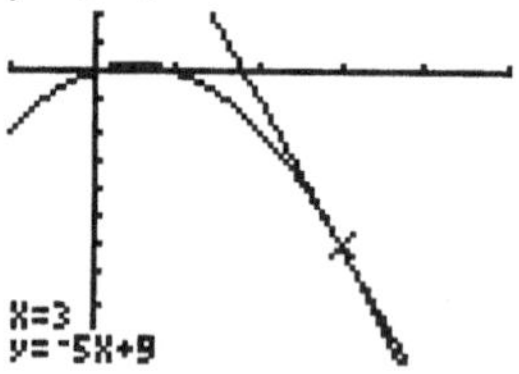

$m=-5$

7. $y=3-\sqrt{x}$

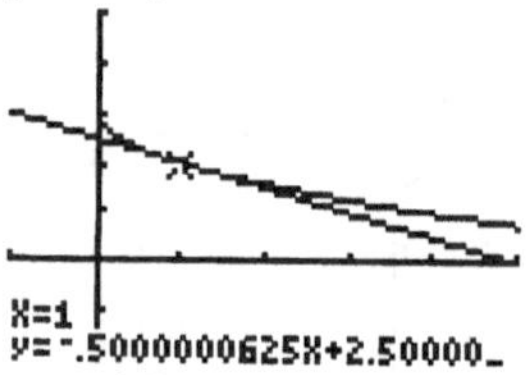

$m=-0.5$

9. $pv=3650 \;\Rightarrow\; v=\dfrac{3650}{p}$

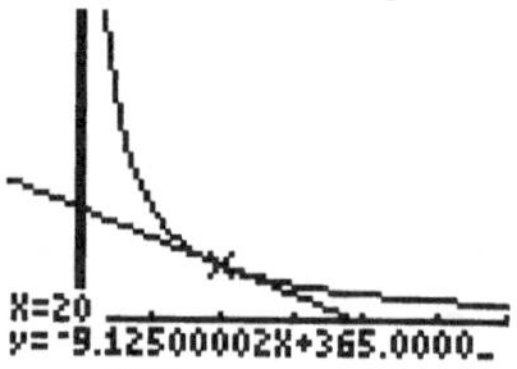

$m=-9.13$

Exercise 23.3 • The Derivative

Use the delta method to find the derivative

1. $y = 3x - 2$

$$y + \Delta y = 3(x + \Delta x) - 2 = 3x + 3\Delta x - 2$$
$$(y + \Delta y) - y = 3x + 3\Delta x - 2 - (3x - 2) = 3x + 3\Delta x - 2 - 3x + 2 = 3\Delta x$$
$$\Delta y = 3\Delta x$$
$$\frac{\Delta y}{\Delta x} = 3$$
$$\frac{dy}{dx} = \lim_{\Delta x \to 0} \frac{\Delta y}{\Delta x} = 3$$

3. $y = 2x + 5$

$$y + \Delta y = 2(x + \Delta x) + 5 = 2x + 2\Delta x + 5$$
$$(y + \Delta y) - y = 2x + 2\Delta x + 5 - (2x + 5) = 2x + 2\Delta x + 5 - 2x - 5 = 2\Delta x$$
$$\Delta y = 2\Delta x$$
$$\frac{\Delta y}{\Delta x} = 2$$
$$\frac{dy}{dx} = \lim_{\Delta x \to 0} \frac{\Delta y}{\Delta x} = 2$$

5. $y = x^2 + 1$

$$y + \Delta y = (x + \Delta x)^2 + 1 = x^2 + 2x\Delta x + (\Delta x)^2 + 1$$
$$(y + \Delta y) - y = x^2 + 2x\Delta x + (\Delta x)^2 + 1 - \left(x^2 + 1\right) = x^2 + 2x\Delta x + (\Delta x)^2 + 1 - x^2 - 1 = 2x\Delta x + (\Delta x)^2$$
$$\Delta y = 2x\Delta x + (\Delta x)^2$$
$$\frac{\Delta y}{\Delta x} = \frac{2x\Delta x + (\Delta x)^2}{\Delta x} = \frac{\Delta x(2x + \Delta x)}{\Delta x} = 2x + \Delta x$$
$$\frac{dy}{dx} = \lim_{\Delta x \to 0} \frac{\Delta y}{\Delta x} = \lim_{\Delta x \to 0} (2x + \Delta x) = 2x + 0 = 2x$$

7. $y = x^3$

$$y + \Delta y = (x + \Delta x)^3 = x^3 + 3x^2\Delta x + 3x(\Delta x)^2 + (\Delta x)^3$$
$$(y + \Delta y) - y = x^3 + 3x^2\Delta x + 3x(\Delta x)^2 + (\Delta x)^3 - \left(x^3\right) = 3x^2\Delta x + 3x(\Delta x)^2 + (\Delta x)^3$$
$$\Delta y = 3x^2\Delta x + 3x(\Delta x)^2 + (\Delta x)^3$$
$$\frac{\Delta y}{\Delta x} = \frac{3x^2\Delta x + 3x(\Delta x)^2 + (\Delta x)^3}{\Delta x} = \frac{\Delta x\left(3x^2 + 3x(\Delta x) + (\Delta x)^2\right)}{\Delta x} = 3x^2 + 3x\Delta x + (\Delta x)^2$$
$$\frac{dy}{dx} = \lim_{\Delta x \to 0} \frac{\Delta y}{\Delta x} = \lim_{\Delta x \to 0} \left(3x^2 + 3x\Delta x + (\Delta x)^2\right) = 3x^2 + 0 + 0 = 3x^2$$

9. $y=\frac{3}{x}$

$$y+\Delta y=\frac{3}{x+\Delta x}$$

$$(y+\Delta y)-y=\frac{3}{x+\Delta x}-\frac{3}{x}=\frac{3x-3(x+\Delta x)}{x(x+\Delta x)}=\frac{-3\Delta x}{x(x+\Delta x)}$$

$$\Delta y=\frac{-3\Delta x}{x(x+\Delta x)}$$

$$\frac{\Delta y}{\Delta x}=\frac{\frac{-3\Delta x}{x(x+\Delta x)}}{\Delta x}=\frac{-3\Delta x}{x(x+\Delta x)}\left(\frac{1}{\Delta x}\right)=-\frac{3}{x(x+\Delta x)}$$

$$\frac{dy}{dx}=\lim_{\Delta x\to 0}\frac{\Delta y}{\Delta x}=\lim_{\Delta x\to 0}\left(-\frac{3}{x(x+\Delta x)}\right)=-\frac{3}{x(x+0)}=-\frac{3}{x^2}$$

11. $y=\sqrt{3-x}$

$$y+\Delta y=\sqrt{3-(x+\Delta x)}=\sqrt{3-x-\Delta x}$$

$$(y+\Delta y)-y=\sqrt{3-x-\Delta x}-\sqrt{3-x}$$

$$\Delta y=\sqrt{3-x-\Delta x}-\sqrt{3-x}$$

$$\frac{\Delta y}{\Delta x}=\frac{\sqrt{3-x-\Delta x}-\sqrt{3-x}}{\Delta x}\left(\frac{\sqrt{3-x-\Delta x}+\sqrt{3-x}}{\sqrt{3-x-\Delta x}+\sqrt{3-x}}\right)=\frac{3-x-\Delta x-3+x}{\Delta x\left(\sqrt{3-x-\Delta x}+\sqrt{3-x}\right)}=\frac{-1}{\sqrt{3-x-\Delta x}+\sqrt{3-x}}$$

$$\frac{dy}{dx}=\lim_{\Delta x\to 0}\frac{\Delta y}{\Delta x}=\lim_{\Delta x\to 0}\left(\frac{-1}{\sqrt{3-x-\Delta x}+\sqrt{3-x}}\right)=\frac{-1}{\sqrt{3-x-0}+\sqrt{3-x}}=\frac{-1}{2\sqrt{3-x}}$$

Slope of the tangent or rate of change at a given value of x

13. $y=\frac{1}{x^2}$ at $x=1$

$$y+\Delta y=\frac{1}{(x+\Delta x)^2}=\frac{1}{x^2+2x\Delta x+(\Delta x)^2}$$

$$(y+\Delta y)-y=\frac{1}{x^2+2x\Delta x+(\Delta x)^2}-\frac{1}{x^2}=\frac{x^2-\left(x^2+2x\Delta x+(\Delta x)^2\right)}{x^2\left(x^2+2x\Delta x+(\Delta x)^2\right)}=\frac{x^2-x^2-2x\Delta x-(\Delta x)^2}{x^2\left(x^2+2x\Delta x+(\Delta x)^2\right)}$$

$$\frac{\Delta y}{\Delta x}=\frac{\frac{-2x\Delta x-(\Delta x)^2}{x^2\left(x^2+2x\Delta x+(\Delta x)^2\right)}}{\Delta x}=\frac{\Delta x(-2x-\Delta x)}{x^2\left(x^2+2x\Delta x+(\Delta x)^2\right)}\left(\frac{1}{\Delta x}\right)=\frac{-2x-\Delta x}{x^2\left(x^2+2x\Delta x+(\Delta x)^2\right)}$$

$$\frac{dy}{dx}=\lim_{\Delta x\to 0}\frac{\Delta y}{\Delta x}=\lim_{\Delta x\to 0}\left(\frac{-2x-\Delta x}{x^2\left(x^2+2x\Delta x+(\Delta x)^2\right)}\right)=\frac{-2x-0}{x^2\left(x^2+0+0\right)}=\frac{-2x}{x^4}=-\frac{2}{x^3}$$

$$\left.\frac{dy}{dx}\right|_{x=1}=-\frac{2}{(1)^3}=-2$$

15. $y = x + \frac{1}{x}$ at $x = 2$

$$y + \Delta y = (x + \Delta x) + \frac{1}{(x + \Delta x)}$$

$$(y + \Delta y) - y = x + \Delta x + \frac{1}{x + \Delta x} - x - \frac{1}{x} = \Delta x + \frac{1}{x + \Delta x} - \frac{1}{x} = \frac{x\Delta x(x + \Delta x) + x - (x + \Delta x)}{x(x + \Delta x)}$$

$$= \frac{x^2\Delta x + x(\Delta x)^2 + x - x - \Delta x}{x(x + \Delta x)} = \frac{x^2\Delta x + x(\Delta x)^2 - \Delta x}{x(x + \Delta x)}$$

$$\frac{\Delta y}{\Delta x} = \frac{\frac{\Delta x\left(x^2 + x\Delta x - 1\right)}{x(x + \Delta x)}}{\Delta x} = \frac{\Delta x\left(x^2 + x\Delta x - 1\right)}{x(x + \Delta x)}\left(\frac{1}{\Delta x}\right) = \frac{x^2 + x\Delta x - 1}{x(x + \Delta x)}$$

$$\frac{dy}{dx} = \lim_{\Delta x \to 0} \frac{\Delta y}{\Delta x} = \lim_{\Delta x \to 0}\left(\frac{x^2 + x\Delta x - 1}{x(x + \Delta x)}\right) = \frac{x^2 + 0 - 1}{x(x + 0)} = \frac{x^2 - 1}{x^2}$$

$$\left.\frac{dy}{dx}\right|_{x=2} = \frac{(2)^2 - 1}{(2)^2} = \frac{3}{4}$$

17. $y = 2x - 3$ at $x = 3$

$$y + \Delta y = 2(x + \Delta x) - 3 = 2x + 2\Delta x - 3$$

$$(y + \Delta y) - y = 2x + 2\Delta x - 3 - (2x - 3) = 2x + 2\Delta x - 3 - 2x + 3 = 2\Delta x$$

$$\frac{\Delta y}{\Delta x} = \frac{2\Delta x}{\Delta x} = 2$$

$$\frac{dy}{dx} = \lim_{\Delta x \to 0} \frac{\Delta y}{\Delta x} = 2$$

$$\left.\frac{dy}{dx}\right|_{x=3} = 2$$

19. $y = 2x^2 - 6$ at $x = 3$

$$y + \Delta y = 2(x + \Delta x)^2 - 6 = 2x^2 + 4x\Delta x + 2(\Delta x)^2 - 6$$

$$(y + \Delta y) - y = 2x^2 + 4x\Delta x + 2(\Delta x)^2 - 6 - \left(2x^2 - 6\right) = 2x^2 + 4x\Delta x + 2(\Delta x)^2 - 6 - 2x^2 + 6 = 4x\Delta x + 2(\Delta x)^2$$

$$\frac{\Delta y}{\Delta x} = \frac{4x\Delta x + 2(\Delta x)^2}{\Delta x} = \frac{\Delta x(4x + 2\Delta x)}{\Delta x} = 4x + 2\Delta x$$

$$\frac{dy}{dx} = \lim_{\Delta x \to 0} \frac{\Delta y}{\Delta x} = \lim_{\Delta x \to 0}(4x + 2\Delta x) = 4x + 2(0) = 4x$$

$$\left.\frac{dy}{dx}\right|_{x=3} = 4(3) = 12$$

Graphics Calculator

21. $y = 2x^2 - 6$ at $x = 3$

```
nDeriv(2X²-6,X,3
)
                12
```

Other Symbols for the Derivative

23. $y = 2x^2 - 3 \Rightarrow y'(x) = 4x$

$y'(3) = 4(3) = 12$

25. $f(x) = 7 - 4x^2$

$$f'(x) = \lim_{\Delta x \to 0} \frac{f(x+\Delta x) - f(x)}{\Delta x} = \lim_{\Delta x \to 0} \frac{7 - 4(x+\Delta x)^2 - (7 - 4x^2)}{\Delta x} = \lim_{\Delta x \to 0} \frac{7 - 4x^2 - 8x\Delta x - 4(\Delta x)^2 - 7 + 4x^2}{\Delta x}$$

$$= \lim_{\Delta x \to 0} \frac{-8x\Delta x - 4(\Delta x)^2}{\Delta x} = \lim_{\Delta x \to 0} \frac{\Delta x(-8x - 4\Delta x)}{\Delta x} = \lim_{\Delta x \to 0} (-8x - 4\Delta x) = -8x - 0 = -8x$$

27. $f(x) = 7 - 4x^2 \Rightarrow f'(x) = -8x$

$f'(-3) = -8(-3) = 24$

Operator Notation $\quad \frac{dy}{dx} = \frac{d}{dx}(y) = D_x(y) = D(y) = y' = y'(x) = f'(x) = \lim_{\Delta x \to 0} \frac{f(x+\Delta x) - f(x)}{\Delta x}$

29. $$\frac{d}{dx}(x^2 - 1) = \lim_{\Delta x \to 0} \frac{(x+\Delta x)^2 - 1 - (x^2 - 1)}{\Delta x} = \lim_{\Delta x \to 0} \frac{x^2 + 2x\Delta x + (\Delta x)^2 - 1 - x^2 + 1}{\Delta x} = \lim_{\Delta x \to 0} \frac{2x\Delta x + (\Delta x)^2}{\Delta x}$$

$$= \lim_{\Delta x \to 0} \frac{\Delta x(2x + \Delta x)}{\Delta x} = \lim_{\Delta x \to 0} (2x + \Delta x) = 2x + 0 = 2x$$

31. $$D_x(x^2) = \lim_{\Delta x \to 0} \frac{(x+\Delta x)^2 - x^2}{\Delta x} = \lim_{\Delta x \to 0} \frac{x^2 + 2x\Delta x + (\Delta x)^2 - x^2}{\Delta x} = \lim_{\Delta x \to 0} \frac{2x\Delta x + (\Delta x)^2}{\Delta x}$$

$$= \lim_{\Delta x \to 0} \frac{\Delta x(2x + \Delta x)}{\Delta x} = \lim_{\Delta x \to 0} (2x + \Delta x) = 2x + 0 = 2x$$

33. $D(x^2 - 1) = \frac{d}{dx}(x^2 - 1) = 2x$ Found in Problem 29.

\

Exercise 23.4 • Rules for Derivatives

Derivative of a Constant

1. $y = 8 \Rightarrow \frac{dy}{dx} = 0$

3. $y = a^2 \Rightarrow \frac{dy}{dx} = 0$

Derivative of a Constant Times a Power Function

5. $y = x \Rightarrow \dfrac{dy}{dx} = 1x^{1-1} = 1x^0 = 1(1) = 1$

7. $y = x^7 \Rightarrow \dfrac{dy}{dx} = 7x^{7-1} = 7x^6$

9. $y = 3x^2 \Rightarrow \dfrac{dy}{dx} = 3(2)x^{2-1} = 6x$

Power Function with Negative Exponent

11. $y = x^{-5} \Rightarrow \dfrac{dy}{dx} = -5x^{-5-1} = -5x^{-6}$

13. $y = \dfrac{1}{x} = x^{-1} \Rightarrow \dfrac{dy}{dx} = -1x^{-1-1} = -1x^{-2}$ or $-\dfrac{1}{x^2}$

15. $y = \dfrac{3}{x^3} = 3x^{-3}$

$\dfrac{dy}{dx} = 3(-3)x^{-3-1} = -9x^{-4}$ or $-\dfrac{9}{x^4}$

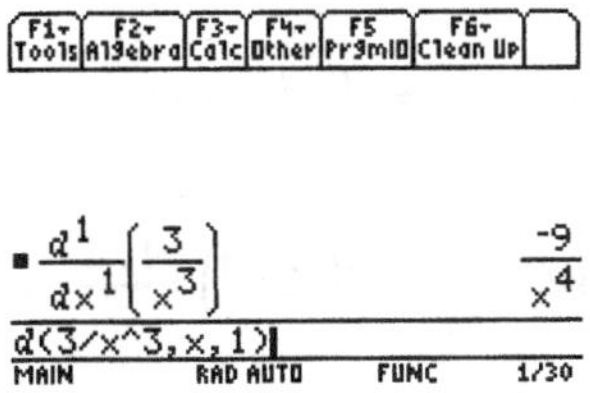

Power Function with Fractional Exponent

17. $y = 7.5x^{1/3} \Rightarrow \dfrac{dy}{dx} = 7.5\left(\dfrac{1}{3}\right)x^{1/3-1} = 2.5x^{1/3-3/3} = 2.5x^{-2/3}$

19. $y = 4\sqrt{x} = 4x^{1/2} \Rightarrow \dfrac{dy}{dx} = 4\left(\dfrac{1}{2}\right)x^{1/2-1} = 2x^{1/2-2/2} = 2x^{-1/2}$ or $\dfrac{2}{\sqrt{x}}$

21. $y = -17\sqrt{x^3} = -17\left(x^3\right)^{1/2} = -17x^{3/2} \Rightarrow \dfrac{dy}{dx} = -17\left(\dfrac{3}{2}\right)x^{3/2-1} = -\dfrac{51}{2}x^{3/2-2/2} = -\dfrac{51}{2}x^{1/2}$ or $-\dfrac{51\sqrt{x}}{2}$

Derivative of a Sum

23. $y = 3 - 2x \Rightarrow \dfrac{dy}{dx} = 0 - 2 = -2$

25. $y = 3x - x^3 \Rightarrow \dfrac{dy}{dx} = 3 - 3x^2$

27. $y = 3x^3 + 7x^2 - 2x + 5 \Rightarrow \dfrac{dy}{dx} = 3\left(3x^2\right) + 7(2x) - 2 + 0 = 9x^2 + 14x - 2$

29. $y = ax + b \Rightarrow \dfrac{dy}{dx} = a + 0 = a$

31. $y = \frac{x^2}{2} - \frac{x^7}{7} = \frac{1}{2}x^2 - \frac{1}{7}x^7$

$\frac{dy}{dx} = \frac{1}{2}(2x) - \frac{1}{7}(7x^6) = x - x^6$

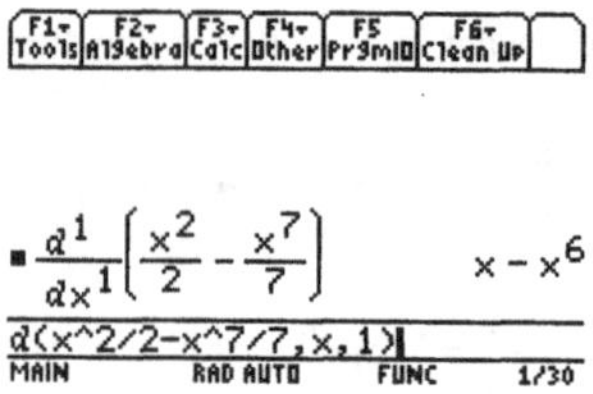

33. $y = 2x^{3/4} + 4x^{-1/4} \Rightarrow \frac{dy}{dx} = 2\left(\frac{3}{4}x^{3/4-1}\right) + 4\left(-\frac{1}{4}x^{-1/4-1}\right) = \frac{3}{2}x^{-1/4} - x^{-5/4}$

35. $y = 2x^{4/3} - 3x^{2/3} \Rightarrow \frac{dy}{dx} = 2\left(\frac{4}{3}x^{4/3-1}\right) - 3\left(\frac{2}{3}x^{2/3-1}\right) = \frac{8}{3}x^{1/3} - 2x^{-1/3}$

Other Symbols for the Derivative

37. $y = 2x^3 - 3 \Rightarrow y' = 2(3x^2) = 6x^2$

39. $\frac{d}{dx}(3x^5 + 2x) = 3(5x^4) + 2 = 15x^4 + 2$

41. $D_x(7.8 - 5.2x^{-2}) = -5.2(-2x^{-3}) = \frac{10.4}{x^3}$

43. $D(3x^2 + 2x) = 3(2x) + 2 = 6x + 2$

Derivative at a Given Point

45 $y = x^3 - 5$

$y' = 3x^2$

$y'(1) = 3(1)^2 = 3$

47. $f(x) = 2.75x^2 - 5.02x$

$f'(x) = 2.75(2x) - 5.02 = 5.50x - 5.02$

$f'(3.36) = 5.50(3.36) - 5.02 = 13.5$

49. $y = x^2 - 2 \Rightarrow y' = 2x$

The slope of the tangent at $x = 2$ is

$y'(2) = 2(2) = 4$.

51. $y = 5x^3 \Rightarrow y' = 15x^2$

The rate of change at $x = 0.50$ is

$y'(0.50) = 15(0.50)^2 = 3.75$.

Functions with Other Variables

53. $v = 5t^2 - 3t + 4 \Rightarrow \frac{dv}{dt} = 10t - 3$

55. $s = 58.3t^3 - 63.8t \Rightarrow \frac{ds}{dt} = 175t^2 - 63.8$

57. $y = \sqrt{5w^3} = \sqrt{5}w^{3/2} \Rightarrow \frac{dy}{dw} = \sqrt{5}\left(\frac{3}{2}\right)w^{1/2} = \frac{3\sqrt{5w}}{2}$

59. $v = \frac{85.3}{t^4} = 85.3t^{-4} \Rightarrow \frac{dv}{dt} = -341t^{-5} = -\frac{341}{t^5}$

An Application

61. $s = 5t^2 + 3t \Rightarrow v = \dfrac{ds}{dt} = 10t + 3$

$$v(3.55) = \left.\frac{ds}{dt}\right|_{t=3.55} = 10(3.55) + 3 = 38.5 \text{ in./s}$$

Exercise 23.5 • Derivative of a Function Raised to a Power

1. $y = (2x+1)^5 \Rightarrow \dfrac{dy}{dx} = 5(2x+1)^{5-1}(2) = 10(2x+1)^4$

3. $y = \left(3x^2+2\right)^4 - 2x \Rightarrow \dfrac{dy}{dx} = 4\left(3x^2+2\right)^3(6x) - 2 = 24x\left(3x^2+2\right)^3 - 2$

5. $y = (2-5x)^{3/5} \Rightarrow \dfrac{dy}{dx} = \dfrac{3}{5}(2-5x)^{-2/5}(-5) = -\dfrac{3}{(2-5x)^{2/5}}$

7. $y = \dfrac{2.15}{x^2+a^2} = 2.15\left(x^2+a^2\right)^{-1} \Rightarrow \dfrac{dy}{dx} = -2.15\left(x^2+a^2\right)^{-2}(2x) = -\dfrac{4.30x}{\left(x^2+a^2\right)^2}$

9. $y = \dfrac{3}{x^2+2} = 3\left(x^2+2\right)^{-1}$

$$\frac{dy}{dx} = -3\left(x^2+2\right)^{-2}(2x) = -\frac{6x}{\left(x^2+2\right)^2}$$

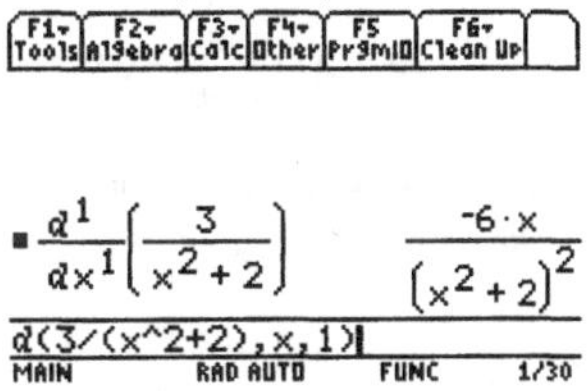

11. $y = \left(a - \dfrac{b}{x}\right)^2 = \left(a - bx^{-1}\right)^2 \Rightarrow \dfrac{dy}{dx} = 2\left(a - \dfrac{b}{x}\right)\left(bx^{-2}\right) = \dfrac{2b}{x^2}\left(a - \dfrac{b}{x}\right)$

13. $y = \sqrt{1-3x^2} = \left(1-3x^2\right)^{1/2} \Rightarrow \dfrac{dy}{dx} = \dfrac{1}{2}\left(1-3x^2\right)^{-1/2}(-6x) = -3x\left(1-3x^2\right)^{-1/2} = -\dfrac{3x}{\sqrt{1-3x^2}}$

15. $y = \sqrt{1-2x} = (1-2x)^{1/2} \Rightarrow \dfrac{dy}{dx} = \dfrac{1}{2}(1-2x)^{-1/2}(-2) = -\dfrac{1}{\sqrt{1-2x}}$

17. $y = \sqrt[3]{4-9x} = (4-9x)^{1/3} \Rightarrow \dfrac{dy}{dx} = \dfrac{1}{3}(4-9x)^{-2/3}(-9) = -\dfrac{3}{(4-9x)^{2/3}}$

19. $y = \dfrac{1}{\sqrt{x+1}} = (x+1)^{-1/2} \Rightarrow \dfrac{dy}{dx} = -\dfrac{1}{2}(x+1)^{-3/2}(1) = -\dfrac{1}{2(x+1)^{3/2}} = -\dfrac{1}{2\sqrt{(x+1)^3}}$

21. $\dfrac{d}{dx}\left(3x^5+2x\right)^2 = 2\left(3x^5+2x\right)\left(15x^4+2\right)$

23. $D_x\left(4.8-7.2x^{-2}\right)^2=2\left(4.8-7.2x^{-2}\right)\left(14.4x^{-3}\right)=\dfrac{28.8\left(4.8-7.2x^{-2}\right)}{x^3}$

25. $v=\left(5t^2-3t+4\right)^2 \Rightarrow \dfrac{dv}{dt}=2\left(5t^2-3t+4\right)\left(10t-3\right)$

27. $s=\left(8.3t^3-3.8t\right)^{-2} \Rightarrow \dfrac{ds}{dt}=-2\left(8.3t^3-3.8t\right)^{-3}\left(24.9t^2-3.8\right)=\dfrac{7.6-49.8t^2}{\left(8.3t^3-3.8t\right)^3}$

29. $y=\left(4.82x^2-8.25x\right)^3 \Rightarrow y'=3\left(4.82x^2-8.25x\right)^2(9.64x-8.25)$

$y'(3.77)=3\left(4.82(3.77)^2-8.25(3.77)\right)^2(9.64(3.77)-8.25)=118{,}000$

31. $y=\left(x^2-x\right)^3 \Rightarrow y'=3\left(x^2-x\right)^2(2x-1)$

$y'(3)=3\left((3)^2-(3)\right)^2(2(3)-1)=540$

An Application

33. $v=3.45\left(t^2+2\right)^2 \Rightarrow a(t)=\dfrac{dv}{dt}=6.9\left(t^2+2\right)(2t)=13.8t\left(t^2+2\right)$

$a(1.00)=\left.\dfrac{dv}{dt}\right|_{t=1.00}=13.8(1.00)\left((1.00)^2+2\right)=41.4 \text{ ft/s}^2$

Exercise 23.6 • Derivatives of Products and Quotients

Products

1. $y=x\left(x^2-3\right) \Rightarrow \dfrac{dy}{dx}=x\dfrac{d}{dx}\left(x^2-3\right)+\left(x^2-3\right)\dfrac{d}{dx}(x)=x(2x)+\left(x^2-3\right)=2x^2+x^2-3=3x^2-3$

3. $y=x\left(x^2-2\right)^2 \Rightarrow \dfrac{dy}{dx}=x(2)\left(x^2-2\right)(2x)+\left(x^2-2\right)^2(1)=4x^4-8x^2+x^4-4x^2+4=5x^4-12x^2+4$

5. $y=(5+3x)(3+7x) \Rightarrow \dfrac{dy}{dx}=(5+3x)(7)+(3+7x)(3)=35+21x+9+21x=42x+44$

7. $y=(x+3)(5x-6) \Rightarrow \dfrac{dy}{dx}=(x+3)(5)+(5x-6)(1)=5x+15+5x-6=10x+9$

9. $y=x^3\left(8.24x-6.24x^3\right)$

$\dfrac{dy}{dx}=x^3\left(8.24-18.7x^2\right)+\left(8.24x-6.24x^3\right)\left(3x^2\right)=8.24x^3-18.7x^5+24.7x^3-18.7x^5=33.3x^3-37.4x^5$

11. $y = 3x\sqrt{5+x^2} = 3x\left(5+x^2\right)^{1/2}$

$$\frac{dy}{dx} = 3x\left(\frac{1}{2}\right)\left(5+x^2\right)^{-1/2}(2x) + \left(5+x^2\right)^{1/2}(3)$$

$$= \frac{3x^2}{\sqrt{5+x^2}} + 3\sqrt{5+x^2}$$

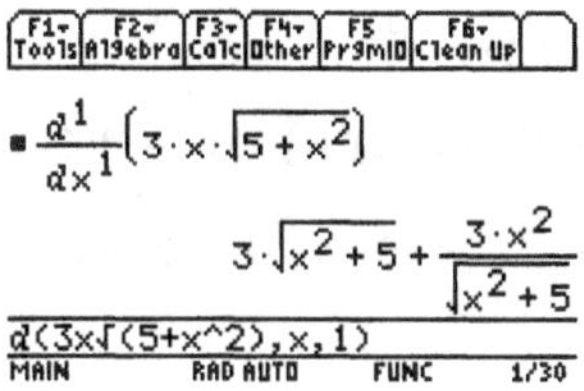

13. $y = \sqrt{x}\left(3x^2+2x-3\right) = (x)^{1/2}\left(3x^2+2x-3\right)$

$$\frac{dy}{dx} = (x)^{1/2}(6x+2) + \left(3x^2+2x-3\right)\left(\frac{1}{2}x^{-1/2}\right) = (6x+2)\sqrt{x} + \frac{3x^2+2x-3}{2\sqrt{x}}$$

$$= \frac{(6x+2)2x + 3x^2+2x-3}{2\sqrt{x}} = \frac{12x^2+4x+3x^2+2x-3}{2\sqrt{x}} = \frac{15x^2+6x-3}{2\sqrt{x}}$$

15. $y = \left(2x^2-3\right)\sqrt[3]{3x+5} = \left(2x^2-3\right)(3x+5)^{1/3}$

$$\frac{dy}{dx} = \left(2x^2-3\right)\left(\frac{1}{3}\right)(3x+5)^{-2/3}(3) + (3x+5)^{1/3}(4x) = \frac{2x^2-3}{(3x+5)^{2/3}} + 4x\sqrt[3]{3x+5}$$

17. $$\frac{d}{dx}\left[\left(2x^5+5x\right)^2(x-3)\right] = \left(2x^5+5x\right)^2(1) + (x-3)2\left(2x^5+5x\right)\left(10x^4+5\right)$$

$$= \left(2x^5+5x\right)^2 + (2x-6)\left(2x^5+5x\right)\left(10x^4+5\right)$$

19. $D_x\left[\left(x^2-2\right)(x-6)\right] = \left(x^2-2\right)(1) + (x-6)(2x) = x^2-2+2x^2-12x = 3x^2-12x-2$

21. $y = \left(x^2-1\right)\sqrt{x+7} = \left(x^2-1\right)(x+7)^{1/2}$

$$y' = \left(x^2-1\right)\left(\frac{1}{2}\right)(x+7)^{-1/2}(1) + (x+7)^{1/2}(2x) = \frac{x^2-1}{2\sqrt{x+7}} + 2x\sqrt{x+7}$$

The rate of change at $x=3$ is $y'(3) = \dfrac{(3)^2-1}{2\sqrt{(3)+7}} + 2(3)\sqrt{(3)+7} = 20.2$.

Constant Times a Function

23. $y = 6(x-9) \Rightarrow \dfrac{dy}{dx} = 6\dfrac{d}{dx}(x-9) = 6(1) = 6$

25. $y = \pi(2x-4)^3 \Rightarrow \dfrac{dy}{dx} = \pi(3)(2x-4)^2(2) = 6\pi(2x-4)^2$

Products with More Than Two Factors

27. $y = x(x-7)(x+1)$

$$\frac{dy}{dx} = x(x-7)(1) + x(x+1)(1) + (x-7)(x+1)(1) = x^2-7x+x^2+x+x^2-6x-7 = 3x^2-12x-7$$

29. $y = x(x+1)^2(x-2)^3$

$$\frac{dy}{dx} = x(x+1)^2 3(x-2)^2(1) + x(x-2)^3 2(x+1)(1) + (x+1)^2(x-2)^3(1)$$
$$= 3x(x+1)^2(x-2)^2 + \left(2x^2+2x\right)(x-2)^3 + (x+1)^2(x-2)^3$$

Quotients

31. $y = \dfrac{x}{x+2}$

$$\frac{dy}{dx} = \frac{(x+2)\frac{d}{dx}(x) - x\frac{d}{dx}(x+2)}{(x+2)^2}$$
$$= \frac{x+2-x}{(x+2)^2} = \frac{2}{(x+2)^2}$$

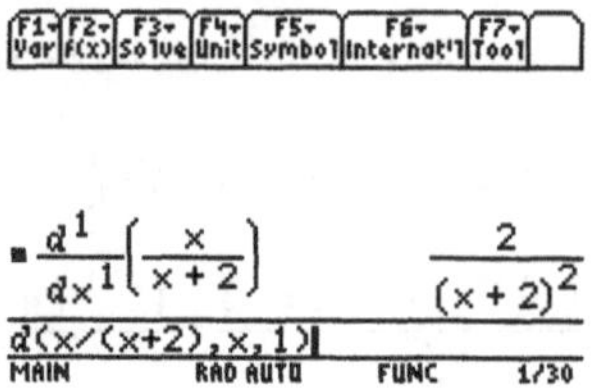

33. $y = \dfrac{x^2}{4-x^2} \Rightarrow \dfrac{dy}{dx} = \dfrac{\left(4-x^2\right)(2x) - x^2(-2x)}{\left(4-x^2\right)^2} = \dfrac{8x-2x^3+2x^3}{\left(4-x^2\right)^2} = \dfrac{8x}{\left(4-x^2\right)^2}$

35. $y = \dfrac{x+2}{x-3} \Rightarrow \dfrac{dy}{dx} = \dfrac{(x-3)(1)-(x+2)(1)}{(x-3)^2} = \dfrac{x-3-x-2}{(x-3)^2} = \dfrac{-5}{(x-3)^2}$

37. $y = \dfrac{x^{1/2}}{x^{1/2}+1} \Rightarrow \dfrac{dy}{dx} = \dfrac{\left(x^{1/2}+1\right)\left(\frac{1}{2}x^{-1/2}\right) - x^{1/2}\left(\frac{1}{2}x^{-1/2}\right)}{\left(x^{1/2}+1\right)^2} = \dfrac{\frac{1}{2}+\frac{1}{2}x^{-1/2}-\frac{1}{2}}{\left(\sqrt{x}+1\right)^2} = \dfrac{1}{2\sqrt{x}\left(\sqrt{x}+1\right)^2}$

39. a is a constant.

$$w = \frac{z}{\sqrt{z^2-a^2}} = \frac{z}{\left(z^2-a^2\right)^{1/2}} \Rightarrow \frac{dw}{dz} = \frac{\left(z^2-a^2\right)^{1/2}(1) - z\left(\frac{1}{2}\left(z^2-a^2\right)^{-1/2}(2z)\right)}{\left[\left(z^2-a^2\right)^{1/2}\right]^2} = \frac{\sqrt{z^2-a^2} - \frac{z^2}{\sqrt{z^2-a^2}}}{z^2-a^2}$$
$$= \frac{\frac{z^2-a^2-z^2}{\sqrt{z^2-a^2}}}{z^2-a^2} = \frac{-a^2}{\left(z^2-a^2\right)^{3/2}}$$

41. $y = \dfrac{\sqrt{16+3x}}{x} = \dfrac{(16+3x)^{1/2}}{x} \Rightarrow y' = \dfrac{x\frac{1}{2}(16+3x)^{-1/2}(3) - (16+3x)^{1/2}}{x^2} = \dfrac{\frac{3x}{2\sqrt{16+3x}} - \sqrt{16+3x}}{x^2}$

The slope of the tangent at $x=3$ is $y'(3) = \dfrac{\frac{3(3)}{2\sqrt{16+3(3)}} - \sqrt{16+3(3)}}{(3)^2} = \dfrac{\frac{9}{10}-5}{9} = -0.456$.

43. $y=\dfrac{x}{\sqrt{8-x^2}}=\dfrac{x}{\left(8-x^2\right)^{1/2}} \Rightarrow y'=\dfrac{\left(8-x^2\right)^{1/2}(1)-x\frac{1}{2}\left(8-x^2\right)^{-1/2}(-2x)}{\left[\left(8-x^2\right)^{1/2}\right]^2}=\dfrac{\sqrt{8-x^2}+\dfrac{x^2}{\sqrt{8-x^2}}}{8-x^2}$

$$y'(2)=\frac{\sqrt{8-(2)^2}+\dfrac{(2)^2}{\sqrt{8-(2)^2}}}{8-(2)^2}=\frac{2+\dfrac{4}{2}}{4}=1$$

An Application

45. $T=(t+3)\sqrt{t+1}=(t+3)(t+1)^{1/2}$

$$T'(t)=(t+3)\left(\frac{1}{2}\right)(t+1)^{-1/2}(1)+(t+1)^{1/2}(1)=\frac{t+3}{2\sqrt{t+1}}+\sqrt{t+1}$$

$$T'(2.35)=\frac{(2.35)+3}{2\sqrt{(2.35)+1}}+\sqrt{(2.35)+1}=3.29\,^\circ\text{F/h}$$

Exercise 23.7 • Other Variables, Implicit Relations, and Differentials

Derivatives with Respect to Other Variables

1. $y=2u^3 \Rightarrow \dfrac{dy}{dw}=2\left(3u^2\dfrac{du}{dw}\right)=6u^2\dfrac{du}{dw}$

3. $w=y^2+u^3 \Rightarrow \dfrac{dw}{du}=2y\dfrac{dy}{du}+3u^2$

5. $\dfrac{d}{dx}\left(x^3y^2\right)=x^3\dfrac{d}{dx}\left(y^2\right)+y^2\dfrac{d}{dx}\left(x^3\right)=x^3\left(2y\dfrac{dy}{dx}\right)+y^2\left(3x^2\right)=2x^3y\dfrac{dy}{dx}+3x^2y^2$

7. $\dfrac{d}{dt}\sqrt{3z^2+5}=\dfrac{d}{dt}\left(3z^2+5\right)^{1/2}=\dfrac{1}{2}\left(3z^2+5\right)^{-1/2}\left(6z\dfrac{dz}{dt}\right)=\dfrac{3z}{\sqrt{3z^2+5}}\dfrac{dz}{dt}$

9. $x=y^2-7y \Rightarrow \dfrac{dx}{dy}=2y-7$

11. $x=(y-2)(y+3)^5$

$$\frac{dx}{dy}=(y-2)5(y+3)^4(1)+(y+3)^5(1)=(y+3)^4\left[5(y-2)+(y+3)\right]$$

$$=(y+3)^4\left[5y-10+y+3\right]=(6y-7)(y+3)^4$$

Derivatives of Implicit Relations

13. $5x-2y=7$

$5-2y'=0$

$y'=\dfrac{5}{2}$

15. $xy=5$

$xy'+y=0$

$y'=-\dfrac{y}{x}$

17. $y^2 = 4ax$

$2yy' = 4a$

$y' = \frac{2a}{y}$

19. $x^3 + y^3 - 3axy = 0$

$3x^2 + 3y^2y' - 3axy' - 3ay = 0$

$3y'\left(y^2 - ax\right) = 3\left(ay - x^2\right)$

$y' = \frac{ay - x^2}{y^2 - ax}$

21. $y + y^3 = x + x^3$

$y' + 3y^2y' = 1 + 3x^2$

$y'\left(1 + 3y^2\right) = 1 + 3x^2$

$y' = \frac{1 + 3x^2}{1 + 3y^2}$

23. $y^3 - 4x^2y^2 + y^4 = 9$

$3y^2y' - 4x^2(2yy') - 4y^2(2x) + 4y^3y' = 0$

$yy'\left(3y - 8x^2 + 4y^2\right) = 8xy^2$

$y' = \frac{8xy}{3y - 8x^2 + 4y^2}$

25. $x^2 + y^2 = 25$

$2x + 2yy' = 0$

$y' = -\frac{x}{y}$

at $x = 2,\ 4 + y^2 = 25 \Rightarrow y = \sqrt{25 - 4} = \sqrt{21}$

slope of tangent $y'\big|_{(2,\sqrt{21})} = -\frac{2}{\sqrt{21}} = -0.436$

27. $2x^2 + 2y^3 - 9xy = 0$

$4x + 6y^2y' - 9xy' - 9y = 0$

$y'\left(6y^2 - 9x\right) = 9y - 4x$

$y' = \frac{9y - 4x}{6y^2 - 9x}$

slope of the tangent $y'\big|_{(1,2)} = \frac{9(2) - 4(1)}{6(2)^2 - 9(1)} = \frac{18 - 4}{24 - 9} = \frac{14}{15}$

Differentials

29. $y = x^3$

$\frac{dy}{dx} = 3x^2$

$dy = 3x^2 dx$

31. $y = \frac{x - 1}{x + 1}$

$\frac{dy}{dx} = \frac{(x+1) - (x-1)}{(x+1)^2} = \frac{x + 1 - x + 1}{(x+1)^2} = \frac{2}{(x+1)^2}$

$dy = \frac{2dx}{(x+1)^2}$

33. $y = x^3 + 3x$

$\dfrac{dy}{dx} = 3x^2 + 3$

$dy = (3x^2 + 3)dx$

35. $3x^2 - 2xy + 2y^2 = 3$

$6x - 2x\dfrac{dy}{dx} - 2y + 4y\dfrac{dy}{dx} = 0$

$2\dfrac{dy}{dx}(2y - x) = 2y - 6x$

$\dfrac{dy}{dx} = \dfrac{y - 3x}{2y - x}$

$dy = \dfrac{y - 3x}{2y - x}dx$

37. $2x^2 + 3xy + 4y^2 = 20$

$4x + 3x\dfrac{dy}{dx} + 3y + 8y\dfrac{dy}{dx} = 0$

$\dfrac{dy}{dx}(3x + 8y) = -(4x + 3y)$

$\dfrac{dy}{dx} = -\dfrac{4x + 3y}{3x + 8y}$

$dy = -\dfrac{4x + 3y}{3x + 8y}dx$

An Application

39. $x^2 + y^2 = (8.25)^2$

$2x + 2yy' = 0$

$y' = -\dfrac{x}{y}$

at $x = 6.25,\ (6.25)^2 + y^2 = (8.25)^2 \Rightarrow y = \sqrt{8.25^2 - 6.25^2} = 5.39$

slope of the tangent $y'|_{(6.25,5.39)} = -\dfrac{6.25}{5.39} = -1.16$

Exercise 23.8 • Higher-Order Derivatives

1. $y = 2x^3 \Rightarrow y' = 6x^2 \Rightarrow y'' = 12x$

3. $y = 4x^3 + 3x^2 \Rightarrow y' = 12x^2 + 6x \Rightarrow y'' = 24x + 6$

5. $y = 3x^4 - x^3 + 5x \Rightarrow y' = 12x^3 - 3x^2 + 5$

$y'' = 36x^2 - 6x$

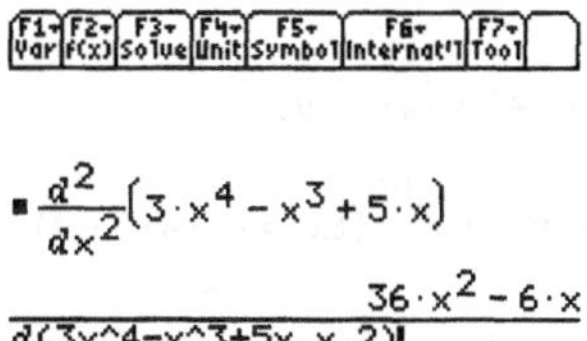

7. $y=\dfrac{x^2}{x+2} \Rightarrow y'=\dfrac{(x+2)2x-x^2(1)}{(x+2)^2}=\dfrac{2x^2+4x-x^2}{(x+2)^2}=\dfrac{x^2+4x}{(x+2)^2}$

$$y''=\frac{(x+2)^2(2x+4)-(x^2+4x)2(x+2)(1)}{\left[(x+2)^2\right]^2}=\frac{(x+2)\left[2x^2+8x+8-2x^2-8x\right]}{(x+2)^4}=\frac{8}{(x+2)^3}$$

9. $y=\sqrt{5-4x^2}=\left(5-4x^2\right)^{1/2}$

$$y'=\frac{1}{2}\left(5-4x^2\right)^{-1/2}(-8x)=-4x\left(5-4x^2\right)^{-1/2}$$

$$y''=-4x\left(-\frac{1}{2}\right)\left(5-4x^2\right)^{-3/2}(-8x)+\left(5-4x^2\right)^{-1/2}(-4)=\frac{-16x^2}{\left(5-4x^2\right)^{3/2}}-\frac{4}{\left(5-4x^2\right)^{1/2}}\frac{\left(5-4x^2\right)}{\left(5-4x^2\right)}$$

$$=\frac{-16x^2-20+16x^2}{\left(5-4x^2\right)^{3/2}}=\frac{-20}{\left(5-4x^2\right)^{3/2}}$$

11. $y=3x^3+2x^2$

$y'=9x^2+4x$

$y''=18x+4$

$y''(2)=18(2)+4=40$

An Application

13. $s(t)=4.55t^3+2.85t^2+5.22$

$v(t)=s'(t)=13.65t^2+5.7t$

$a(t)=s''(t)=27.3t+5.7$

$v(1.55)=13.65(1.55)^2+5.7(1.55)=41.6 \text{ cm/s}$

$a(1.55)=27.3(1.55)+5.7=48.0 \text{ cm/s}^2$

Chapter 23 • Review Problems

1. $\lim\limits_{x\to1}\dfrac{x^2+2-3x}{x-1}=\lim\limits_{x\to1}\dfrac{(x-1)(x-2)}{x-1}=\lim\limits_{x\to1}(x-2)=1-2=-1$

3. $f(x)=\sqrt{4x^2+9}=\left(4x^2+9\right)^{1/2}$

$$f'(x)=\frac{1}{2}\left(4x^2+9\right)^{-1/2}(8x)=4x\left(4x^2+9\right)^{-1/2}=\frac{4x}{\sqrt{4x^2+9}} \Rightarrow f'(2)=\frac{4(2)}{\sqrt{4(2)^2+9}}=\frac{8}{5}$$

5. $\frac{x^2}{4}+\frac{y^2}{9}=1$

$\frac{2x}{4}+\frac{2y}{9}\frac{dy}{dx}=0$

$\frac{2y}{9}\frac{dy}{dx}=-\frac{x}{2}$

$\frac{dy}{dx}=-\frac{x}{2}\frac{9}{2y}=-\frac{9x}{4y}$

7. $y=5x-3x^2$

$y+\Delta y=5(x+\Delta x)-3(x+\Delta x)^2=5x+5\Delta x-3x^2-6x\Delta x-3(\Delta x)^2$

$(y+\Delta y)-y=5x+5\Delta x-3x^2-6x\Delta x-3(\Delta x)^2-(5x-3x^2)=5x+5\Delta x-3x^2-6x\Delta x-3(\Delta x)^2-5x+3x^2$

$=5\Delta x-6x\Delta x-3(\Delta x)^2$

$\frac{\Delta y}{\Delta x}=\frac{\Delta x(5-6x-3\Delta x)}{\Delta x}=5-6x-3\Delta x$

$\frac{dy}{dx}=\lim_{\Delta x\to 0}\frac{\Delta y}{\Delta x}=\lim_{\Delta x\to 0}(5-6x-3\Delta x)=5-6x-3(0)=5-6x$

9. $\lim_{x\to 0}\frac{e^x}{x}=$ Does Not Exist

However, $\lim_{x\to 0^+}\frac{e^x}{x}=+\infty$

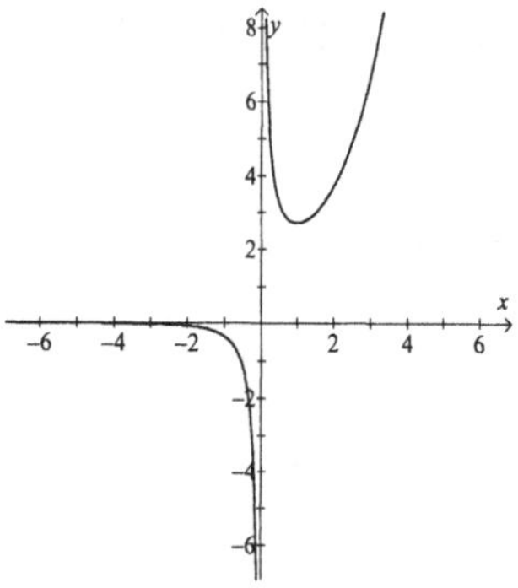

11. $\lim_{x\to -7}\frac{x^2+6x-7}{x+7}=\lim_{x\to -7}\frac{(x+7)(x-1)}{x+7}=\lim_{x\to -7}(x-1)=-7-1=-8$

13. $D(3x^4+2)^2=2(3x^4+2)(12x^3)=72x^7+48x^3$

15. $v=5t^2-3t+4\ \Rightarrow\ \frac{dv}{dt}=10t-3$

17. $\lim_{x\to 5}\frac{25-x^2}{x-5}=\lim_{x\to 5}\frac{(5-x)(5+x)}{x-5}=\lim_{x\to 5}-(x+5)=-(5+5)=-10$

19. $D_x\left[(21.7x+19.1)(64.2-17.9x^{-2})^2\right]=(21.7x+19.1)(2)(64.2-17.9x^{-2})(35.8x^{-3})+(64.2-17.9x^{-2})^2(21.7)$

$=71.6x^{-3}(21.7x+19.1)(64.2-17.9x^{-2})+21.7(64.2-17.9x^{-2})^2$

21. $\dfrac{d}{dx}\left(4x^3 - 3x + 2\right) = 12x^2 - 3$

23. $y = (3x+2)\left(x^2 - 7\right) \Rightarrow \dfrac{dy}{dx} = (3x+2)(2x) + \left(x^2 - 7\right)(3) = 6x^2 + 4x + 3x^2 - 21 = 9x^2 + 4x - 21$

25. $y = \dfrac{2x}{x^2 - 9} \Rightarrow \dfrac{dy}{dx} = \dfrac{\left(x^2 - 9\right)(2) - (2x)(2x)}{\left(x^2 - 9\right)^2} = \dfrac{2x^2 - 18 - 4x^2}{\left(x^2 - 9\right)^2} = \dfrac{-2x^2 - 18}{\left(x^2 - 9\right)^2}$

27. $y = \left(2x^3 - 4\right)^2 \Rightarrow \dfrac{dy}{dx} = 2\left(2x^3 - 4\right)\left(6x^2\right) = 24x^5 - 48x^2$

29. $y = x^{2/5} + 2x^{1/3} \Rightarrow \dfrac{dy}{dx} = \dfrac{2}{5}x^{-3/5} + \dfrac{2}{3}x^{-2/3}$

31. $f(x) = (2x+1)\left(5x^2 - 2\right) \Rightarrow f'(x) = (2x+1)(10x) + \left(5x^2 - 2\right)(2) = 20x^2 + 10x + 10x^2 - 4 = 30x^2 + 10x - 4$

$f''(x) = 60x + 10$

33. $y = (2x-5)^2 \Rightarrow \dfrac{dy}{dx} = 2(2x-5)(2) = 8x - 20 \Rightarrow dy = (8x - 20)dx$

35. $x^3 - y - 2x = 5$

$3x^2 - \dfrac{dy}{dx} - 2 = 0$

$\dfrac{dy}{dx} = 3x^2 - 2$

$dy = \left(3x^2 - 2\right)dx$

Chapter 24: Graphical Applications of the Derivative

Exercise 24.1 • Equations of Tangents and Normals

1. $y = x^2 + 2 \Rightarrow y(1) = 1^2 + 2 = 3$

$y' = 2x$

Tangent

$m_{\text{tangent}} = y'(1) = 2(1) = 2$

$\frac{y-3}{x-1} = 2$

$y - 3 = 2x - 2$

$2x - y + 1 = 0$

Normal

$m_{\text{normal}} = -\frac{1}{m_{\text{tangent}}} = -\frac{1}{2}$

$\frac{y-3}{x-1} = -\frac{1}{2}$

$-2y + 6 = x - 1$

$x + 2y - 7 = 0$

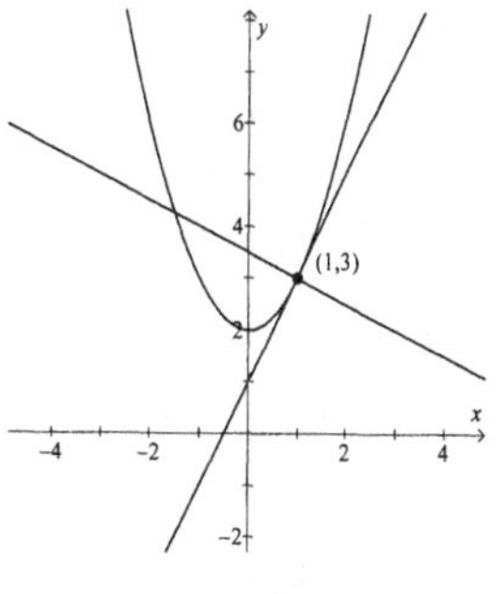

3. $y = 3x^2 - 1 \Rightarrow y(2) = 3(2)^2 - 1 = 11$

$y' = 6x$

Tangent

$m_{\text{tangent}} = y'(2) = 6(2) = 12$

$\frac{y-11}{x-2} = 12$

$y - 11 = 12x - 24$

$12x - y - 13 = 0$

Normal

$m_{\text{normal}} = -\frac{1}{12}$

$\frac{y-11}{x-2} = -\frac{1}{12}$

$-12y + 132 = x - 2$

$x + 12y - 134 = 0$

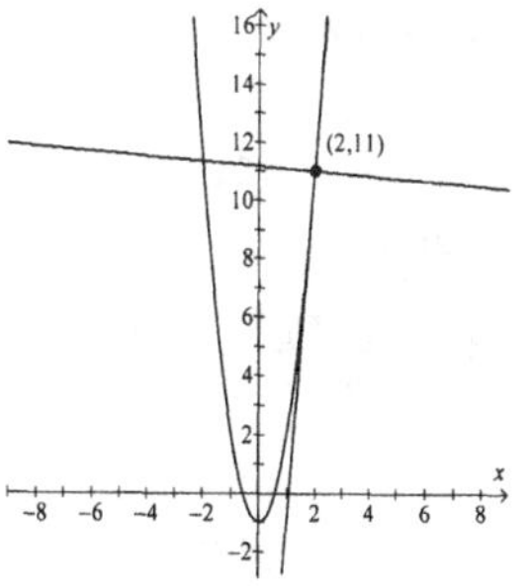

5. $x^2 + y^2 = 25 \Rightarrow 2x + 2yy' = 0 \Rightarrow y' = -\frac{x}{y}$

Tangent

$m_{\text{tangent}} = y'\big|_{(3,4)} = -\frac{3}{4}$

$\frac{y-4}{x-3} = -\frac{3}{4}$

$4y - 16 = -3x + 9$

$3x + 4y - 25 = 0$

Normal

$m_{\text{normal}} = -\frac{1}{(-3/4)} = \frac{4}{3}$

$\frac{y-4}{x-3} = \frac{4}{3}$

$3y - 12 = 4x - 12$

$4x - 3y = 0$

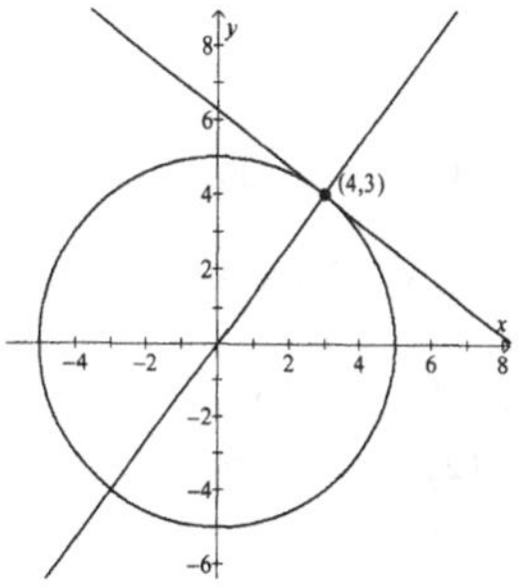

7. $y = x^3 - 3x^2 \Rightarrow y' = 3x^2 - 6x$

$y' = 9 \Rightarrow 3x^2 - 6x = 9$

$x^2 - 2x - 3 = 0$

$(x-3)(x+1) = 0$

$x = 3$ or $x = -1$

We want the first quadrant point, so use $x = 3$.

$y(3) = 3^3 - 3(3)^2 = 0$

First quadrant point is $(3,0)$.

9. $x^2 + y^2 = 25 \Rightarrow 2x + 2yy' = 0$

$$y' = -\frac{x}{y}$$

$$y' = \frac{3}{4} \Rightarrow -\frac{x}{y} = \frac{3}{4} \Rightarrow y = -\frac{4}{3}x$$

$$x^2 + \left(-\frac{4}{3}x\right)^2 = 25$$

$$x^2 + \frac{16}{9}x^2 = 25 \Rightarrow \frac{25}{9}x^2 = 25$$

$$x^2 = 9 \Rightarrow x = \pm 3$$

The points of contact are $(-3,4)$ and $(3,-4)$.

Angles between Curves

11. $y = 2 - x \Rightarrow y' = -1$

$m_1 = -1$

$y = x^2 \Rightarrow y' = 2x$

$m_2 = 2(1) = 2$

$$\tan\phi = \frac{m_2 - m_1}{1 + m_1 m_2} = \frac{2-(-1)}{1+(-1)(2)} = -3$$

$\phi = \tan^{-1}(3) = 71.6^\circ$

or $180^\circ - 71.6^\circ = 108.4^\circ$

13. $y = x^2 + x - 2 \Rightarrow y' = 2x + 1$

$m_1 = 2(1) + 1 = 3$

$y = x^2 - 5x + 4 \Rightarrow y' = 2x - 5$

$m_2 = 2(1) - 5 = -3$

$$\tan\phi = \frac{m_2 - m_1}{1 + m_1 m_2} = \frac{-3-3}{1+(-3)(3)} = \frac{3}{4}$$

$\phi = \tan^{-1}\left(\frac{3}{4}\right) = 36.9^\circ$

or $180^\circ - 36.9^\circ = 143.1^\circ$

Exercise 24.2 • Maximum, Minimum, and Inflection Points

Increasing and Decreasing Functions

1. $y = 3x^2 - 4 \Rightarrow y' = 6x$

$y'(2) = 12 > 0$ increasing at $x = 2$

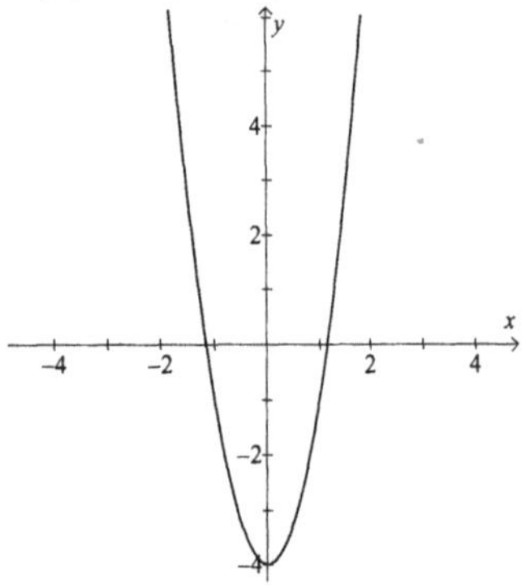

3. $y = 4x^2 - x \Rightarrow y' = 8x - 1$

$y'(-2) = -17 < 0$ decreasing at $x = -2$

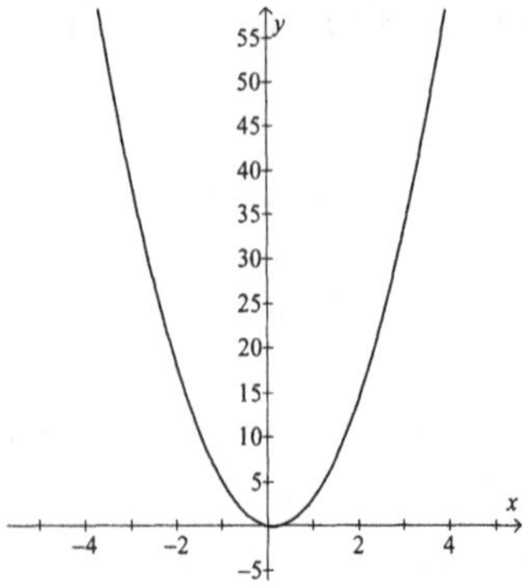

5. $y = 3 - 2x + x^2 \Rightarrow y' = 2x - 2$
$y'(0) = -2 < 0$ decreasing at $x = 0$

7. $y = x^3 + x^2 \Rightarrow y' = 3x^2 + 2x$
$y'(-2) = 3(-2)^2 + 2(-2) = 8 > 0$ increasing at $x = -2$

9. $y = 3x + 5 \Rightarrow y' = 3$
increasing for all x

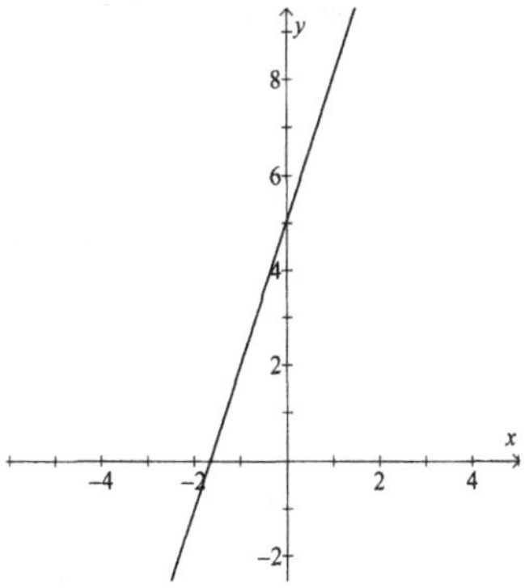

11. $y = x^3 - 3 \Rightarrow y' = 3x^2$
$3x^2 = 0 \Rightarrow x = 0$
$y'(-1) = 3(-1)^2 = 3 > 0$ increasing for $x < 0$
$y'(1) = 3(1)^2 = 3 > 0$ increasing for $x > 0$
increasing for all x

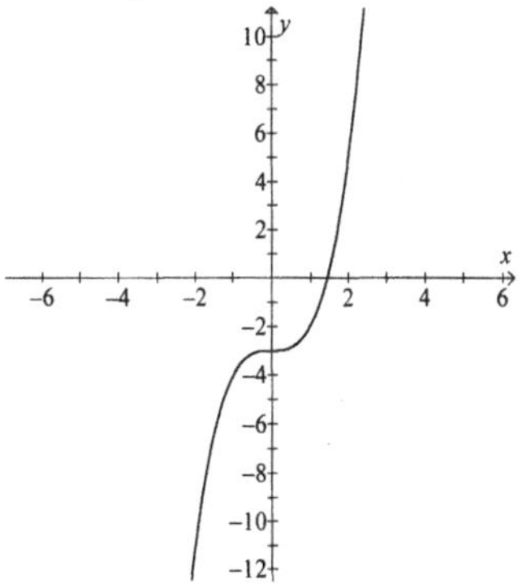

13. $y = 2x + x^2 \Rightarrow y' = 2x + 2 \qquad 2x + 2 = 0 \Rightarrow x = -1$
$y'(-2) = 2(-2) + 2 = -2 > 0$ decreasing for $x < -1$
$y'(0) = 2(0) + 2 = 2 > 0$ increasing for $x > -1$

15. $y = 2x + x^2 \Rightarrow y' = 2x + 2 \qquad 2x + 2 = 0 \Rightarrow x = -1$
$y'(-2) = 2(-2) + 2 = -2 > 0$ decreasing for $x < -1$
$y'(0) = 2(0) + 2 = 2 > 0$ increasing for $x > -1$

Concavity

17. $y = x^4 + x^2$

$y' = 4x^3 + 2x$

$y'' = 12x^2 + 2$

$y''(2) = 12(2)^2 + 2 = 50 > 0$

so concave upward at $x = 2$

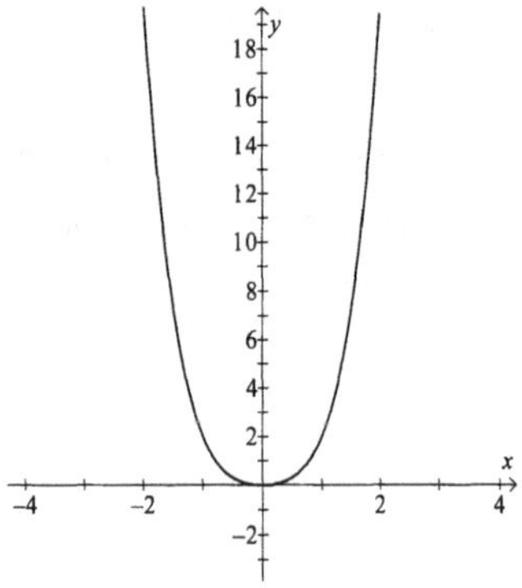

19. $y = -2x^3 - 2\sqrt{x+2} = -2x^3 - 2(x+2)^{1/2}$

$y' = -6x^2 - (x+2)^{-1/2}$

$y'' = -12x + \frac{1}{2}(x+2)^{-3/2}$

$y''\left(\frac{1}{4}\right) = -12\left(\frac{1}{4}\right) + \frac{1}{2}\left(\left(\frac{1}{4}\right)+2\right)^{-3/2} = -2.85 < 0$

so concave downward at $x = \frac{1}{4}$

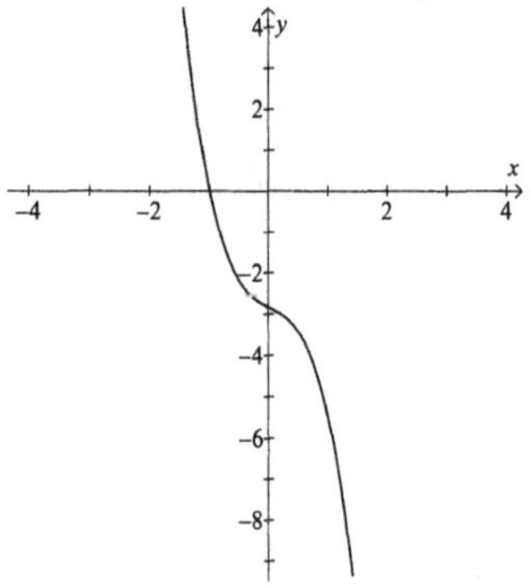

21. $y = 2x + x^2 \Rightarrow y' = 2 + 2x \Rightarrow y'' = 2$ $\quad y''(1) = 2 > 0$ so concave upward at $x = 1$

23. $y = x + x^3 \Rightarrow y' = 1 + 3x^2 \Rightarrow y'' = 6x$ $\quad y''(-1) = 6(-1) = -6 < 0$ so concave downward at $x = -1$

Maximum and Minimum Points

25. $y = x^2 \Rightarrow y' = 2x$

$2x = 0 \Rightarrow x = 0$

$y'' = 2$

$y''(0) = 2 > 0 \qquad y(0) = 0$

$(0,0)$ is a minimum point.

27. $y = x^3 - 7x^2 + 36 \Rightarrow y' = 3x^2 - 14x$

$3x^2 - 14x = 0 \Rightarrow x(3x - 14) = 0$

$x = 0$ or $x = \frac{14}{3} = 4.67$

$y'' = 6x - 14$

$y''(0) = -14 < 0 \qquad y(0) = 36$

$y''\left(\frac{14}{3}\right) = 14 > 0 \qquad y(4.67) = -14.8$

$(0,36)$ is a maximum point.

$(4.67, -14.8)$ is a minimum point.

29. $y = 2x^2 - x^4 \Rightarrow y' = 4x - 4x^3$

$4x - 4x^3 = 0 \Rightarrow 4x\left(1 - x^2\right) = 0$

$x = 0 \text{ or } x = \pm 1$

$y'' = 4 - 12x^2$

$y''(-1) = -8 < 0 \qquad y(-1) = 1$

$y''(0) = 4 > 0 \qquad y(0) = 0$

$y''(1) = -8 < 0 \qquad y(1) = 1$

$(-1,1)$ and $(1,1)$ are maximum points.

$(0,0)$ is a minimum point.

31. $y = x^4 - 4x \Rightarrow y' = 4x^3 - 4$

$4x^3 - 4 = 0 \Rightarrow 4\left(x^3 - 1\right) = 0$

$x = 1$

$y'' = 12x^2$

$y''(1) = 12 > 0 \qquad y(1) = -3$

$(1,-3)$ is a minimum point.

33. $y = 3x^4 - 4x^3 - 12x^2$

$y' = 12x^3 - 12x^2 - 24x$

$12x^3 - 12x^2 - 24x = 0$

$12x\left(x^2 - x - 2\right) = 0$

$12x(x-2)(x+1) = 0$

$x = -1,\ x = 0, \text{ or } x = 2$

$y(-1) = -5;\ y(0) = 0;\ y(2) = -32$

$(-1,-5)$ and $(2,-32)$ are minimum points.

$(0,0)$ is a maximum point.

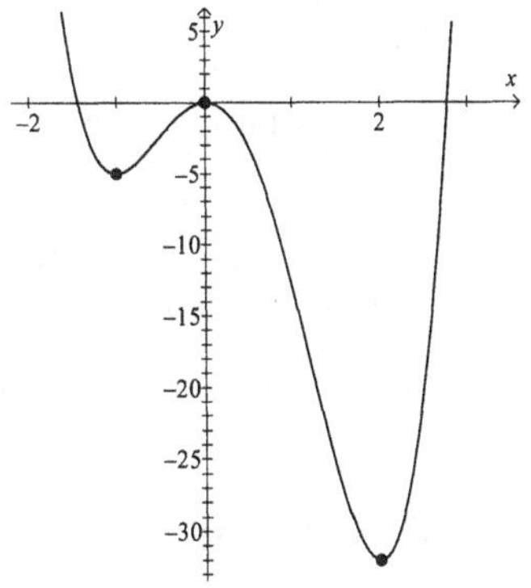

35. $y = x^3 + 3x^2 - 9x + 5$

$y' = 3x^2 + 6x - 9$

$3x^2 + 6x - 9 = 0$

$3\left(x^2 + 2x - 3\right) = 0$

$3(x+3)(x-1) = 0$

$x = -3 \text{ or } x = 1$

$y'' = 6x + 6$

$y''(-3) = -12 < 0 \qquad y(-3) = 32$

$y''(1) = 12 > 0 \qquad y(1) = 0$

$(-3,32)$ is a maximum point.

$(1,0)$ is a minimum point.

Implicit Relations

37. $4x^2+9y^2=36$

$8x+18yy'=0 \Rightarrow y'=-\dfrac{4x}{9y}$

$-\dfrac{4x}{9y}=0 \Rightarrow x=0$

when $x=0$: $9y^2=36$

$y^2=4 \Rightarrow y=\pm 2$

$$y''=-\frac{9y(4)-4x(9y')}{(9y)^2}=\frac{36\left[x\left(-\frac{4x}{9y}\right)-y\right]}{81y^2}=\frac{-16x^2-36y^2}{81y^3}$$

$$y''\big|_{(0,2)}=\frac{-16(0)^2-36(2)^2}{81(2)^3}=-0.222<0$$

$$y''\big|_{(0,-2)}=\frac{-16(0)^2-36(-2)^2}{81(-2)^3}=0.222>0$$

$(0,2)$ is a maximum point. $(0,-2)$ is a minimum point.

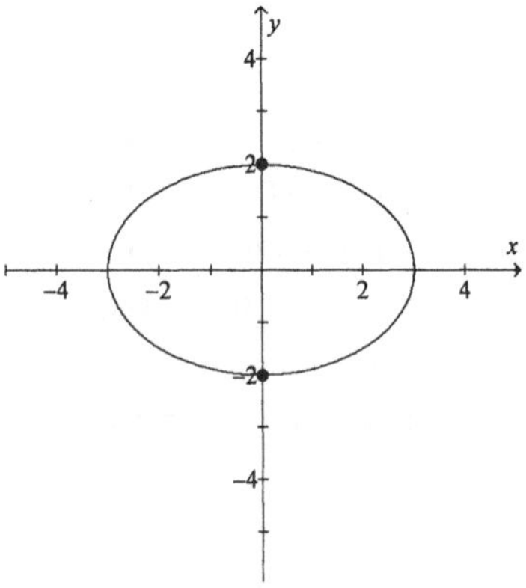

39. $x^2-x-2y^2+36=0$

$2x-1-4yy'=0 \Rightarrow y'=\dfrac{2x-1}{4y}$

$\dfrac{2x-1}{4y}=0 \Rightarrow x=\dfrac{1}{2}$

when $x=\dfrac{1}{2}$: $\dfrac{1}{4}-\dfrac{1}{2}-2y^2+36=0$

$y^2=17.875 \Rightarrow y=\pm 4.23$

$$y''=\frac{(4y)(2)-(2x-1)(4y')}{(4y)^2}=\frac{8y-(2x-1)4\left(\frac{2x-1}{4y}\right)}{16y^2}$$

$$=\frac{8y^2-(2x-1)^2}{16y^3}$$

$$y''\big|_{(0.5,-4.23)}=\frac{8(-4.23)^2-(2(0.5)-1)^2}{16(-4.23)^3}=-0.12<0$$

$$y''\big|_{(0.5,4.23)}=\frac{8(4.23)^2-(2(0.5)-1)^2}{16(4.23)^3}=0.12>0$$

$(0.5,-4.32)$ is a maximum point. $(0.5,4.32)$ is a minimum point.

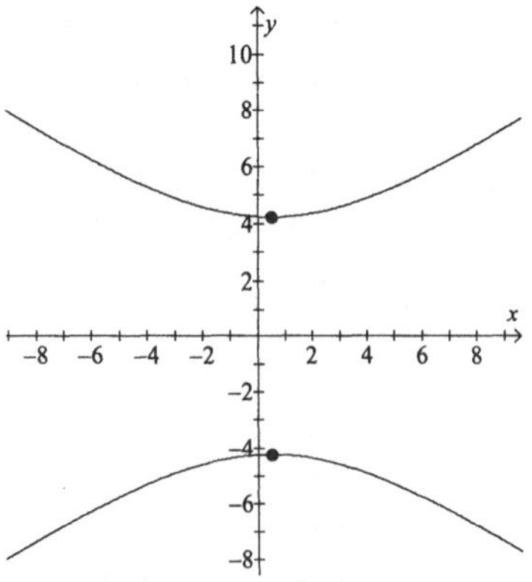

41. $y = x^4 + x \;\Rightarrow\; y' = 4x^3 + 1 \;\Rightarrow\; y'' = 12x^2$

$4x^3 + 1 = 0 \;\Rightarrow\; x^3 = -\frac{1}{4} \;\Rightarrow\; x = \sqrt[3]{\frac{1}{4}} \approx -0.630$

$y''(-0.630) = 12(-0.630)^2 > 0 \qquad y(-0.630) = (-0.630)^4 + (-0.630) - 0.472$

$(-0.630, -0.472)$ is a minimum point.

43. $y = 5 - x^2 + 2 = 3 - x^2 \;\Rightarrow\; y' = 2x \;\Rightarrow\; y'' = 2$

$2x = 0 \;\Rightarrow\; x = 0$

$y''(0) = 2 > 0 \qquad y(0) = 5 = (0)^2 + 2 = 7$

$(0,7)$ is a minimum point.

Inflection Points

45. $y = x^3 + 3$

$y' = 3x^2$

$y'' = 6x$

$6x = 0 \;\Rightarrow\; x = 0$

$y''(-1) = 6(-1) = -6 < 0$ concave downward for $x < 0$

$y''(1) = 6(1) = 6 > 0$ concave upward for $x > 0$

$y(0) = 3$

$(0,3)$ is an inflection point.

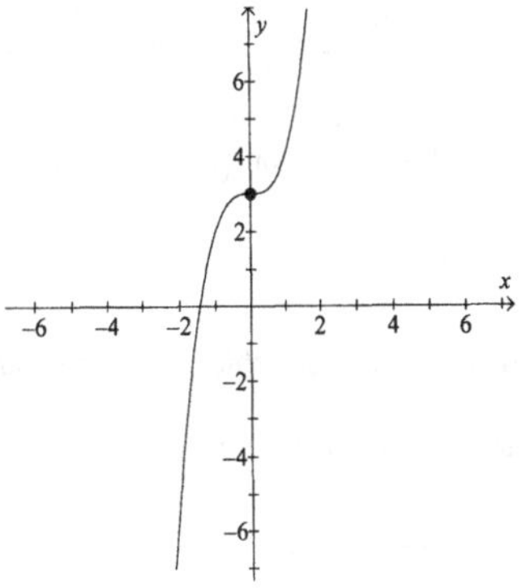

47. $y = x^4 - x^3 + 1$

$y' = 4x^3 - 3x^2$

$y'' = 12x^2 - 6x$

$12x^2 - 6x = 0 \;\Rightarrow\; 6x(2x-1) = 0 \;\Rightarrow\; x = 0$ or $x = \frac{1}{2}$

$y''(-1) = 18 > 0$ concave upward for $x < 0$

$y''\left(\frac{1}{4}\right) = -0.75 < 0$ concave downward for $0 < x < \frac{1}{2}$

$y''(1) = 6 > 0$ concave upward for $x > \frac{1}{2}$

$y(0) = 1;\; y\left(\frac{1}{2}\right) = \frac{15}{16}$

$(0,1)$ and $\left(\frac{1}{2}, \frac{15}{16}\right)$ are inflection points.

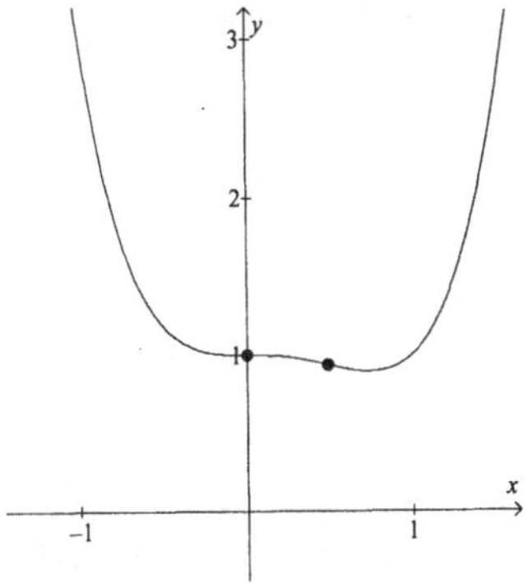

49. $y = x^3 - x \Rightarrow y' = 3x^2 - 1 \Rightarrow y'' = 6x \qquad 6x = 0 \Rightarrow x = 0$

$y''(-1) = 6(-1) = -6 < 0$ concave downward for $x < 0$

$y''(1) = 6(1) = 6 > 0$ concave upward for $x > 0$

$y(0) = (0)^3 - (0)^2 = 0$

$(0,0)$ is an inflection point.

51. $y = 5x^3 - 2x^2 + 1 \Rightarrow y' = 15x^2 - 4x \Rightarrow y'' = 30x - 4 \qquad 30x - 4 = 0 \Rightarrow x = \frac{2}{15} \approx 0.133$

$y''(0) = 30(0) - 4 = -4 < 0$ concave downward for $x < \frac{2}{15}$

$y''(1) = 30(1) - 4 = 26 > 0$ concave upward for $x > \frac{2}{15}$

$y\left(\frac{2}{15}\right) = 5\left(\frac{2}{15}\right)^3 - 2\left(\frac{2}{15}\right)^2 + 1 = \frac{659}{675} \approx 0.976$

$\left(\frac{2}{15}, \frac{659}{675}\right)$ or $(0.133, 0.976)$ is an inflection point.

Exercise 24.3 • Sketching, Verifying, and Interpreting Graphs

Make a complete graph of each function. Locate all features of interest.

1. $y = 4x^2 - 5 \Rightarrow y' = 8x \Rightarrow y'' = 8$

$y' = 0 \Rightarrow 8x = 0 \Rightarrow x = 0$

$y''(0) = 8 > 0 \quad y(0) = -5$

$(0,-5)$ is a minimum point.

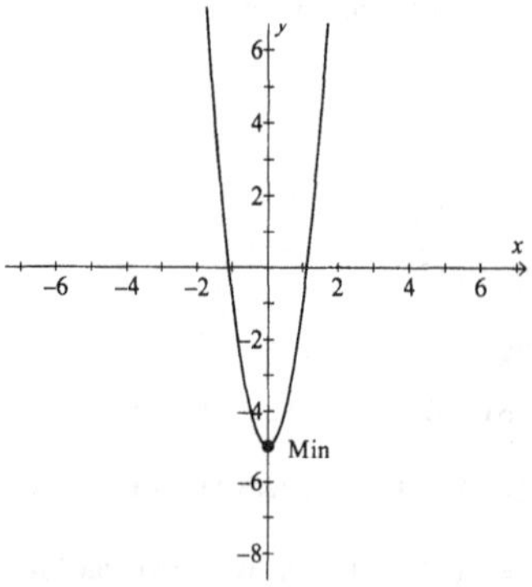

3. $y = 5 - x^{-1} \Rightarrow y' = x^{-2}$

No points of interest, but there is a vertical asymptote at $x = 0$ and a horizontal asymptote at $y = 5$.

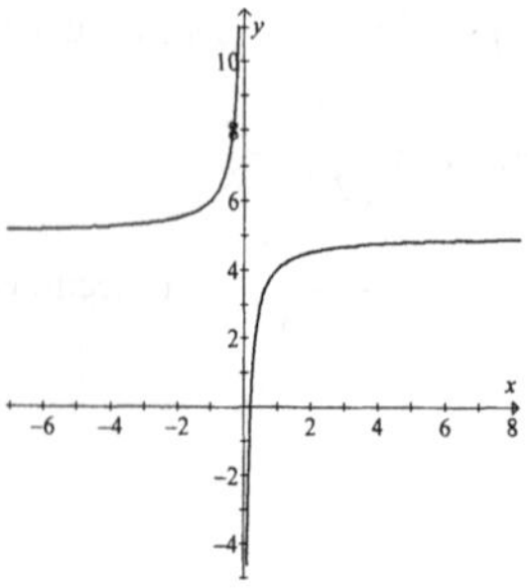

5. $y = x^4 - 8x^2 \Rightarrow y' = 4x^3 - 16x$

$4x^3 - 16x = 0 \Rightarrow 4x(x^2 - 4) = 0$

$x = 0, \pm 2$

$y'' = 12x^2 - 16 \Rightarrow 12x^2 - 16 = 0 \Rightarrow 4(3x^2 - 4) = 0$

$x^2 = \frac{4}{3} \Rightarrow x = \pm\frac{2}{\sqrt{3}} = \pm 1.15$

$y''(-2) = 32 > 0$ concave up for $x < -1.15$

$y''(0) = -16 < 0$ concave down for $-1.15 < x < 1.15$

$y''(2) = 32 > 0$ concave up for $x > 1.15$

$y(-2) = -16;\ y(0) = 0;\ y(2) = -16$

$y\left(-\frac{2}{\sqrt{3}}\right) = -8.89;\ y\left(\frac{2}{\sqrt{3}}\right) = -8.89$

$(-2,-16)$ and $(2,-16)$ are minimum points; $(0,0)$ is a maximum point;

$\left(-\frac{2}{\sqrt{3}}, -8.89\right)$ and $\left(\frac{2}{\sqrt{3}}, -8.89\right)$ are inflection points.

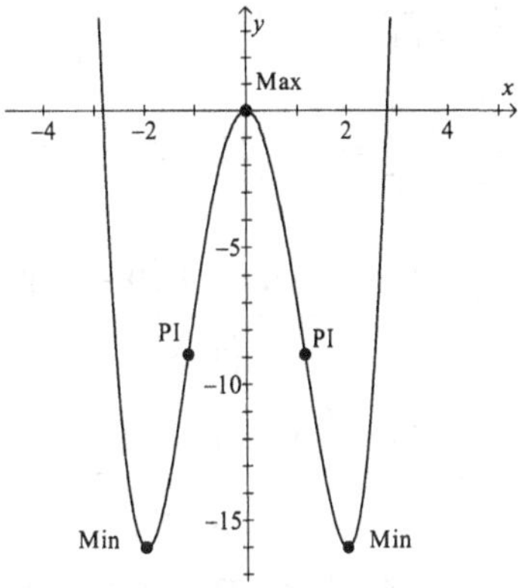

7. $y = x^3 - 9x^2 + 24x - 7 \Rightarrow y' = 3x^2 - 18x + 24$

$3x^2 - 18x + 24 = 0 \Rightarrow 3(x^2 - 6x + 8) = 0$

$3(x-2)(x-4) = 0 \Rightarrow x = 2 \text{ or } 4$

$y'' = 6x - 18 \Rightarrow 6x - 18 = 0 \Rightarrow x = 3$

$y''(2) = -6 < 0$ concave down for $x < 3$

$y''(4) = 6 > 0$ concave up for $x > 3$

$y(2) = 13;\ y(3) = 11;\ y(4) = 9$

$(4,9)$ is a minimum point; $(2,13)$ is a maximum point; $(3,11)$ is an inflection point.

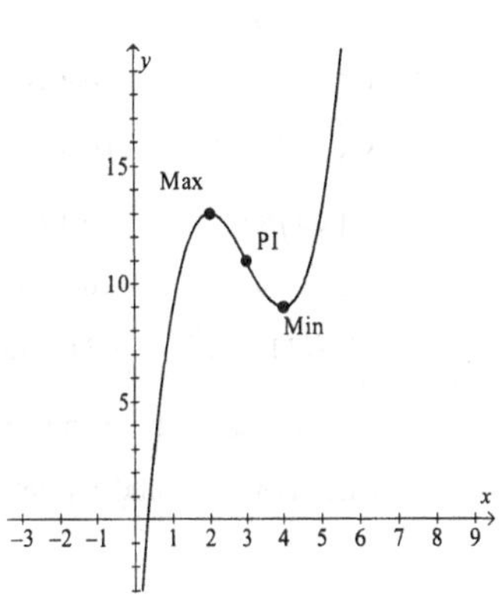

9. $y = 5x - x^5 \Rightarrow y' = 5 - 5x^4$

$5 - 5x^4 = 0 \Rightarrow 5(1 - x^4) = 0 \Rightarrow x = \pm 1$

$y'' = -20x^3 \Rightarrow -20x^3 = 0 \Rightarrow x = 0$

$y''(-1) = 20 > 0$ concave up for $x < 0$

$y''(1) = -20 < 0$ concave down for $x > 0$

$y(-1) = -4;\ y(0) = 0;\ y(1) = 4$

$(-1,-4)$ is a minimum point; $(1,4)$ is a maximum point;

$(0,0)$ is an inflection point.

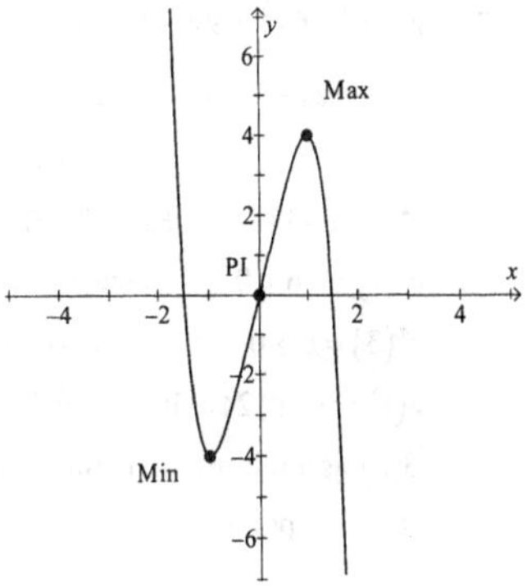

11. $y = \dfrac{6x}{3+x^2} \Rightarrow y' = \dfrac{6(3+x^2)-6x(2x)}{(3+x^2)^2} = \dfrac{18+6x^2-12x^2}{(3+x^2)^2}$

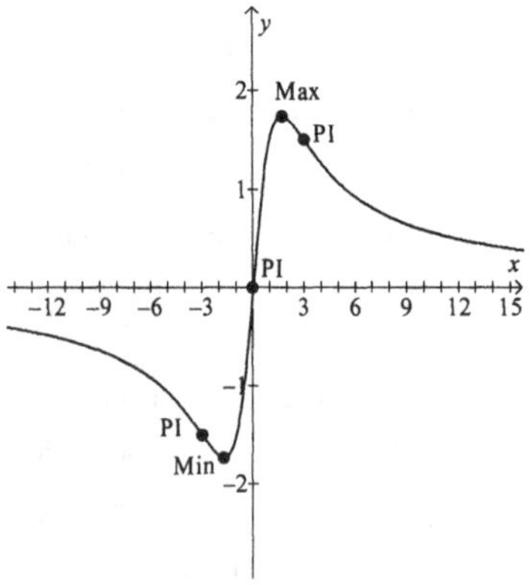

$\dfrac{18-6x^2}{(3+x^2)^2} = 0 \Rightarrow 18-6x^2 = 0 \Rightarrow x = \pm\sqrt{3}$

$y'' = \dfrac{(3+x^2)^2(-12x)-4x(18-6x^2)(3+x^2)}{(3+x^2)^4}$

$= \dfrac{(3+x^2)\left[(3+x^2)(-12x)-4x(18-6x^2)\right]}{(3+x^2)^4}$

$= \dfrac{-36x-12x^3-72x+24x^3}{(3+x^2)^3} = \dfrac{12x^3-108x}{(3+x^2)^3}$

$\dfrac{12x(x^2-9)}{(3+x^2)^3} = 0 \Rightarrow 12x(x^2-9) = 0 \Rightarrow x = 0, \pm 3$

$y''(-4) < 0$ concave down for $x < -3$

$y''(-\sqrt{3}) > 0$ concave up for $-3 < x < 0$

$y''(\sqrt{3}) < 0$ concave down for $0 < x < 3$

$y''(4) > 0$ concave up for $x > 3$

$y(-\sqrt{3}) = -\sqrt{3};\ y(\sqrt{3}) = \sqrt{3};\ y(-3) = -1.5;\ y(0) = 0;\ y(3) = 1.5$

$(\sqrt{3},\sqrt{3})$ is a maximum point; $(-\sqrt{3},-\sqrt{3})$ is a minimum point; $(-3,-1.5)$, $(0,0)$, and $(3,1.5)$ are inflection points.

There is a horizontal asymptote at $y = 0$.

13. $y = x^3-6x^2+9x+3 \Rightarrow y' = 3x^2-12x+9$

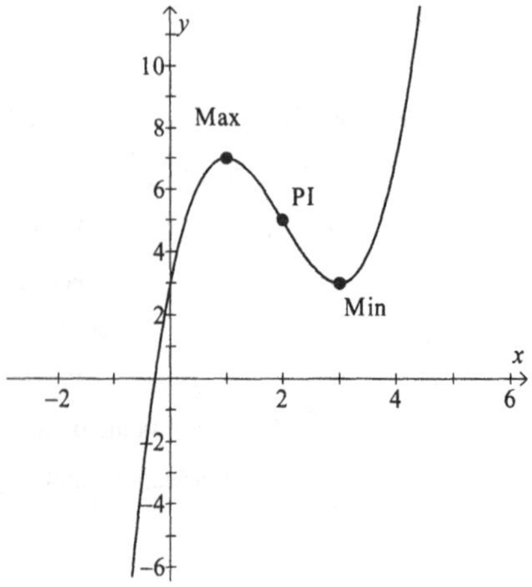

$3x^2-12x+9 = 0 \Rightarrow 3(x^2-4x+3) = 0$

$3(x-1)(x-3) = 0 \Rightarrow x = 1 \text{ or } 3$

$y'' = 6x-12 \Rightarrow 6x-12 = 0 \Rightarrow x = 2$

$y''(1) = -6 < 0$ concave down for $x < 2$

$y''(3) = 6 > 0$ concave up for $x > 2$

$y(1) = 7;\ y(2) = 5;\ y(3) = 3$

$(3,3)$ is a minimum point; $(1,7)$ is a maximum point; $(2,5)$ is an inflection point.

15. $y = \dfrac{96x - 288}{x^2 + 2x + 1}$

$y' = \dfrac{96\left(x^2 + 2x + 1\right) - (96x - 288)(2x + 2)}{\left(x^2 + 2x + 1\right)^2}$

$= \dfrac{-96x^2 + 576x + 672}{\left(x^2 + 2x + 1\right)^2}$

$\dfrac{-96x^2 + 576x + 672}{\left(x^2 + 2x + 1\right)^2} = 0 \;\Rightarrow\; -96(x+1)(x-7) = 0$

$x = -1$ or 7

-1 is not in the domain.

$y'' = \dfrac{-192\left[-x^3 + 9x^2 + 21x + 11\right]}{\left(x^2 + 2x + 1\right)^3}$

$-192\left[-x^3 + 9x^2 + 21x + 11\right] = 0 \;\Rightarrow\; x = -1 \text{ or } 11$

$y''(7) < 0$ concave down for $x < 11$

$y''(12) > 0$ concave up for $x > 11$

$y(7) = 6;\; y(11) = 5.33$

$(7, 6)$ is a maximum point; $(11, 5.33)$ is an inflection point.

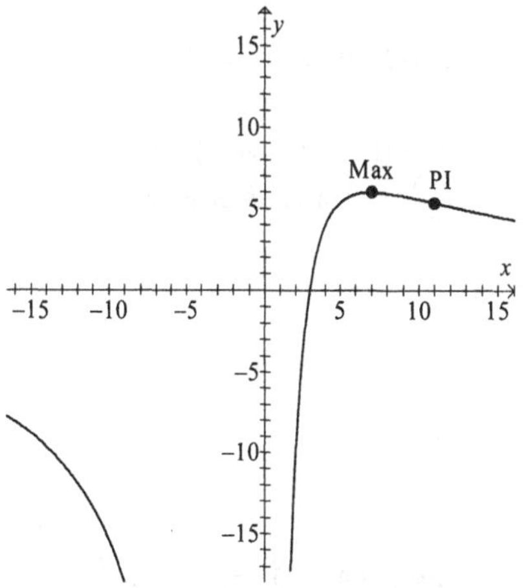

17. $y = \dfrac{x}{\sqrt{x^2 + 1}} = \dfrac{x}{\left(x^2 + 1\right)^{1/2}} \;\Rightarrow\; y' = \dfrac{\left(x^2 + 1\right)^{1/2} - x^2\left(x^2 + 1\right)^{-1/2}}{x^2 + 1}$

$y' = \dfrac{x^2 + 1 - x^2}{\left(x^2 + 1\right)^{3/2}} = \dfrac{1}{\left(x^2 + 1\right)^{3/2}} \neq 0$

$y'' = \left(-\dfrac{3}{2}\right)\left(x^2 + 1\right)^{-5/2}(2x) = \dfrac{-3x}{\left(x^2 + 1\right)^{5/2}}$

$\dfrac{-3x}{\left(x^2 + 1\right)^{5/2}} = 0 \;\Rightarrow\; -3x = 0 \;\Rightarrow\; x = 0$

$y''(-1) > 0$ concave up for $x < 0$

$y''(1) < 0$ concave down for $x > 0$

$y(0) = 0$

$(0, 0)$ is an inflection point. There are horizontal asymptotes at $y = \pm 1$.

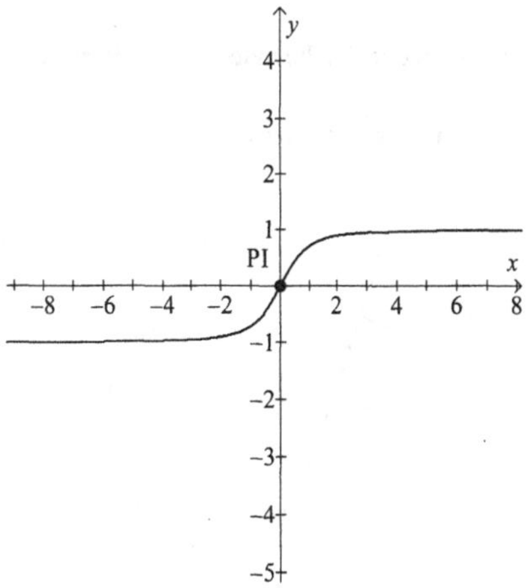

19. $y = x^2 + 2x \quad \Rightarrow \quad y' = 2x + 2$

$2x + 2 = 0 \quad \Rightarrow \quad x = -1$

$y'' = 2$

$y''(-1) = 2 > 0 \quad y(-1) = -1$

$(-1,-1)$ is a minimum point.

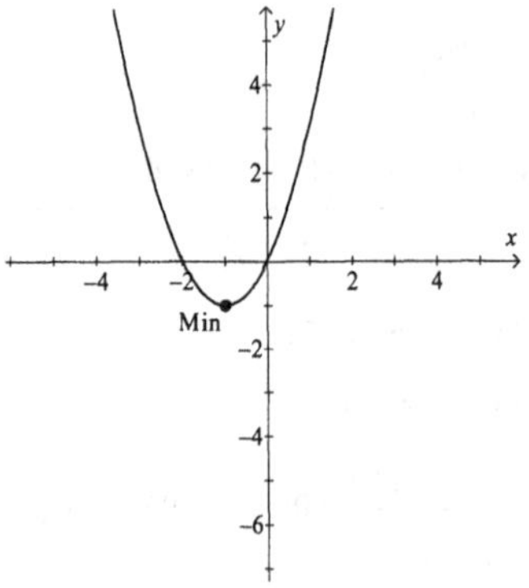

21. $y = x^3 + 4x^2 - 5 \quad \Rightarrow \quad y' = 3x^2 + 8x$

$3x^2 + 8x = 0 \quad \Rightarrow \quad x(3x+8) = 0 \quad \Rightarrow \quad x = 0 \text{ or } -\frac{8}{3}$

$y'' = 6x + 8 \quad \Rightarrow \quad 6x + 8 = 0 \quad \Rightarrow \quad x = -\frac{4}{3}$

$y''(-2) = -4 < 0$ concave down for $x < -\frac{4}{3}$

$y''(0) = 8 > 0$ concave up for $x > -\frac{4}{3}$

$y\left(-\frac{8}{3}\right) = 4.48; \; y(0) = -5; \; y\left(-\frac{4}{3}\right) = -0.26$

$(0,-5)$ is a minimum point; $\left(-\frac{8}{3}, 4.48\right)$ is a maximum point; $\left(-\frac{4}{3}, -0.26\right)$ is an inflection point.

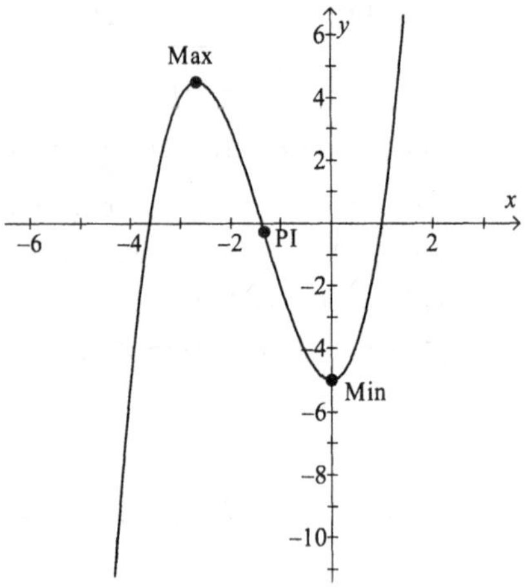

Graph the region bounded by the given curves.

23. $y = 3x^2$ and $y = 2x$

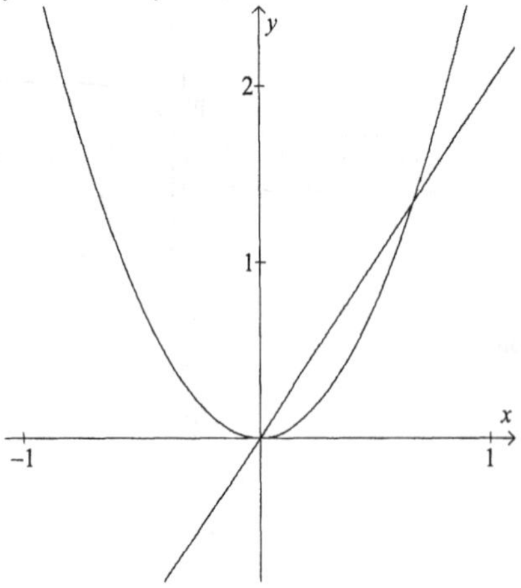

25. $y = 5x^2 - 2x$, $y = 4$ and the y axis, in the second quadrant

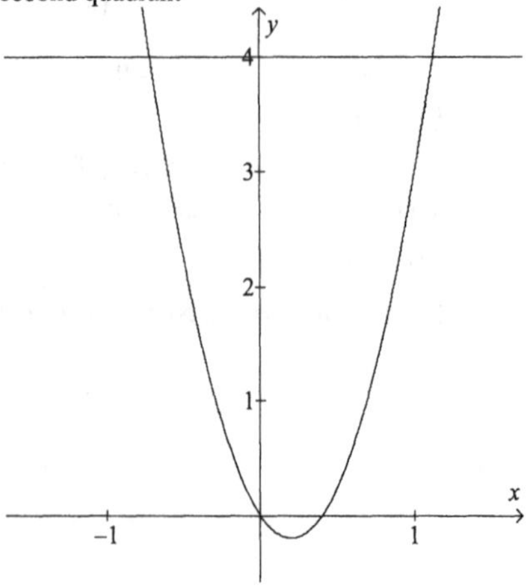

Chapter 24 • Review Problems

1. $y = 4x^2 - 2x + 5 \Rightarrow y' = 8x - 2$

$8x - 2 = 0 \Rightarrow x = \dfrac{1}{4}$

$y'' = 8$

$y''\left(\dfrac{1}{4}\right) = 8 > 0 \qquad y\left(\dfrac{1}{4}\right) = 4.75$

$(0.250, 4.75)$ is a minimum point.

3. $y = 3\sqrt[3]{x} - x = 3x^{1/3} - x \Rightarrow y' = x^{-2/3} - 1$

$x^{-2/3} - 1 = 0 \Rightarrow x = \pm 1$

$y'' = -\dfrac{2}{3}x^{-5/3}$

$y''(-1) = \dfrac{2}{3} > 0 \qquad y(-1) = -2$

$y''(1) = -\dfrac{2}{3} < 0 \qquad y(1) = 2$

$(-1, -2)$ is a minimum point.

$(1, 2)$ is a maximum point.

5. $y = \dfrac{1 + 2x}{3 - x} \Rightarrow y' = \dfrac{(3-x)(2) - (1+2x)(-1)}{(3-x)^2} = \dfrac{6 - 2x + 1 + 2x}{(3-x)^2} = \dfrac{7}{(3-x)^2}$

Tangent

$m_{\text{tangent}} = y'(2) = \dfrac{7}{(3-2)^2} = 7$

$\dfrac{y-5}{x-2} = 7$

$y - 5 = 7x - 14$

$7x - y - 9 = 0$

Normal

$m_{\text{normal}} = -\dfrac{1}{7}$

$\dfrac{y-5}{x-2} = -\dfrac{1}{7}$

$-7y + 35 = x - 2$

$x + 7y - 37 = 0$

7. $y = \sqrt{x^2 + 7} = (x^2 + 7)^{1/2} \Rightarrow y' = \dfrac{1}{2}(x^2 + 7)^{-1/2}(2x) = \dfrac{x}{\sqrt{x^2 + 7}}$

$m_{\text{tangent}} = y'(3) = \dfrac{3}{\sqrt{9 + 7}} = \dfrac{3}{4}$

$\dfrac{y-4}{x-3} = \dfrac{3}{4} \Rightarrow 4y - 16 = 3x - 9 \Rightarrow 3x - 4y + 7 = 0$

x intercept $(y = 0)$: $3x + 7 = 0 \Rightarrow x = -\dfrac{7}{3}$

9. $y = \sqrt{4x} = 2x^{1/2} \Rightarrow y' = x^{-1/2}$

$x > 0$

$y'(1) = (1)^{-1/2} > 0$ rising for $x > 0$

curve never falls

11. $y = x^2 - x^5 \Rightarrow y' = 2x - 5x^4 \Rightarrow y'' = 2 - 20x^3$

$y''(1) = 2 - 20(1)^3 = -18 < 0$

so concave downward at $x = 1$

13. $y = 3x^3 - 2x + 4 \Rightarrow y(2) = 3(2)^3 - 2(2) + 4 = 24$

$y' = 9x^2 - 2$

Tangent

$m_{\text{tangent}} = y'(2) = 9(2)^2 - 2 = 34$

$\frac{y - 24}{x - 2} = 34$

$y - 24 = 34x - 68$

$34x - y - 44 = 0$

Normal

$m_{\text{normal}} = -\frac{1}{34}$

$\frac{y - 24}{x - 2} = -\frac{1}{34}$

$-34y + 816 = x - 2$

$x + 34y - 818 = 0$

15. First find the x-coordinate of the point of intersection for the curves $y = \frac{x^2}{4}$ and $y = \frac{2}{x}$

$\frac{x^2}{4} = \frac{2}{x} \Rightarrow x^3 = 8 \Rightarrow x = 2$

$y_1 = \frac{x^2}{4} \Rightarrow y_1' = \frac{1}{2}x$

$m_1 = \frac{1}{2}(2) = 1$

$y_2 = \frac{2}{x} = 2x^{-1} \Rightarrow y_2' = -2x^{-2} = \frac{-2}{x^2}$

$m_2 = \frac{-2}{2^2} = -\frac{1}{2}$

$$\tan\phi = \frac{m_2 - m_1}{1 + m_1 m_2} = \frac{-\frac{1}{2} - 1}{1 + (1)\left(-\frac{1}{2}\right)} = -3$$

$\phi = \tan^{-1}(|-3|) = 71.6°$,

or $180° - 71.6° = 108.4°$

Chapter 25: More Applications of the Derivative

Exercise 25.1 • Rate of Change

Rate of Change with Respect to Time

1. $T = 55.6t^2 + 28.2t + 44.8\ °\text{F}$

$$\frac{dT}{dt} = 111.2t + 28.2\ °\text{F/h}$$

$$\left.\frac{dT}{dt}\right|_{t=2.00\text{ h}} = 111.2(2) + 28.2 = 251\ °\text{F/h}$$

3. $q = \dfrac{44.5}{\sqrt{t^2 - 26.4}} = 44.5\left(t^2 - 26.4\right)^{-1/2}\ \text{ft}^3/\text{min}$

$$\frac{dq}{dt} = \frac{-44.5t}{\left(t^2 - 26.4\right)^{-3/2}}\ \text{ft}^3/\text{min}^2$$

$$\left.\frac{dq}{dt}\right|_{t=15.3\text{ min}} = \frac{-44.5(15.3)}{\left((15.3)^2 - 26.4\right)^{-3/2}} = -0.227\ \text{ft}^3/\text{min}^2$$

Electric Current

5. $i\big|_{t=5.92\text{ s}} = 6.96(5.92) - 1.64 = 39.6\ \text{A}$

7. $P = vi = (33.54t - 7.90)(6.96t - 1.64) = 233.4t^2 - 110.0t + 12.96$

$$P\big|_{t=4.88\text{ s}} = 233.4(4.88)^2 - 110.0(4.88) + 12.96 = 5040\ \text{W}$$

Current in a Capacitor

9. $v = 6.27t^2 - 15.3t + 52.2$

$$\frac{dv}{dt} = 12.54t - 15.3$$

$$i = C\frac{dv}{dt} = \left(33.5\times10^{-6}\right)(12.54t - 15.3)$$

$$i\big|_{t=5.50\text{ s}} = \left(33.5\times10^{-6}\right)\left(12.54(5.50) - 15.3\right) = 0.0018\ \text{A}=1.80\ \text{mA}$$

Voltage Across an Inductor

11. $i = 5.22t^2 - 4.02t$

$$\frac{di}{dt} = 10.44t - 4.02$$

$$v = L\frac{di}{dt} = 1.44(10.44t - 4.02)$$

$$v\big|_{t=2.00\text{ s}} = 1.44\left(10.44(2.00) - 4.02\right) = 24.3\ \text{V}$$

Rate of Change with Respect to Another Variable

13. $pv = k \Rightarrow k = 25.5(146) = 3723$

$$p = \frac{k}{v} = kv^{-1}$$

$$\frac{dp}{dv} = -kv^{-2} = -\frac{k}{v^2}$$

$$\frac{dp}{dv} = -\frac{3723}{(146)^2} = -0.175 \text{ lb/in.}^2/\text{in.}^3$$

15. $v = \frac{4}{3}\pi r^3$

$$\frac{dv}{dr} = 4\pi r^2$$

$$\frac{dv}{dr} = 4\pi(1.00)^2 = 12.6 \text{ m}^3/\text{min}$$

\

17. $P = 0.324\sqrt{L} = 0.324L^{1/2}$

$$\frac{dP}{dL} = 0.162L^{-1/2} = \frac{0.162}{\sqrt{L}}$$

$$\frac{dP}{dL} = \frac{0.162}{\sqrt{9.00}} = 0.0540 \text{ s/in.}$$

Beam Deflection

19. $y = \frac{wx^2}{24EI}\left(x^2 + 6L^2 - 4Lx\right)$

$$= \frac{w}{24EI}\left(x^4 + 6L^2x^2 - 4Lx^3\right)$$

$$\frac{dy}{dx} = \frac{w}{24EI}\left(4x^3 + 12L^2x - 12Lx^2\right)$$

$$= \frac{wx}{6EI}\left(x^2 + 3L^2 - 3Lx\right)$$

Exercise 25.2 • Motion of a Point

Straight-Line Motion

1. $s = 32t - 8t^2$

$$v = \frac{ds}{dt} = 32 - 16t$$

$$v(2.00) = 32 - 16(2.00) = 0$$

$$a = \frac{dv}{dt} = -16$$

$$a(2.00) = -16.0$$

3. $s = t^2 + t^{-1} + 3$

$v = \frac{ds}{dt} = 2t - t^{-2}$

$v(0.500) = 2(0.500) - (0.500)^{-2} = -3.00$

$a = \frac{dv}{dt} = 2 + 2t^{-3}$

$a(0.500) = 2 + 2(0.500)^{-3} = 18.0$

5. $s = 120t - 16t^2$

$v = \frac{ds}{dt} = 120 - 32t$

$v(4.00) = 120 - 32(4.00) = -8.00$

$a = \frac{dv}{dt} = -32$

$a(4.00) = -32.0$

7. $s = 40t + 16t^2$

$v = \frac{ds}{dt} = 40 + 32t \quad \Rightarrow \quad v(2.00) = 40 + 32(2.00) = 104 \text{ ft/s}$

$a = \frac{dv}{dt} = 32 \quad \Rightarrow \quad a(2.00) = 32.0 \text{ ft/s}^2$

9. $s = 6t^2$

$v = \frac{ds}{dt} = 12t \quad \Rightarrow \quad v(5.00) = 12(5.00) = 60.0 \text{ ft/s}$

11. a) $s = 2000t - 16t^2$

$v = \frac{ds}{dt} = 2000 - 32t \quad \Rightarrow \quad v(0) = 2000 \text{ ft/s}$

b) $0 = 2000 - 32t \quad \Rightarrow \quad t = \frac{2000}{32} = 62.5 \text{ s} \quad \Rightarrow \quad s(62.5) = 2000(62.5) - 16(62.5)^2 = 62{,}500 \text{ ft}$

c) $v(10.0) = 2000 - 32(10.0) = 1680 \text{ ft/s}$

13. $s = 16t^2 - 64t + 64$

Point comes to a rest when $v = 0$

$v = \frac{ds}{dt} = 32t - 64 \quad \Rightarrow \quad 32t - 64 = 0 \quad \Rightarrow \quad t = 2 \text{ s}$

$s(2) = 16(2)^2 - 64(2) + 64 = 0$

$a = \frac{dv}{dt} = 32 \text{ units/s}^2$

Curvilinear Motion

15. $y = x^4 + x^2$

$y' = 4x^3 + 2x$

$y'(2.55) = 4(2.55)^3 + 2(2.55) = 71.4$

direction of travel is $\theta = \arctan(71.4) = 89.2°$

components of velocity

$v_x = 1.25\cos(89.2°) = 0.0175$ in./s

$v_y = 1.25\sin(89.2°) = 1.25$ in./s

Equations Given in Parametric Form

17. magnitude $v = \sqrt{v_x^2 + v_y^2} = \sqrt{(32.3)^2 + (-27.3)^2} = 42.3$ cm/s

$\phi = \arctan\left(\frac{v_y}{v_x}\right) = \arctan\left(\frac{-27.3}{32.3}\right) = -40.2°$

x component is positive and y component is negative means resultant is in quadrant IV

direction $\theta = 360° - 40.2° = 320°$

Trajectories

19. a) components for initial velocity

$v_{0x} = v_0 \cos\theta = 6350\cos 43.0° = 4644$ $\quad v_{0y} = v_0 \sin\theta = 6350\sin 43.0° = 4331$

horizontal and vertical positions

$x = v_{0x}t = 4644t \Rightarrow x(7.00) = 4644(7.00) = 32{,}500$ ft

$y = v_{0y}t - \frac{g}{2}t^2 = 4331t - \frac{32}{2}t^2 \Rightarrow y(7.00) = 4331(7.00) - 16(7.00)^2 = 29{,}500$ ft

b) horizontal and vertical velocities

$v_x = 4644 \Rightarrow v_x(7.00) = 4640$ ft/s

$v_y = 4331 - 32t \Rightarrow v_y(7.00) = 4331 - 32(7.00) = 4110$ ft/s

Rotation

21. $\theta = 44.8t^3 + 29.3t^2 + 81.5$

$\omega = \frac{d\theta}{dt} = 134.4t^2 + 58.6t \Rightarrow \omega(4.25) = 134.4(4.25)^2 + 58.6(4.25) = 2680$ rad/s

23. $\theta = 184 + 271t^3$

a) $\omega = \frac{d\theta}{dt} = 813t^2 \Rightarrow \omega(1.25) = 813(1.25)^2 = 1270$ rad/s

b) $\alpha = \frac{d\omega}{dt} = 1626t \Rightarrow \alpha(1.25) = 1626(1.25) = 2030\ \text{rad/s}^2$

Exercise 25.3 • Related Rates

One Moving Object

1. $\frac{dx}{dt} = 100$ m/s; at $t = 1$ min $= 60$ s, $x = 100(60) = 6000$ m

$$z^2 = x^2 + (8000)^2 \Rightarrow z = \sqrt{6000^2 + 8000^2} = 10000 \text{ m}$$

$$2z\frac{dz}{dt} = 2x\frac{dx}{dt}$$

$$\frac{dz}{dt} = \frac{x\frac{dx}{dt}}{z} = \frac{6000(100)}{10000} = 60 \text{ m/s}$$

The distance from the pond to the plane is increasing at a rate of 60 m/s.

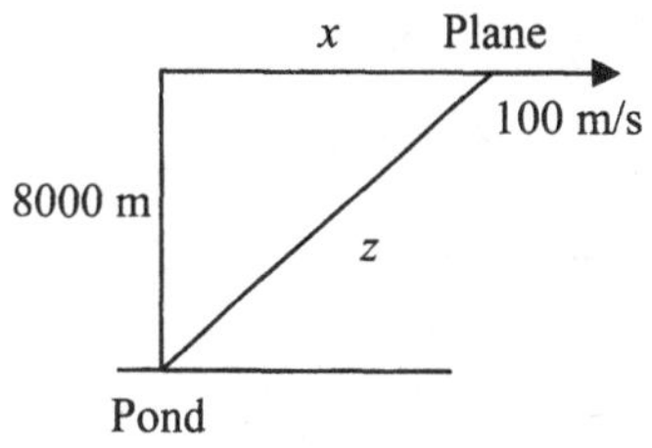

3. $\frac{dx}{dt} = -8.00$ mi/h; $x = 50.0$ ft $= 0.00947$ mi

$$100 \text{ ft}\left(\frac{1}{5280}\right) = 0.01894 \text{ mi}$$

$$z^2 = x^2 + (0.01894)^2 \Rightarrow z = \sqrt{0.00947^2 + 0.01894^2} = 0.02118 \text{ mi}$$

$$2z\frac{dz}{dt} = 2x\frac{dx}{dt}$$

$$\frac{dz}{dt} = \frac{x\frac{dx}{dt}}{z} = \frac{0.00947(-8.00)}{0.02118} = -3.58 \text{ mi/h}$$

The runner is approaching the top of the tower at a rate of 3.58 mi/h.

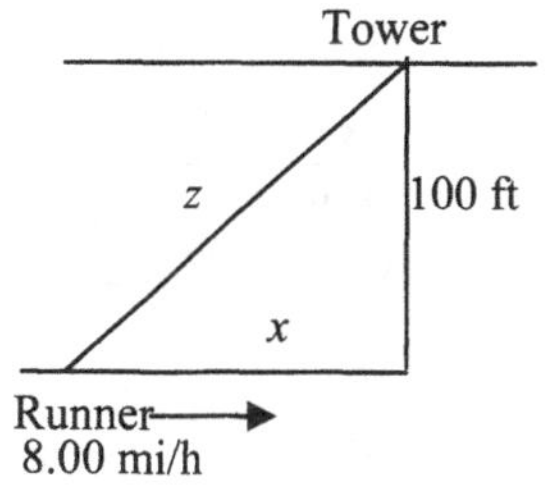

Ropes and Cables

5. $\frac{dx}{dt} = 4.00$ m/s; $z = 50.0$ m

$$z^2 = x^2 + (40.0)^2 \Rightarrow x = \sqrt{50.0^2 - 40.0^2} = 30.0 \text{ m}$$

$$2z\frac{dz}{dt} = 2x\frac{dx}{dt}$$

$$\frac{dz}{dt} = \frac{x\frac{dx}{dt}}{z} = \frac{30.0(4.00)}{50.0} = 2.40 \text{ m/s}$$

The cable is leaving the boat at a rate of 2.40 m/s.

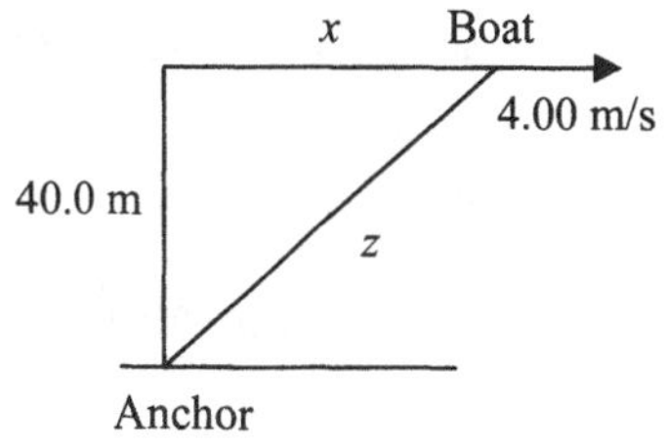

7. $\frac{dx}{dt} = -4.00$ ft/s;

When AB is horizontal a right triangle is formed

$$(30.0)^2 = x^2 + y^2 \Rightarrow 0 = 2x\frac{dx}{dt} + 2y\frac{dy}{dt}$$

Also, $x = \sqrt{30.0^2 - 20.0^2} = 22.36$ ft

$$\frac{dy}{dt} = -\frac{x\frac{dx}{dt}}{y} = -\frac{22.36(-4.00)}{20.0} = 4.47 \text{ ft/s}$$

B is rising at a rate of 4.47 ft/s.

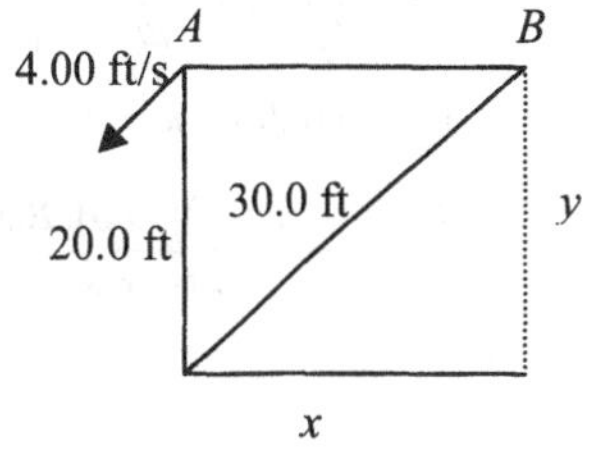

9. $\frac{dx}{dt} = 12.0 \text{ in./s}$

$z^2 = x^2 + (54.0)^2 \Rightarrow z = \sqrt{80.0^2 + 54.0^2} = 96.52 \text{ in.}$

$2z\frac{dz}{dt} = 2x\frac{dx}{dt}$

$\frac{dz}{dt} = -\frac{x\frac{dx}{dt}}{z} = \frac{80.0(12.0)}{96.52} = 9.95 \text{ in./s}$

The bucket is rising at a rate of 9.95 in./s.

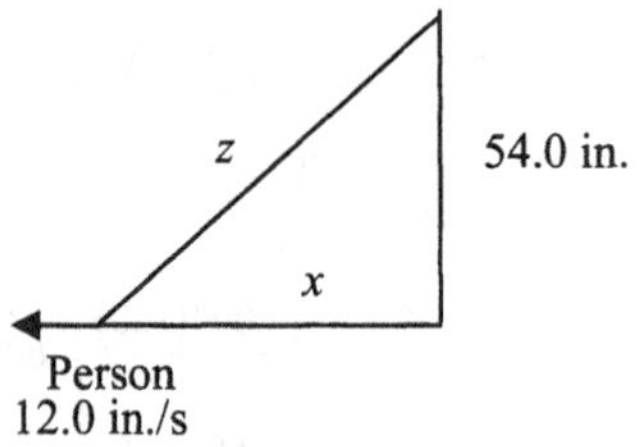

11. $\frac{dx}{dt} = 5.00 \text{ ft/s}$

$\frac{s}{30.0} = \frac{6.00}{x}$

$s = \frac{6.00(30.0)}{x} = 180x^{-1}$

$\frac{ds}{dt} = -180x^{-2}\frac{dx}{dt} = \frac{-180}{(15.0)^2}(5.00) = -4.00 \text{ ft/s}$

The shadow is decreasing at a rate of 4.00 ft/s.

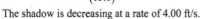

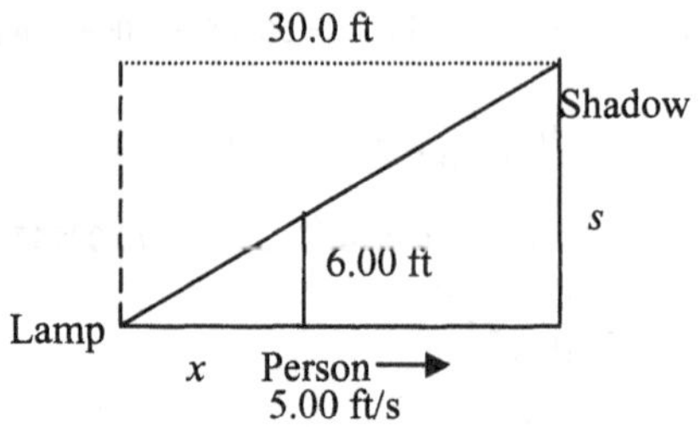

Expansion and Contraction

13. $\frac{ds}{dt} = 0.00500 \text{ in./s}$

at $t = 20$ s, $s = 10 + 20(0.00500) = 10.1$ in.

$A = s^2$

$\frac{dA}{dt} = 2s\frac{ds}{dt}$

$= 2(10.1)(0.00500) = 0.101 \text{ in.}^2\text{/s}$

15. $\frac{dV}{dt} = 10.0 \text{ in.}^3\text{/min}$

$V = s^3$

$V = 125 \text{ in}^3 \Rightarrow 125 = s^3 \Rightarrow s = \sqrt[3]{125} = 5 \text{ in.}$

$\frac{dV}{dt} = 3s^2\frac{ds}{dt}$

$10.0 = 3(5)^2\frac{ds}{dt} \Rightarrow \frac{ds}{dt} = 0.133 \text{ in./min}$

17. $\frac{dx}{dt} = 1.00 \text{ in./s}, \ x = 10.0 \text{ in.}, \ y = 20.0 \text{ in.}, \ \frac{dV}{dt} = 0$

$V = \frac{\pi x^2 y}{4} \Rightarrow \frac{dV}{dt} = \frac{2\pi x}{4}\frac{dx}{dt}y + \frac{\pi x^2}{4}\frac{dy}{dt}$

$0 = 2\pi(10.0)(1.00)(20.0) + \pi(10.0)^2\frac{dy}{dt}$

$\frac{dy}{dt} = -\frac{2\pi(10.0)(1.00)(20.0)}{\pi(10.0)^2} = -2(1.00)(2) = -4.00 \text{ in./min}$

Fluid Flow

19. $\frac{dV}{dt}=15.0\text{ m}^3/\text{min}$

$\frac{r}{h}=\frac{5.00}{20.0}\Rightarrow r=\frac{1}{4}h\Rightarrow\frac{dr}{dt}=\frac{1}{4}\frac{dh}{dt}$

when $h=8.00$ m, $r=\frac{1}{4}(8.00)=2.00$ m

$V=\frac{1}{3}\pi r^2h\Rightarrow\frac{dV}{dt}=\frac{2}{3}\pi r\frac{dr}{dt}h+\frac{1}{3}\pi r^2\frac{dh}{dt}$

$\frac{2}{3}\pi(2.00)\left(\frac{1}{4}\frac{dh}{dt}\right)(8.00)+\frac{1}{3}\pi(2.00)^2\frac{dh}{dt}=15.0$

$\frac{8.00}{3}\pi\frac{dh}{dt}+\frac{4.00}{3}\pi\frac{dh}{dt}=15.0$

$12\pi\frac{dh}{dt}=45.0\Rightarrow\frac{dh}{dt}=1.19\text{ m/min}$

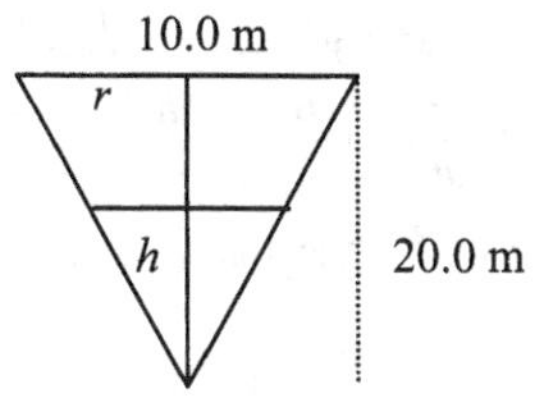

21. $\frac{dV}{dt}=8.00\text{ ft}^3/\text{min},\ l=10.0\text{ ft}$

$V=\frac{1}{2}(2h)(h)l=h^2(10.0)$

$\frac{dV}{dt}=20.0h\frac{dh}{dt}$

$20.0(2.00)\frac{dh}{dt}=8.00$

$\frac{dh}{dt}=\frac{8.00}{40.0}=0.200\text{ ft/min}$

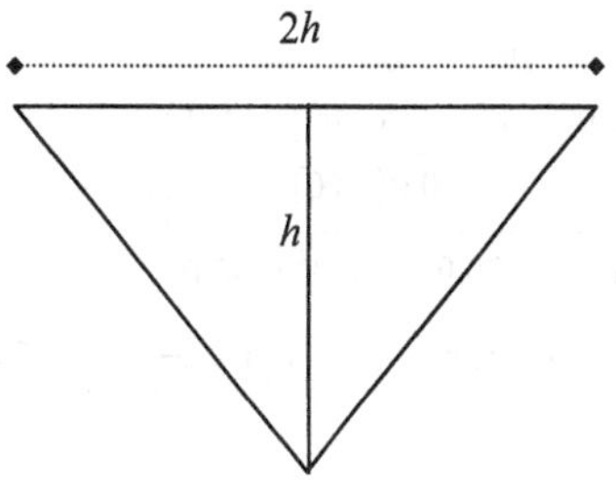

Gas Laws

23. $p=50.0\text{ lb/in.}^2,\ v=10.0\text{ ft}^3,\ \frac{dv}{dt}=-1.00\text{ ft}^3/\text{s}$

$pv^{1.4}=C\Rightarrow 1.4pv^{0.4}\frac{dv}{dt}+v^{1.4}\frac{dp}{dt}=0$

$1.4(50.0)(10.0)^{0.4}(-1.00)+(10.0)^{1.4}\frac{dp}{dt}=0$

$\frac{dp}{dt}=\frac{1.4(50.0)}{10.0}=7.0\text{ lb/in.}^2/\text{s}$

Two Moving Objects [See Example 18]

25. $\frac{dx}{dt} = 100$ km/h, $\frac{dy}{dt} = 150$ km/h

at 2 p.m., $x = 100(2) = 200$ km and $y = 150$ km

$$z^2 = x^2 + y^2 \Rightarrow z = \sqrt{200^2 + 150^2} = 250 \text{ km}$$

$$2z\frac{dz}{dt} = 2x\frac{dx}{dt} + 2y\frac{dy}{dt}$$

$$250\frac{dz}{dt} = 200(100) + 150(150)$$

$$\frac{dz}{dt} = \frac{42500}{250} = 170 \text{ km/h}$$

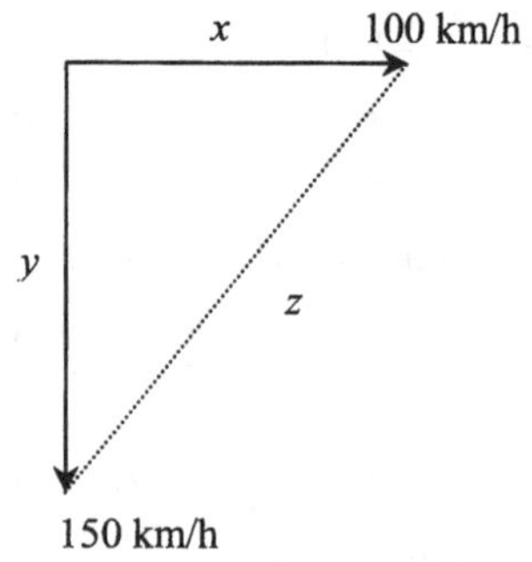

27. $h = 30.0$ ft, $\frac{dx}{dt} = 5.00$ ft/s, $\frac{dy}{dt} = 10.0$ ft/s

$$s^2 = x^2 + y^2 + 30.0^2$$

$$2s\frac{ds}{dt} = 2x\frac{dx}{dt} + 2y\frac{dy}{dt} + 0$$

after 3.00 s, $x = 5.00(3.00) = 15.0$ m

$y = 10.0(3.00) = 30.0$ m, so

$$s = \sqrt{15.0^2 + 30.0^2 + 30.0^2} = 45 \text{ ft}$$

$$45\frac{ds}{dt} = 15.0(5.00) + 30.0(10.0) \Rightarrow \frac{ds}{dt} = \frac{375}{45} = 8.33 \text{ ft/s}$$

Miscellaneous

29. $y = \frac{Px^3}{3EI}$

$$\frac{dy}{dt} = \frac{P3x^2}{3EI}\frac{dx}{dt} = \frac{Px^2}{EI}\frac{dx}{dt}$$

$P = 165,\ E = 1320000,\ I = 10.9$

$x = 75.0,\ \frac{dx}{dt} = 25.0$

$$\frac{dy}{dt} = \frac{165(75.0)^2}{1320000(10.9)}(25.0) = 1.61 \text{ in./s}$$

Exercise 25.4 • Optimization

Number Puzzles

1. $N(x)=x+\frac{1}{2}\left(\frac{1}{x^2}\right)=x+\frac{1}{2}x^{-2}$

$N'(x)=1-x^{-3}=1-\frac{1}{x^3}$

$N''(x)=3x^{-4}=\frac{3}{x^4}$

$N'(x)=0 \Rightarrow 1-\frac{1}{x^3}=0$

$x^3=1$

$x=1$

$N''(1)=\frac{3}{1}>0$ minimum at $x=1$

The smallest sum is when $x=1$.

3. $x+y=20 \Rightarrow y=20-x$

$P=yx^2$

$P(x)=(20-x)x^2=20x^2-x^3$

$P'(x)=40x-3x^2$

$P''(x)=40-6x$

$P'(x)=0 \Rightarrow 40x-3x^2=0$

$x(40-3x)=0$

$x=0$ or $\frac{40}{3}$

$P''(0)=40>0$ minimum at $x=0$

$P''\left(\frac{40}{3}\right)=-40<0$ maximum at $x=\frac{40}{3}$

$x=\frac{40}{3} \Rightarrow y=20-\frac{40}{3}=\frac{20}{3}$

The two numbers are $6\frac{2}{3}$ and $13\frac{1}{3}$.

Minimum Perimeter

5. Area $=xy=432 \Rightarrow y=\frac{432}{x}$

Cost $=2x+y+\frac{1}{2}y=2x+\frac{3}{2}y$

$C(x)=2x+\frac{3}{2}\left(\frac{432}{x}\right)=2x+648x^{-1}$

$C'(x)=2-648x^{-2}=2-\frac{648}{x^2}$

$C''(x)=1296x^{-3}=\frac{1296}{x^3}$

$C'(x)=0 \Rightarrow 2-\frac{648}{x^2}=0$

$2x^2=648$

$x^2=324$

$x=18$ (Exclude -18)

$C''(18)>0$ minimum at $x=18$

$x=18\text{ m} \Rightarrow y=\frac{432}{18}=24\text{ m}$

The dimensions should be $18\text{ m}\times 24\text{ m}$.

7. Area $=xy=162 \Rightarrow y=\frac{162}{x}$

Take x to be the length of the short side

Perimeter $=2x+y$

$P(x)=2x+\frac{162}{x}=2x+162x^{-1}$

$P'(x)=2-162x^{-2}=2-\frac{162}{x^2}$

$P''(x)=324x^{-3}=\frac{324}{x^3}$

$P'(x)=0 \Rightarrow 2-\frac{162}{x^2}=0$

$2x^2=162$

$x^2=81$

$x=9$ (Exclude -9)

$P''(9)>0$ minimum at $x=9$

$x=9\text{ yd} \Rightarrow y=\frac{162}{9}=18\text{ yd}$

The dimensions should be $9\text{ yd}\times 18\text{ yd}$.

Maximum Volume of Containers

9. Volume $= \pi r^2 h$

Surface Area $= 2r\pi h + \pi r^2 = 100 \Rightarrow \quad h = \dfrac{100 - \pi r^2}{2r\pi}$

$$V(r) = \pi r^2\left(\frac{100 - \pi r^2}{2r\pi}\right) = 50r - \frac{1}{2}\pi r^3 \Rightarrow V'(r) = 50 - \frac{3}{2}\pi r^2 \Rightarrow V''(r) = -3\pi r$$

$$V'(r) = 0 \Rightarrow 50 - \frac{3}{2}\pi r^2 = 0$$

$$r = \frac{10}{\sqrt{3\pi}} = 3.257$$

$V''(3.257) < 0$ maximum at $r = 3.257$

$d = 2(3.257) = 6.51$ cm and $h = \dfrac{100 - \pi(3.257)^2}{2(3.257)\pi} = 3.26$ cm

The maximum volume occurs when the diameter of the cylinder is 6.51 cm and the height is 3.26 cm.

Maximum Area of Plane Figures

11. Area $= xy$

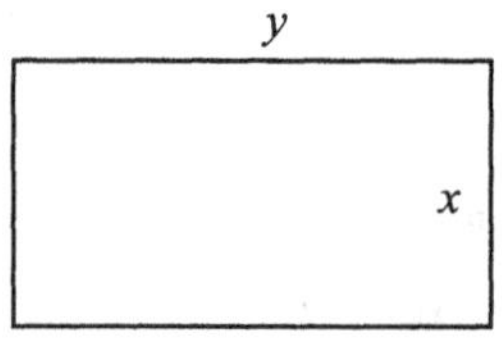

Perimeter $= 2x + 2y = 20.0 \Rightarrow y = 10 - x$

$A(x) = x(10 - x) = 10x - x^2$

$A'(x) = 10 - 2x$

$A''(x) = -2$

$A'(x) = 0 \Rightarrow 10 - 2x = 0$

$x = 5$

$A''(5) = -2 < 0$ maximum at $x = 5$

$y = 10 - 5 = 5$

Maximum area is $5(5) = 25$ in.^{2}.

13. Area $= xy$

$x_1 = \dfrac{h}{2}$ and $x_2 = \dfrac{30 - h}{3}$

$$b = x_2 - x_1 = \frac{30 - h}{3} - \frac{h}{2} = \frac{60 - 2h - 3h}{6} = \frac{60 - 5h}{6}$$

$$A(h) = h\left(\frac{60 - 5h}{6}\right) = 10h - \frac{5}{6}h^2$$

$$A'(h) = 10 - \frac{5}{3}h$$

$$A''(h) = -\frac{5}{3}$$

$$A'(h) = 0 \Rightarrow 10 - \frac{5}{3}h = 0$$

$$h = 6$$

$A''(6) < 0$ maximum at h = 6

The area is a maximum for a y value of 6.

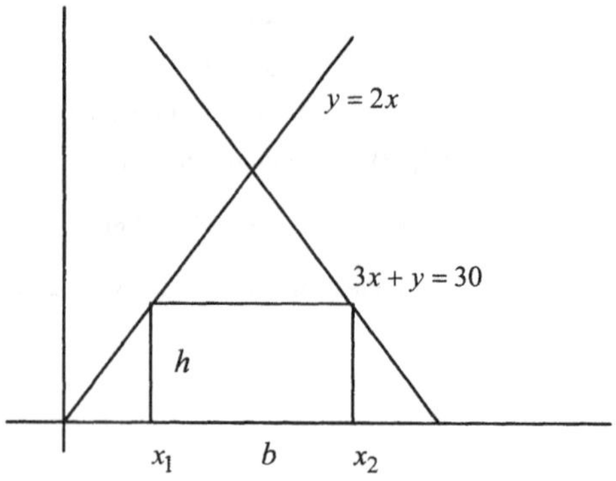

Maximum Cross-Sectional Area

15 $A(x) = 4.00x + 2\left(\frac{1}{2}x\sqrt{16-x^2}\right) = 4.00x + x\left(16-x^2\right)^{1/2}$

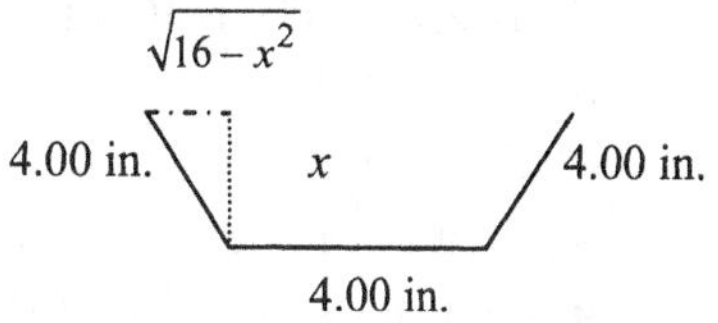

$$A'(x) = 4.00 + x\left(\frac{1}{2}\right)\left(16-x^2\right)^{-1/2}(-2x) + \left(16-x^2\right)^{1/2}$$

$$= 4 - x^2\left(16-x^2\right)^{-1/2} + \left(16-x^2\right)^{1/2}$$

$$A''(x) = -x^3\left(16-x^2\right)^{-3/2} - 3x\left(16-x^2\right)^{-1/2}$$

$$A'(x) = 0 \Rightarrow 4 - x^2\left(16-x^2\right)^{-1/2} + \left(16-x^2\right)^{1/2} = 0$$

multiply both sides by $\left(16-x^2\right)^{1/2}$:

$$4\left(16-x^2\right)^{1/2} - x^2 + 16 - x^2 = 0$$

$$2\left(16-x^2\right)^{1/2} = x^2 - 8$$

square both sides

$$4\left(16-x^2\right) = x^4 - 16x^2 + 64 \Rightarrow x^4 - 12x^2 = 0$$

$$x^2\left(x^2-12\right) = 0 \Rightarrow x = 0 \text{ or } \pm\sqrt{12} \quad \left(\text{Exclude } 0 \text{ and } -\sqrt{12}\right)$$

$A''\left(\sqrt{12}\right) < 0$ maximum at $x = \sqrt{12} = 3.46$

The gutter should be 3.46 in. deep.

Minimum Distance

17. $y = \frac{x^2}{2}$

Minimizing the square of the distance will minimize the distance, so the calculations can be made easier by not working with the radicals.

$$d^2 = (x-4)^2 + \left(\frac{x^2}{2} - 1\right)^2$$

$$2d(d') = 2(x-4) + 2\left(\frac{x^2}{2} - 1\right)x$$

$$d' = 0 \Rightarrow 2x - 8 + x^3 - 2x = 0$$

$$x^3 - 8 = 0 \Rightarrow x = 2$$

$$y = \frac{2^2}{2} = 2$$

The point is $(2,2)$.

Inscribed Plane Figures

19. Area = xy

The altitude of the triangle is $h = \sqrt{10^2 - 5^2} = 8.66$

By similar triangles, $\frac{8.66}{10} = \frac{8.66 - x}{y} \Rightarrow y = 10 - 1.155x$

$A(x) = x(10 - 1.155x) = 10x - 1.155x^2$

$A'(x) = 10 - 2.31x$

$A''(x) = -2.31$

$A'(x) = 0 \Rightarrow 10 - 2.31x = 0 \Rightarrow x = 4.33$

$A''(4.33) < 0$ maximum at $x = 4.33$

$y = 10 - 1.155(4.33) = 5.00$

The dimensions of the rectangle are $5.00 \text{ in.} \times 4.33 \text{ in.}$

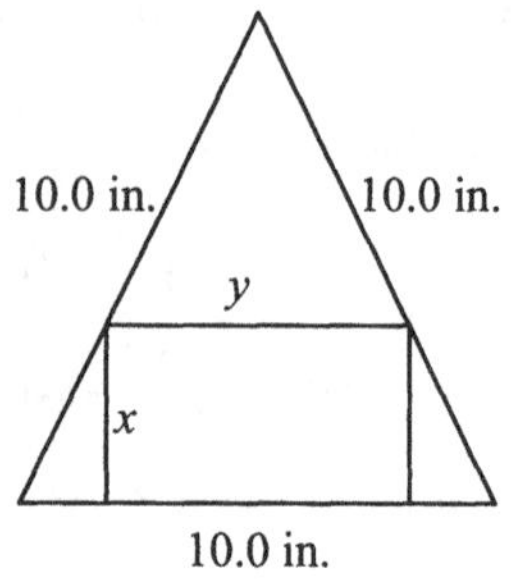

21. Area = wh

$\frac{x^2}{100} + \frac{y^2}{49} = 1 \Rightarrow y = \pm\sqrt{49\left(1 - \frac{x^2}{100}\right)} = \pm\frac{7}{10}\sqrt{100 - x^2}$

$h = 2\left(\frac{7}{10}\right)\sqrt{100 - \left(\frac{w}{2}\right)^2} = \frac{7}{10}\sqrt{400 - w^2}$

$A(w) = w\left(\frac{7}{10}\right)\left(400 - w^2\right)^{1/2}$

$A'(w) = \frac{7}{10}w\left(\frac{1}{2}\right)\left(400 - w^2\right)^{-1/2}(-2w) + \left(\frac{7}{10}\right)\left(400 - w^2\right)^{1/2}$

$= \frac{-\frac{7}{10}w^2 + \frac{7}{10}\left(400 - w^2\right)}{\sqrt{400 - w^2}} = \frac{-\frac{14}{10}w^2 + 280}{\sqrt{400 - w^2}}$

$A'(w) = 0 \Rightarrow -\frac{14}{10}w^2 + 280 = 0$

$w = \pm 14.14$ (exclude -14.14)

Graph $A(x)$ to verify maximum at $w = 14.14$

$h = \frac{7}{10}\sqrt{400 - (14.14)^2} = 9.90$

The dimensions of the rectangle are $9.90 \text{ units} \times 14.1 \text{ units}$.

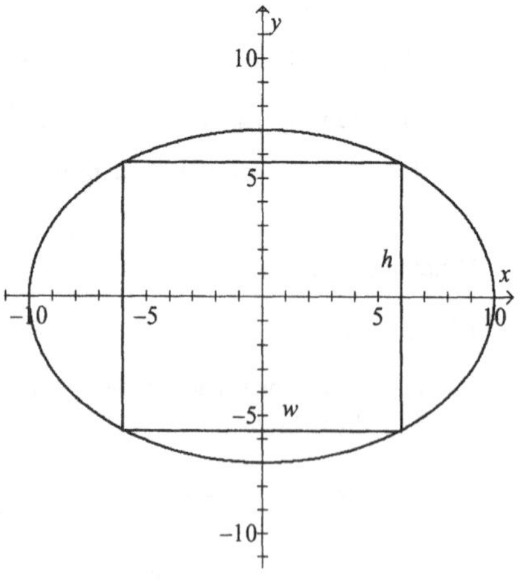

Inscribed Volumes

23. $h^2+(2r)^2=10.0^2 \;\Rightarrow\; r^2=\frac{1}{4}\left(100-h^2\right)$

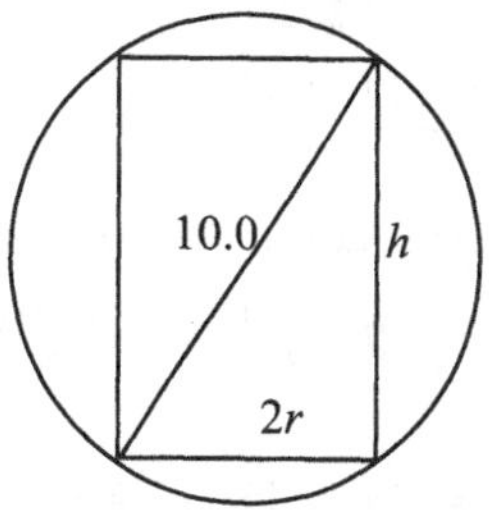

Volume $=\pi r^2 h$

$V(h)=\pi\left[\frac{1}{4}\left(100-h^2\right)\right]h=\frac{100\pi}{4}h-\frac{\pi}{4}h^3$

$V'(h)=\frac{100\pi}{4}-\frac{3\pi}{4}h^2$

$V''(h)=-\frac{3\pi}{2}h$

$V'(h)=0 \;\Rightarrow\; \frac{100\pi}{4}-\frac{3\pi}{4}h^2=0$

$h=\sqrt{\frac{100}{3}}\pm 5.77$ (exclude -5.77)

$V''(5.77)<0$ maximum at h = 5.77

$r=\sqrt{\frac{1}{4}\left(100-5.77^2\right)}=4.08$

The cylinder has radius of 4.08 cm and height of 5.77 cm.

25. $r^2=(9.00)^2-(h-9.00)^2=81-h^2+18h-81=18h-h^2$

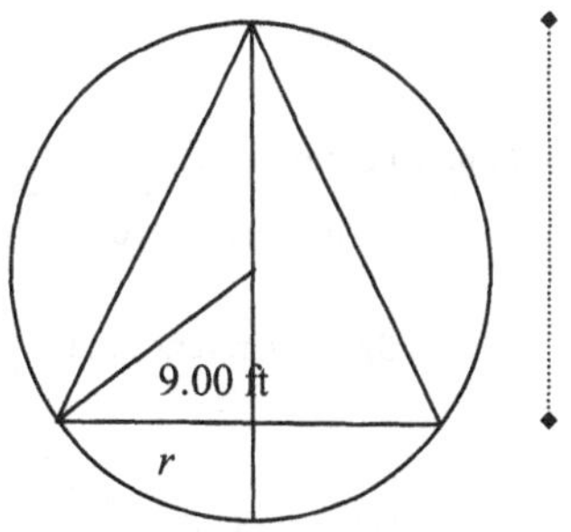

Volume $=\frac{1}{3}\pi r^2 h$

$V(h)=\frac{1}{3}\pi\left(18h-h^2\right)h=6\pi h^2-\frac{\pi}{3}h^3$

$V'(h)=12\pi h-\pi h^2$

$V''(h)=12\pi-2\pi h$

$V'(h)=0 \;\Rightarrow\; 12\pi h-\pi h^2=0$

$\pi h(12-h)=0$

$h=0$ or 12 (exclude 0)

$V''(h)<0$ maximum at $h=12$

The cone has a height of 12.0 ft.

Most Economical Dimensions of Containers

27. $V = \pi r^2 h + \left(\frac{1}{2}\right)\frac{4}{3}\pi r^3 = 755$

$$h = \frac{755 - (2/3)\pi r^3}{\pi r^2}$$

$$SA = \pi r^2 + 2\pi rh + 2\pi r^2 = 3\pi r^2 + 2\pi rh$$

$$S(r) = 3\pi r^2 + 2\pi r\left(\frac{755 - (2/3)\pi r^3}{\pi r^2}\right) = 3\pi r^2 + 1510r^{-1} - \frac{4}{3}\pi r^2 = 1.67\pi r^2 + 1510r^{-1}$$

$$S'(r) = 3.34\pi r - 1510r^{-2} = 3.34\pi r - \frac{1510}{r^2}$$

$$S''(r) = 3.34\pi + 3020r^{-3} = 3.34\pi + \frac{3020}{r^3}$$

$$S'(r) = 0 \Rightarrow 3.34\pi r - \frac{1510}{r^2} = 0$$

$$3.34\pi r^3 - 1510 = 0$$

$$r = \sqrt[3]{\frac{1510}{3.34\pi}} = 5.24$$

$S''(5.24) > 0$ minimum at $r = 5.24$

$$h = \frac{755 - (2/3)\pi(5.24)^3}{\pi(5.24)^2} = 5.26$$

The silo has a radius of 5.24 m and a total height of 5.24 + 5.26 = 10.50 m with the roof or 5.26 m without.

Minimum Travel Time

28. Shore distance $= \sqrt{10.0^2 - 6.00^2} = 8.00$ Rowing distance $= \sqrt{x^2 + (6.00)^2}$ Walking distance $= 8 - x$

At 3.00 mi/hr, rowing time $= \frac{\sqrt{x^2+36}}{3.00}$. At 4.00 mi/hr, walking time $= \frac{8-x}{4}$.

$$\text{Total time: } T(x) = \frac{\sqrt{x^2+36}}{3.00} + \frac{8-x}{4}$$

$$T'(x) = \frac{1}{3}\left(\frac{1}{2}\right)\left(x^2+36\right)^{-1/2}(2x) + \frac{1}{4}(-1) = \frac{x}{3\sqrt{x^2+36}} - \frac{1}{4}$$

$$T'(x) = 0 \Rightarrow \frac{x}{3\sqrt{x^2+36}} - \frac{1}{4} = 0$$

$$4x = 3\sqrt{x^2+36}$$

$$16x^2 = 9\left(x^2+36\right)$$

$$7x^2 = 324 \Rightarrow x = 6.80$$

Check the graph of $T(x)$ to verify minimum.

He should land at 6.80 mi.

Beam Problem

29. $x^2+y^2=18.0^2 \Rightarrow y^2=324-x^2$

$S=kxy^2$

$S(x)=kx\left(324-x^2\right)=324kx-kx^3 \Rightarrow S'(x)=324k-3kx^2 \Rightarrow S''(x)=-6x$

$S'(x)=0 \Rightarrow 324k-3kx^2=0$

$x=\sqrt{108}=10.4$

$S''(10.4)<0$ maximum at $r=2.643$

$y=\sqrt{324-(10.4)^2}=14.7$

The dimensions are $10.4 \text{ in.} \times 14.7 \text{ in.}$

Light

31. Illumination from M: $E_M=\dfrac{kI}{x^2}$ Illumination from N: $E_N=\dfrac{k\left(\frac{1}{3}I\right)}{(100-x)^2}$

$$E(x)=\frac{kI}{x^2}+\frac{k\left(\frac{1}{3}I\right)}{(100-x)^2}=kIx^{-2}+\frac{1}{3}kI(100-x)^{-2}$$

$$E'(x)=-2kIx^{-3}-\frac{2}{3}kI(100-x)^{-3}(-1)=\frac{-2kI}{x^3}+\frac{2kI}{3(100-x)^3}$$

$$E'(x)=0 \Rightarrow \frac{2kI}{x^3}=\frac{2kI}{3(100-x)^3}$$

$$x^3=3(100-x)^3$$

$$x=\sqrt[3]{3}(100-x)$$

$$x+x\sqrt[3]{3}=100\sqrt[3]{3}$$

$$x=\frac{100\sqrt[3]{3}}{1+\sqrt[3]{3}}=59.1$$

Graph $E(x)$ to verify minimum at $x=59.1$.

The illumination is at a minimum at a point 59.1 in. from M.

Electrical

33. $P(i)=30i-2i^2 \Rightarrow P'(i)=30-4i \Rightarrow P''(i)=-4$

$P'(i)=0 \Rightarrow 30-4i=0$

$i=7.5$

$P''(7.5)<0$ maximum at $i=7.5$

The current must be 7.5 A to deliver the maximum power.

35. $E(i) = \dfrac{115i - 25 - i^2}{115i} = 1 - \dfrac{25}{115}i^{-1} - \dfrac{1}{115}i$

$E'(i) = \dfrac{25}{115}i^{-2} - \dfrac{1}{115} = \dfrac{25}{115i^2} - \dfrac{1}{115}$ and $E''(i) = -\dfrac{50}{115}i^{-3} = -\dfrac{50}{115i^3}$

$E'(i) = 0 \Rightarrow \dfrac{25}{115i^2} = \dfrac{1}{115}$

$i^2 = 25 \Rightarrow i = 5$

$E''(5) < 0$ maximum at $i = 5$.

The efficiency is a maximum at a current of 5.0 A.

Mechanisms

37. $E(x) = \dfrac{x - \mu x^2}{x + \mu} \Rightarrow E'(x) = \dfrac{(x+\mu)(1-2\mu x) - (x - \mu x^2)}{(x+\mu)^2}$

$E'(x) = 0 \Rightarrow (x+\mu)(1-2\mu x) = (x - \mu x^2)$

$x - 2\mu x^2 + \mu - 2\mu^2 x = x - \mu x^2$

$\mu x^2 + 2\mu^2 x - \mu = 0$

$x^2 + 2\mu x - 1 = 0$

For $\mu = 0.45$, $x^2 + 0.9x - 1 = 0 \Rightarrow x = \dfrac{-0.9 \pm \sqrt{0.81+4}}{2} = 0.65$ (Exclude $x = -1.55$)

Graphical Solution

39. $V = \pi r^2 h = 10000 \Rightarrow h = \dfrac{10000}{\pi r^2}$

$\text{Cost} = 2(4.23)\pi r^2 + 3.81(2r\pi h) + 5.85(2(2r\pi) + h)$ [h is for the seam in the cylindrical siding of the can]

$C(r) = 8.46\pi r^2 + 7.62 r\pi\left(\dfrac{10000}{\pi r^2}\right) + 23.4 r\pi + 5.85\left(\dfrac{10000}{\pi r^2}\right) = 8.46\pi r^2 + \dfrac{76200}{r} + 23.4 r\pi + \dfrac{58500}{\pi r^2}$

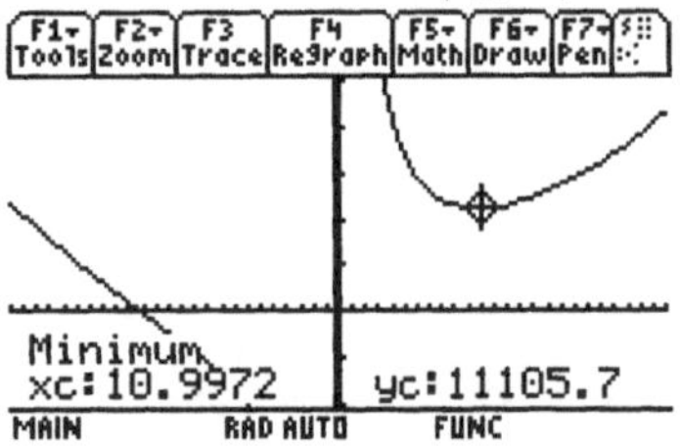

The minimum cost of \$11,106 will occur when $r = 11.0$ ft and $h = \dfrac{10000}{\pi(11.0)^2} = 26.3$ ft .

Chapter 23 • Review Problems

1. $\frac{dx}{dt}=160\text{ ft/s},\ \frac{dy}{dt}=120.0\text{ ft/s}$

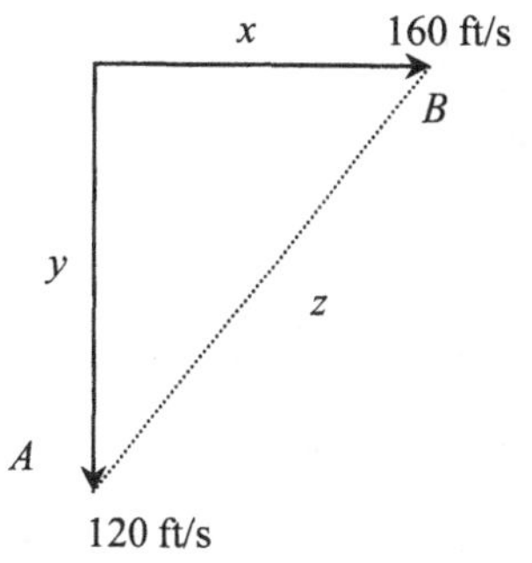

$$x=(160\text{ ft/s})(12\text{ min})\left(\frac{60\text{ s}}{1\text{ min}}\right)=115{,}200\text{ ft}$$

$$y=(120\text{ ft/s})(24\text{ min})\left(\frac{60\text{ s}}{1\text{ min}}\right)=172{,}800\text{ ft}$$

$$z^2=x^2+y^2 \Rightarrow z=\sqrt{115200^2+172800^2}=207{,}700\text{ ft}$$

$$2z\frac{dz}{dt}=2x\frac{dx}{dt}+2y\frac{dy}{dt}$$

$$207{,}700\frac{dz}{dt}=115{,}200(160)+172{,}800(120)$$

$$\frac{dz}{dt}=189\text{ ft/s}$$

3. $\frac{dx}{dt}=-5.0\text{ km/h}=-5000\text{ m/h};\ x=80\text{ m}$

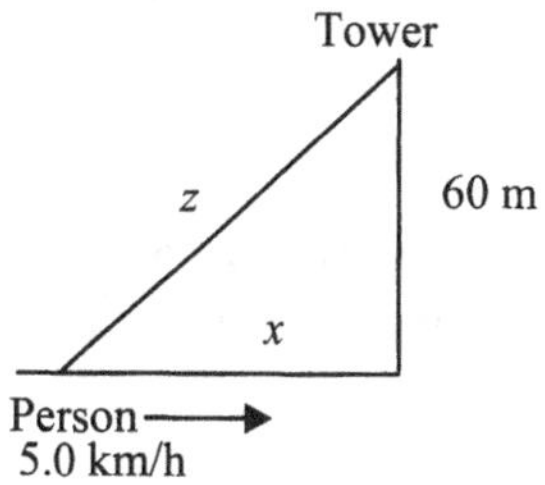

$$z^2=x^2+60^2 \Rightarrow z=\sqrt{80^2+60^2}=100\text{ m}$$

$$2z\frac{dz}{dt}=2x\frac{dx}{dt}$$

$$\frac{dz}{dt}=\frac{x\frac{dx}{dt}}{z}=\frac{80(-5000)}{100}=-4000\text{ m/h}=-4.0\text{ km/h}$$

5. $r=\frac{h}{\sqrt{3}};\ \frac{dV}{dt}=-5.00\text{ cm}^3/\text{min}$

$$V=\frac{1}{3}\pi r^2 h=\frac{1}{3}\pi\left(\frac{h}{\sqrt{3}}\right)^2 h=\frac{\pi}{9}h^3$$

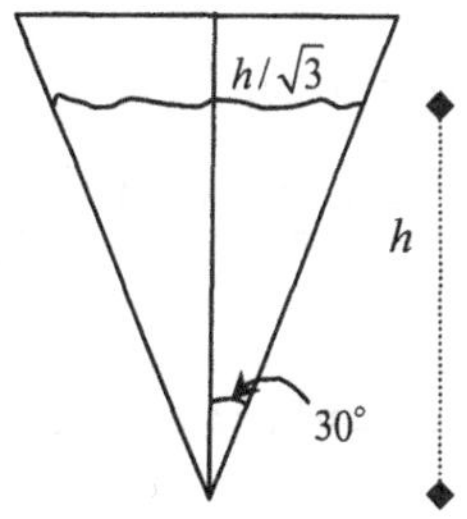

$$\frac{dV}{dt}=\frac{\pi}{3}h^2\frac{dh}{dt} \Rightarrow \frac{dh}{dt}=\frac{\frac{dV}{dt}}{\frac{\pi}{3}h^2}=\frac{-5.00}{\frac{\pi}{3}h^2}$$

$$A=\pi r\sqrt{r^2+h^2}=\pi\left(\frac{h}{\sqrt{3}}\right)\sqrt{\frac{h^2}{3}+h^2}=\pi\left(\frac{h}{\sqrt{3}}\right)h\sqrt{\frac{4}{3}}$$

$$=\pi\frac{h^2}{\sqrt{3}}\left(\frac{2}{\sqrt{3}}\right)=\frac{2}{3}\pi h^2$$

$$\frac{dA}{dt}=\frac{4}{3}\pi h\frac{dh}{dt}=\frac{4}{3}\pi h\left(\frac{-5.00}{\frac{\pi}{3}h^2}\right)=\frac{-20}{h}$$

$$\left.\frac{dA}{dt}\right|_{h=6.00\text{ cm}}=\frac{-20}{6.00}=-3.33\text{ cm}^2/\text{min}$$

The inner surface area of the tank is being exposed at a rate of $3.33\text{ cm}^2/\text{min}$.

7. $P = k\left(sv - v^2\right)$

$$\frac{dP}{dv} = ks - 2kv$$

$$\frac{dP}{dv} = 0 \quad \Rightarrow \quad ks = 2kv \quad \Rightarrow \quad v = \frac{s}{2}$$

9. $h^2 + (2r)^2 = 12^2 \;\Rightarrow\; r^2 = \frac{1}{4}\left(144 - h^2\right)$

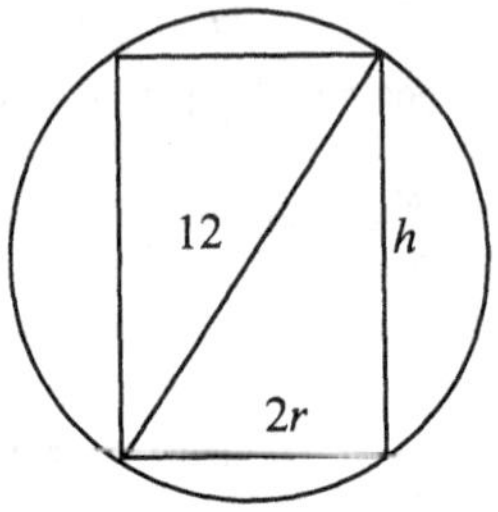

Volume $= \pi r^2 h$

$$V(h) = \pi\left[\frac{1}{4}\left(144 - h^2\right)\right]h = \frac{144\pi}{4}h - \frac{\pi}{4}h^3$$

$$V'(h) = \frac{144\pi}{4} - \frac{3\pi}{4}h^2$$

$$V''(h) = -\frac{3\pi}{2}h$$

$$V'(h) = 0 \;\Rightarrow\; \frac{144\pi}{4} - \frac{3\pi}{4}h^2 = 0$$

$$h = \sqrt{\frac{144}{3}} = 6.93$$

$V''(6.93) < 0$ maximum at $h = 6.93$

The cylinder has a height of 6.93.

11. $p = 40\text{ lb/in.}^2,\; v = \left(5.0\text{ ft}^3\right)\left(\frac{12\text{ in.}}{1\text{ ft}}\right)^3 = 8640\text{ in.}^3$

$$\frac{dv}{dt} = \left(0.20\text{ ft}^3/\text{s}\right)\left(\frac{12\text{ in.}}{1\text{ ft}}\right)^3 = 345.6\text{ in.}^3/\text{s}$$

$$pv^{1.41} = k \;\Rightarrow\; 1.41pv^{0.41}\frac{dv}{dt} + v^{1.41}\frac{dp}{dt} = 0$$

$$1.41(40)(8640)^{0.41}(345.6) + (8640)^{1.41}\frac{dp}{dt} = 0$$

$$\frac{dp}{dt} = \frac{-1.41(40)(8640)^{0.41}(345.6)}{(8640)^{1.41}} = -2.26\text{ lb/in.}^2 \text{ per second}$$

13. $s = -t^3 + 3t^2 + 24t + 28 \;\Rightarrow\; v = \frac{ds}{dt} = -3t^2 + 6t + 24$

Point at rest when $v = 0 \;\Rightarrow\; -3t^2 + 6t + 24 = 0$

$$t^2 - 2t - 8 = 0$$

$$(t-4)(t+2) = 0 \;\Rightarrow\; t = -2 \text{ or } 4$$

$t = -2 \;\Rightarrow\; s = 0$ and $t = 4 \;\Rightarrow\; s = 108$

15. Area = wh

$$A(x) = 2x\left(\frac{8}{x^2+4}\right) = \frac{16x}{x^2+4}$$

$$A'(x) = \frac{(x^2+4)16 - 16x(2x)}{(x^2+4)^2} = \frac{64-16x^2}{(x^2+4)^2}$$

$$A'(x) = 0 \Rightarrow 64 - 16x^2 = 0 \Rightarrow x = \pm 2 \text{ (exclude } -2)$$

$$A(2) = \frac{16(2)}{(2)^2+4} = 4$$

The largest rectangle has an area of 4 square units.

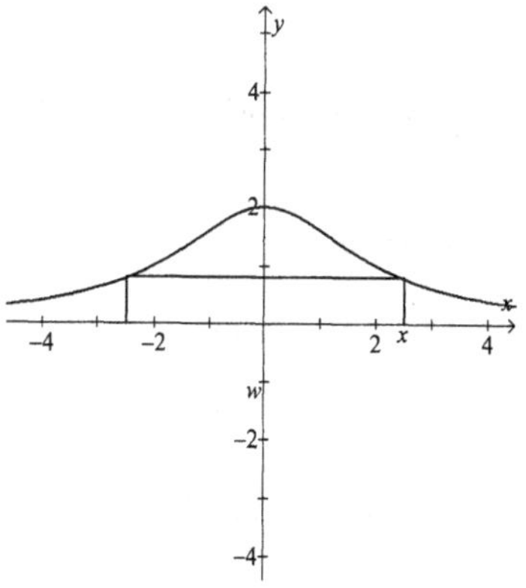

17. $x + y = 10 \Rightarrow y = 10 - x$

$$P = y^2x^3 \Rightarrow P(x) = (10-x)^2x^3 = 100x^3 - 20x^4 + x^5$$

$$P'(x) = 300x^2 - 80x^3 + 5x^4 \quad \text{and} \quad P''(x) = 600x - 240x^2 + 20x^3$$

$$P'(x) = 0 \Rightarrow 300x^2 - 80x^3 + 5x^4 = 0$$

$$5x^2(x^2 - 16x + 60) = 0$$

$$5x^2(x-6)(x-10) = 0$$

$$x = 0, 6, \text{ or } 10$$

$P''(6) < 0$ maximum at $x = 6$

$x = 6 \Rightarrow y = 10 - 6 = 4.$ The two numbers are 4 and 6.

19. $V = x^2y$

$$A = x^2 + 4xy = 300 \Rightarrow y = \frac{300-x^2}{4x}$$

$$V(x) = x^2\left(\frac{300-x^2}{4x}\right) = 75x - \frac{1}{4}x^3 \Rightarrow V'(x) = 75 - \frac{3}{4}x^2 \Rightarrow V''(x) = -\frac{3}{2}x$$

$$V'(x) = 0 \Rightarrow 75 = \frac{3}{4}x^2 \Rightarrow x = 10$$

$V''(10) = -15 < 0$ maximum at $x = 10$

$y = \frac{300-100}{4(10)} = 5$ The dimensions are 10 in. × 10 in. × 5 in.

21. $\theta = 18.5t^2 + 12.8t + 14.8$

a) $\omega = \frac{d\theta}{dt} = 37t + 12.8 \Rightarrow \omega(3.50) = 37(3.50) + 12.8 = 142$ rad/s

b) $\alpha = \frac{d\omega}{dt} = 37 \Rightarrow \alpha(3.50) = 37.0$ rad/s^2

23. $q = 2.84t^2 + 6.25t^3 \Rightarrow i = \frac{dq}{dt} = 5.68t + 18.75t^2$

$i(1.25) = 5.68(1.25) + 18.75(1.25)^2 = 36.4$ A

Chapter 26: Integration

Exercise 1 ◊ The Indefinite Integral

1. $\int dx = x + C$

3. $\int 6\,dx = 6\int dx = 6x + C$

5. $\int 5\,dx = 5\int dx = 5x + C$

7. $\int \pi\,dx = \pi\int dx = \pi x + C$

9. $\int x\,dx = \dfrac{x^{1+1}}{1+1} = \dfrac{x^2}{2} = \dfrac{1}{2}x^2 + C$

11. $\int x^4 dx = \dfrac{x^{4+1}}{4+1} = \dfrac{x^5}{5} = \dfrac{1}{5}x^5 + C$

13. $\int 5x\,dx = 5\int x\,dx = 5\left(\dfrac{x^{1+1}}{1+1}\right) = 5\left(\dfrac{x^2}{2}\right) = \dfrac{5}{2}x^2 + C$

15. $\int 8x^4 dx = 8\int x^4 dx = \dfrac{8}{5}x^5 + C$

17. $\int (x+4)\,dx = \int x\,dx + \int 4\,dx = \dfrac{x^2}{2} + 4x + C$

19. $\int (x^2 + 4x)\,dx = \int x^2 dx + 4\int x\,dx = \dfrac{x^3}{3} + 2x^2 + C$

21. $\int (3x^2 - 24x + 4)\,dx = 3\int x^2 dx - 24\int x\,dx + 4\int dx = x^3 - 12x^2 + 4x + C$

23. $\int x^{1/2} dx = \dfrac{x^{1/2+1}}{(1/2)+1} = \dfrac{x^{3/2}}{(3/2)} = \dfrac{2x^{3/2}}{3} + C$

25. $\int x^{4/3} dx = \dfrac{3x^{7/3}}{7} + C$

27. $\int \sqrt{5x}\,dx = \sqrt{5}\int x^{1/2} dx = \dfrac{2}{3}\sqrt{5}x^{3/2} + C$

29. $\int 9\sqrt[5]{2x}\,dx = \dfrac{15}{2^{4/5}}x^{6/5} + C$

31. $\int \dfrac{1}{x^2}dx = \int x^{-2} dx = \dfrac{x^{-2+1}}{-2+1} = \dfrac{x^{-1}}{-1} = -\dfrac{1}{x} + C$

33. $\int \dfrac{dx}{x^4} = \int x^{-4} dx = -\dfrac{1}{3x^3} + C$

35. $\int \dfrac{dx}{\sqrt{x}} = \int x^{-1/2} dx = \dfrac{x^{-1/2+1}}{-(1/2)+1} = \dfrac{x^{1/2}}{(1/2)} = 2x^{1/2} + C = 2\sqrt{x} + C$

Simplify and integrate

37. $\int \sqrt{x}(3x-2)\,dx = \int x^{1/2}(3x-2)\,dx = \int 3x^{3/2} - 2x^{1/2} dx = \dfrac{6}{5}x^{5/2} - \dfrac{4}{3}x^{3/2} + C$

38. $\int (x+1)^2 dx = \int (x^2 + 2x + 1)dx = \dfrac{x^3}{3} + x^2 + x + C$

39. $\int \frac{4x^2 - 2\sqrt{x}}{x} dx = 4\int \frac{x^2}{x} dx - 2\int \frac{x^{1/2}}{x} dx = 2x^2 - 4\sqrt{x} + C$

40. $\int (x+2)(x-3) dx = \int (x^2 - x - 6) dx = \frac{x^3}{3} - \frac{x^2}{2} - 6x + C$

41. $\int (1-x)^3 dx = \int (1 - 3x + 3x^2 - x^3) dx = x - \frac{3x^2}{2} + x^3 - \frac{x^4}{4} + C = -\frac{x^4}{4} + x^3 - \frac{3x^2}{2} + x + C$

42. $\int \frac{x^3 + 2x^2 - 3x - 6}{x+2} dx = \int \frac{(x+2)(x^2-3)}{x+2} dx = \frac{x^3}{3} - 3x + C$

Exercise 2 ◊ Rules for Finding Integrals
Integral of a Power Function

1. Let $u = x^4 + 1$; $du = 4x^3 dx$

$\int (x^4+1)^3 4x^3 dx \Rightarrow \int u^3 du = \frac{u^4}{4} = \frac{1}{4}(x^4+1)^4 + C$

3. Let $u = x^3 + 1$; $du = 3x^2 dx$

$\int 9(x^3+1)^2 x^2 dx \Rightarrow 3\int u^2 du = u^3 = (x^3+1)^3 + C$

5. Let $u = x^2 + 2x$; $du = (2x+2)dx$

$\int 2(x^2+2x)^3 (2x+2) dx \Rightarrow 2\int u^3 du = \frac{2u^4}{4} = \frac{1}{2}(x^2+2x)^4 + C$

7. Let $u = 1 - x$; $du = dx$

$\int \frac{-dx}{(1-x)^2} = \int (1-x)^{-2}(-dx) = -\int u^{-2} du = u^{-1} = \frac{1}{1-x} + C$

9. Let $u = x^2 - 2$; $du = 2x\,dx$

$\int x\sqrt{x^2-2}dx = \int (x^2-2)^{1/2} x\,dx \Rightarrow \frac{1}{2}\int u^{1/2} du = \frac{1}{2}\left(\frac{u^3}{3/2}\right) = \frac{1}{3}(x^2-2)^{3/2} + C$

11. Let $u = 1 - x^3$; $du = -3x^2 dx$

$\int \frac{x^2 dx}{\sqrt{1-x^3}} = -\frac{1}{3}\int (1-x^3)^{-1/2}(-3x^2) dx = -\frac{1}{3}\int u^{-1/2} du = -\frac{1}{3}\left(\frac{u^{1/2}}{1/2}\right) = -\frac{2}{3}\sqrt{1-x^3} + C$

Other Variables

13. $\int 5t^4 dt = 5\int t^4 dt = 5\left(\frac{t^{4+1}}{4+1}\right) = 5\left(\frac{t^5}{5}\right) = t^5 + C$

15. Let $u = z^3 + 3z$; $du = (3z^2+3)dz$

$\int (z^3+3z)^3 (z^2+1) dz \Rightarrow \frac{1}{3}\int u^3 du = \frac{1}{3}\left(\frac{u^4}{4}\right) = \frac{(z^3+3z)^4}{12} + C$

17. $\int \frac{t^2+9}{t} dt = \int t\,dt + \int \frac{9}{t} dt = \frac{t^2}{2} + 9\ln|t| + C$

Integral of *du/u*

19. $\int \frac{3}{x} dx = 3\ln|x| + C$

21. Let $u = z^3 - 3$; $du = 3z^2 dz$

$$\int \frac{5z^2}{z^3-3}dz \Rightarrow \frac{5}{3}\int \frac{du}{u} = \frac{5}{3}\ln\left|z^3-3\right| + C$$

$$\int \frac{tdt}{6-t^2} \Rightarrow -\frac{1}{2}\int \frac{du}{u} = -\frac{1}{2}\ln\left|6-t^2\right| + C$$

23. $\int \frac{x+1}{x}dx = \int dx + \int \frac{1}{x}dx = x + \ln|x| + C$

Miscellaneous Rules from a Table of Integrals

25. Rule 57: $\int \frac{du}{u^2-a^2} = \frac{1}{2a}\ln\left|\frac{u-a}{u+a}\right| + C$; Let $u = 3t$ and $a = 2$; $du = 3dt$

$$\int \frac{dt}{4-9t^2} = -\int \frac{dt}{9t^2-4} = -\frac{1}{3}\int \frac{du}{u^2-a^2} \Rightarrow -\frac{1}{12}\ln\left|\frac{3t-2}{3t+2}\right| + C$$

27. Rule 69: $\int \sqrt{a^2-u^2}\,du = \frac{u}{2}\sqrt{a^2-u^2} + \frac{a^2}{2}\sin^{-1}\frac{u}{a} + C$; Let $u = 3x$; $du = 3dx$; $a = 5$

$$\int \sqrt{25-9x^2}\,dx \Rightarrow \frac{x}{2}\sqrt{25-9x^2} + \frac{25}{6}\sin^{-1}\frac{3x}{5} + C$$

29. Rule 56: $\int \frac{du}{a^2+b^2u^2} = \frac{1}{ab}\tan^{-1}\frac{bu}{a} + C$; Let $u = x$; $du = dx$; $a = 3$, $b = 1$

$$\int \frac{dx}{x^2+9} \Rightarrow \frac{1}{3}\tan^{-1}\frac{x}{3} + C$$

31. Rule 56: $\int \frac{du}{u(a+bu)} = \frac{-1}{a}\ln\left|\frac{a+bu}{u}\right| + C$

$$\int \frac{dx}{16x^2+9} \Rightarrow \frac{1}{12}\tan^{-1}\frac{4x}{3} + C$$

33. Rule57: $\int \frac{du}{u^2-a^2} = \frac{1}{2a}\ln\left|\frac{u-a}{u+a}\right|$; Let $u = 2x$, $a = 3$, $du = 2\,dx$

$$\int \frac{dx}{9-4x^2} \Rightarrow -\frac{1}{12}\ln\left|\frac{2x-3}{2x+3}\right| + C$$

35. Rule57: $\int \frac{du}{u^2-a^2} = \frac{1}{2a}\ln\left|\frac{u-a}{u+a}\right|$; Let $u = x, a = 2$

$$\int \frac{dx}{x^2-4} \Rightarrow \frac{1}{4}\ln\left|\frac{x-2}{x+2}\right| + C$$

37. Rule66: $\int \sqrt{u^2 \pm a^2}\,du = \frac{u}{2}\sqrt{u^2 \pm a^2} \pm \frac{a^2}{2}\ln\left|u + \sqrt{u^2 \pm a^2}\right| + C$; Let $u = x$, $a = 2$

$$\int \sqrt{\frac{x^2}{4}-1}\,dx = \frac{1}{2}\int \sqrt{x^2-4}dx = \frac{x}{4}\sqrt{x^2-4} - \ln\left|x + \sqrt{x^2-4}\right| + C$$

39. Rule61: $\int \frac{du}{\sqrt{a^2-u^2}} = \sin^{-1}\frac{u}{a} + C$; Let $a = 1$, $u = x^2$, $du = 2x\,dx$

$$\int \frac{5xdx}{\sqrt{1-x^4}} \Rightarrow \frac{5}{2}\sin^{-1}x^2 + C$$

Exercise 3 ◊ Simple Differential Equations
Simple First –Order Differential Equations

1. $\frac{dy}{dx}=4x^2 \Rightarrow dy=4x^2dx \Rightarrow \int dy=\int 4x^2dx \Rightarrow y=\frac{4x^3}{3}+C$

3. $\frac{dy}{dx}=x^{-3} \Rightarrow dy=x^{-3}dx \Rightarrow \int dy=\int x^{-3}dx \Rightarrow y=-\frac{1}{2x^2}+C$

5. $\frac{ds}{dt}=\frac{1}{2}t^{-2/3} \Rightarrow ds=\frac{1}{2}t^{-2/3}dt \Rightarrow \int ds=\frac{1}{2}\int t^{-2/3}dt \Rightarrow s=\frac{1}{2}\left(\frac{t^{1/3}}{1/3}\right)+C=\frac{3}{2}\sqrt[3]{t}+C$

Finding the Constant of Integration

7. $y'=3x \Rightarrow \int dy=3\int x\,dx \Rightarrow y=\frac{3x^2}{2}+C$; at $(2,6)$, $y=6=\frac{12}{2}+C \therefore C=0$ and $y=\frac{3x^2}{2}$

9. $y'=\sqrt{x} \Rightarrow \int dy=\int x^{1/2}dx \Rightarrow y=\frac{x^{3/2}}{3/2}+C$; at $(2,4)$, $y=4=\frac{2^{3/2}}{3/2}+C \therefore C=2.12 \Rightarrow 2x^{3/2}-3y+6.34=0$

11. $\frac{dy}{dx}=\sqrt{2x} \Rightarrow dy=(2x)^{1/2}dx \Rightarrow \int dy=\frac{1}{2}\int(2x)^{1/2}2dx \Rightarrow y=\frac{\frac{1}{2}(2x)^{3/2}}{(3/2)}+C=\frac{1}{3}(2x)^{3/2}+C$

at $\left(\frac{1}{2},\frac{1}{3}\right)$, $C=0$ and $\frac{3}{2}y=\frac{1}{2}(2x)^{3/2}$; when $x=2$, $y=\frac{8}{3}$

Simple Second-Order Differential Equations

13. $y''=4 \Rightarrow dy'=4dx \Rightarrow \int dy'=4\int dx \Rightarrow y'=4x+C$

Slope of 3 at $(2,6) \Rightarrow y'=\frac{dy}{dx}=3=8+C \therefore C=-5$ and $dy=(4x-5)dx$

$y'=\frac{dy}{dx}=4x-5 \Rightarrow \int dy=\int(4x-5)dx \Rightarrow y=2x^2-5x+C$

at $(2,6)$, $6=8-10+C \therefore C=8$ and $y=2x^2-5x+8$

15. $y''=12x^2-6 \Rightarrow dy'=(12x^2-6)dx \Rightarrow \int dy'=\int(12x^2-6)dx \Rightarrow y'=\frac{12x^3}{3}-6x+C$

Slope of 20 at $(2,4) \Rightarrow y'=20=32-12+C \therefore C=0$ and $dy=(4x^3-6x)dx$

$y'=\frac{dy}{dx}=4x^3-6x \Rightarrow \int dy=\int(4x^3-6x)dx \Rightarrow y=x^4-3x^2+C$; at $(2,4)$, $4=16-12+C$

$\therefore C=0$ and $y=x^4-3x^2$

17. $\frac{dv}{dt}=21.5 \Rightarrow \int dv=\int 21.5dt \Rightarrow v=21.5t+C$

at $t=0$, $v=27.6\text{m/sec}$, so $C=27.6$ and $v=21.5t+27.6\text{m/sec}$

$v=\frac{ds}{dt}=\int ds=\int vdt=\int(21.5t+27.6)dt=10.75t^2+27.6t+K$

at $t=o$, $s=44.3$ so $K=44.3$ and $s=10.75t^2+27.6t+44.3$

Exercise 4 ◊ The Definite Integral

1. $\int_1^2 x\,dx=\frac{1}{2}x^2\Big|_1^2=2-\frac{1}{2}=\frac{3}{2}=1.50$

3. $\int_1^3 7x^2dx=\frac{7}{3}x^3\Big|_1^3=63-2.3=60.7$

5. $\int_0^4(x^2+2x)dx=\int_0^4 x^2dx+2\int_0^4 x\,dx=\frac{1}{3}x^3\Big|_0^4+x^2\Big|_0^4=21.3-0+16-0=37.3$

7. Let $u = x + 3$; $du = dx$

$$\int_2^4 (x+3)^2 dx \Rightarrow \int_5^7 u^2 du \Rightarrow \frac{u^3}{3}\Bigg|_5^7 \Rightarrow \frac{(x+3)^3}{3}\Bigg|_2^4 = 114.3 - 41.6 = 72.7$$

9. $\int_1^{10} \frac{dx}{x} = \ln|u|\Big|_1^{10} = \ln 10 - \ln 1 = 2.30 - 0 = 2.30$

11. Let $u = 4 + x^2$; $du = 2x\,dx$

$$\int_0^1 \frac{x\,dx}{\sqrt{4+x^2}} \Rightarrow \frac{1}{2}\int_4^5 u^{-1/2} du \Rightarrow \frac{(1/2)u^{1/2}}{(1/2)}\Bigg|_0^1 = \sqrt{4+x^2}\Big|_0^1 = \sqrt{5} - \sqrt{4} = 0.236$$

13. Let $u = 3 + 2x$; $du = -2dx$

$$\int_0^1 \frac{dx}{\sqrt{3-2x}} \Rightarrow -\frac{1}{2}\int_3^1 u^{-1/2} du \Rightarrow -\frac{(1/2)u^{1/2}}{(1/2)}\Bigg|_3^1 = -\sqrt{3-2x}\Big|_0^1 = -\left(\sqrt{1} - \sqrt{3}\right) = 0.732$$

An Application

15. $\int_0^{48.0} 12\pi x\,dx = 12\pi \int_0^{48.0} x\,dx = 6\pi x^2\Big|_0^{48.0} = 43{,}400 in^3$

Exercise 5 ◊ Approximate Area under a Curve

Estimation of Areas

1. $y = x^2 + 4$

x^*	1.5	2.5	3.5	Total
$f(x^*)\Delta x$	6.25	10.25	16.25	32.75

3. $y = 2 - x^2$

x^*	−0.25	0.25	0.75	Total
$f(x^*)\Delta x$	0.969	0.969	0.719	2.66

5. $y = (x-2)^2$

x^*	0.5	1.5	2.5	3.5	Total
$f(x^*)\Delta x$	2.25	0.25	0.25	2.25	5.33

7. $y = x^2 + 1$ from 0 to 8

x^*	1	3	5	7	Total
$f(x^*)\Delta x$	4	20	52	100	179

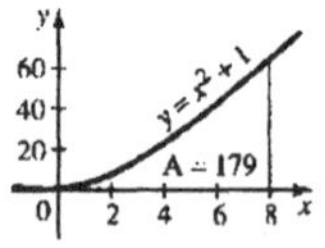

9. $y = \frac{1}{x}$ from 2 to 10

x^*	3	5	7	9	Total
$f(x^*)\Delta x$	0.667	0.40	0.286	0.222	1.58

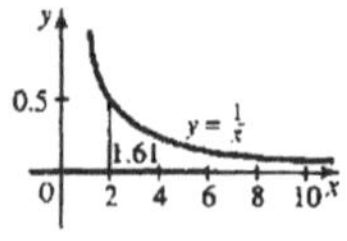

Sigma Notation

11. $\sum_{n=1}^{5} n = 1 + 2 + 3 + 4 + 5 = 15$

13. $\sum_{n=1}^{7} 3n = 3 + 6 + 9 + 12 + 15 + 18 + 21 = 84$

15. $\sum_{n=1}^{5} n(n-1) = 0+2+6+12+20 = 40$

Approximate Areas by Midpoint Method

17. Find the area under $y = x^2 + 1$ from $x = 0$ to 8

x^*	1	3	5	7	Total
$f(x^*)\Delta x$	4	20	52	100	176

19. Find the area under $y = \dfrac{1}{x}$ from $x = 2$ to 10

x^*	3	5	7	9	Total
$f(x^*)\Delta x$	0.66	0.40	0.29	0.22	1.56

Exercise 6 ◊ Exact Area under a Curve

1. $y = 2x$ from $x = 0$ to 10

$$A = \int_0^{10} 2x\,dx = x^2\Big|_0^{10} = 100 - 0 = 100$$

3. $y = 3 + x^2$ from $x = -5$ to 5

$$A = \int_{-5}^{5} (3 + x^2)\,dx = 3x\Big|_{-5}^{5} + \frac{x^3}{3}\Big|_{-5}^{5} = 15 + 15 + 41\frac{2}{3} + 41\frac{2}{3} = 113\frac{1}{3}$$

5. $y = x^3$ from $x = 0$ to 4

$$A = \int_0^4 x^3\,dx = \frac{x^4}{4}\Big|_0^4 = 64 - 0 = 64$$

7. $y = \dfrac{1}{\sqrt{x}}$ from $x = \dfrac{1}{2}$ to 8

$$A = \int_{1/2}^{8} x^{-1/2}\,dx = 2x^{1/2}\Big|_{1/2}^{8} = 5.657 - 1.414 = 4.24$$

9. $y = x^2 + x + 1$ from $x = 2$ to 3

$$A = \int_2^3 (x^2 + x + 1)\,dx = \frac{x^3}{3}\Big|_2^3 + \frac{x^2}{2}\Big|_2^3 + x\Big|_2^3 = \left(\frac{27}{3} - \frac{8}{3}\right) + \left(\frac{9}{2} - \frac{4}{2}\right) + (3 - 2) = 9.83$$

11. $y = 2x + \dfrac{1}{x^2}$ from $x = 1$ to 4

$$A = \int_1^4 \left(2x + \frac{1}{x^2}\right) dx = x^2\Big|_1^4 - \frac{1}{x}\Big|_1^4 = 16 - 1 - \frac{1}{4} + 1 = 15.75$$

An Application

13. $A = 2\int_{-30}^{23} \left(6.5 - \dfrac{13x^2}{1800}\right) dx = 2\left[6.5x\Big|_{-30}^{23} - \dfrac{13x^3}{5400}\Big|_{-30}^{23}\right] = 2\left[(149.5 + 195) - (29 + 65)\right] = 500 \text{ ft}^2$

CHAPTER 26 REVIEW PROBLEMS

Perform each integration

1. $\int \dfrac{dx}{\sqrt[3]{x}} = \int x^{-1/3}\,dx = \dfrac{3x^{2/3}}{2} + C$

3. $\int 3.1y^2\,dy = 1.03y^3 + C$

5. Let $u = 4x$; $du = 4\,dx$

$$\int\sqrt{4x}dx \Rightarrow \frac{1}{4}\int u^{1/2}du = \frac{1}{4}\left(\frac{u^{3/2}}{3/2}\right) = \frac{4x\sqrt{x}}{3} + C$$

7. Let $u = x^4 - 2x^3$; $du = \left(4x^3 - 6x^2\right)dx$

$$\int\left(x^4 - 2x^3\right)\left(2x^3 - 3x^2\right)dx \Rightarrow \frac{1}{2}\int u\,du = \frac{1}{2}\left(\frac{u^2}{2}\right) = \frac{1}{4}\left(x^4 - 2x^3\right)^2 + C$$

9. Let $u = x + 5$; $du = dx$

$$\int\frac{dx}{x+5} \Rightarrow \int\frac{du}{u} = \ln|x+5| + C$$

11. $\int_2^7\left(x^2 - 2x + 3\right)dx = \frac{x^3}{3} - x^2 + 3x\Big|_2^7 = \frac{343}{3} - \frac{8}{3} - (49-4) + 3(7-2) = 81\frac{2}{3}$

13. $\int_0^a\left(\sqrt{a} - \sqrt{x}\right)^2 dx = \int_0^a\left(a - 2\sqrt{a}\sqrt{x} + x\right)dx = ax - \frac{4\sqrt{a}}{3}x^{3/2} + \frac{x^2}{2}\Big|_0^a = a^2 - \frac{4a^2}{3} + \frac{a^2}{2} = \frac{a^2}{6}$

15. Rule 66; $u = 5x$, $du = 5dx$, $a = 3$

$$\int_0^3\sqrt{9+25x^2}\,dx = \frac{1}{5}\left|\frac{5x}{2}\sqrt{25x^2+9} + \frac{9}{2}\ln\left|5x + \sqrt{25x^2+9}\right|\right|_0^3 = \frac{1}{5}\left(\frac{15}{2}(234)^{1/2} + \frac{9}{2}\ln\left|15 + (234)^{1/2}\right|\right) - \frac{1}{5}\left(\frac{9}{2}\ln|3|\right) = 25.03$$

17. Rule 52; $u = x$, $a = 1$, $b = 5$

$$\int_1^4 x\sqrt{1+5x}\,dx = \frac{2(15x-2)}{15(5)^2}(1+5x)^{3/2}\Big|_1^4 = 28.75$$

19. Rule 57; $u = 2x$, $a = 3$, $du = 2dx$

$$\int\frac{dx}{9-4x^2} = -\int\frac{dx}{4x^2-9} = -\frac{1}{2}\left(\frac{1}{6}\ln\left|\frac{2x-3}{2x+3}\right|\right) + C = -\frac{1}{12}\ln\left|\frac{2x-3}{2x+3}\right| + C$$

21. Rule 56; $u = x$, $a = 3$, $b = 2$

$$\int\frac{dx}{9+4x^2} = \frac{1}{6}\tan^{-1}\left(\frac{2x}{3}\right) + C$$

23. Rule 62; $u = 3x$, $du = 3dx$, $a = 2$

$$\int_2^6\frac{dx}{\sqrt{4+9x^2}} = \frac{1}{3}\left|\ln\left|3x + \sqrt{9x^2+4}\right|\right|_2^6 = 0.3583$$

25. $\frac{dN}{dt} = 0.5N \Rightarrow \frac{dN}{0.5N} = dt \Rightarrow \int\frac{dN}{0.5N} = \int dt; 2\ln N = t + C \Rightarrow \ln N = \frac{t}{2} + \frac{C}{2}$

$N = e^{(t/2)+(C/2)} = e^{t/2}e^{C/2} = ke^{t/2}$; since $N = 100$ when $t = 0$, $100 = ke^0 = k \therefore N = 100e^{t/2}$

27. $\sum_{d=1}^{4}\frac{d^2}{d+1} = \frac{1}{2} + \frac{4}{3} + \frac{9}{4} + \frac{16}{5} = 7.283$

29. $A = \sqrt{8}\int_0^2 x^{1/2}dx = \sqrt{8}\left(\frac{x^{3/2}}{3/2}\right)\Big|_0^2 = \frac{16}{3}$

Chapter 27: Applications of the Integral

Exercise 1 ◊ Applications to Motion
Displacement

1. $v = \frac{ds}{dt} = 11.6t + 21.4 \text{ cm/s} \Rightarrow ds = (11.6t + 21.4)dt$

$\int ds = \int (11.6t + 21.4)dt \therefore s = 5.80t^2 + 21.4t + C$

Since the initial displacement (at $t = 0$) is 12.6 cm, $C = 12.6$ cm and $s = 5.80t^2 + 21.4t + 12.6$ cm
At $t = 7.00$s, displacement is 447 cm

3. $v = \frac{ds}{dt} = \frac{1}{8}t^2 \text{ ft/s} \Rightarrow ds = \frac{1}{8}t^2 dt \Rightarrow \int ds = \frac{1}{8}\int t^2 dt$

Velocity

5. $a = \frac{dv}{dt} = 1.41t^2 + 5.28 \text{ ft/s}^2 \Rightarrow \int dv = \int (1.41t^2 + 5.28)dt$

$v = \frac{1.41t^3}{3} + 5.28t + C$; At $t = 0$, $v = 2.58$ ft/s $\therefore C = 2.58$

$v = 0.470t^3 + 5.28t + 2.58$; At $t = 1.00s$, $v = 8.33$ ft/s

7. $a = t^3 - 25.8 \text{ cm/s}^2$

$v = \int (t^3 - 25.8)dt = \frac{1}{4}t^4 - 25.8t + C$; At $t = 0$, $v = 15.8\text{cm/s} \therefore C = 15.8\text{cm/s}$

$v = \frac{1}{4}t^4 - 25.8t + 15.8$; At $t = 5.00s$, $v = 156.25 - 129 + 15.8 = 43.1\text{cm}$

Freely Falling Body

9. (a) $a = 32.2 \text{ ft/s}^2\text{t}$

(b) $v = 1.77 + 32.2t^2$ ft/s

(c) $s = 0 + 1.77t + 16.1t^2$ ft

(d) $a = 32.2 \text{ ft/s}^2$; $v = 291.6$ ft/s; $s = 150$ ft

11. $a = -32 \text{ ft/s}^2 = \frac{dv}{dt} \Rightarrow v = \int -32\,dt = -32t + v_0 \therefore v_0 = 20$

$v = \frac{ds}{dt} \Rightarrow ds = -32\int (t + 20)dt \Rightarrow s = -16t^2 + 20t + C$

At $t = 0$, $C = 0 \therefore s = 20t - 16t^2$

Motion along a Curve

13. $v_x = \int a_x\,dt = 5.00\int t^2 dt = \frac{5.00}{3}t^3 + C; C = 0 \Rightarrow v_x = \frac{5.00}{3}t^3$

$v_y = \int a_y\,dt = 2.00\int t\,dt = t^2 + C$; $C = 0 \Rightarrow v_y = t^2$

At $t = 10$, $v_x = 1670$ cm/s, $v_y = 100$ cm/s

15. $a_x = 3t$; $a_y = 2t^2$; initial location (9,1)

$v_x = \int a_x dt = 3\int t\,dt = \frac{3}{2}t^2 + C$; $C = 6.00 \text{ cm/s} \therefore v_x = \frac{3}{2}t^2 + 6$

$v_y = \int a_y dt = 2\int t^2 dt = \frac{2}{3}t^3 + C$; $C = 2.00 \text{ cm/s} \therefore v_y = \frac{2}{3}t^3 + 2$

At $t = 15$, $v_x = \frac{3}{2}(15)^2 + 6 = 344$ cm; $v_y = \frac{2}{3}(15)^3 + 2 = 2250$ cm

Rotation

17. $\omega = \int \alpha dt = 8.5\int t^2 dt = \frac{8.5}{3}t^3 + C$ but $C = 0$

$\theta = \int \omega dt = \frac{8.5}{3}\int t^3 dt = \frac{8.5}{3}\left(\frac{t^4}{4}\right) + C$ but $C = 0$

At $t = 20$, angular displacement $\theta = \frac{8.5\times(20)^4}{12} = 113300 \text{ rad} \Rightarrow \frac{113300}{2\pi} = 18{,}000$ revolutions

19. $\omega = \int \alpha\, dt = 7.24\int dt = 7.24t + C \Rightarrow 1.25 = (7.24)(0) + C \therefore C = 1.25$

At $t = 2$, angular velocity $\omega = 7.24(2) + 1.25 = 15.7$ rad/s

Exercise 2 ◊ Applications to Electric Circuits

Charge

1. $q = \int i\,dt = \int (2t+3)dt = t^2 + 3t + k;\ k = 8.13$

At $t = 1.00\text{s},\ q = (1.00)^2 + (3)(1.00) + 8.13 = 12.1\text{C}$

3. $q = \int i\,dt = \int (3.25 + t^3)dt = 3.25t + \frac{t^4}{4} + k;\ k = 16.8$

At $t = 3.75\text{s},\ q = (3.75)(3.25) + \frac{(3.75)^4}{4} + 16.8 = 78.4$ C

Voltage Across a Capacitor

5. Let $u = 5 + t^2;\ du = 2t\,dt$

$$v = \frac{1}{C}\int i\,dt = \frac{1}{15.2}\int t\sqrt{5+t^2}\,dt \Rightarrow \frac{1}{15.2}\left(\frac{1}{2}\right)\int u^{1/2}du = \frac{1}{15.2}\left(\frac{1}{2}\right)\left(\frac{u^{3/2}}{3/2}\right) + k;$$

$$v = \frac{1}{15.2}\frac{\left(5+t^2\right)^{3/2}}{3} + k;\ k = 2.00 - \frac{(5)^{3/2}}{45.6} = 2.00 - 0.245 = 1.755$$

At $t = 1.755\text{s},\ v = \frac{\left(5 + (1.755)^2\right)^{3/2}}{45.6} + 1.755 = 2.26$ V

Current in an Inductor

7. $i = \frac{1}{L}\int v\,dt = \frac{1}{1.05}\int \sqrt{23t}\,dt = \frac{1}{1.05}\int (23t)^{1/2}dt = \frac{1}{1.05}\left(\frac{1}{23}\right)\left(\frac{(23t)^{3/2}}{3/2}\right) + k;\ k = 0$

At $t = 1.25\text{s},\ i = \frac{2(23\times1.25)^{3/2}}{(23)(3.15)} = 4.25$ A

9. $i = \frac{1}{L}\int v\,dt = \frac{1}{15.0}\int \left(28.5 + \sqrt{6t}\right)dt = \frac{1}{15.0}\int 28.5\,dt + \frac{1}{15.0}\int (6t)^{1/2}dt \Rightarrow \frac{28.5t}{15.0} + \frac{1}{15}\left(\frac{1}{6}\right)\left(\frac{(6t)^{3/2}}{3/2}\right) + k$

At $t = 2.5\text{s},\ i = \frac{(28.5)(2.5)}{15.0} + \frac{2(6\times2.5)^{3/2}}{(6)45.0} + 15 = 20.2$ A

Exercise 3 ◊ Finding Areas by Integration

NOTE: Each of the functions in the problems of Exercise 3 should be sketched or graphed by calculator prior to solving them. This helps avoid problems or errors.

Area Bounded by the *x* Axis

1. $y = 3x^2 + 2x$ from $x = 1$ to 3

$$A=\int_1^3\left(3x^2+2x\right)dx=x^3+x^2\Big|_1^3=36-2=34$$

3. $y=3\sqrt{x}$ from $x=1$ to 5

$$A=3\int_1^5 x^{1/2}dx=\frac{3}{(3/2)}x^{3/2}\Big|_1^5=2x^{3/2}\Big|_1^5=22.4-2=20.4$$

5. $y^2=16-x$ from $x=0$ to 16; Let $u=16-x$ and $du=-dx$

$$A=\int_{16}^0 u^{1/2}du=\int_{16}^0\frac{2}{3}u^{3/2}du=\frac{2}{3}(16-x)^{3/2}\Big|_0^{16}=42\frac{2}{3}$$

7. Horizontal elements are better for this problem. Curve intersects the first quadrant axes at (2, 0) and (0, 1)
$x+y+y^2=2\Rightarrow x=2-y-y^2$

$$A=\int_0^1\left(2-y-y^2\right)dy=2y-\frac{y^2}{2}-\frac{y^3}{3}\Big|_0^1=1\frac{1}{6}$$

9. $10y=x^2-80\Rightarrow y=\dfrac{x^2}{10}-8$ from $x=1$ to 6

$$A=\int_1^6 0-\left(\frac{x^2}{10}-8\right)dx=8x-\frac{x^3}{30}\Big|_1^6=\left(48-\frac{216}{30}\right)-\left(8-\frac{1}{30}\right)=32.83$$

11. $y=x^3-4x^2+3x$ from $x=1$ to 3

$$A=\int_1^3 0-\left(x^3-4x^2+3x\right)dx=0-\left(\frac{x^4}{4}-\frac{4x^3}{3}+\frac{3x^2}{2}\right)\Big|_1^3=0-\left(\frac{81}{4}-\frac{108}{3}+\frac{27}{2}\right)-\left(\frac{1}{4}-\frac{4}{3}+\frac{3}{2}\right)=2\frac{2}{3}$$

Areas Bounded by the y Axis

13. $y=x^2+2$ from $y=3$ to $5\Rightarrow x^2=y-2\Rightarrow x=(y-2)^{1/2}$

$$dA=x\,dy;\ A=\int_3^5(y-2)^{1/2}dy=\frac{2(y-2)^{3/2}}{3}\Big|_3^5=3.464-\frac{2}{3}=2.797$$

15. $y^3=4x$ from $y=0$ to $4\Rightarrow x=\dfrac{y^3}{4}$

$$dA=x\,dy;\ A=\frac{1}{4}\int_0^4 y^3dy=\frac{1}{16}y^4\Big|_0^4=16$$

Area between Two Curves

17. Using the point-slope method, the equation of the chord is $y=-2+2x$
$f(x)=6+4x-x^2;\ g(x)=-2+2x;\ f(x)-g(x)=6+4x-x^2-(-2+2x)=8+2x-x^2$

$$A=\int_{-2}^4\left(8+2x-x^2\right)dx=8x+x^2-\frac{x^3}{3}\Big|_{-2}^4=\left(32+16-21\frac{1}{3}\right)-\left(-16+4+2\frac{2}{3}\right)=36$$

19. Using the point-slope method, the equation of the chord is $y=\dfrac{x+4}{3}$
$y^3=x^2\Rightarrow y=x^{2/3}$; the area is below the chord and above the curve which intersect at $x=-1$ and $x=8$

$$A=\int_{-1}^8\left(\frac{x+4}{3}-x^{2/3}\right)dx=\int_{-1}^8\left(-x^{2/3}+\frac{x}{3}+\frac{4}{3}\right)dx=-\frac{3}{5}x^{5/3}+\frac{x^2}{6}+\frac{4}{3}x\Big|_{-1}^8=2.70$$

Area Enclosed by Two Curves

21. Horizontal elements are better for this problem. Curves intersect at (1, –2) and (4, 4)

$$y^2 = 4x \Rightarrow x = \frac{y^2}{4};\ 2x - y = 4 \Rightarrow x = \frac{y+4}{2}$$

$$A = \int_{-2}^{4}\left(\frac{y+4}{2} - \frac{y^2}{4}\right)dy = \int_{-2}^{4}\left(-\frac{y^2}{4} + \frac{y}{2} + 2\right)dy = -\frac{y^3}{12} + \frac{y^2}{4} + 2y\Big|_{-2}^{4} = 9$$

23. $A = \int_0^2\left[(2x)^{1/2} - \frac{x^2}{2}\right]dx = 2^{1/2}\left(\frac{2}{3}x^{3/2}\right) - \frac{x^3}{6}\Big|_0^2 = \frac{4}{3}$

Areas of Geometric Figures

25. $A = \int_0^a a\,dx = ax\big|_0^a = a^2 - 0 = a^2$

27. $A = \int_0^b\left(-\frac{h}{b}x + h\right)dx = -\frac{hx^2}{2b} + hx\big|_0^b = -\frac{hb^2}{2b} + hb = \frac{1}{2}hb$

29. Use Rule 69 in Appendix C

$$A = 4\int_0^r \sqrt{r^2 - x^2}\,dx = 4\left[\frac{x}{2}\sqrt{r^2 - x^2} + \frac{r^2}{2}\sin^{-1}\frac{x}{r}\right]_0^r = 4\left[0 + \frac{r^2}{2}\sin^{-1}1\right] - \left[0 + \frac{r}{2}\sin^{-1}0\right] = 4\left[\frac{r^2}{2}\left(\frac{\pi}{2}\right) - 0\right] = \pi r^2$$

31. $\frac{x^2}{a^2} + \frac{y^2}{b^2} = 1 \Rightarrow y^2 = b^2\left(1 - \frac{x^2}{a^2}\right) = \frac{b^2\left(a^2 - x^2\right)}{a^2} \Rightarrow y = \pm\frac{b}{a}\sqrt{a^2 - x^2}$

$$A = 4\int_0^a \frac{b}{a}\sqrt{a^2 - x^2}\,dx = \frac{b}{a}\left(4\int_0^a \sqrt{a^2 - x^2}\right)dx = \frac{b}{a}\left((4)\frac{x}{2}\sqrt{a^2 - x^2} + \frac{a^2}{2}\sin^{-1}\frac{x}{a}\Big|_0^a\right)$$

$$A = \frac{b}{a}\left(\pi a^2\right) = \pi ab$$

Applications

33. $\frac{x^2}{9^2} + \frac{y^2}{4.5^2} = 1 \Rightarrow x^2 = 9^2\left(1 - \frac{y^2}{4.5^2}\right) = \frac{9^2}{4.5^2}\left(4.5^2 - y^2\right) = 2^2\left(4.5^2 - y^2\right) \Rightarrow x = \pm 2\sqrt{\left(4.5^2 - y^2\right)}$

$$A = 2\int_{-4.5}^{1.5} 2\sqrt{\left(4.5^2 - y^2\right)}\,dy = 4\left[\frac{y}{2}\sqrt{4.5^2 - y^2} + \frac{4.5^2}{2}\sin^{-1}\frac{y}{4.5}\right]_{-4.5}^{1.5} = 90.1\text{ ft}^2$$

35. Volume = cross sectional area × height

$$\frac{x^2}{16^2} + \frac{y^2}{9^2} = 1$$

$$A = 4\int_0^{16} 9\sqrt{\left(1 - \frac{x^2}{16^2}\right)}\,dx = \frac{9}{4}\int_0^{16}\sqrt{16^2 - x^2}\,dx$$

$$A = \frac{9}{4}\left[\frac{x}{2}\sqrt{16^2 - x^2} + \frac{16^2}{2}\arcsin\frac{x}{16}\right]_0^{16} = \frac{9}{4}(16)(8)\left(\frac{\pi}{2}\right) = 9(16)\pi\text{ in}^2$$

$$V = 9(16)(\pi)(95) = 42{,}977\text{ in}^3 = 24.9\text{ ft}^3$$

37. $y = 6.5 - kx^2 \Rightarrow 0 = 6.5 - k(-30)^2 \Rightarrow k = \frac{13}{1800}$

$$A = 2\int_{-30}^{23}\left(6.5 - \frac{13x^2}{1800}\right)dx = 2\left(6.5 - \frac{13x^2}{1800}\right)\Bigg|_{-30}^{23}$$

$$A = 2\left[6.5(23) - \frac{(13)(23)^3}{5400} + 6.5(30) - \frac{(13)(30)^3}{5400}\right] = 500\text{ ft}^2$$

39. $y = 6 - px^2 \Rightarrow 0 = 6 - 36p \Rightarrow p = \frac{1}{6} \Rightarrow y_1 = 6 - \frac{1}{6}x^2$

$$x^2+(y-k)^2=12^2 \Rightarrow -k^2=12^2-6^2=108 \Rightarrow k=\pm\sqrt{108} \Rightarrow \left(y+\sqrt{108}\right)^2=12^2-x^2$$

$$y_2=-\sqrt{108}+\sqrt{144-x^2}$$

$$A=2\int_0^6 (y_1-y_2)dx=2\int_0^6\left[\left(6-\frac{1}{6}x^2\right)+\sqrt{108}-\sqrt{144-x^2}\right]dx$$

$$A=2\left[\left(6+\sqrt{108}\right)x-\frac{x^3}{18}\right]_0^6-2\int_0^6\sqrt{144-x^2}\,dx$$

$$\int_0^6\sqrt{144-x^2}\,dx=\left[\frac{x}{2}\sqrt{144-x^2}+\frac{144}{2}\sin^{-1}\frac{x}{12}\right]_0^6=68.876$$

$$A=2\left[\left(6+\sqrt{108}\right)(6)-\frac{6^3}{18}\right]-2(68.876)=34.9\text{ ft}^2$$

Exercise 4 ◊ Volumes by Integration
NOTE: Each of the functions in the problems of Exercise 4 should be sketched or graphed by calculator prior to solving them. This helps avoid problems or errors.

Rotation about the *x* Axis

1. $y=x^3$ and $x=2$

$$dV=\pi y^2\,dx=\pi x^6\,dx \Rightarrow V=\pi\int_0^2 x^6dx=\pi\left(\frac{x^7}{7}\right)\Bigg|_0^2=57.4$$

3. $y=\dfrac{x^{3/2}}{2}$ and $x=2$

$$dV=\pi\left(\frac{x^{3/2}}{2}\right)^2dx=\frac{\pi}{4}x^3dx \Rightarrow V=\pi\int_0^2\left(\frac{x^{3/2}}{2}\right)^2dx=\frac{\pi}{4}\int_0^2 x^3dx=\frac{\pi x^4}{16}\Bigg|_0^2=\pi$$

5. $y^2(2-x)=x^3$ and $x=1 \Rightarrow y^2=\dfrac{x^3}{2-x}=-x^2-2x-4+\dfrac{8}{2-x}$

$$dV=\pi y^2dx=\pi\left(-x^2-2x-4+\frac{8}{2-x}\right)dx$$

$$V=\pi\int_0^1\left(-x^2-2x-4+\frac{8}{2-x}\right)dx=\pi\left(-\frac{x^3}{3}-x^2-4x-8\ln|2-x|\right)\Bigg|_0^1$$

$$V=\pi\left(-\frac{16}{3}+8\ln 2\right)=0.666$$

7. $x^{2/3}+y^{2/3}=1$ from $x=0$ to 1

$$dV=\pi y^2dx=\pi\left(1-x^{2/3}\right)^3dx=\pi\left(1-3x^{2/3}+3x^{4/3}-x^2\right)dx$$

$$V=\pi\int_0^1\left(1-3x^{2/3}+3x^{4/3}-x^2\right)dx=\pi\left(x-\frac{9}{5}x^{5/3}+\frac{9}{7}x^{7/3}-\frac{x^3}{3}\right)\Bigg|_0^1=0.479$$

Rotation about the *y* Axis

9. $y=x^3$, the x axis, and $y=8$

$$V=\pi\int_0^8 x^2dy=\pi\int_0^8 y^{2/3}dy=\frac{3\pi y^{5/3}}{5}\Bigg|_0^8=\frac{3\pi}{5}8^{5/3}=60.3$$

11. $9x^2+16y^2=144 \Rightarrow x^2=16\left(1-\dfrac{y^2}{9}\right)$

$$V = \pi\int_0^3 x^2 dy = 16\pi\int_0^3\left(1-\frac{y^2}{9}\right)dy = 16\pi\left(y-\frac{y^3}{27}\right)\Bigg|_0^3 = 32\pi$$

13. $y^2 = 4x$ and $y = 4$

$$dV = \pi x^2 dy = \pi\frac{y^4}{16}dy$$

$$V = \pi\int_0^4 \frac{y^4}{16}dy = \frac{\pi y^5}{16\times 5}\Bigg|_0^4 = \frac{64\pi}{5} = 40.2$$

Solid of Revolution with a Hole

15. $y = 2\sqrt{x}$ and $x = 3$, about the y axis $\Rightarrow x^{1/2} = \frac{y}{2} \Rightarrow x^2 = \frac{y^4}{16} x^2 = 9$

$$V = \pi\int_0^{2\sqrt{3}}\left(r_o^2 - r_i^2\right)dy = \pi\int_0^{2\sqrt{3}}\left(9-\frac{y^4}{16}\right)dy = \pi\left(9y-\frac{y^5}{80}\right)\Bigg|_0^{2\sqrt{3}} = \frac{72\sqrt{3}\pi}{5}$$

17. $y = 3\sqrt{x}$ and $y = 2$, about the x axis $\Rightarrow y^2 = 9x \Rightarrow y^2 = 4$; $x = \frac{4}{9}$ when $y = 2$

$$V = \pi\int_0^{4/9}(4-9x)dx = \pi\left(4x-\frac{9x^2}{2}\right)\Bigg|_0^{4/9} = \frac{8}{9}\pi$$

Rotation about a Noncoordinate Axis

19. above $y = 3$ and below $y = 4x - x^2$, about $y = 3$

$$dV = \pi(y-3)^2 dx = \pi\left(-3+4x-x^2\right)^2 dx = \pi\left(x^4 - 8x^3 + 22x^2 - 24x + 9\right)dx$$

$$V = \pi\int_1^3\left(x^4 - 8x^3 + 22x^2 - 24x + 9\right)dx = \pi\left(\frac{x^5}{5} - 2x^4 + \frac{22}{3}x^3 - 12x^2 + 9x\right)\Bigg|_1^3 = 3.35$$

21. $y = 4$ and $y = 4 + 6x - 2x^2$, about $y = 4$

$$dV = \pi(y-4)^2 dx = \pi\left(6x-2x^2\right)^2 dx = \pi\left(36x^2 - 24x^3 + 4x^4\right)dx$$

$$V = 4\pi\int_0^3\left(9x^2 - 6x^3 + x^4\right)dx = 4\pi\left(\frac{9x^3}{3} - \frac{6x^4}{4} + \frac{x^5}{5}\right)\Bigg|_0^3 = 102$$

Volumes of Familiar Solids

23. $V = \pi\int_0^h \frac{r^2x^2}{h^2}dx = \frac{\pi r^2}{h^2}\frac{x^3}{3}\Bigg|_0^h = \frac{\pi r^2 h}{3}$

Applications

25. $x = ky^2$; $48 = k(24)^2 \therefore k = \frac{1}{12}$; $12x = y^2$

$$dV = \pi y^2 dx = 12\pi x\,dx$$

$$V = 12\pi\int_0^{48} x\,dx = 6\pi x^2\Big|_0^{48.0} = 43{,}429 in^3 = 25.1 \text{ ft}^3$$

27. Cylinder: $V_1 = \pi 12(5)^2 = 942 \text{ mm}^3$

Paraboloid: $y = kx^2 \Rightarrow 14 = k(5)^2 \therefore k = \dfrac{14}{25}$

$$dV = \pi x^2 dy = \pi\left(\frac{25}{14}y\right)dy;\ V_2 = \pi\int_0^{14}\left(\frac{25}{14}y\right)dy = \pi\left(\frac{25y^2}{28}\right)\Bigg|_0^{14} = 550\text{ mm}^3$$

$$V_1 + V_2\left(11.3\text{ g/cm}^3\right) = (0.942 + 0.550)\left(11.3\text{ g/cm}^3\right) = 1.492\text{ cm}^3\left(11.3\text{ g/cm}^3\right) = 16.9\text{ g}$$

CHAPTER 27 REVIEW PROBLEMS

1. $4x = 5 - x \Rightarrow 5x = 5 \Rightarrow x = 1$ and $y = \pm 2$

$$V = \pi\int_0^1 4x\,dx + \pi\int_1^5 (5-x)\,dx = \pi 2x^2\Big|_0^1 - \pi\frac{(5-x)^2}{2}\Bigg|_1^5 = \pi(2-(-8)) = 10\pi = 31.4$$

3. $x^2 = 8y;\ y = \dfrac{64}{x^2+16}$; by Rule 56: $\displaystyle\int\frac{du}{a^2+b^2u^2} = \frac{1}{ab}\tan^{-1}\frac{bu}{a} + C$

$$y = \frac{64}{x^2+16} = \frac{64}{8y+16} \Rightarrow 8y^2 + 16y - 64 = 0 \Rightarrow y^2 + 2y - 8 = 0$$

$(y+4)(y-2) = 0$ and $y = 2; y \ge 0$ as $x^2 = 8y$

$$A = 2\int_0^4\left(\frac{64}{x^2+16} - \frac{x^2}{8}\right)dx = 128\int_0^4\frac{dx}{x^2+16} - \frac{1}{4}\int_0^4 x^2\,dx$$

$$A = \frac{128}{4}\tan^{-1}\frac{x}{4}\Bigg|_0^4 - \frac{1}{4}\frac{x^3}{3}\Bigg|_0^4 = 19.8$$

5. $A = 2\displaystyle\int_0^4 x^{5/2}\,dx = \frac{4}{5}x^{5/2}\Bigg|_0^4 = 25.6$

7. $A = \displaystyle\int_1^2\sqrt{x^3 - x^2}\,dx = 2\int_1^2 x\sqrt{x-1}\,dx$

$$A = 2\int_1^2\left[(x-1)+1\right](x-1)^{1/2}\,dx$$

$$A = 2\int_1^2\left[(x-1)^{3/2} + (x-1)^{1/2}\right]dx$$

$$A = 2\left(\frac{2}{5}(x-1)^{5/2} + \frac{2}{3}(x-1)^{3/2}\right)\Bigg|_1^2 = 2.13$$

9. $\alpha = 7.25t^2$ rad/s^2

$$\omega = 7.25\int t^2\,dt = \frac{7.25}{3}t^3 + \omega_0 = 2.417t^3\text{ rad/s}^2$$

$$\theta = 2.417\int t^3\,dt = \frac{2.417}{4}t^4 + \theta_0 = 0.604t^4;\text{NOTE: }\omega_0 = 0$$

$$\omega(20.0) = 2.417(20.0)^3 = 19{,}300\text{ rad/s}$$

$$\theta(20.0) = 0.604(20.0)^4 = 96{,}600\text{ rad} = 15{,}400\text{ rev}$$

11. $a_x = \dfrac{dv_x}{dt} = t \Rightarrow dv_x = t\,dt;\ v_x = \dfrac{t^2}{2} + v_{x_0}$; at $t = 0,\ v_x = 4 \Rightarrow v_{x_0} = 4;\ v_x = \dfrac{t^2}{2} + 4$

$$v_x = \frac{ds_x}{dt} = \frac{1}{2}t^2 + 4 \Rightarrow ds_x = \left(\frac{1}{2}t^2 + 4\right)dt; s_x = \frac{t^3}{6} + 4t + s_{x0}; at\ t = 0, s_x = 1 \Rightarrow s_x = \frac{t^3}{6} + 4t + 1$$

$$a_y = \frac{dv_y}{dt} = 5t \Rightarrow dv_y = 5t\,dt;\ v_y = \frac{5t^2}{2} + v_{y_0};\text{ at } t = 0,\ v_y = 15 \Rightarrow v_{y_0} = 15;\ v_y = \frac{5t^2}{2} + 15$$

$$v_y = \frac{ds_y}{dt} = \frac{5}{2}t^2 + 15 \Rightarrow ds_y = \left(\frac{5}{2}t^2 + 15\right)dt;\ s_y = \frac{5t^3}{6} + 15t + s_{y_0};\text{ at } t = 0, s_y = 1 \Rightarrow s_{y_0} = 1;\ s_y = \frac{5t^3}{6} + 15t + 1$$

13. $y^2 = 9\left(1-\dfrac{x^2}{16}\right) = \dfrac{9}{16}\left(16-x^2\right)$

$$V = 2\int_0^4 \pi y^2 dx = 2\pi\int_0^4 \frac{9}{16}\left(16-x^2\right)dx = \frac{9}{8}\pi\left(16x-\frac{x^3}{3}\right)\Bigg|_0^4 = 48\pi = 151$$

15. $V_{\text{Cylinder}} = \pi(8)^2 4 = 804$

$$V_{\text{Area}} = 2\pi\int_0^8 (8-y)x\,dy = 2\pi\int_0^8 (8-y)y^{2/3}dy = 2\pi\int_0^8\left(8y^{2/3}-y^{5/3}\right)dy$$

$$= 2\pi\left(\frac{24}{5}y^{5/3}-\frac{3}{8}y^{8/3}\right)\Bigg|_0^8 = 362 \therefore 804-362 = 442$$

17. $A = 4\int_0^4 3\sqrt{1-\dfrac{x^2}{16}}\,dx = 3\int_0^4 \sqrt{16-x^2}\,dx = 3\left(\dfrac{x}{2}\sqrt{16-x^2}+\dfrac{16}{2}\sin^{-1}\dfrac{x}{4}\right)\Bigg|_0^4 = 3\left(\dfrac{8\pi}{2}\right) = 12\pi$

19. $v = 8.9+(3t)^{1/2}$ V

$$i = \frac{1}{L}\int v\,dt = \frac{1}{L}\int\left[8.9+(3t)^{1/2}\right]dt = \frac{1}{25.0}\left(8.9t+\frac{\sqrt{3}t^{3/2}}{(3/2)}\right)+i_0$$

$$= 0.356t+0.0462t^{3/2}+1$$

$$i(5.00) = 0.356(5.00)+0.0462(5.00)^{3/2}+1 = 3.30\text{ A}$$

Chapter 28: More Applications of the Integral

Exercise 1 ◊ Length of Arc

1. $y = x^{3/2} \Rightarrow \frac{dy}{dx} = \frac{3}{2}x^{1/2} \Rightarrow \left(\frac{dy}{dx}\right)^2 = \frac{9}{4}x$

$s = \int_0^{5/9} \sqrt{1+\frac{9x}{4}}\,dx;\ \text{Let } u = 1+\frac{9x}{4},\ du = \frac{9}{4}dx$

$s = \frac{4}{9}\int_0^{9/4} u^{1/2}\,du \Rightarrow \frac{4}{9}\int_0^{9/4} \frac{u^{3/2}}{(3/2)}du = \frac{4}{9}\left(\frac{2}{3}u^{3/2}\right)\Big|_0^{9/4} = \frac{8}{27}\left(1+\frac{9}{4}x\right)^{3/2}\Big|_0^{5/9} = \frac{8}{27}(3.375-1) = 0.704$

3. $y = \left(36-x^2\right)^{1/2} \Rightarrow \frac{dy}{dx} = \frac{1}{2}\left(36-x^2\right)^{-1/2}(-2x) \Rightarrow \left(\frac{dy}{dx}\right)^2 = \frac{x^2}{36-x^2}$

NOTE: This is the circle $x^2+y^2=36$

$s = \int_0^6 \sqrt{1+\frac{x^2}{36-x^2}}\,dx = \int_0^6 \sqrt{\frac{36-x^2+x^2}{36-x^2}}\,dx = \int_0^6 \frac{6}{\sqrt{36-x^2}}dx = 6\int_0^6 \frac{dx}{\sqrt{36-x^2}}$

By Rule 61, Appendix C $\Rightarrow 6\left(\sin^{-1}\left(\frac{x}{6}\right)\right)\Big|_0^6 = 6\left(\frac{\pi}{2}-0\right) = 3\pi = 9.42$

5. $y = \frac{x^2}{4} \Rightarrow \frac{dy}{dx} = \frac{x}{2} \Rightarrow \left(\frac{dy}{dx}\right)^2 = \frac{x^2}{4}$

$s = \int_0^4 \sqrt{1+\left(\frac{x}{2}\right)^2}\,dx = \frac{1}{2}\int_0^4 \sqrt{4+x^2}\,dx$

By Rule 66, Appendix C $\Rightarrow \frac{1}{2}\left(\frac{x}{2}\sqrt{4+x^2}+\frac{4}{2}\ln\left|x+\sqrt{4+x^2}\right|\right)\Big|_0^4 = 5.92$

7. $y = 4x-x^2 \Rightarrow \frac{dy}{dx} = 4-2x \Rightarrow \left(\frac{dy}{dx}\right)^2 = (4-2x)^2$

$s = \int_0^4 \sqrt{1+(4-2x)^2}\,dx = -\frac{1}{2}\int_0^4 \sqrt{1+(4-2x)^2}\,(-2)dx$

By Rule 66, Appendix C $\Rightarrow -\frac{1}{2}\left(\frac{4-2x}{2}\sqrt{1+(4-2x)^2}+\frac{1}{2}\ln\left|4-2x+\sqrt{1+(4-2x)^2}\right|\right)\Big|_0^4 = 9.29$

9. $x = \frac{1}{8}y^2 \Rightarrow \frac{dx}{dy} = \frac{y}{4} \Rightarrow \left(\frac{dx}{dy}\right)^2 = \frac{y^2}{16}$

$s = \int_0^4 \sqrt{1+\frac{y^2}{16}}\,dy = \frac{1}{4}\int_0^4 \sqrt{16+y^2}\,dy$

By Rule 66, Appendix C $\Rightarrow \frac{1}{4}\left(\frac{y}{2}\sqrt{16+y^2}+\frac{16}{2}\ln\left|y+\sqrt{16+y^2}\right|\right)\Big|_0^4 = 4.59$

Applications

11. $ky = x^2;\ 200k = (500)^2\ \therefore k = 1250 \Rightarrow y = \frac{x^2}{1250} \Rightarrow \frac{dy}{dx} = \frac{x}{625} \Rightarrow \left(\frac{dy}{dx}\right)^2 = \frac{x^2}{(625)^2}$

$\sqrt{1+\left(\frac{dy}{dx}\right)^2} = \frac{\sqrt{(625)^2+x^2}}{625} \Rightarrow s = 2\int_0^{500}\sqrt{1+\left(\frac{dy}{dx}\right)^2}\,dx = \frac{2}{625}\int_0^{500}\sqrt{(625)^2+x^2}\,dx$

By Rule 66, Appendix C $\Rightarrow s = \frac{2}{625}\left(\frac{x}{2}\sqrt{(625)^2+x^2}+\frac{(625)^2}{2}\ln\left|x+\sqrt{(625)^2+x^2}\right|\right)\Big|_0^{500} = 1096\text{ ft}$

13. $y = 0.0625x^2 - 5x + 100; \frac{dy}{dx} = 0.125x - 5; \left(\frac{dy}{dx}\right)^2 = (0.125x - 5)^2$

By Rule 66, Appendix C $\Rightarrow s = 2\int_0^{40} \sqrt{1 + (0.125x - 5)^2}\,dx = \frac{2}{0.125}\int_0^{40} \sqrt{1 + (0.125x - 5)^2}\,(0.125)\,dx$

$$= 16\left(\frac{0.125x - 5}{2}\sqrt{1 + (0.125x - 5)^2} + \frac{1}{2}\ln\left|(0.125x - 5) + \sqrt{1 + (0.125x - 5)^2}\right|\right)\Bigg|_0^{40} = 223 \text{ ft}$$

15. The circular arc intercepts an angle of 60^0 or $\pi/3$ rad. The length is $12(\pi/3) = 12.6$ ft.

$$y = kx^2 \Rightarrow k = \frac{6}{36} = \frac{1}{6} \Rightarrow y = \frac{x^2}{6} \Rightarrow \frac{dy}{dx} = \frac{x}{3} \Rightarrow \left(\frac{dy}{dx}\right)^2 = \frac{x^2}{9}$$

$$s = 2\int_0^6 \sqrt{1 + \frac{x^2}{9}}\,dx = \frac{2}{3}\int_0^6 \sqrt{9 + x^2}\,dx = \frac{1}{3}\left(\frac{x}{2}\sqrt{9 + x^2} + \frac{9}{2}\ln\left|x + \sqrt{9 + x^2}\right|\right)\Bigg|_0^6 = 17.7 \text{ ft}$$

Perimeter is 12.6 ft + 17.7 ft = 30.3 ft

Exercise 2 ◊ Area of Surface of Revolution

Rotation About the *x* Axis

1. $y = \frac{x^3}{9}$ from $x = 0$ to $3 \Rightarrow \frac{dy}{dx} = \frac{3}{9}x^2 \Rightarrow \left(\frac{dy}{dx}\right)^2 = \frac{x^4}{9}$

$$S = 2\pi\int_0^3 \frac{x^3}{9}\sqrt{1 + \frac{x^4}{9}}\,dx = \frac{2\pi}{9}\int_0^3 \frac{\sqrt{9 + x^4}}{3}(x^3 dx)$$

$$S = \frac{2\pi}{27}\left(\frac{1}{4}\right)\int_0^3 (9 + x^4)^{1/2}(4x^3 dx) = \frac{\pi}{54}(9 + x^4)^{3/2}\frac{2}{3}\Big|_0^3 = 32.1$$

3. $y = 3\sqrt{x}$ from $x = 0$ to $4 \Rightarrow \frac{dy}{dx} = \frac{3}{2}x^{-1/2} \Rightarrow \left(\frac{dy}{dx}\right)^2 = \frac{9}{4}x^{-1}$

$$S = 2\pi\int_0^4 (3\sqrt{x})\sqrt{1 + \frac{9}{4}x^{-1}}\,dx = 6\pi\int_0^4 \sqrt{x + \frac{9}{4}}dx == 6\pi\left(\frac{2}{3}\right)\left(x + \frac{9}{4}\right)^{3/2}\Bigg|_0^4 = 49\pi = 154$$

5. $y = \sqrt{4 - x}$ in the first quadrant $\Rightarrow \frac{dy}{dx} = \frac{1}{2}(4 - x)^{-1/2}(-1) \Rightarrow \left(\frac{dy}{dx}\right)^2 = \frac{1}{16 - 4x}$

$$S = 2\pi\int_0^4 \sqrt{4 - x}\sqrt{\frac{17 - 4x}{16 - 4x}}\,dx = 2\pi\int_0^4 \sqrt{\frac{17 - 4x}{4}}\,dx$$

$$S = -\frac{1}{4}\pi\int_0^4 \sqrt{17 - 4x}(-4dx) = -\frac{\pi}{4}\left(\frac{2}{3}\right)(17 - 4x)^{3/2}\Big|_0^4 = 36.2$$

7. $y = (1 - x^{2/3})^{3/2}$ in the first quadrant $\Rightarrow \frac{dy}{dx} = -\frac{(1 - x^{2/3})^{1/2}}{x^{1/3}} \Longrightarrow \sqrt{1 + \left(\frac{dy}{dx}\right)^2} = \sqrt{1 + \frac{(1 - x^{2/3})}{x^{2/3}}} = \sqrt{\frac{1}{x^{2/3}}} = x^{-1/3}$

$$S = 2\pi\int_0^1 (1 - x^{2/3})^{3/2}(x^{-1/3})dx;\ u = 1 - x^{2/3}, du = -\frac{2}{3}x^{-1/3}dx$$

$$S = 3\pi\left(\frac{2}{5}\right)\left((1 - x^{2/3})^{5/2}\right)\Big|_0^1 = -\frac{6}{5}\pi(0 - 1) = 3.77$$

Rotation About the *y* Axis

9. $y = 3x^2$ from $x = 0$ to $5 \Rightarrow \frac{dy}{dx} = 6x \Rightarrow \left(\frac{dy}{dx}\right)^2 = 36x^2$

$$S = 2\pi\int_0^5 x\sqrt{1 + 36x^2}\,dx = \frac{2\pi}{72}\int_0^5 (1 + 36x^2)^{1/2}(72x\,dx) = \frac{2\pi}{108}(1 + 36x^2)^{3/2}\Big|_0^5 = 1573$$

11. $y = 4 - x^2$ from $x = 0$ to $2 \Rightarrow \dfrac{dy}{dx} = -2x \Rightarrow \left(\dfrac{dy}{dx}\right)^2 = 4x^2$

$$S = 2\pi\int_0^2 x\sqrt{1+4x^2}\,dx = \frac{2\pi}{8}\int_0^2 \left(1+4x^2\right)^{1/2}(8x\,dx) = \frac{\pi}{6}\left(1+4x^2\right)^{3/2}\Big|_0^2 = 36.2$$

Geometric Figures

13. $x^2 + y^2 = r^2 \Rightarrow y = \sqrt{r^2 - x^2} \Rightarrow \dfrac{dy}{dx} = \dfrac{1}{2}\left(r^2 - x^2\right)^{-1/2}(-2x) = \dfrac{x}{\sqrt{r^2 - x^2}}$

$$\sqrt{1+\left(\frac{dy}{dx}\right)^2} = \sqrt{1+\frac{x^2}{r^2-x^2}} = \sqrt{\frac{r^2}{r^2-x^2}} = -\frac{r}{\sqrt{r^2-x^2}}$$

$$s = 4\pi\int_0^r \sqrt{r^2-x^2}\left(\frac{r}{\sqrt{r^2-x^2}}\right)dx = 4\pi\int_0^r r\,dx = 4\pi(rx)\Big|_0^r = 4\pi r^2$$

Applications

15. $x = ky^2 \Rightarrow 4 = k(2)^2 \therefore k = 1 \Rightarrow y^2 = x \Rightarrow y = x^{1/2} \Rightarrow \dfrac{dy}{dx} = \dfrac{1}{2x^{1/2}}$

$$\sqrt{1+\left(\frac{dy}{dx}\right)^2} = \sqrt{1+\frac{1}{4x}}$$

$$S = 2\pi\int_0^4 x^{1/2}\sqrt{1+\frac{1}{4x}}\,dx = 2\pi\int_0^4 \sqrt{x+\frac{1}{4}}\,dx = 2\pi\left(\frac{2}{3}\right)\left(x+\frac{1}{4}\right)^{3/2}\Bigg|_0^4 = 36.2 \text{ ft}^2$$

Exercise 3 ◊ Centroids
Centroids of Simple Shapes

1. $0 + 4 + 3 + (-2) = 5 \Rightarrow \bar{x} = \dfrac{5}{4}$

$$0 + 2 + (-5) + (-3) = -6 \Rightarrow \bar{y} = -\frac{3}{2}$$

3. Area $= 6\left(1\dfrac{1}{2}\right) + 2(3) + 3\dfrac{1}{2}(1) = 9 + 6 + 3\dfrac{1}{2} = 18\dfrac{1}{2}$

$$M_y = 9(3) + 6(3) + 3\frac{1}{2}(3) = 55\frac{1}{2} \Rightarrow \bar{x} = \frac{55\frac{1}{2}}{18\frac{1}{2}} = 3$$

$$M_x = 9\left(4\frac{3}{4}\right) + 6\left(2\frac{1}{2}\right) + 3\frac{1}{2}\left(\frac{1}{2}\right) = 59\frac{1}{2} \Rightarrow \bar{y} = \frac{59\frac{1}{2}}{18\frac{1}{2}} = 3.22$$

Centroids of Areas by Integration

5. $y^2 = 4x \Rightarrow y = \pm 2\sqrt{x}$ and $x = 4$

$$A = 2\int_0^4 2x^{1/2}dx = 4\left(\frac{2}{3}\right)x^{3/2}\Big|_0^4 = \frac{64}{3}$$

$$\bar{x} = \frac{3}{64}\int_0^4 x\left(2x^{1/2} - \left(-2x^{1/2}\right)\right)dx = \frac{3}{64}\int_0^4 4x^{3/2}dx = \frac{3}{64}4\left(\frac{2}{5}\right)x^{5/2}\Big|_0^4 = 2.40;\ \bar{y} = 0$$

7. $y = x^2$, the x axis and $x = 3$

$$A=\int_0^3 x^2dx=\frac{x^3}{3}\bigg|_0^3=9$$

$$\overline{y}=\frac{1}{18}\int_0^3 x^2\left(x^2dx\right)=\frac{1}{18}\left(\frac{x^5}{5}\right)\bigg|_0^3=\frac{27}{10}=2.7$$

Areas Bounded by Two Curves

9. $y=x^3$ and $y=4x\Rightarrow x^3=4x\therefore x=2$ and $x=0$

$$A=\int_0^2\left(4x-x^3\right)dx=2x^2-\frac{x^4}{4}\bigg|_0^2=4$$

$$\overline{x}=\frac{1}{4}\int_0^2 x\left(4x-x^3\right)dx=\frac{1}{4}\int_0^2\left(4x^2-x^4\right)dx=\frac{1}{4}\left(\frac{4}{3}x^3-\frac{1}{5}x^5\right)\bigg|_0^2=1.067$$

11. $y^2=x$ and $x^2=y\Rightarrow x^4=x\Rightarrow x\left(x^3-1\right)=0\Rightarrow x=0$ and $x=1$

$$A=\int_0^1\left(x^{1/2}-x^2\right)dx=\left(\frac{2}{3}x^{3/2}-\frac{x^3}{3}\right)\bigg|_0^1=\frac{1}{3}$$

$$\overline{x}=3\int_0^1 x\left(x^{1/2}-x^2\right)dx=3\int_0^1\left(x^{3/2}-x^3\right)dx=3\left(\frac{2}{5}x^{5/2}-\frac{x^4}{4}\right)\bigg|_0^1=\frac{9}{20}$$

By symmetry, $\overline{y}=\frac{9}{20}$

13. $2y=x^2$ and $y=x^3$

$$A=\int_0^{1/2}\left(\frac{x^2}{2}-x^3\right)dx=\left(\frac{x^3}{6}-\frac{x^4}{4}\right)\bigg|_0^{1/2}=0.00521$$

$$\overline{x}=\frac{1}{0.00521}\int_0^{1/2}x\left(\frac{x^2}{2}-x^3\right)dx=\frac{1}{0.00521}\left(\frac{x^4}{8}-\frac{x^5}{5}\right)\bigg|_0^{1/2}=\frac{3}{10}$$

15. $y=x^2$ and $y=2x+3$

$$A=\int_0^3\left(-x^2+2x+3\right)dx=\left(-\frac{x^3}{3}+x^2+3x\right)\bigg|_0^3=9$$

$$\overline{x}=\frac{1}{9}\int_0^3 x\left(-x^2+2x+3\right)dx=\frac{1}{9}\int_0^3\left(-x^3+2x^2+3x\right)dx=1.25$$

Centroids of Volumes of Revolution

17. $6y=x^2, x=6,$ and the x axis

$$V=\pi\int_0^6\left(\frac{x^2}{6}\right)^2dx=\frac{\pi}{36}\int_0^6 x^4dx=\left(\frac{\pi}{36}\right)\left(\frac{1}{5}x^5\right)\bigg|_0^6=\frac{216\pi}{5}$$

$$\overline{x}V=\pi\int_0^6 x\left(\frac{x^2}{6}\right)^2dx=\frac{\pi}{36}\left(\frac{1}{6}x^6\right)\bigg|_0^6=216\pi;\ \overline{x}=\frac{216\pi}{\left(\frac{216\pi}{5}\right)}=5$$

19. Rotating about the y axis

$y^2 = 4x, y = 6$ and the y axis

$$V = \pi\int_0^6 (x^2)dy = \pi\int_0^6 \frac{y^4}{16}dy = \frac{\pi}{16}\left(\frac{1}{5}y^5\right)\Big|_0^6 = \frac{6^5\pi}{80}$$

$$\bar{y}V = \pi\int_0^6 yx^2dy = \pi\int_0^6 \frac{y^5}{16}dy = \frac{\pi}{16}\left(\frac{1}{6}y^6\right)\Big|_0^6 = \frac{6^5\pi}{16}$$

$$\bar{y} = \frac{\frac{6^5\pi}{16}}{\frac{6^5\pi}{80}} = 5$$

21. $V = \frac{1}{3}\pi r^2 h$

$$\bar{x}V = \pi\int_0^h x\left(r - \frac{r}{h}x\right)^2 dx = \pi r^2\int_0^h\left(x - 2\frac{x^2}{h} + \frac{x^3}{h^2}\right)dx$$

$$\bar{x}V = \pi r^2\left(\frac{x^2}{2} - \frac{2x^3}{3h} + \frac{x^4}{4h^2}\right)\Big|_0^h = \pi r^2\left(\frac{h^2}{2} - \frac{2h^2}{3} + \frac{h^2}{4}\right)\Big|_0^h = \frac{\pi r^2 h^2}{12}$$

$$\bar{x} = \frac{\frac{\pi r^2 h^2}{12}}{\frac{1}{3}\pi r^2 h} = \frac{h}{4}$$

Applications

23. Let the x axis run through the center of Figure 26-21 and the y axis be vertical along the base of the semicircle.

$$A_T = \frac{0.6+3}{2}(6) = 10.8 \text{ ft}^2;\ A_{SC} = \frac{1}{2}\pi(1.5)^2 = 3.534 ft^2;\ A_{TOTAL} = 14.334 \text{ ft}^2$$

$$\bar{x}A = \int_{-6}^0 x(1.5+0.20x)dx + \int_0^{1.5} x\sqrt{2.25-x^2}\,dx = \left(0.75x^2 + 0.05x^3\right)\Big|_{-6}^0 - \frac{1}{2}\left(\frac{2}{3}\right)\left(\left[2.25-x^2\right]^{3/2}\right)\Big|_0^{1.5}$$

$$\bar{x}A = -11.475;\ x_1 = \frac{-15.075}{8.067} = -1.87;\ \bar{x} = 7.50 - 1.50 - 1.87 = 4.13 \text{ ft}$$

25. $y = 18 + kx^2 \Rightarrow 90 = 18 + k(82)^2 \therefore k = 0.0107$

$$V = 2\pi\int_0^{82} x(18+kx^2)dx + \pi(95^2 - 82^2)(90) = 2\pi\int_0^{82}(18x + kx^3)dx + 650{,}592 = 2\pi\left(9x^2 + \frac{kx^4}{4}\right)\Big|_0^{82} + 650{,}592 = 1{,}790{,}729.8$$

$$\bar{y}V = \int_0^{82}\left(\frac{18+kx^2}{2}\right)2\pi x(18+kx^2)dx + (650{,}592)\left(\frac{90}{2}\right) = \pi\int_0^{82}(18^2x + 36kx^3 + k^2x^5)dx + 29{,}276{,}640$$

$$\bar{y}V = \pi\left((9)18x^2 + 9kx^4 + \frac{k^2}{6}x^6\right)\Big|_0^{82} + 29{,}276{,}640 = 82^2\pi\left|162 + 9k(82)^2 + \frac{k^2}{6}(82)^4\right| + 29{,}276{,}640$$

$$\bar{y}V = 35{,}371{,}479 + 29{,}276{,}640 = 64{,}601{,}250;\ \bar{y} = \frac{64{,}601{,}250}{1{,}791{,}440} = 36.1 \text{ mm}$$

Exercise 4 ◊ Fluid Pressure

1. $F = (62.4)(50.0)(\pi)(3.00)^2 = 88{,}200$ lb

3. Depth of centroid = 8.00 + $\frac{1}{2}(6.0) = 11$; Area = 60; $F = (62.4)(11.0)(60.0) = 41{,}200$ lb

5. $A = \frac{1}{2}2\sqrt{3} = \sqrt{3}$

$$A\bar{y}=\int_0^{\sqrt{3}} y\left(\frac{2}{\sqrt{3}}y\right)dy=\frac{2}{\sqrt{3}}\left(\frac{y^3}{3}\right)\Big|_0^{\sqrt{3}}=2;\ \bar{y}=\frac{2}{\sqrt{3}}$$

depth of centroid $=\sqrt{3}-\frac{2}{\sqrt{3}}=\frac{1}{\sqrt{3}};\ F=(62.4)\left(\frac{1}{\sqrt{3}}\right)\sqrt{3}=62.4\text{ lb}$

7. Integrate horizontally; $F=60.0\bar{y}A$

$$\bar{y}A=\int_0^3 y2\left(6\sqrt{1-\frac{y^2}{9}}\right)dy=4\int_0^3 y\sqrt{9-y^2}dy=-2\int_0^3\sqrt{9-y^2}(-2y\,dy)$$

$$\bar{y}A=-2\left(\frac{2}{3}(9-y^2)^{3/2}\right)\Big|_0^3=0+\frac{4}{3}(27)=36;\ F=60.0(36)=2160\text{ lb}$$

Exercise 5 ◊ Work
Springs

1. $W=\int_2^4 Fdx=\int_2^4 xdx=25.0x^2\Big|_2^4=25.0(12)\text{in}\cdot\text{lb}=25.0\text{ ft}\cdot\text{lb}$

3. $W=\int_1^3 Fdx=\int_1^3 8xdx=4x^2\Big|_1^3=32\text{ in}\cdot\text{lb}$

Tanks

5. $V=\pi\int_2^6 x^2dy=\pi\int_2^6(36-y^2)dy=\pi\left(36y-\frac{y^3}{3}\right)\Big|_2^6=74.67\pi=234.6$

$$\bar{y}V=\pi\int_2^6 yx^2dy=\pi\int_2^6 y(36-y^2)dy=\pi\left(18y^2-\frac{y^4}{4}\right)\Big|_2^6=256\pi$$

$$\bar{y}=\frac{256\pi}{74.67\pi}=3.43\text{ ft};\ \text{Work}=(234.6)(62.4)(3.43)=50{,}200\text{ ft}\cdot\text{lb}$$

7. $y=30-5x\ \therefore x=\frac{30-y}{5}=6-\frac{1}{5}y$

$$V=\pi\int_0^{10}x^2dy=\pi\int_0^{10}\left(36-\frac{12}{5}y+\frac{y^2}{25}\right)dy=\pi\left(36y-\frac{6}{5}y^2+\frac{y^3}{75}\right)\Big|_0^{10}=253.3\pi$$

$$\bar{y}V=\pi\int_0^{10}y\left(36-\frac{12}{5}y+\frac{y^2}{25}\right)dy=\pi\left(18y^2-\frac{4}{5}y^3+\frac{y^4}{100}\right)\Big|_0^{10}=1100\pi$$

$\bar{y}=\frac{1100\pi}{253.3\pi}=4.34;$ the centroid will be raised $(10+4.34)=14.34$

$\text{Work}=(253.3\pi)(50)(14.34)=5.71x10^5\text{ ft}\cdot\text{lb}$

Gas Laws

9. $p_1v_1=p_2v_2;v_2=\frac{p_1}{p_2}(v_1)=\frac{15}{80}(200)=37.5ft^3$

$$W=-c\int_{200}^{37.5}\frac{dv}{v}=-c\ln|v|\Big|_{200}^{37.5}=-432{,}000(\ln 37.5-\ln 200)=723{,}000\text{ ft}\cdot\text{lb}$$

Miscellaneous

11. $W=\int_{50}^{100}\frac{k}{x^2}dx=-kx^{-1}\Big|_{50}^{100}=\frac{k}{100}\text{ ft}\cdot\text{lb}$

13. $W=\int_0^{20}\left[200+(500-x)\right]dx=\left(700x-\frac{x^2}{2}\right)\Big|_0^{20}=13{,}800\text{ ft}\cdot\text{lb}$

Exercise 6 ◊ Moment of Inertia
Moment of Inertia of an Area by Integration

1. $I_x = \frac{1}{3}\int_0^1 y^3 dx = \frac{1}{3}\int_0^1 x^3 dx = \frac{1}{3}\left(\frac{x^4}{4}\right)\Bigg|_0^1 = \frac{1}{12}$

3. $I_y = \int_0^4 x^2 y\, dx = \int_0^4 2x^{5/2} dx = \frac{4}{7}\left(x^{7/2}\right)\Bigg|_0^4 = \frac{2^9}{7}$

$$A = \int_0^4 2x^{1/2} dx = \frac{4}{3}x^{3/2}\Big|_0^4 = \frac{2^5}{3};\ r_y = \sqrt{\frac{I_y}{A}} = \sqrt{\frac{\frac{2^9}{7}}{\frac{2^5}{3}}} = 4\sqrt{\frac{3}{7}} = 2.62$$

5. $I_x = \frac{1}{3}\int_0^2 y^2\, dA = \frac{1}{3}\int_0^2 y^2\left(4-x^2\right)dx = \frac{1}{3}\int_0^2\left(4-x^2\right)^3 dx$

$$I_x = \frac{1}{3}\int_0^2\left(64-48x^2+12x^4-x^6\right)dx = \frac{1}{3}\left(64x-16x^3+\frac{12}{5}x^5-\frac{1}{7}x^7\right)\Bigg|_0^2 = 19.5$$

7. $y^3 = 1-x^2$

$$I_x = \frac{1}{3}\int_0^1 y^3 dx = \frac{1}{3}\int_0^1\left(1-x^2\right)dx = \frac{1}{3}\left(x-\frac{x^3}{3}\right)\Bigg|_0^1 = \frac{2}{9}$$

Polar Moment of Inertia

9. $I_x = m\int_0^2 y^2 2\pi y(2-x)dy = m\int_0^2 y^2 2\pi y(2-y)dy = 2\pi m\int_0^2\left(2y^3-y^4\right)dy$

$$I_x = 2\pi m\left(\frac{y^4}{2}-\frac{y^5}{5}\right)\Bigg|_0^2 = 10.1\text{ m}$$

11. $I_x = m\int_0^4 y^2 2\pi y(2-x)dy = 2\pi m\int_0^4 y^3\left(2-y^{1/2}\right)dy$

$$I_x = 2\pi m\int_0^4\left(2y^3-y^{7/2}\right)dy = 2\pi m\left(\frac{y^4}{2}-\frac{2}{9}y^{9/2}\right)\Bigg|_0^4 = 89.4\text{ m}$$

13. $M = \frac{1}{3}\pi r^2 hm;\ I_x = \int_0^r y^2 2\pi y(x)m\,dy$

$$M = 2\pi m\int_0^r y^3(x)dy = \frac{2\pi mh}{r}\int_0^r\left(ry^3-y^4\right)dy = \frac{2\pi mh}{r}\left(\frac{r}{4}y^4-\frac{1}{5}y^5\right)\Bigg|_0^r$$

$$M = \frac{2\pi mh}{r}r^4\left(\frac{r}{4}-\frac{r}{5}\right) = 2\pi mhr^3\left(\frac{r}{20}\right) = \frac{\pi mhr^4}{10};\ \frac{I_x}{M} = \frac{3\pi mhr^4}{10\pi r^2 hm} = \frac{3}{10}r^2;\ I_x = \frac{3}{10}Mr^2$$

15. $V = \int_0^p \pi y^2 dx = \int_0^p \pi(4px)dx = 4\pi p\left(\frac{x^2}{2}\right)_0^p = 2\pi p^3;\ M = 2\pi p^3 m$

$$I_x = \int_0^p \frac{m}{2}\pi y^4 dx = \frac{\pi m}{2}\int_0^p (4px)^2 dx = 8\pi mp^2\int_0^p x^2 dx = \frac{8}{3}m\pi p^2\left(x^3\right)_0^p = \frac{8}{3}m\pi p^5$$

$$\frac{I_x}{M} = \frac{8\pi mp^5}{6\pi p^3 m} = \frac{4}{3}p^2;\ I_x = \frac{4}{3}Mp^2$$

CHAPTER 28 REVIEW PROBLEMS

1. $r = 10;\ A = \frac{1}{2}\pi(10)^2 = 50\pi$

$$\overline{x}=\frac{1}{50\pi}\int_0^{10}x(2y)dx=\frac{2}{50\pi}\int_0^{10}x\sqrt{100-x^2}\,dx=-\frac{1}{50\pi}\int_0^{10}\sqrt{100-x^2}(-2x\,dx)$$

$$\overline{x}=\frac{1}{50\pi}\left(-\frac{2}{3}\left(100-x^2\right)^{3/2}\right)\Bigg|_0^{10}=\frac{(3/2)1000}{50\pi}=4.24$$

3. $A=\frac{1}{4}\pi(3)(4)=3\pi$

$$\overline{x}=\frac{1}{3\pi}\int_0^4 3x\sqrt{1-\frac{x^2}{16}}\,dx=\frac{1}{3\pi}\left(\frac{3}{4}\right)\int_0^4 x\sqrt{16-x^2}dx=-\frac{1}{3\pi}\left(\frac{3}{8}\right)\int_0^4\sqrt{16-x^2}(-2x\,dx)$$

$$\overline{x}=-\frac{1}{3\pi}\left(\frac{3}{8}\right)\left(\frac{2}{3}\right)\left[\left(16-x^2\right)^{3/2}\right]\Bigg|_0^4=1.70$$

$$\overline{y}=\frac{1}{3\pi}\int_0^3 yx\,dy=\frac{1}{3\pi}\int_0^3 y\left(4\sqrt{1-\frac{y^2}{9}}\right)dy=\frac{1}{3\pi}\left(\frac{4}{3}\right)\int_0^3\sqrt{9-y^2}\,y\,dy$$

$$\overline{y}=\frac{4}{9\pi}\frac{1}{2}\int_0^3\left(9-y^2\right)^{1/2}(-2y)dy=\frac{2}{9\pi}\left(\frac{2}{3}\left(9-y^2\right)^{3/2}\right)\Bigg|_0^3=\frac{4}{\pi}=1.27$$

5. $V=\frac{1}{3}\pi(4)^2(12)=64\pi$

$$\overline{y}V=\pi\int_0^{12}yx^2dy=\pi\int_0^{12}\frac{y^3}{9}dy=\left(\frac{\pi}{36}\right)y^4\Bigg|_0^{12}=\frac{12^3\pi}{3}$$

$$\overline{y}=\frac{\frac{12^3\pi}{3}}{64\pi}=9;\ \text{centroid to be raised } (12-9)=3\text{ ft}$$

$$W=(64\pi)(80.0)(3)=48{,}300\text{ ft}\cdot\text{lb}$$

7. $A=\frac{1}{2}\pi(3)^2=\frac{9}{2}\pi$

$$\overline{y}=\frac{2}{9\pi}\int_{-3}^0 y2x\,dy=\frac{4}{9\pi}\int_{-3}^0 y\sqrt{9-y^2}dy=-\frac{2}{9\pi}\int_{-3}^0\sqrt{9-y^2}(-2y\,dy)=-\frac{2}{9\pi}\left[\frac{2}{3}\left(9-y^2\right)^{3/2}\right]\Bigg|_{-3}^0$$

$$\overline{y}=-\frac{(2)(18)}{9\pi}=-\frac{4}{\pi};\quad F=(62.4)\left(\frac{9}{2}\pi\right)\left(\frac{4}{\pi}\right)=1120\text{ lb}$$

9. $W=\int_{1.00}^{3.00}5.45x\,dx=5.45\left(\frac{x^2}{2}\right)\Bigg|_{1.00}^{3.00}=21.8\text{ in}\cdot\text{lb}$

11. $A=\frac{1}{2}\pi(4)^2=8\pi$

$$\overline{y}=\frac{2}{8\pi}\int_{-4}^0 yx\,dy=\frac{2}{8\pi}\int_{-3}^0 y\sqrt{16-y^2}\,dy=-\frac{1}{8\pi}\int_{-4}^0\sqrt{16-y^2}(-2y\,dy)=-\frac{1}{8\pi}\left[\frac{2}{3}\left(16-y^2\right)^{3/2}\right]\Bigg|_{-4}^0$$

$$\overline{y}=-\frac{1}{8\pi}\left[\frac{2}{3}(64)\right]=-\frac{128/3}{8\pi}=-\frac{16}{3\pi};\quad F=(60)(8\pi)\left(\frac{16}{3\pi}\right)=2560\text{ lb}$$

13. $y^2=x+3;\ x=y^2-3;\ \frac{dx}{dy}=2y$

outer surface $= \int_0^2 2\pi y\sqrt{1+(2y)^2}\,dy = \frac{\pi}{4}\int_0^2 \sqrt{1+4y^2}\,(8y\,dy) = \frac{\pi}{4}\left(\frac{2}{3}(1+4y^2)^{3/2}\right)\Big|_0^2 = \frac{\pi}{6}(17^{3/2}-1) = 36.18$

$\frac{y^2}{4} = x;\ \frac{dx}{dy} = \frac{y}{2}$

inner surface $= \int_0^2 2\pi y\sqrt{1+\left(\frac{y}{2}\right)^2}\,dy = \frac{\pi}{2}\int_0^2 \sqrt{4+y^2}\,(2y\,dy) = \frac{\pi}{2}\left[\frac{2}{3}(4+y^2)^{3/2}\right]_0^2 = \frac{\pi}{3}(8^{3/2}-4^{3/2}) = 15.32$

total surface $= 51.5$

15. area of conical surface$=\frac{1}{2}(2\pi)(4)\left(\sqrt{4^2+2^2}\right) = 4\pi\sqrt{20} = 56.199$

area of parabolic surface: $y = \frac{x^2}{4} \Rightarrow \frac{dy}{dx} = \frac{2x}{4} = \frac{x}{2}$

$S = 2\pi\int_0^4 x\sqrt{1+\left(\frac{x}{2}\right)^2}\,dx = \pi\int_0^4 x(4+x^2)^{1/2}dx = \frac{\pi}{2}\int_0^4 (4+x^2)^{1/2}\,2x\,dx$

$S = \left[\frac{\pi}{2}\left(\frac{2}{3}\right)(4+x^2)^{3/2}\right]_0^4 = \frac{\pi}{3}(89.443-8) = 85.287$

total surface area $= 56.199 + 85.287 = 141.486 = 141$

17. area of conical surface $=\frac{1}{2}(2\pi)(8)\left(\sqrt{8^2+2^2}\right) = 8\pi\sqrt{68} = 207.25$

area of parabolic surface: $y = x^3 \Rightarrow \frac{dy}{dx} = 3x^2$

$S = 2\pi\int_0^2 x^3\sqrt{1+(3x^2)^2}\,dx = 2\pi\int_0^2 x^3(1+9x^4)^{1/2}dx = \frac{2\pi}{36}\int_0^2 (1+9x^4)^{1/2}\,36x^3dx$

$S = \frac{\pi}{18}\left(\frac{2}{3}\right)\left[(1+9x^4)^{3/2}\right]_0^2 = \frac{\pi}{27}(1746-1) = 203.04$

total surface area $= 207.25 + 203.04 = 410.29 = 410$

Chapter 29: Trigonometric, Logarithmic, and Exponential Functions

Exercise 29.1 • Derivatives of the Sine and Cosine Functions

First Derivatives

1. $y = \sin x \;\Rightarrow\; y' = \cos x$

3. $y = \cos^3 x \;\Rightarrow\; y' = 3\cos^2 x(-\sin x) = -3\sin x\cos^2 x$

5. $y = \sin 3x \;\Rightarrow\; y' = (\cos 3x)(3) = 2\cos 3x$

7. $y = \sin x\cos x \;\Rightarrow\; y' = (\sin x)(-\sin x) + (\cos x)(\cos x) = \cos^2 x - \sin^2 x$

9. $y = 3.75x\cos x \;\Rightarrow\; y' = 3.75x(-\sin x) + 3.75\cos x = 3.75(\cos x - x\sin x)$

11. $y = \sin^2(\pi - x) \;\Rightarrow\; y' = 2\sin(\pi - x)(\cos(\pi - x))(-1) = -2\sin(\pi - x)\cos(\pi - x)$

13. $y = \sin 2\theta\cos\theta \;\Rightarrow\; y' = (\sin 2\theta)(-\sin\theta) + (\cos\theta)(\cos 2\theta)(2) = 2\cos 2\theta\cos\theta - \sin 2\theta\sin\theta$

15. $y = \sin^2 x\cos x \;\Rightarrow\; y' = \sin^2 x(-\sin x) + \cos x(2\sin x)(\cos x) = 2\sin x\cos^2 x - \sin^3 x$

17. $y = 1.23\sin^2 x\cos 3x$
$y' = 1.23\sin^2 x(-\sin 3x)(3) + \cos 3x(1.23)(2\sin x)(\cos x) = 1.23\left(2\cos 3x\sin x\cos x - 3\sin^2 x\sin 3x\right)$

19. $y = \sqrt{\cos 2t} = (\cos 2t)^{1/2} \;\Rightarrow\; y' = \dfrac{1}{2}(\cos 2t)^{-1/2}(-\sin 2t)(2) = \dfrac{-\sin 2t}{\sqrt{\cos 2t}}$

Second Derivatives

21. $y = \cos x \;\Rightarrow\; y' = -\sin x \;\Rightarrow\; y'' = -\cos x$

23. $y = x\cos x \;\Rightarrow\; y' = x(-\sin x) + \cos x = \cos x - x\sin x$
$y'' = -\sin x - (x\cos x + \sin x) = -2\sin x - x\cos x$

25. $f(x) = x\sin\left(\dfrac{\pi}{2}x\right) \;\Rightarrow\; f'(x) = x\cos\left(\dfrac{\pi}{2}x\right)\left(\dfrac{\pi}{2}\right) + \sin\left(\dfrac{\pi}{2}x\right) = \dfrac{\pi}{2}x\cos\left(\dfrac{\pi}{2}x\right) + \sin\left(\dfrac{\pi}{2}x\right)$
$f''(x) = \dfrac{\pi}{2}x\left(-\sin\left(\dfrac{\pi}{2}x\right)\right)\left(\dfrac{\pi}{2}\right) + \dfrac{\pi}{2}\cos\left(\dfrac{\pi}{2}x\right) + \cos\left(\dfrac{\pi}{2}x\right)\left(\dfrac{\pi}{2}\right) = \pi\cos\left(\dfrac{\pi}{2}x\right) - \dfrac{\pi^2}{4}x\sin\left(\dfrac{\pi}{2}x\right)$
$f''(1) = \pi\cos\left(\dfrac{\pi}{2}\right) - \dfrac{\pi^2}{4}\sin\left(\dfrac{\pi}{2}\right) = -\dfrac{\pi^2}{4}$

Implicit Functions

27. $xy - y\sin x - x\cos y = 0$

$xy' + y - y\cos x - y'\sin x - x(-\sin y)y' - \cos y = 0$

$y'(x\sin y - \sin x + x) = y\cos x + \cos y - y \quad \Rightarrow \quad y' = \dfrac{y\cos x + \cos y - y}{x\sin y - \sin x + x}$

29. $x = \sin(x + y)$

$1 = \cos(x + y)(1 + y')$

$1 = \cos(x + y) + y'\cos(x + y)$

$y'\cos(x + y) = 1 - \cos(x + y) \quad \Rightarrow \quad y' = \dfrac{1 - \cos(x + y)}{\cos(x + y)} = \sec(x + y) - 1$

Tangents

31. $y = \sin x \quad \Rightarrow \quad y' = \cos x \quad \Rightarrow \quad m = y'(2) = \cos(2) = -0.4161$

33. $y = x\sin\dfrac{x}{2} \quad \Rightarrow \quad y' = \dfrac{1}{2}x\cos\dfrac{x}{2} + \sin\dfrac{x}{2} \quad \Rightarrow \quad m = y'(2) = \dfrac{1}{2}(2)\cos\dfrac{(2)}{2} + \sin\dfrac{(2)}{2} = 1.382$

Extreme Values and Inflections Points

35. $y = \sin x \quad \Rightarrow \quad y' = \cos x \quad \Rightarrow \quad y'' = -\sin x$

$y' = 0 \quad \Rightarrow \quad \cos x = 0 \quad \Rightarrow \quad x = \dfrac{\pi}{2}$ and $\dfrac{3\pi}{2}$ in the interval between 0 and 2π

$y'' = 0 \quad \Rightarrow \quad -\sin x = 0 \quad \Rightarrow \quad x = 0, \pi,$ and 2π in the interval between 0 and 2π

$y''\left(\dfrac{\pi}{2}\right) = -\sin\left(\dfrac{\pi}{2}\right) = -1 < 0$ maximum at $x = \dfrac{\pi}{2}$

$y''\left(\dfrac{3\pi}{2}\right) = -\sin\left(\dfrac{3\pi}{2}\right) = 1 > 0$ minimum at $x = \dfrac{3\pi}{2}$

$y\left(\dfrac{\pi}{2}\right) = 1, \; y\left(\dfrac{3\pi}{2}\right) = -1, \; y(0) = 0, \; y(\pi) = 0, \; y(2\pi) = 0$

There is a maximum at $\left(\dfrac{\pi}{2}, 1\right)$, a minimum at $\left(\dfrac{3\pi}{2}, -1\right)$, and inflection points at $(0,0)$, $(\pi,0)$, and $(2\pi,0)$.

37. $y = 3\sin x - 4\cos x \quad \Rightarrow \quad y' = 3\cos x + 4\sin x \quad \Rightarrow \quad y'' = -3\sin x + 4\cos x$

$y' = 0 \quad \Rightarrow \quad -3\cos x = 4\sin x \quad \Rightarrow \quad -\dfrac{3}{4} = \dfrac{\sin x}{\cos x} = \tan x \quad \Rightarrow \quad x = \pi - 0.644 = 2.50$ and $x = 2\pi - 0.644 = 5.64$ in the interval between 0 and 2π

$y'' = 0 \quad \Rightarrow \quad 3\sin x = 4\cos x \quad \Rightarrow \quad \dfrac{4}{3} = \dfrac{\sin x}{\cos x} = \tan x \quad \Rightarrow \quad x = 0.927$ and $x = \pi + 0.927 = 4.07$ in the interval between 0 and 2π

$y''(2.50) < 0$ maximum at $x = 2.50$

$y''(5.64) > 0$ minimum at $x = 5.64$

$y(2.50) = 5, \; y(5.64) = -5, \; y(0.927) = 0, \; y(4.07) = 0$

There is a maximum at $(2.50, 5)$, a minimum at $(5.64, -5)$, and inflection points at $(0.927, 0)$ and $(4.07, 0)$.

Electrical Applications

39. a) $v = 11.5\cos(2.84t + 0.75) \Rightarrow \frac{dv}{dt} = -11.5\sin(2.84t + 0.75)(2.84) = -32.66\sin(2.84t + 0.75)$

$i = C\frac{dv}{dt} = (22.5)(-32.66\sin(2.84t + 0.75)) = -735\sin(2.84t + 0.75)\ \mu\text{A}$

b) $\text{i}(0.2) = -735\sin(2.84(0.2) + 0.75) = -712\ \mu\text{A}$

Exercise 29.2 • Derivatives of the Other Trigonometric Functions

First Derivative

1. $y = \tan 2x \Rightarrow y' = 2\sec^2 2x$

3. $y = 5\csc 3x \Rightarrow y' = 5(-\csc 3x \cot 3x)(3) = -15\csc 3x \cot 3x$

5. $y = 3.25\tan x^2 \Rightarrow y' = 3.25\left(\sec^2 x^2\right)(2x) = 6.50x\sec^2 x^2$

7. $y = 7\csc x^3 \Rightarrow y' = 7\left(-\csc x^3 \cot x^3\right)\left(3x^2\right) = -21x^2 \csc x^3 \cot x^3$

9. $y = x\tan x \Rightarrow y' = x\sec^2 x + \tan x$

11. $y = 5x\csc 6x \Rightarrow y' = 5x(-\csc 6x \cot 6x)(6) + 5\csc 6x = 5\csc 6x - 30x\csc 6x \cot 6x$

13. $w = \sin\theta \tan 2\theta \Rightarrow w' = \sin\theta\left(\sec^2 2\theta\right)(2) + \tan 2\theta\cos\theta = 2\sin\theta\sec^2 2\theta + \tan 2\theta\cos\theta$

15. $v = 5\tan t \csc 3t \Rightarrow v' = 5\tan t(-\csc 3t \cot 3t)(3) + 5\csc 3t \sec^2 t = 5\csc 3t \sec^2 t - 15\tan t \csc 3t \cot 3t$

17. $y = 5.83\tan^2 2x \Rightarrow y' = 5.83(2\tan 2x)\sec^2 2x(2) = 23.32\tan 2x \sec^2 2x$

$y'(1) = 23.32\tan 2\sec^2 2 = \frac{23.32\tan 2}{\sec^2 2} = -294$

19. $f(x) = 3\csc^3 3x \Rightarrow f'(x) = 3\left(3\csc^2 3x\right)(-\csc 3x \cot 3x)(3) = -27\csc^3 3x \cot 3x$

$f'(3) = -27\csc^3 9 \cot 9 = \frac{-27}{\sin^3 9 \tan 9} = 853$

Second Derivative

21. $y = 3\tan x \Rightarrow y' = 3\sec^2 x \Rightarrow y'' = 3(2\sec x)\sec x \tan x = 6\sec^2 x \tan x$

23. $y = 3\csc 2\theta \quad \Rightarrow \quad y' = 3(-\csc 2\theta \cot 2\theta)(2) = -6\csc 2\theta \cot 2\theta$

$y'' = -6\csc 2\theta\left(-\csc^2 2\theta\right)(2) - 6\cot 2\theta(-\csc 2\theta \cot 2\theta)(2) = 12\csc^3 2\theta + 12\cot^2 2\theta \csc 2\theta$

$y''(1) = 12\csc^3 2 + 12\cot^2 2\csc 2 = \dfrac{12}{\sin^3 2} + \dfrac{12}{\tan^2 2\sin 2} = 18.73$

Implicit Functions

25. $y\tan x = 2$

$y\sec^2 x + y'\tan x = 0 \quad \Rightarrow \quad y' = -\dfrac{y\sec^2 x}{\tan x} = -y\dfrac{1}{\cos^2 x}\dfrac{\cos x}{\sin x} = -y\sec x\csc x$

27. $\sec(x+y) = 7$

$\sec(x+y)\tan(x+y)(1+y') = 0 \quad \Rightarrow \quad y'(\sec(x+y)\tan(x+y)) = -\sec(x+y)\tan(x+y)$

$y' = \dfrac{-\sec(x+y)\tan(x+y)}{\sec(x+y)\tan(x+y)} = -1$

Tangents

29. $y = \tan x \quad \Rightarrow \quad y' = \sec^2 x$

$m = y'(1) = \sec^2(1) = \dfrac{1}{\cos^2 1} = 3.43 \qquad y(1) = \tan(1) = 1.56$

$\dfrac{y-1.56}{x-1} = 3.43 \quad \Rightarrow \quad y - 1.56 = 3.43x - 3.43 \quad \Rightarrow \quad 3.43x - y = 1.87$

Extreme Values and Inflections Points

31. $y = 2x - \tan x \quad \Rightarrow \quad y' = 2 - \sec^2 x \quad \Rightarrow \quad y'' = -2\sec x(\sec x\tan x) = -2\sec^2 x\tan x$

$y' = 0 \quad \Rightarrow \quad 2 - \sec^2 x = 0 \quad \Rightarrow \quad \cos^2 x = \dfrac{1}{2} \quad \Rightarrow \quad \cos x = \pm\dfrac{1}{\sqrt{2}}$

$x = \dfrac{\pi}{4}$ and $\dfrac{3\pi}{4}$ in the interval between 0 and π

$y'' = 0 \quad \Rightarrow \quad -2\sec^2 x\tan x = 0 \quad \Rightarrow \quad \tan x = 0 \quad \Rightarrow \quad x = 0$ and π in the interval between 0 and π

$(\sec x \neq 0)$

$y''\left(\dfrac{\pi}{4}\right) = -4 < 0$ maximum at $x = \dfrac{\pi}{4}$

$y''\left(\dfrac{3\pi}{4}\right) = 4 > 0$ minimum at $x = \dfrac{3\pi}{4}$

$y\left(\dfrac{\pi}{4}\right) = 0.571, \; y\left(\dfrac{3\pi}{2}\right) = 5.71, \; y(0) = 0, \; y(\pi) = 2\pi$

There is a maximum at $\left(\dfrac{\pi}{4}, 0.571\right)$, a minimum at $\left(\dfrac{3\pi}{4}, 5.71\right)$, and inflection points at $(0,0)$ and (π,π).

Rate of Change

33. $y = 6\sin 4t$

$$v = y' = 6\cos 4t(4) = 24\cos 4t \quad \Rightarrow \quad v(0.0500) = 24\cos 4(0.0500) = 23.5 \text{ cm/s}$$

$$a = y'' = 24(-\sin 4t)(4) = -96\sin 4t \quad \Rightarrow \quad a(0.0500) = -96\sin 4(0.0500) = -19.1 \text{ cm/s}^2$$

Related Rates

35. $h' = 3.00 \text{ in./min} = \frac{1}{4} \text{ ft/min}, \quad \text{h} = 4.00 \text{ ft}$

$$\frac{\theta}{2} = \tan^{-1}\left(\frac{4.00}{4.00}\right) = 45^\circ \quad \Rightarrow \quad \theta = 90^\circ$$

$$\tan\frac{\theta}{2} = \frac{4.00}{h}$$

implicit differentiate with respect to the variable t

$$\left(\sec^2\frac{\theta}{2}\right)\left(\frac{1}{2}\theta'\right) = \frac{-4}{h^2}h'$$

$$\theta' = \frac{-8\cos^2\frac{\theta}{2}}{h^2}h' = \frac{-8\cos^2(45^\circ)}{(4.00)^2}\left(\frac{1}{4}\right) = -0.0625 \text{ rad/min} = -3.58 \text{ deg/min}$$

Optimization

37. $l = \frac{x + 2.67}{\cos\theta}$ and $x = 9\cot\theta \quad \Rightarrow \quad l = \frac{9\cot\theta + 2.67}{\cos\theta}$

$$\frac{dl}{d\theta} = \frac{\cos\theta(-9\csc^2\theta) - (-\sin\theta)(9\cot\theta + 2.67)}{\cos^2\theta}$$

$$\frac{dl}{d\theta} = 0 \quad \Rightarrow \quad (9\cot\theta + 2.67)\sin\theta = 9\cos\theta\csc^2\theta$$

$$\left(9 - \frac{9}{\sin^2\theta}\right)\cos\theta + 2.67\sin\theta = 0$$

$$9\left(\frac{\sin^2\theta - 1}{\sin^2\theta}\right)\cos\theta + 2.67\sin\theta = 0$$

$$9\left(\frac{-\cos^2\theta}{\sin^2\theta}\right)\cos\theta + 2.67\sin\theta = 0$$

$$2.67\sin^3\theta - 9\cos^2\theta = 0$$

$$2.67\tan^3\theta = 9$$

$$\tan^3\theta = \frac{9}{2.67} \quad \Rightarrow \quad \theta = 0.983 \text{ rad}$$

$$x = 9\cot(0.983) = \frac{9}{\tan 0.983} = 6.00 \text{ ft}$$

$$l = \frac{6.00 + 2.67}{\cos 0.983} = 15.6 \text{ ft}$$

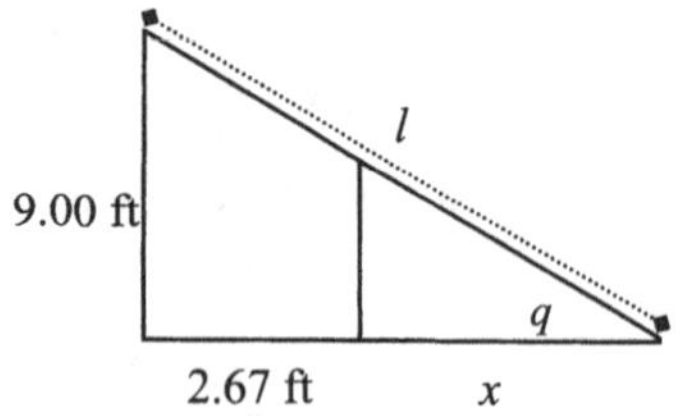

39. $F = \dfrac{fW}{f\sin\theta + \cos\theta} = \dfrac{0.60W}{0.60\sin\theta + \cos\theta} = 0.60W(0.60\sin\theta + \cos\theta)^{-1}$

$$\frac{dF}{d\theta} = -0.60W(0.60\sin\theta + \cos\theta)^{-2}(0.60\cos\theta - \sin\theta) = \frac{0.60W(\sin\theta - 0.60\cos\theta)}{(0.60\sin\theta + \cos\theta)^2}$$

$$\frac{dF}{d\theta} = 0 \Rightarrow 0.60\cos\theta = \sin\theta$$

$$\tan\theta = 0.60 \Rightarrow \theta = 31.0^\circ$$

41. $\tan\theta = \dfrac{12.8}{x} \Rightarrow x = \dfrac{12.8}{\tan\theta}$

$$\sin\theta = \frac{x+5.4}{L} \Rightarrow L = \frac{x+5.4}{\sin\theta}$$

$$L = \frac{(12.8/\tan\theta) + 5.4}{\sin\theta} = 12.8\sec\theta + 5.4\csc\theta$$

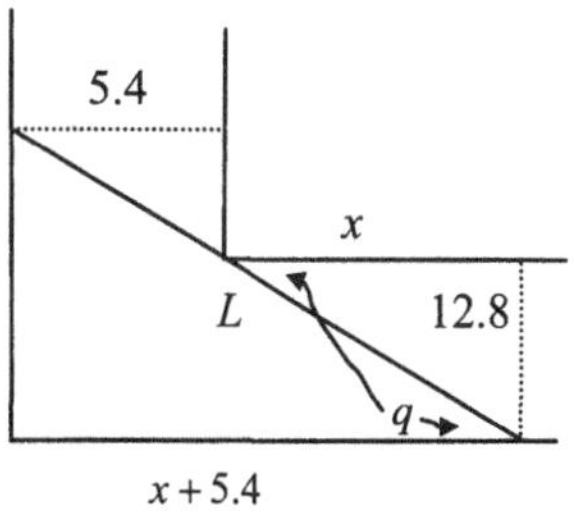

$$\frac{dL}{d\theta} = 12.8\sec\theta\tan\theta - 5.4\csc\theta\cot\theta = \frac{12.8\sin\theta}{\cos^2\theta} - \frac{5.4\cos\theta}{\sin^2\theta}$$

$$\frac{dL}{d\theta} = 0 \Rightarrow 12.8\sin^3\theta = 5.4\cos^3\theta \Rightarrow \tan^3\theta = \frac{5.4}{12.8}$$

$$\tan\theta = \sqrt[3]{\frac{5.4}{12.8}}$$

$$\theta = 36.87^\circ$$

$$L = \frac{12.8}{\cos 36.87^\circ} + \frac{5.4}{\sin 36.87^\circ} = 25.0 \text{ ft}$$

Exercise 29.3 • Derivatives of the Inverse Trigonometric Functions

1. $y = x\,\mathrm{Sin}^{-1}x \Rightarrow y' = x\dfrac{1}{\sqrt{1-x^2}}(1) + (1)\mathrm{Sin}^{-1}x = \mathrm{Sin}^{-1}x + \dfrac{x}{\sqrt{1-x^2}}$

3. $y = \mathrm{Cos}^{-1}\dfrac{x}{a} \Rightarrow y' = \dfrac{-1}{\sqrt{1-\left(\dfrac{x}{a}\right)^2}}\left(\dfrac{1}{a}\right) = \dfrac{-1}{\sqrt{\dfrac{a^2-x^2}{a^2}}}\left(\dfrac{1}{a}\right) = \dfrac{-1/a}{1/a\sqrt{a^2-x^2}} = \dfrac{-1}{\sqrt{a^2-x^2}}$

5. $y = \mathrm{Sin}^{-1}\left(\dfrac{\sin x - \cos x}{\sqrt{2}}\right)$

$$y' = \frac{1}{\sqrt{1-\left(\frac{1}{\sqrt{2}}(\cos x - \sin x)\right)^2}}\left(\frac{1}{\sqrt{2}}(\cos x + \sin x)\right) = \frac{\frac{1}{\sqrt{2}}(\cos x + \sin x)}{\sqrt{\frac{2-(\cos x - \sin x)^2}{2}}}$$

$$= \frac{\frac{1}{\sqrt{2}}(\cos x + \sin x)}{\frac{1}{\sqrt{2}}\sqrt{2 - \cos^2 x + 2\cos x\sin x - \sin^2 x}} = \frac{\cos x + \sin x}{\sqrt{2 + \underbrace{2\cos x\sin x}_{\sin 2x} - \underbrace{(\cos^2 x + \sin^2 x)}_{1}}} = \frac{\cos x + \sin x}{\sqrt{1 + \sin 2x}}$$

7. $y = t^2 \operatorname{Cos}^{-1} t \Rightarrow y' = t^2 \dfrac{-1}{\sqrt{1-(t)^2}}(1) + (2t)\operatorname{Cos}^{-1} t = -\dfrac{t^2}{\sqrt{1-(t)^2}} + 2t\operatorname{Cos}^{-1} t$

9. $y = \operatorname{Arctan}(1+2x) \Rightarrow y' = \dfrac{1}{1+(1+2x)^2}(2) = \dfrac{2}{1+1+4x+4x^2} = \dfrac{2}{2(1+2x+2x^2)} = \dfrac{1}{1+2x+2x^2}$

11. $y = \operatorname{Arccot}\dfrac{x}{a} \Rightarrow y' = \dfrac{-1}{1+\left(\dfrac{x}{a}\right)^2}\left(\dfrac{1}{a}\right) = \dfrac{-1/a}{1/a^2\left(a^2+x^2\right)} = \dfrac{-a}{a^2+x^2}$

13. $y = \operatorname{Arccsc} 2x \Rightarrow y' = \dfrac{-1}{2x\sqrt{(2x)^2-1}}(2) = \dfrac{-1}{x\sqrt{4x^2-1}}$

15. $y = t^2 \operatorname{Arcsin}\dfrac{t}{2} \Rightarrow y' = t^2 \dfrac{1}{\sqrt{1-\left(\dfrac{t}{2}\right)^2}}\left(\dfrac{1}{2}\right) + 2t\operatorname{Arcsin}\dfrac{t}{2} = 2t\operatorname{Arcsin}\dfrac{t}{2} + \dfrac{1/2t^2}{1/2\sqrt{4-t^2}} = 2t\operatorname{Arcsin}\dfrac{t}{2} + \dfrac{t^2}{\sqrt{4-t^2}}$

17. $y = \operatorname{Sec}^{-1}\dfrac{a}{\sqrt{a^2-x^2}}$

$$y' = \frac{1}{\dfrac{a}{\sqrt{a^2-x^2}}\sqrt{\left(\dfrac{a}{\sqrt{a^2-x^2}}\right)^2-1}}\left(\frac{\left(a^2-x^2\right)^{1/2}(0) - a\left(\dfrac{1}{2}\right)\left(a^2-x^2\right)^{-1/2}(-2x)}{a^2-x^2}\right)$$

$$= \frac{1}{\dfrac{a}{\sqrt{a^2-x^2}}\sqrt{\dfrac{a^2-a^2+x^2}{a^2-x^2}}}\left(\frac{ax}{\left(a^2-x^2\right)^{3/2}}\right) = \frac{1}{\sqrt{a^2-x^2}}$$

19. $y = \dfrac{\operatorname{Arctan} x}{x} \Rightarrow y' = \dfrac{x\left(\dfrac{1}{1+x^2}\right) - (1)\operatorname{Arctan} x}{x^2} = \dfrac{1}{x\left(1+x^2\right)} - \dfrac{\operatorname{Arctan} x}{x^2}$

$m = y'(1) = \dfrac{1}{1\left(1+1^2\right)} - \dfrac{\operatorname{Arctan} 1}{1^2} = -0.285$

21. $y = \sqrt{x}\operatorname{Arccot}\dfrac{x}{4} \Rightarrow y' = \sqrt{x}\dfrac{-1}{1+\left(\dfrac{x}{4}\right)^2}\left(\dfrac{1}{4}\right) + \left(\dfrac{1}{2\sqrt{x}}\right)\operatorname{Arccot}\dfrac{x}{4} = \dfrac{-4\sqrt{x}}{16+x^2} + \left(\dfrac{1}{2\sqrt{x}}\right)\operatorname{Arctan}\dfrac{4}{x}$

$m = y'(4) = \dfrac{-4(2)}{16+16} + \left(\dfrac{1}{2(2)}\right)\operatorname{Arctan} 1 = -0.0537$

Exercise 29.4 • Derivatives of Logarithmic Functions

Derivative of $\log_b u$

1. $y = \log 7x \quad \Rightarrow \quad y' = \frac{1}{7x}(\log e)(7) = \frac{1}{x}\log e$

3. $y = \log_b x^3 \quad \Rightarrow \quad y' = \frac{1}{x^3 \ln b}\left(3x^2\right) = \frac{3}{x \ln b}$

5. $y = \log\left(x\sqrt{5+6x}\right)$

$$y' = \frac{1}{\left(x\sqrt{5+6x}\right)}\log e\left(x\frac{1}{2}(5+6x)^{-1/2}(6) + (1)(5+6x)^{1/2}\right) = \frac{1}{\left(x\sqrt{5+6x}\right)}\frac{3x+5+6x}{\sqrt{5+6x}}\log e = \frac{(5+9x)\log e}{5x+6x^2}$$

7. $y = x\log\frac{2}{x} \quad \Rightarrow \quad y' = x\frac{1}{2/x}(\log e)\left(-\frac{2}{x^2}\right) + (1)\log\frac{2}{x} = \log\frac{2}{x} - \log e$

Derivative of $\ln u$

9. $y = \ln 3x \quad \Rightarrow \quad y' = \frac{1}{3x}(3) = \frac{3}{3x} = \frac{1}{x}$

11. $y = \ln\left(x^2 - 3x\right) \quad \Rightarrow \quad y' = \frac{2x-3}{x^2-3x}$

13. $y = 2.75x \ln 1.02x^3 \quad \Rightarrow \quad y' = 2.75x\left(\frac{3.06x^2}{1.02x^3}\right) + 2.75\ln 1.02x^3 = 8.25 + 2.75\ln 1.02x^3$

15. $y = \frac{\ln(x+5)}{x^2} \quad \Rightarrow \quad y' = \frac{x^2\left(\frac{1}{x+5}\right) - 2x\ln(x+5)}{x^4} = \frac{1}{x^2(x+5)} - \frac{2\ln(x+5)}{x^3}$

17. $s = \ln\sqrt{t-5} = \ln(t-5)^{1/2} = \frac{1}{2}\ln(t-5) \quad \Rightarrow \quad s' = \left(\frac{1}{2}\right)\frac{1}{t-5} = \frac{1}{2t-10}$

With Trigonometric Functions

19. $y = \ln \sin x \quad \Rightarrow \quad y' = \frac{\cos x}{\sin x} = \cot x$

21. $y = \sin x \ln \sin x \quad \Rightarrow \quad y' = \sin x\frac{\cos x}{\sin x} + \cos x \ln \sin x = \cos x\left(1 + \ln \sin x\right)$

Implicit Relations

23. $y \ln y + \cos x = 0$

$$y\frac{y'}{y} + y'\ln y - \sin x = 0 \quad \Rightarrow \quad y'(1+\ln y) = \sin x \quad \Rightarrow \quad y' = \frac{\sin x}{1+\ln y}$$

25. $x-y=\ln(x+y)$

$$1-y'=\frac{1+y'}{x+y} \Rightarrow x+y-y'(x+y)=1+y' \Rightarrow y'(1+x+y)=x+y-1 \Rightarrow y'=\frac{x+y-1}{x+y+1}$$

27. $\ln y+x=10$

$$\frac{y'}{y}+1=0 \Rightarrow \frac{y'}{y}=-1 \Rightarrow y'=-y$$

Logarithmic Differentiation

29. $y=\dfrac{\sqrt{a^2-x^2}}{x}$

$$\ln y=\frac{1}{2}\ln\left(a^2-x^2\right)-\ln x$$

$$\frac{1}{y}\frac{dy}{dx}=\frac{1}{2}\left(\frac{-2x}{a^2-x^2}\right)-\frac{1}{x}=\frac{-x}{\left(a^2-x^2\right)}-\frac{1}{x}=\frac{-x^2-a^2+x^2}{x\left(a^2-x^2\right)}=\frac{-a^2}{x\left(a^2-x^2\right)}$$

$$\frac{dy}{dx}=\frac{-a^2}{x\left(a^2-x^2\right)}y=\frac{-a^2}{x\left(a^2-x^2\right)}\left(\frac{\sqrt{a^2-x^2}}{x}\right)=\frac{-a^2}{x^2\sqrt{a^2-x^2}}$$

31. $y=x^{\sin x}$

$$\ln y=\sin x\ln x$$

$$\frac{1}{y}\frac{dy}{dx}=\sin x\left(\frac{1}{x}\right)+(\cos x)\ln x=\frac{\sin x}{x}+\cos x\ln x$$

$$\frac{dy}{dx}=\left(\frac{\sin x}{x}+\cos x\ln x\right)y=x^{\sin x}\left(\frac{\sin x}{x}+\cos x\ln x\right)$$

33. $y=\left(\mathrm{Cos}^{-1}x\right)^x$

$$\ln y=x\ln\left(\mathrm{Cos}^{-1}x\right)$$

$$\frac{1}{y}\frac{dy}{dx}=x\left(\frac{1}{\mathrm{Cos}^{-1}x}\right)\left(\frac{-1}{\sqrt{1-x^2}}\right)+(1)\ln\left(\mathrm{Cos}^{-1}x\right)=\ln\left(\mathrm{Cos}^{-1}x\right)-\frac{x}{\left(\mathrm{Cos}^{-1}x\right)\sqrt{1-x^2}}$$

$$\frac{dy}{dx}=\left(\ln\left(\mathrm{Cos}^{-1}x\right)-\frac{x}{\left(\mathrm{Cos}^{-1}x\right)\sqrt{1-x^2}}\right)y=\left(\mathrm{Cos}^{-1}x\right)^x\left(\ln\left(\mathrm{Cos}^{-1}x\right)-\frac{x}{\left(\mathrm{Cos}^{-1}x\right)\sqrt{1-x^2}}\right)$$

Tangent to a Curve

35. $y=\ln x \Rightarrow$ when $y=0,\ 0=\ln x \Rightarrow x=1$ $\quad y'=\dfrac{1}{x} \Rightarrow m=y'(1)=\dfrac{1}{1}=1$

37. $y=\log(4x-3) \Rightarrow y'=\dfrac{4}{4x-3}\log e$ $\quad m=y'(2)=\dfrac{4}{4(2)-3}\log e=0.3474$

Angle of Intersection

39. $y=\ln(x+1) \Rightarrow y'=\dfrac{1}{x+1}$ $\qquad m_1=\dfrac{1}{(2)+1}=\dfrac{1}{3}$

$y=\ln(7-2x) \Rightarrow y'=\dfrac{-2}{7-2x}$ $\qquad m_2=\dfrac{-2}{7-2(2)}=-\dfrac{2}{3}$

$$\tan\phi=\frac{m_2-m_1}{1+m_1m_2}=\frac{-\frac{2}{3}-\frac{1}{3}}{1+\left(\frac{1}{3}\right)\left(-\frac{2}{3}\right)}=\frac{-1}{1-\frac{2}{9}}=-\frac{9}{7}$$

$$\phi=\tan^{-1}\left(\frac{9}{7}\right)=52^\circ \text{ or } 180^\circ-52^\circ=128^\circ$$

Extreme Values and Points of Inflection

41. $y=x\ln x \Rightarrow y'=x\left(\dfrac{1}{x}\right)+(1)\ln x=1+\ln x \Rightarrow y''=\dfrac{1}{x}$

$y'=0 \Rightarrow 1+\ln x=0 \Rightarrow \ln x=-1 \Rightarrow x=e^{-1}=\dfrac{1}{e}$

$y''\neq 0$ no inflection points

$y''\left(\dfrac{1}{e}\right)=e>0$ minimum at $x=\dfrac{1}{e}$ $\qquad y\left(\dfrac{1}{e}\right)=-\dfrac{1}{e}$

There is a minimum at $\left(\dfrac{1}{e},-\dfrac{1}{e}\right)$.

43. $y=\dfrac{x}{\ln x} \Rightarrow y'=\dfrac{(\ln x)(1)-x\left(\frac{1}{x}\right)}{(\ln x)^2}=\dfrac{\ln x-1}{(\ln x)^2}$

$$y''=\frac{(\ln x)^2\left(\frac{1}{x}\right)-(\ln x-1)2\ln x\frac{1}{x}}{(\ln x)^4}=\frac{\ln x-2\ln x+2}{x(\ln x)^3}=\frac{2-\ln x}{x(\ln x)^3}$$

$y'=0 \Rightarrow \ln x-1=0 \Rightarrow \ln x=1 \Rightarrow x=e$

$y''=0 \Rightarrow 2-\ln x=0 \Rightarrow \ln x=2 \Rightarrow x=e^2$

$y''(e)>0$ minimum at $x=e$ $\left(\text{also concave up for } x<e^2\right)$

$y''\left(e^3\right)<0$ concave down for $x<e^2$

$y(e)=e,\ y\left(e^2\right)=\dfrac{e^2}{2}$

There is a minimum at (e,e), and an inflection point at $\left(e^2,\dfrac{e^2}{2}\right)$.

Roots

45. $x - 10\log x = 0 \;\Rightarrow\; x = 1.37$

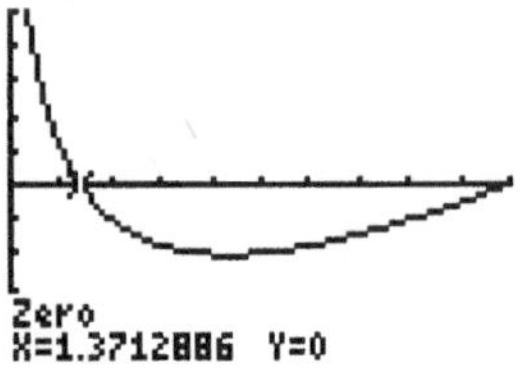

Applications

47. $S = x^2 \ln\left(\dfrac{1}{x}\right) = x^2 \ln\left(x^{-1}\right) = -x^2 \ln x$

$S' = -x^2\left(\dfrac{1}{x}\right) - 2x\ln x = -x\left(1 + 2\ln x\right)$

$S' = 0 \;\Rightarrow\; -x\left(1 + 2\ln x\right) = 0 \;\Rightarrow\; \ln x = -\dfrac{1}{2} \;\Rightarrow\; x = e^{-1/2} = 0.607 \text{ (Exclude } x = 0)$

49. $C = 20 \times 10^{-5}$ moles/liter, $\dfrac{dC}{dt} = 5.5 \times 10^{-5}$ per min

$\text{pH} = -\log_{10} C$

$\dfrac{d\text{pH}}{dt} = -\dfrac{1}{C\ln 10}\dfrac{dC}{dt} = -\dfrac{1}{\left(20 \times 10^{-5}\right)\ln 10}\left(5.5 \times 10^{-5}\right) = 0.12$ per min

51. $P_1 = 2.0$ W, $P_2 = 400$ W, $\dfrac{dP_2}{dt} = -0.50$ W/day

$\text{dB} = 10\log\left(\dfrac{P_2}{P_1}\right) = 10\log\left(\dfrac{P_2}{2}\right) \;\Rightarrow\; \dfrac{d\text{dB}}{dP_2} = 10\left(\dfrac{1}{(P_2/2)\ln 10}\right)\left(\dfrac{1}{2}\right) = \dfrac{10}{P_2 \ln 10}$

$\dfrac{d\text{dB}}{dt} = \dfrac{d\text{dB}}{dP_2} \cdot \dfrac{dP_2}{dt} = \dfrac{10}{(400)\ln 10}(-0.50) = -0.0054$ dB/day

Exercise 29.5 • Derivatives of the Exponential Functions

Derivative of b^u

1. $y = 3^{2x} \;\Rightarrow\; y' = 3^{2x}(2)\ln 3 = 2\left(3^{2x}\right)\ln 3$

3. $y = (x)\left(10^{2x+3}\right) \;\Rightarrow\; y' = x\left[2\left(10^{2x+3}\right)\ln 10\right] + \left(10^{2x+3}\right)(1) = 10^{2x+3}\left(1 + 2x\ln 10\right)$

5. $y = 2^{x^2} \;\Rightarrow\; y' = 2^{x^2}(2x)\ln 2 = 2x\left(2^{x^2}\right)\ln 2$

Derivative of e^u

7. $y = e^{2x} \Rightarrow y' = 2e^{2x}$

9. $y = e^{e^x} \Rightarrow y' = e^{e^x} e^x = e^{e^x + x}$

11. $y = e^{\sqrt{1-x^2}} \Rightarrow y' = e^{\sqrt{1-x^2}} \left(\frac{1}{2}\left(1-x^2\right)^{-1/2}(-2x) \right) = -\frac{xe^{\sqrt{1-x^2}}}{\sqrt{1-x^2}}$

13. $y = \frac{2}{e^x} = 2e^{-x} \Rightarrow y' = 2e^{-x}(-1) = -2e^{-x}$ or $-\frac{2}{e^x}$

15. $y = x^2 e^{3x} \Rightarrow y' = x^2 e^{3x}(3) + e^{3x}(2x) = xe^{3x}(3x+2)$

17. $y = \frac{e^x}{x} \Rightarrow y' = \frac{xe^x - e^x(1)}{x^2} = \frac{e^x(x-1)}{x^2}$

19. $y = \frac{e^x - e^{-x}}{x^2} \Rightarrow y' = \frac{x^2\left(e^x + e^{-x}\right) - \left(e^x - e^{-x}\right)2x}{\left(x^2\right)^2} = \frac{x^2 e^x + x^2 e^{-x} - 2xe^x + 2xe^{-x}}{x^4} = \frac{(x-2)e^x + (x+2)e^{-x}}{x^3}$

21. $y = \left(x + e^x\right)^2 \Rightarrow y' = 2\left(x + e^x\right)\left(1 + e^x\right) = 2\left(x + xe^x + e^x + e^{2x}\right)$

23. $y = \frac{\left(1 + e^x\right)^2}{x} \Rightarrow y' = \frac{x(2)\left(1+e^x\right)\left(e^x\right) - \left(1+e^x\right)^2(1)}{x^2} = \frac{\left(1+e^x\right)\left(2xe^x - e^x - 1\right)}{x^2}$

With Trigonometric Functions

25. $y = \sin^3 e^x \Rightarrow y' = 3\left(\sin^2 e^x\right)\left(\cos e^x\right)\left(e^x\right) = 3e^x \sin^2 e^x \cos e^x$

27. $y = e^\theta \cos 2\theta \Rightarrow y' = e^\theta(-\sin 2\theta)(2) + (\cos 2\theta)e^\theta = e^\theta(\cos 2\theta - 2\sin 2\theta)$

Implicit Relations

29. $e^x + e^y = 1$

$e^x + e^y(y') = 0 \Rightarrow y' = \frac{-e^x}{e^y} = -e^{x-y}$

31. $e^y = \sin(x+y)$

$e^y(y') = \cos(x+y)(1+y') \Rightarrow y'\left(e^y - \cos(x+y)\right) = \cos(x+y) \Rightarrow y' = \frac{\cos(x+y)}{e^y - \cos(x+y)}$

33. $f(t)=e^{\sin t}\cos t \;\Rightarrow\; f'(t)=e^{\sin t}(-\sin t)+(\cos t)e^{\sin t}(\cos t)=e^{\sin t}\left(\cos^2 t-\sin t\right)$

$f''(t)=e^{\sin t}\left(2\cos t(-\sin t)-\cos t\right)+\left(\cos^2 t-\sin t\right)e^{\sin t}(\cos t)=e^{\sin t}\left(\cos^3 t-3\cos t\sin t-\cos t\right)$

$f''(0)=e^{\sin 0}\left(\cos^3 0-3\cos 0\sin t-\cos 0\right)=1(1-0-1)=0$

With Logarithmic Functions

35. $y=e^x\ln x \;\Rightarrow\; y'=e^x\left(\frac{1}{x}\right)+(\ln x)e^x=e^x\left(\ln x+\frac{1}{x}\right)$

37. $y=\ln x^{e^x}=e^x\ln x \;\Rightarrow\; y'=e^x\left(\frac{1}{x}\right)+(\ln x)e^x=e^x\left(\frac{1}{x}+\ln x\right)$

39. $y=e^t\cos t$

$y'=e^t(-\sin t)+(\cos t)e^t=e^t(\cos t-\sin t)$

$y''=e^t(-\sin t-\cos t)+(\cos t-\sin t)e^t=e^t(-2\sin t)=-2e^t\sin t$

41. $y=e^x\sin x$

$y'=e^x(\cos x)+(\sin x)e^x=e^x(\cos x+\sin x)$

$y''=e^x(-\sin x+\cos x)+(\cos x+\sin x)e^x=e^x(2\cos x)=2e^x\cos x$

Approximate Solution

43. $e^x+x-3=0 \;\Rightarrow\; x=0.79$

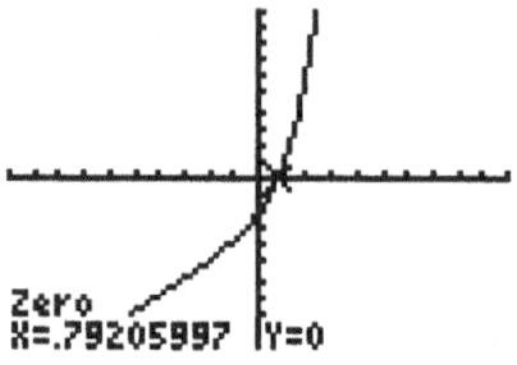

45. $5e^{-x}+x-5=0 \;\Rightarrow\; x=4.97$

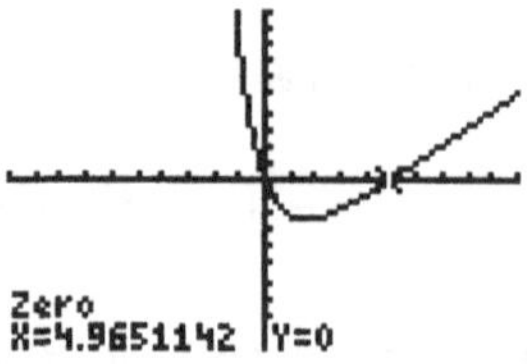

Maximum, Minimum, and Inflection Points

47. $y=e^{2x}+5e^{-2x} \;\Rightarrow\; y'=2e^{2x}-10e^{-2x} \;\Rightarrow\; y''=4e^{2x}+20e^{-2x}$

$y'=0 \;\Rightarrow\; 2e^{2x}=10e^{-2x} \;\Rightarrow\; e^{4x}=5 \;\Rightarrow\; \ln e^{4x}=\ln 5 \;\Rightarrow\; x=\frac{\ln 5}{4}=0.402$

$y''=0 \;\Rightarrow\; 4e^{2x}=20e^{-2x} \;\Rightarrow\; e^{4x}=5 \;\Rightarrow\; x=\frac{\ln 5}{4}=0.402$

$y'(0)=2-10<0$ decreasing on $x<0.402$ and $y'(1)=13.4>0$ increasing on $x>0.402$

minimum at $x=0.402$

$y(0.402)=4.472$

There is a minimum at $(0.402, 4.472)$.

49. $y = 10e^{-x}\sin x \quad \Rightarrow \quad y' = 10e^{-x}\cos x - 10e^{-x}\sin x = 10e^{-x}(\cos x - \sin x)$

$y'' = 10e^{-x}(-\sin x - \cos x) - 10e^{-x}(\cos x - \sin x) = 10e^{-x}(-2\cos x) = -20e^{-x}\cos x$

$y' = 0 \quad \Rightarrow \quad 10e^{-x}(\cos x - \sin x) = 0 \quad \Rightarrow \quad \cos x = \sin x \quad \Rightarrow \quad x = \frac{\pi}{4}$ and $\frac{5\pi}{4}$

$y''\left(\frac{\pi}{4}\right) < 0$ maximum at $x = \frac{\pi}{4}$ and $y''\left(\frac{5\pi}{4}\right) > 0$ minimum at $x = \frac{5\pi}{4}$

$y\left(\frac{\pi}{4}\right) = 3.224,\ y\left(\frac{5\pi}{4}\right) = -0.139$

There is a maximum at $\left(\frac{\pi}{4}, 3.224\right)$ and a minimum point at $\left(\frac{5\pi}{4}, -0.139\right)$.

Applications

51. $P = 9000e^{0.08t} \quad \Rightarrow \quad P' = 9000(0.08)e^{0.08t} = 720e^{0.08t}$

$P'(3) = 720e^{0.08(3)} = \$915.30$ per year

53. $y = \frac{1}{2}\left(e^{x} + e^{-x}\right) \quad \Rightarrow \quad y' = \frac{1}{2}\left(e^{x} - e^{-x}\right)$

$m = y'(5) = \frac{1}{2}\left(e^{5} - e^{-5}\right) = 74.2$

55. $N = 1855e^{-0.5t} \quad \Rightarrow \quad N' = 1855(-0.5)e^{-0.5t} = -927.5e^{-0.5t}$

$N'(1) = -927.5e^{-0.5(1)} = -563$ rev/min^2

57. $N = 10{,}000e^{0.1t} \quad \Rightarrow \quad N' = 10{,}000(0.1)e^{0.1t} = 1000e^{0.1t}$

(a) $N'(0) = 1000e^{0.1(0)} = 1000$ bacteria/h

(b) $N'(100) = 1000e^{0.1(100)} = 2.2 \times 10^{7}$ bacteria/h

59. $p = 29.92e^{-h/5} \quad \Rightarrow \quad \frac{dp}{dt} = 29.92\left(-\frac{1}{5}\right)e^{-h/5}\frac{dh}{dt} = -5.984e^{-h/5}\frac{dh}{dt}$

$h = 18.0$ mi, $\frac{dh}{dt} = 1500$ m/h

$\frac{dp}{dt} = -5.984e^{-(18.0)/5}(1500) = -245$ in. Hg/h

61. $d = 64.0e^{0.00676h} \quad \Rightarrow \quad \frac{dp}{dh} = 64.0(0.00676)e^{0.00676h} = 0.43264e^{0.00676h}$

$\left.\frac{dp}{dh}\right|_{h=1.00 \text{ mi}} = 0.43264e^{0.00676(1)} = 0.436$ lb/ft^3/mi

Electrical Applications

63. $v = 448\left(1 - e^{-t/122}\right) \quad \Rightarrow \quad \frac{dv}{dt} = 448\left(-\left(-\frac{1}{122}\right)e^{-t/122}\right) = 3.67e^{-t/122}$

a) $i = C\frac{dv}{dt} = (185)\left(3.67e^{-t/122}\right) = 679e^{-t/122}\ \mu\text{A}$

b) $i(150) = 679e^{-(150)/122} = 199\ \mu\text{A}$

65. $i = 225e^{-t/128} \quad \Rightarrow \quad \frac{di}{dt} = 225\left(-\frac{1}{128}\right)e^{-t/128} = -1.758e^{-t/128}$

a) $v = L\frac{di}{dt} = (37.2)\left(-1.758e^{-t/128}\right) = -65.4e^{-t/128}\ \text{V}$

b) $v(155) = -65.4e^{-(155)/128} = -19.5\ \text{V}$

Exercise 29.6 • Integrals of the Exponential and Logarithmic Functions

Exponential Functions

1. $\int a^{5x}dx = \frac{1}{5}\int a^{5x}(5)dx = \frac{1}{5}\left(\frac{1}{\ln a}a^{5x}\right) + C = \frac{a^{5x}}{5\ln a} + C$

3. $\int 5^{7x}dx = \frac{1}{7}\int 5^{7x}(7)dx = \frac{1}{7}\left(\frac{1}{\ln 5}5^{7x}\right) + C = \frac{5^{7x}}{7\ln 5} + C$

5. $\int a^{3y}dy = \frac{1}{3}\int a^{3y}(3)dy = \frac{1}{3}\left(\frac{1}{\ln a}a^{3y}\right) + C = \frac{a^{3y}}{3\ln a} + C$

7. $\int 4e^{x}dx = 4\int e^{x}dx = 4e^{x} + C$

9. $\int xe^{x^2}dx = \frac{1}{2}\int e^{x^2}(2x)dx = \frac{1}{2}e^{x^2} + C$

11. $\int_1^2 xe^{3x^2}dx = \frac{1}{6}\int_1^2 e^{3x^2}(6x)dx = \frac{1}{6}e^{3x^2}\Big|_1^2 = \frac{1}{6}\left[e^{3(4)} - e^{3(1)}\right] = 27{,}122$

13. $\int \frac{e^{\sqrt{x}}}{\sqrt{x}}dx = \int e^{x^{1/2}}\left(x^{-1/2}\right)dx = 2\int e^{x^{1/2}}\left(\frac{1}{2}x^{-1/2}\right)dx = 2e^{\sqrt{x}} + C$

15. $\int_0^1\left(e^{x} - 1\right)^2 dx = \int_0^1\left(e^{2x} - 2e^{x} + 1\right)dx = \left(\frac{1}{2}e^{2x} - 2e^{x} + x\right)\Big|_0^1 = \left(\frac{1}{2}e^{2} - 2e + 1\right) - \left(\frac{1}{2}(1) - 2(1) + 0\right) = 0.7580$

17. $\int \frac{e^{\sqrt{x-2}}}{\sqrt{x-2}}dx = \int e^{(x-2)^{1/2}}\left((x-2)^{-1/2}\right)dx = 2\int e^{(x-2)^{1/2}}\left(\frac{1}{2}(x-2)^{-1/2}\right)dx = 2e^{\sqrt{x-2}} + C$

19. $\int\left(e^{x/a}+e^{-x/a}\right)dx = a\int e^{x/a}\left(\frac{1}{a}\right)dx+(-a)\int e^{-x/a}\left(-\frac{1}{a}\right)dx = ae^{x/a}-ae^{-x/a}+C$

Logarithmic Functions

21. $\int \ln 3x\,dx = \frac{1}{3}\int \ln 3x\,(3)\,dx = \frac{1}{3}(3x)(\ln 3x-1)+C = x(\ln 3x-1)+C$ or $x\ln 3x - x + C$

23. $\int_1^2 x\ln x^2\,dx = \frac{1}{2}\int_1^2 \ln x^2\,(2x)\,dx = \frac{1}{2}\left(x^2\right)\left(\ln x^2-1\right)\Big|_1^2 = \frac{1}{2}(4)(\ln 4-1)-\frac{1}{2}(1)(\ln 1-1) = 1.273$

25. $\int_2^4 x\log\left(x^2+1\right)dx = \frac{1}{2}\int_3^4 \frac{\ln\left(x^2+1\right)}{\ln 10}(2x)\,dx = \frac{1}{2\ln 10}\left(x^2+1\right)\left(\ln\left(x^2+1\right)-1\right)\Big|_2^4$

$= \frac{1}{2\ln 10}\left[17(\ln(17)-1)-5(\ln(5)-1)\right] = 6.106$

27. $y=e^{2x} \Rightarrow A=\int_1^3 e^{2x}dx = \frac{1}{2}\int_1^3 e^{2x}(2)\,dx = \frac{1}{2}e^{2x}\Big|_1^3 = \frac{1}{2}\left[e^{2(3)}-e^{2(1)}\right] = 198.0$

29. $dV = 2\pi(1-x)\,y\,dx = 2\pi(1-x)e^x dx$

$V = 2\pi\int_0^1\left(e^x - xe^x\right)dx$

$= 2\pi\left(e^x - e^x(x-1)\right)\Big|_0^1$ [Use Rule 37 in the Table of Integrals]

$= 2\pi\left(2e^x - xe^x\right)\Big|_0^1$

$= 2\pi\left[(2e-e)-(2)\right]$

$= 2\pi(e-2) = 4.51$

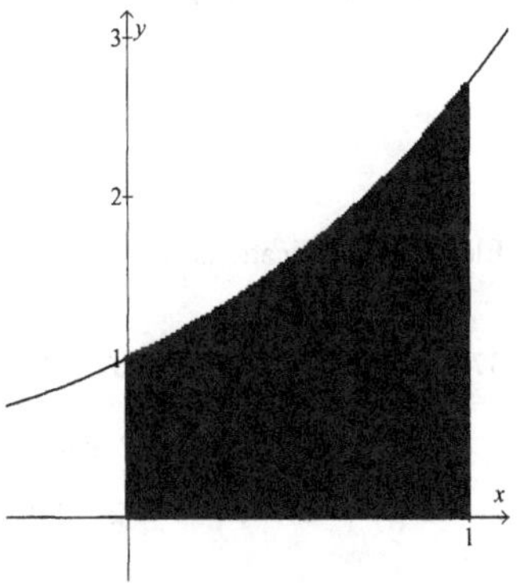

31. $y=e^{-x} \Rightarrow y' = -e^{-x}$

$S = 2\pi\int_0^{100} y\sqrt{1+(y')^2}\,dx = 2\pi\int_0^{100} e^{-x}\sqrt{1+\left(-e^{-x}\right)^2}\,dx$

$= -2\pi\int_0^{100}\left(\sqrt{\left(e^{-x}\right)^2+1}\right)\left(-e^{-x}\right)dx$

$= -2\pi\left(\frac{e^{-x}}{2}\sqrt{e^{-2x}+1}+\frac{1}{2}\ln\left|e^{-x}+\sqrt{e^{-2x}+1}\right|\right)\Big|_0^{100}$ [Use Rule 66]

$= 7.21$

33. $A=\int_0^1 e^x\,dx = e^x\Big|_0^1 = e-1$

$\bar{y}=\dfrac{1}{2A}\int_0^1\left[\left(e^x\right)^2-0\right]dx$

$=\dfrac{1}{2A}\left(\dfrac{1}{2}\right)\int_0^1 e^{2x}(2)\,dx$

$=\dfrac{1}{4A}e^{2x}\Big|_0^1=\dfrac{1}{4A}\left(e^2-1\right)$

$=\dfrac{e^2-1}{4(e-1)}=\dfrac{e+1}{4}=0.930$

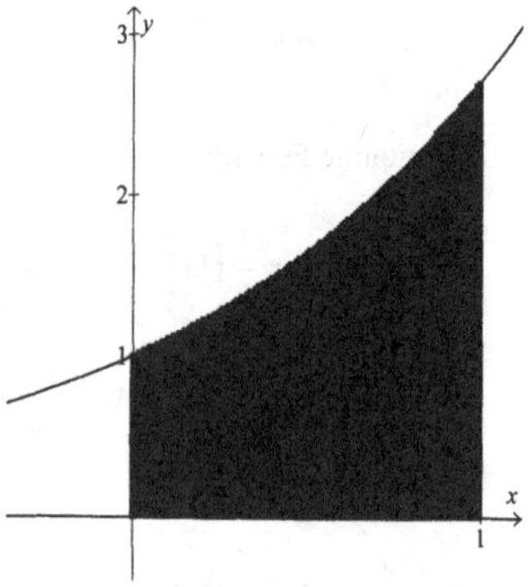

35. $I_x=\dfrac{1}{3}\int_0^1\left(e^x\right)^3 dx=\dfrac{1}{3}\left(\dfrac{1}{3}\right)\int_0^1 e^{3x}(3)\,dx=\dfrac{1}{9}e^{3x}\Big|_0^1=\dfrac{1}{9}\left(e^3-1\right)=2.12$

$I_y=\int_0^1 x^2e^x\,dx=\left[e^x\left(x^2-2x+2\right)\right]\Big|_0^1$ [Use Rule 38]

$=e(1-2+2)-1(2)$

$=e-2=0.718$

The moment is $(2.12, 0.718)$.

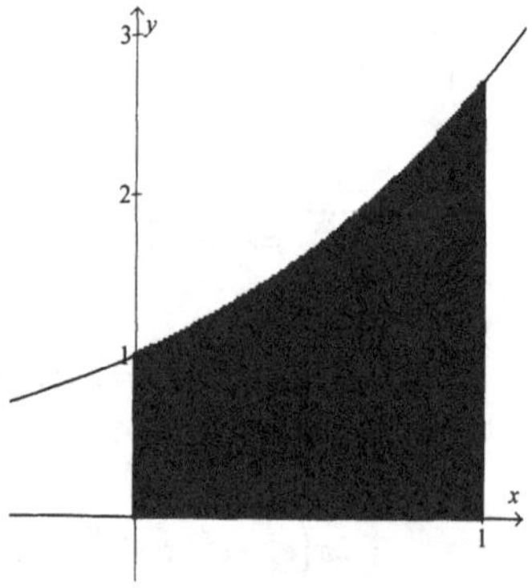

Electrical Applications

37. $V=\dfrac{1}{C}\int i\,dt=\dfrac{1}{3.85}\int 84.6e^{-t/127}\,dt=\dfrac{84.6}{3.85}(-127)\int e^{-t/127}\left(-\dfrac{1}{127}\right)dt=-2790e^{-t/127}+k$

$V(0)=0 \;\Rightarrow\; 0=-2790(1)+k \;\Rightarrow\; k=2790$

$V(t)=-2790e^{-t/127}+2790$ V

Exercise 29.7 • Integrals of the Trigonometric Functions

1. $\int \sin 3x\,dx=\dfrac{1}{3}\int \sin 3x\,(3)\,dx=\dfrac{1}{3}(-\cos 3x)+C=-\dfrac{1}{3}\cos 3x+C$

3. $\int \tan 5\theta\,d\theta=\dfrac{1}{5}\int \tan 5\theta(5)\,d\theta=\dfrac{1}{5}\left(-\ln\left|\cos 5\theta\right|\right)+C=-\dfrac{1}{5}\ln\left|\cos 5\theta\right|+C$

5. $\int \sec 4x\,dx=\dfrac{1}{4}\int \sec 4x(4)\,dx=\dfrac{1}{4}\ln\left|\sec 4x+\tan 4x\right|+C$

7. $\int 3\tan 9\theta\,d\theta=\dfrac{3}{9}\int \tan 9\theta(9)\,d\theta=\dfrac{1}{3}\left(-\ln\left|\cos 9\theta\right|\right)+C=-\dfrac{1}{3}\ln\left|\cos 9\theta\right|+C$

9. $\int x\sin x^2\,dx = \frac{1}{2}\int \sin x^2\,(2x)\,dx = \frac{1}{2}\left(-\cos x^2\right) + C = -\frac{1}{2}\cos x^2 + C$

11. $\int \theta^2 \tan\theta^3\,d\theta = \frac{1}{3}\int \tan\theta^3\left(3\theta^2\right)d\theta = \frac{1}{3}\left(-\ln\left|\cos\theta^3\right|\right) + C = -\frac{1}{3}\ln\left|\cos\theta^3\right| + C$

13. $\int \sin(x+1)\,dx = -\cos(x+1) + C$

15. $\int \tan(4-5\theta)\,d\theta = -\frac{1}{5}\int \tan(4-5\theta)(-5)\,d\theta = -\frac{1}{5}\left(-\ln\left|\cos(4-5\theta)\right|\right) + C = \frac{1}{5}\ln\left|\cos(4-5\theta)\right| + C$

17. $\int x\sec\left(4x^2-3\right)dx = \frac{1}{8}\int \sec\left(4x^2-3\right)(8x)\,dx = \frac{1}{8}\ln\left|\sec\left(4x^2-3\right) + \tan\left(4x^2-3\right)\right| + C$

19. $\int x\cos x^2\,dx = \frac{1}{2}\int \cos x^2\,(2x)\,dx = \frac{1}{2}\sin x^2 + C$ or $\frac{\sin x^2}{2} + C$

21. $\int_0^\pi \sin\phi\,d\phi = -\cos\phi\Big|_0^\pi = -\cos\pi - (-\cos 0) = -1(-1) + 1 = 2$

23. $\int_0^\pi \cos\frac{\theta}{2}\,d\theta = 2\int_0^\pi \cos\frac{\theta}{2}\left(\frac{1}{2}\right)d\theta = 2\sin\frac{\theta}{2}\Big|_0^\pi = 2\sin\frac{\pi}{2} - 2\sin 0 = 2(1) - 0 = 2$

25. $y = \sin x \;\Rightarrow\; A = \int_0^\pi \sin x\,dx = -\cos x\Big|_0^\pi = -(\cos\pi - \cos 0) = -(-2) = 2$

27. $y = 2\sin\left(\frac{1}{2}\pi x\right)$

$$A = \int_0^2 2\sin\left(\frac{1}{2}\pi x\right)dx = 2\left(\frac{2}{\pi}\right)\int_0^2 \sin\left(\frac{1}{2}\pi x\right)\left(\frac{\pi}{2}\right)dx = \frac{4}{\pi}\left(-\cos\left(\frac{1}{2}\pi x\right)\right)\Bigg|_0^2 = -\frac{4}{\pi}(\cos\pi - \cos 0) = -\frac{4}{\pi}(-2) = \frac{8}{\pi}$$

29. $A = \int_0^{\pi/2}(\cos x - 0)\,dx + \int_{\pi/2}^{3\pi/2}(0 - \cos x)\,dx$

$= \sin x\Big|_0^{\pi/2} - \sin x\Big|_{\pi/2}^{3\pi/2}$

$= \left(\sin\frac{\pi}{2} - 0\right) - \left(\sin\frac{3\pi}{2} - \sin\frac{\pi}{2}\right)$

$= 1 - (-1 - 1)$

$= 3$

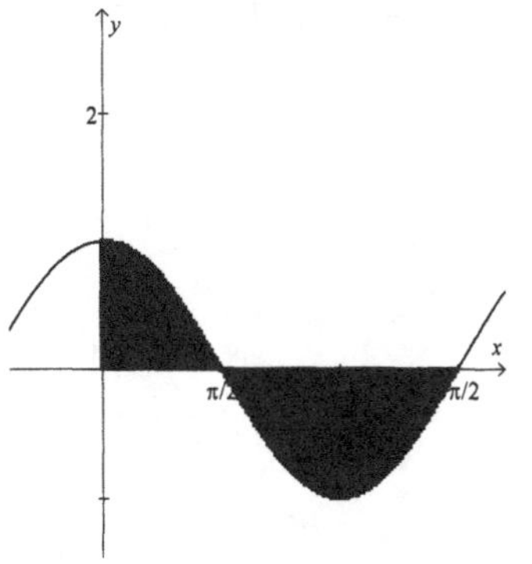

31. $y = \sin x \;\Rightarrow\; y' = \cos x$

$$S = 2\pi\int_0^{\pi} y\sqrt{1+(y')^2}\,dx = -2\pi\int_0^{\pi} -\sin x\sqrt{1+(\cos x)^2}\,dx$$

$$= -2\pi\left(\frac{\cos x}{2}\sqrt{\cos^2 x+1}+\frac{1}{2}\ln\left|\cos x+\sqrt{\cos^2 x+1}\right|\right)\Bigg|_0^{\pi} \quad \text{[Use Rule 66]}$$

$$= -2\pi\left[\left(\frac{-1}{2}\sqrt{(-1)^2+1}+\frac{1}{2}\ln\left|-1+\sqrt{(-1)^2+1}\right|\right)-\left(\frac{1}{2}\sqrt{1+1}+\frac{1}{2}\ln\left|1+\sqrt{1+1}\right|\right)\right]$$

$$= -2\pi\left[\left(-\frac{\sqrt{2}}{2}+\frac{1}{2}\ln\left|\sqrt{2}-1\right|\right)-\left(\frac{\sqrt{2}}{2}+\frac{1}{2}\ln\left|\sqrt{2}+1\right|\right)\right]$$

$$= -2\pi\left[-\sqrt{2}+\frac{1}{2}\ln\left|\sqrt{2}-1\right|-\frac{1}{2}\ln\left|\sqrt{2}+1\right|\right]$$

$$= -14.4$$

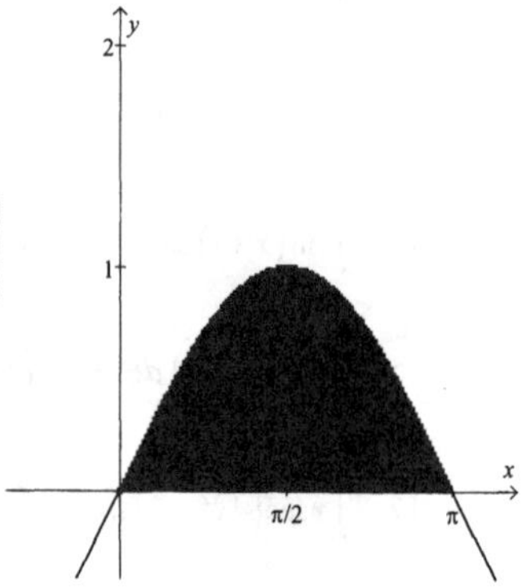

33. $A = \int_0^{\pi}\sin x\,dx = -\cos x\big|_0^{\pi} = -(-1)+1 = 2$

$$\bar{x} = \frac{1}{A}\int_0^{\pi} x[\sin x - 0]dx = \frac{1}{2}\int_0^{\pi} x\sin x\,dx = \frac{1}{2}(\sin x - x\cos x)\big|_0^{\pi} = \frac{1}{2}\left[(0-\pi(-1))-0\right] = \frac{\pi}{2}$$

$$\bar{y} = \frac{1}{2A}\int_0^{\pi}\left[\left(e^x\right)^2-0\right]dx = \frac{1}{4}\int_0^{\pi}\left[(\sin x)^2-0\right]dx = \frac{1}{4}\int_0^{\pi}\sin^2 x\,dx = \frac{1}{4}\left(\frac{x}{2}-\frac{\sin 2x}{4}\right)\Bigg|_0^{\pi} = \frac{1}{4}\left(\frac{\pi}{2}-0\right) = \frac{\pi}{8}$$

The coordinates of the centroid are $\left(\frac{\pi}{2},\frac{\pi}{8}\right)$.

Electrical Applications

35. a) $q = \int i\,dt = \int 84.3\sin(11.5t+5.48)\,dt = 84.3\left(\frac{1}{11.5}\right)\int(\sin(11.5t+5.48))(11.5)dt = -7.33\cos(11.5t+5.48)+k$

$q(0) = 0 \;\Rightarrow\; 0 = -7.33\cos(5.48)+k \;\Rightarrow\; k = 7.33\cos 5.48 = 5.09$

$q(t) = -7.33\cos(11.5t+5.48)+5.09$ C

b) $q(2) = -7.33\cos(11.5(2)+5.48)+5.09 = 12.3$ C

Exercise 29.8 • Average and Root Mean Square Values

Find the average ordinate for each function in the given interval

1. $y = x^2 \Rightarrow y_{\text{avg}} = \dfrac{1}{6-0}\displaystyle\int_0^6 x^2\,dx = \dfrac{1}{6}\left(\dfrac{1}{3}x^3\right)\Bigg|_0^6 = \dfrac{1}{18}\left(6^3 - 0\right) = 12$

3. $y = \sqrt{1+2x} = (1+2x)^{1/2}$

$$y_{\text{avg}} = \frac{1}{12-4}\int_4^{12}(1+2x)^{1/2}\,dx = \frac{1}{8}\left(\frac{1}{2}\right)\int_4^{12}(1+2x)^{1/2}(2)\,dx = \frac{1}{16}\left(\frac{2}{3}(1+2x)^{3/2}\right)\Bigg|_4^{12} = \frac{1}{24}\left(25^{3/2} - 9^{3/2}\right)$$

$$= \frac{1}{24}(125-27) = 4.08$$

5. $y = \sin^2 x$ [Use Rule 16 in the Table of Integrals]

$$y_{\text{avg}} = \frac{1}{(\pi/2)-0}\int_0^{\pi/2}\sin^2 x\,dx = \frac{2}{\pi}\left(\frac{x}{2} - \frac{\sin 2x}{4}\right)\Bigg|_0^{\pi/2} = \frac{2}{\pi}\left[\left(\frac{\pi/2}{2} - \frac{\sin\pi}{4}\right) - (0)\right] = \frac{2}{\pi}\left(\frac{\pi}{4}\right) = \frac{1}{2}$$

Find the rms value for each function in the given interval.

7. $y = 2x+1$

$$\text{rms}^2 = \frac{1}{6-0}\int_0^6 (2x+1)^2\,dx = \frac{1}{6}\int_0^6 4x^2 + 4x + 1\,dx = \frac{1}{6}\left(\frac{4}{3}x^3 + 2x^2 + x\right)\Bigg|_0^6$$

$$= \frac{1}{6}\left(\frac{4}{3}(6^3) + 2(6)^2 + 6 - (0)\right) = \frac{1}{6}(288 + 72 + 6) = 61$$

$\text{rms} = \sqrt{61} = 7.81$

9. $y = x + 2x^2$

$$\text{rms}^2 = \frac{1}{4-1}\int_1^4 \left(x+2x^2\right)^2\,dx = \frac{1}{3}\int_1^4 4x^4 + 4x^3 + x^2\,dx = \frac{1}{3}\left(\frac{4}{5}x^5 + x^4 + \frac{1}{3}x^3\right)\Bigg|_1^4$$

$$= \frac{1}{3}\left[\left(\frac{4}{5}(4^5) + 4^4 + \frac{1}{3}(4^3)\right) - \left(\frac{4}{5} + 1 + \frac{1}{3}\right)\right] = 365$$

$\text{rms} = \sqrt{365} = 19.1$

11. $y = 2\cos x$ [Use Rule 17 in the Table of Integrals]

$$\text{rms}^2 = \frac{1}{(\pi/2)-(\pi/6)}\int_{\pi/6}^{\pi/2} 4\cos^2 x\,dx = \frac{3}{\pi}(4)\left(\frac{x}{2} + \frac{\sin 2x}{4}\right)\Bigg|_{\pi/6}^{\pi/2} = \frac{12}{\pi}\left[\left(\frac{\pi}{4} + \frac{\sin\pi}{4}\right) - \left(\frac{\pi}{12} + \frac{\sin(\pi/3)}{4}\right)\right] = 1.17$$

$\text{rms} = \sqrt{1.17} = 1.08$

12. $y = 5\sin 2x$ [Use Rule 16 in the Table of Integrals]

$$\text{rms}^2 = \frac{1}{(\pi/6)-0}\int_0^{\pi/6} 25\sin^2 2x\,dx = \frac{6}{\pi}\left(\frac{25}{2}\right)\int_0^{\pi/6}\sin^2 2x(2)\,dx = \frac{75}{\pi}\left(\frac{2x}{2} - \frac{\sin 4x}{4}\right)\Bigg|_0^{\pi/6}$$

$$= \frac{75}{\pi}\left(\frac{\pi}{6} - \frac{\sin(2\pi/3)}{4} - (0)\right) = 7.33$$

$\text{rms} = \sqrt{7.33} = 2.71$

Chapter 29 • Review Problems

1. $y=\frac{a}{2}\left(e^{x/a}-e^{-x/a}\right) \Rightarrow y'=\frac{a}{2}\left(e^{x/a}\left(\frac{1}{a}\right)-e^{-x/a}\left(-\frac{1}{a}\right)\right)=\frac{1}{2}\left(e^{x/a}+e^{-x/a}\right)$

3. $y=8\tan\sqrt{x}=8\tan x^{1/2} \Rightarrow y'=8\left(\sec^2 x^{1/2}\right)\left(\frac{1}{2}x^{-1/2}\right)=\frac{4}{\sqrt{x}}\sec^2\sqrt{x}$

5. $y=x\,\text{Arctan}\,4x \Rightarrow y'=x\frac{1}{1+(4x)^2}(4)+(1)\,\text{Arctan}\,4x=\frac{4x}{16x^2+1}+\text{Arctan}\,4x$

7. $y^2=\sin 2x$

   $2yy'=(\cos 2x)(2) \Rightarrow y'=\frac{\cos 2x}{y}$

9. $y=x\sin x \Rightarrow y'=x(\cos x)+(1)\sin x=x\cos x+\sin x$

11. $y=x^3\cos x \Rightarrow y'=x^3(-\sin x)+\left(3x^2\right)\cos x=3x^2\cos x-x^3\sin x$

13. $y=\frac{\sin x}{x} \Rightarrow y'=\frac{x(\cos x)-(1)\sin x}{x^2}=\frac{x\cos x-\sin x}{x^2}$

15. $y=\log\left[x\left(1+x^2\right)\right]=\log\left(x+x^3\right) \Rightarrow y'=\frac{1+3x^2}{x^3+x}(\log e)$

17. $y=\ln\left(x+\sqrt{x^2+a^2}\right) \Rightarrow y'=\frac{1+(1/2)\left(x^2+a^2\right)^{-1/2}(2x)}{x+\sqrt{x^2+a^2}}=\frac{\sqrt{x^2+a^2}+x}{\sqrt{x^2+a^2}\left(x+\sqrt{x^2+a^2}\right)}=\frac{1}{\sqrt{x^2+a^2}}$

19. $y=\csc 3x \Rightarrow y'=(-\csc 3x\cot 3x)(3)=-3\csc 3x\cot 3x$

21. $y=\frac{\sin x}{\cos x}=\tan x \Rightarrow y'=\sec^2 x$

23. $y=\ln\left(2x^3+x\right) \Rightarrow y'=\frac{6x^2+1}{2x^3+x}$

25. $y=x^2\,\text{Arccos}\,x \Rightarrow y'=x^2\left(-\frac{1}{\sqrt{1-x^2}}\right)(1)+(2x)\,\text{Arccos}\,x=2x\text{Arccos}\,x-\frac{x^2}{\sqrt{1-x^2}}$

27. $xy = 4\log\frac{x}{2} \quad \Rightarrow \quad y = \frac{4\log\frac{x}{2}}{x}$

$$y' = 4\left(\frac{x\left(\frac{1/2}{x/2}\right)\log e - (1)\log\frac{x}{2}}{x^2}\right) = 4\left(\frac{\log e - \log\frac{x}{2}}{x^2}\right)$$

$$y'' = 4\left(\frac{-x^2\left(\frac{1/2}{x/2}\right)\log e - 2x\left(\log e - \log\frac{x}{2}\right)}{x^4}\right) = 4\left(\frac{-x\log e - 2x\log e + 2x\log\frac{x}{2}}{x^4}\right) = 4\left(\frac{2\log\frac{x}{2} - 3\log e}{x^3}\right)$$

$y'' = 0 \quad \Rightarrow \quad 2\log\frac{x}{2} = 3\log e \quad \Rightarrow \quad \log\frac{x}{2} = \frac{3\log e}{2} \quad \Rightarrow \quad \log x - \log 2 = \frac{3\log e}{2} \quad \Rightarrow \quad \log x = \frac{3\log e}{2} + \log 2$

$\log x = 0.9525 \quad \Rightarrow \quad x = 10^{0.9525} = 8.96$

$y''(1) < 0$ concave down for $x < 8.96$ and $y''(10) > 0$ concave up for $x > 8.96$

so inflection point at $x = 8.96$ with $y(8.96) = 0.291$

There is an inflection point at $(8.96, 0.291)$.

29. $\int_0^1 3x\sin x^2\, dx = 3\left(\frac{1}{2}\right)\int_0^1 \sin x^2\,(2)x dx = -\frac{3}{2}\cos x^2\Big|_0^1 = -\frac{3}{2}[\cos 1 - \cos 0] = 0.690$

31. $\int 6x\cos 2x^2\, dx = 6\left(\frac{1}{4}\right)\int \cos 2x^2\,(4x)\, dx = \frac{3}{2}\sin 2x^2 + C$

33. $\int_0^{\pi/2} 3\cos 5x\, dx = 3\left(\frac{1}{5}\right)\int_0^{\pi/2} \cos 5x(5)\, dx = \frac{3}{5}\sin 5x\Big|_0^{\pi/2} = \frac{3}{5}\left[\sin\left(5\left(\frac{\pi}{2}\right)\right) - 0\right] = \frac{3}{5}(1) = \frac{3}{5}$

35. $\int x\ln x^2\, dx = \frac{1}{2}\int \ln x^2\,(2)x\, dx = \frac{1}{2}\left(x^2\right)\left(\ln x^2 - 1\right) + C = \frac{x^2}{2}\left(\ln x^2 - 1\right) + C$

37. $\int x\ln\left(3x^2 - 2\right) dx = \frac{1}{6}\int \ln\left(3x^2 - 2\right)(6)x\, dx$

$= \frac{1}{6}\left(3x^2 - 2\right)\left(\ln\left(3x^2 - 2\right) - 1\right) + C$ or $\frac{1}{6}\left(3x^2 - 2\right)\ln\left(3x^2 - 2\right) - \frac{x^2}{2} + C$

39. $\int_0^{\pi} x^2\cos\left(x^3 + 5\right) dx = \left(\frac{1}{3}\right)\int_0^{\pi} \cos\left(x^3 + 5\right)(3)x^2\, dx = \frac{1}{3}\sin\left(x^3 + 5\right)\Big|_0^{\pi} = \frac{1}{3}\left[\sin\left(\pi^3 + 5\right) - \sin 5\right] = -0.0112$

41. $y = \sin^2 x$

$y_{\text{avg}} = \frac{1}{2\pi - 0}\int_0^{2\pi} \sin^2 x\, dx = \frac{1}{2\pi}\left(\frac{x}{2} - \frac{\sin 2x}{4}\right)\Big|_0^{2\pi}$ [Use Rule 16 in the Table of Integrals]

$= \frac{1}{2\pi}\left[\left(\frac{2\pi}{2} - \frac{\sin 4\pi}{4}\right) - (0)\right] = \frac{1}{2\pi}(\pi) = \frac{1}{2}$

43. $y = 2\tan x - \tan^2 x \Rightarrow y' = 2\sec^2 x - 2\tan x \sec^2 x = 2\sec^2 x(1 - \tan x)$

$y' = 0 \Rightarrow \tan x = 1 \Rightarrow x = \frac{\pi}{4}$

$y'(0) > 0$ increasing for $x < \frac{\pi}{4}$ and $y'(1) < 0$ decreasing for $x > \frac{\pi}{4}$

so maximum at $x = \frac{\pi}{4}$

45. $y = \frac{\text{Arcsec}\,2x}{\sqrt{x}} \Rightarrow y' = \frac{x^{1/2}\left(\frac{2}{2x\sqrt{4x^2-1}}\right) - \frac{1}{2}x^{-1/2}\text{Arcsec}\,2x}{x} = \frac{\frac{1}{\sqrt{4x^2-1}} - \frac{1}{2}\text{Arcsec}\,2x}{x^{3/2}}$

$y'(1) == \frac{\frac{1}{\sqrt{4-1}} - \frac{1}{2}\text{Arcsec}\,2}{1} = \frac{1}{\sqrt{3}} - \frac{1}{2}\text{Arccos}\frac{1}{2} = \frac{1}{\sqrt{3}} - \frac{1}{2}\left(\frac{\pi}{3}\right) = 0.0538$

47. $y = \sin x \Rightarrow y' = \cos x \Rightarrow m_{\tan} = y'\left(\frac{\pi}{6}\right) = \cos\left(\frac{\pi}{6}\right) = \frac{\sqrt{3}}{2}$

$y - \sin\frac{\pi}{6} = \frac{\sqrt{3}}{2}\left(x - \frac{\pi}{6}\right) \Rightarrow y - \frac{1}{2} = \frac{\sqrt{3}}{2}\left(x - \frac{\pi}{6}\right)$

49. $y = 2x + 5 \Rightarrow m = 2$

$y = \tan x \Rightarrow y' = \sec^2 x$

$\sec^2 x = 2 \Rightarrow \cos^2 x = \frac{1}{2} \Rightarrow \cos x = \frac{1}{\sqrt{2}} \Rightarrow x = \frac{\pi}{4}$

51. $2\sin\frac{1}{2}x - \cos 2x = 0$

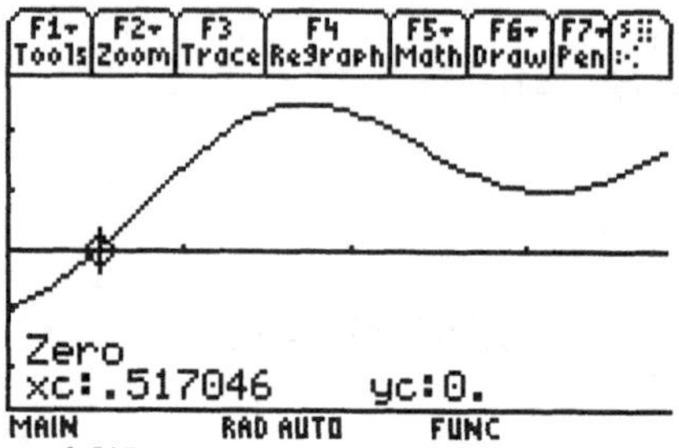

$x = 0.517$

53. $T = 100t + 1500e^{-0.2t}$

$T' = 100 + 1500(-0.2)e^{-0.2t} = 100 - 300e^{-0.2t}$

$T'' = -300(-0.2)e^{-0.2t} = 60e^{-0.2t}$

$T' = 0 \Rightarrow 300e^{-0.2t} = 100 \Rightarrow e^{-0.2t} = \frac{1}{3} \Rightarrow -0.2t = \ln\frac{1}{3} \Rightarrow t = \frac{\ln(1/3)}{-0.2} = 5.49 \text{ h}$

$T''(5.49) > 0$ so mininimum at $t = 5.49$

$T(5.49) = 100(5.49) + 1500e^{-0.2(5.49)} = 1050^\circ\text{F}$

The minimum temperature of 1050°F occurs at 5.49 h.

55. a) $v = 22.5\cos(48.3t + 0.95) \Rightarrow \frac{dv}{dt} = -22.5(48.3)\sin(48.3t + 0.95) = -1086.75\sin(48.3t + 0.95)$

$i = C\frac{dv}{dt} = (184)(-1086.75\sin(48.3t + 0.95)) = -200000\sin(48.3t + 0.95)\ \mu\text{A} = -200\sin(48.3t + 0.95)\text{ mA}$

b) $i(0.1) = -200\sin(48.3(0.1) + 0.95) = -96.4\ \text{mA}$

57. a) $q = \int i\,dt = \int 735\sin(33.6t + 0.73)\,dt = 735\left(\frac{1}{33.6}\right)\int(\sin(33.6t + 0.73))(33.6)dt$

$= -21.9\cos(33.6t + 0.73) + k$

$q(0) = 0 \Rightarrow 0 = -21.9\cos(0.73) + k \Rightarrow k = 21.9\cos 0.73 = 16.3$

$q(t) = -21.9\cos(33.6t + 0.73) + 16.3\ \text{C}$

b) $q(2.00) = -21.9\cos(33.6(2.00) + 0.73) + 16.3 = 8.06\ \text{C}$

59. a) $v = 274\left(1 - e^{-t/335}\right) \Rightarrow \frac{dv}{dt} = -274\left(-\frac{1}{335}\right)e^{-t/335} = 0.8179e^{-t/335}$

$i = C\frac{dv}{dt} = (482)\left(0.8179e^{-t/335}\right) = 394e^{-t/335}\ \mu\text{A}$

b) $i(200) = 394e^{-200/335} = 217\ \mu\text{A}$

61. $V = \frac{1}{C}\int i\,dt = \frac{1}{834}\int 63.5e^{-t/77.3}\,dt = \frac{63.5}{834}(-77.3)\int e^{-t/77.3}\left(-\frac{1}{77.3}\right)dt = -5.89e^{-t/77.3} + k$

$V(0) = 0 \Rightarrow 0 = -5.89(1) + k \Rightarrow k = 5.89$

$V(t) = -5.89e^{-t/77.3} + 5.89\ \text{V}$

Chapter 30: First-Order Differential Equations

Exercise 1 ◊ Definitions

1. $\frac{dy}{dx}+3xy=5$ (a) First order (b) first degree (c) ordinary

3. $D^3y-4Dy=2xy$ (a) Third order (b) first degree (c) ordinary

5. $3(y'')^4-5y'=3y$ (a) Second order (b) fourth degree (c) ordinary

7. $\frac{dy}{dx}=7x$
   $dy=7x\,dx$
   $y=\int 7x\,dx$
   $y=\frac{7x^2}{2}+C$

9. $4x-3y'=5$
   $4x\,dx-3\,dy=5$ [multiply by dx]
   $\int 4x\,dx-\int 3\,dy=\int 5\,dx$
   $2x^2-3y=5x+C$
   $3y=2x^2-5x+C$ [now solve for y]
   $y=\frac{2x^2}{3}-\frac{5x}{3}+\frac{C}{3}\Rightarrow y=\frac{2x^2}{3}-\frac{5x}{3}+C\left[C \text{ is a constant just as } \frac{C}{3} \text{ is a constant}\right]$

11. $dy=x^2\,dx$
   $\int dy=\int x^2\,dx$
   $y=\frac{x^3}{3}+C$

13. $y'=\frac{2y}{x},\ y=Cx^2$
   $y=Cx^2\Rightarrow y'=2Cx$
   Substituting $y=Cx^2$ and $y'=2Cx$ into $y'=\frac{2y}{x}$
   $2Cx=\frac{2(Cx^2)}{x}\Rightarrow 2Cx=2Cx$; thus $y=Cx^2$ is a solution.

15. $Dy=\frac{2y}{x},\ y=Cx^2$
   $y=Cx^2\Rightarrow Dy=2Cx$
   Substituting $y=Cx^2$ and $y'=2Cx$ into $Dy=\frac{2y}{x}$
   $2Cx=\frac{2(Cx^2)}{x}\Rightarrow 2Cx=2Cx$; thus $y=Cx^2$ is a solution.

Exercise 2 ◊ Solving a DE by Calculator, Graphically, and Numerically
Calculator Solution

1. $y'=\frac{x}{y}$

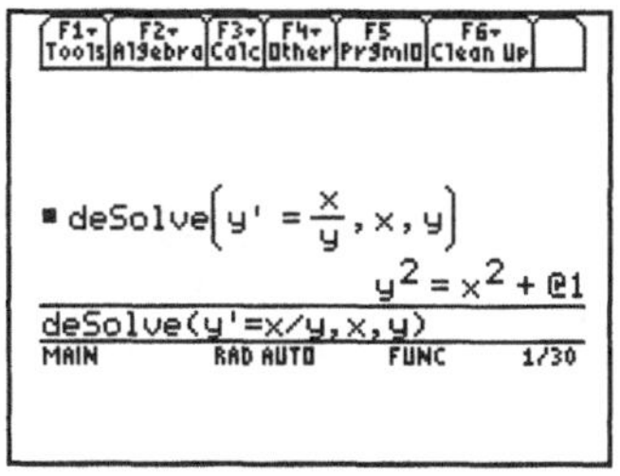

3. $y' = \dfrac{x^2}{y^3}$

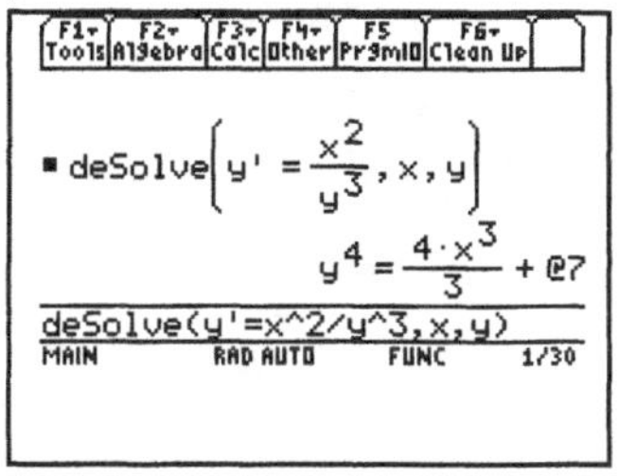

5. $y' = \dfrac{x^2}{y^2}$ $y(0) = 1$

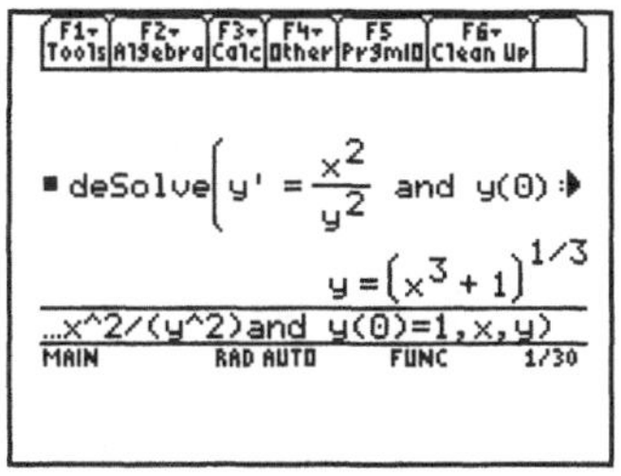

Graphical and Numerical Solution

7. $y' = x$ Start at (0, 1). Find $y(2)$.

x	y	m
0.00	1.00	0.00
0.10	1.00	0.10
0.20	1.01	0.20
0.30	1.03	0.30
0.40	1.06	0.40
0.50	1.10	0.50
0.60	1.15	0.60
0.70	1.21	0.70
0.80	1.28	0.80
0.90	1.36	0.90
1.00	1.45	1.00
1.10	1.55	1.10
1.20	1.66	1.20
1.30	1.78	1.30
1.40	1.91	1.40
1.50	2.05	1.50
1.60	2.20	1.60

y –value of 1.01 calculated from previous y value (1.00) + step value (0.10) times previous slope (0.10)

y –value of 1.36 calculated from previous y value (1.28) + step value (0.10) times previous slope (0.80)

1.70	2.36	1.70
1.80	2.53	1.80
1.90	2.71	1.90
2.00	**2.90**	2.00
2.10	3.10	2.10

$y(2) = 3$

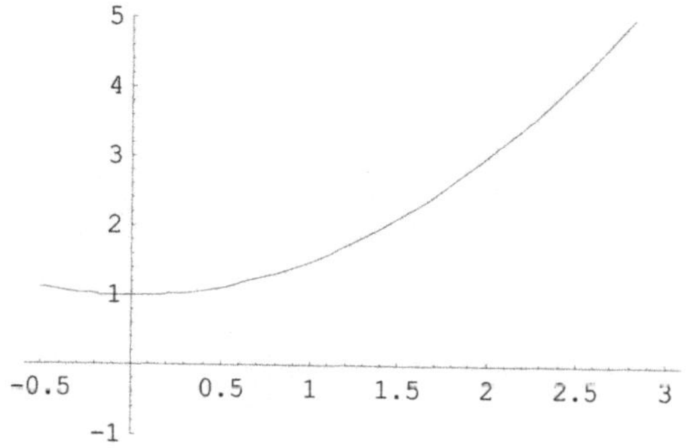

9. $y' = x - 2y$ Start at (0, 4). Find $y(3)$.

x	y	m
0.00	4.00	-8.00
0.10	3.20	-6.30
0.20	2.57	-4.94
0.30	2.08	-3.85
0.40	1.69	-2.98
0.50	1.39	-2.29
0.60	1.16	-1.73
0.70	0.99	-1.28
0.80	0.86	-0.93
0.90	0.77	-0.64
1.00	0.71	-0.41
1.10	0.67	-0.23
1.20	0.64	-0.08
1.30	0.63	0.03
1.40	0.64	0.13
1.50	0.65	0.20
1.60	0.67	0.26
1.70	0.70	0.31
1.80	0.73	0.35
1.90	0.76	0.38
2.00	0.80	0.40
2.10	0.84	0.42
2.20	0.88	0.44
2.30	0.93	0.45
2.40	0.97	0.46
2.50	1.02	0.47
2.60	1.06	0.47
2.70	1.11	0.48

2.80	1.16	0.48
2.90	1.21	0.49
3.00	**1.26**	0.49
3.10	1.30	0.49

$y(3) = 1.26$

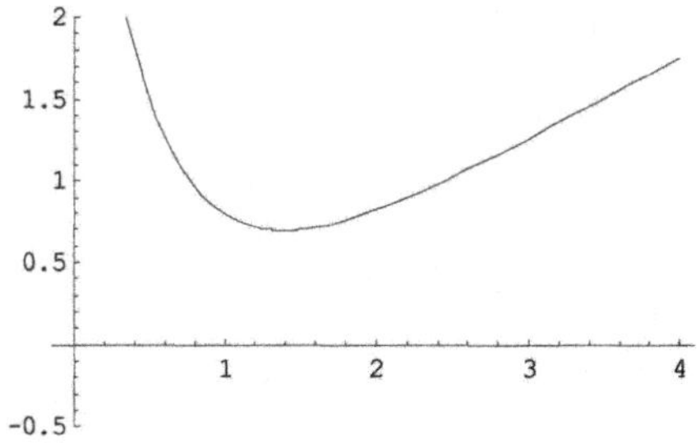

Exercise 3 ◊ First-Order DE: Variables Separable
General Solution

1. $y' = \frac{x}{y}$

$\frac{dy}{dx} = \frac{x}{y} \Rightarrow y\,dy = x\,dx$[separate variables]

$\int y\,dy = \int x\,dx$ [integrate]

$\frac{y^2}{2} = \frac{x^2}{2} + C_1$

$y^2 = x^2 + 2C_1$ [solve for y]

$y = \pm\sqrt{x^2 + C}$

By calculator

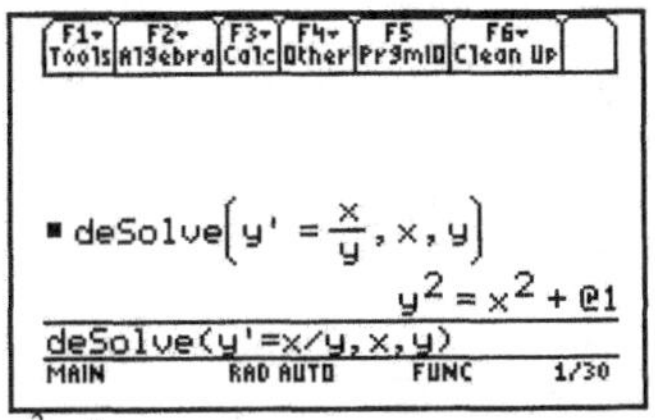

3. $dy = x^2 y\,dx$

$\frac{dy}{y} = x^2\,dx$

$\int \frac{1}{y}\,dy = \int x^2\,dx$

$\ln|y| = \frac{x^3}{3} + C_1$

$3\ln|y| = x^3 + C$

$\ln|y^3| = x^3 + C$

By calculator

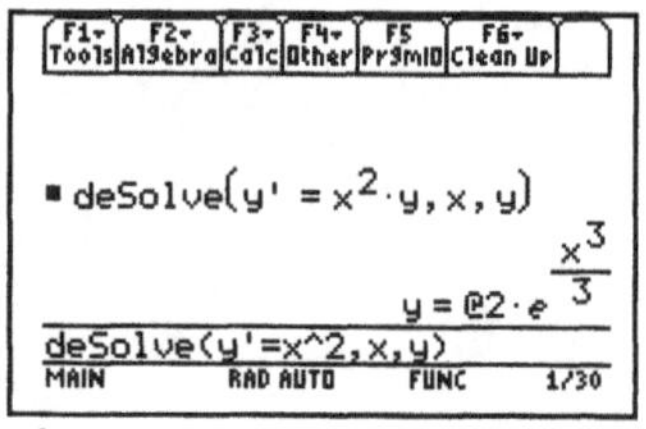

5. $y' = \dfrac{x^2}{y^3}$

$\dfrac{dy}{dx} = \dfrac{x^2}{y^3}$

$y^3\,dy = x^2\,dx$ [separate variables]

$\int y^3\,dy = \int x^2\,dx$

$\dfrac{y^4}{4} = \dfrac{x^3}{3} + C_1$

$4x^3 - 3y^4 = C$

7. $xy\,dx - (x^2+1)\,dy = 0$

$xy\,dx = \left(x^2+1\right)dy \Rightarrow x\,dx = \dfrac{\left(x^2+1\right)dy}{y} \Rightarrow \dfrac{x\,dx}{x^2+1} = \dfrac{dy}{y}$

$\int \dfrac{x\,dx}{x^2+1} = \int \dfrac{dy}{y} \Rightarrow \dfrac{1}{2}\int \dfrac{2x\,dx}{x^2+1} = \int \dfrac{dy}{y}$

$\dfrac{1}{2}\ln\left|x^2+1\right| = \ln|y| + C_1$

$\ln\left(x^2+1\right)^{1/2} = \ln|y| + C_1$

$\ln\left(x^2+1\right)^{1/2} - \ln|y| = C_1$

$\ln\dfrac{\sqrt{x^2+1}}{y} = C_1$

$e^{\ln\frac{\sqrt{x^2+1}}{y}} = e^{C_1}$ [exponentiate] $\Rightarrow \dfrac{\sqrt{x^2+1}}{y} = C_1 \Rightarrow y = \dfrac{\sqrt{x^2+1}}{C}$

9. $(1+x^2)\,dy + (y^2+1)\,dx = 0$

$\dfrac{dy}{y^2+1} + \dfrac{dx}{x^2+1} = 0$ [separate variables]

$\int \dfrac{1}{y^2+1}dy + \int \dfrac{1}{x^2+1}dx + C_1 = 0$

$\arctan y + \arctan x + C_1 = 0 \Rightarrow \arctan y + \arctan x = C$

11. $\sqrt{1+x^2}\,dy + xy\,dx = 0$

$\dfrac{dy}{y} + \dfrac{x\,dx}{\sqrt{1+x^2}} = 0$ [separate variables]

$\int \dfrac{dy}{y} + \dfrac{1}{2}\int 2x\left(1+x^2\right)^{-1/2}dx = 0$

$\ln y + \left(1+x^2\right)^{1/2} + C = 0$

$\ln y + \sqrt{1+x^2} = C$

13. $(y^2+1)\,dx = (x^2+1)\,dy$

$$\frac{dx}{x^2+1}=\frac{dy}{y^2+1}\text{ [separate variables]}$$

$$\int\frac{dx}{x^2+1}=\int\frac{dy}{y^2+1}$$

$$\arctan x=\arctan y+C$$

By calculator

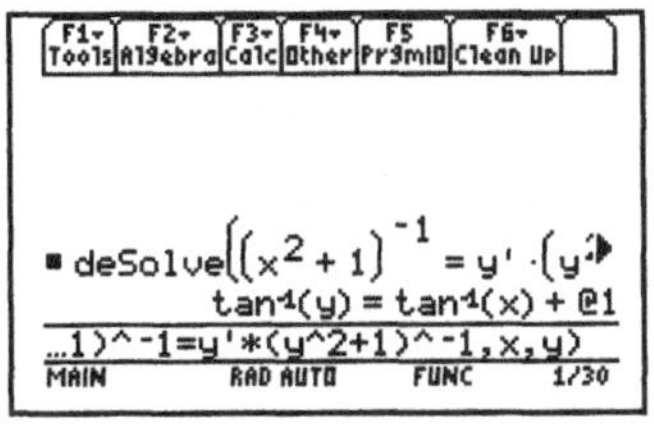

15. $(2+y)\,dx+(x-2)\,dy=0$

$$(2+y)dx=-(x-2)dy$$

$$\frac{dx}{x-2}=-\frac{dy}{2+y}\text{ [separate variables]}$$

$$\int\frac{dx}{x-2}=-\int\frac{dy}{2+y}$$

$$\ln|x-2|=-\ln|2+y|+C_1\Rightarrow\ln|x-2|+\ln|2+y|=C_1$$

$$\ln|(x-2)(2+y)|=C_1$$

$$e^{\ln|(x-2)(2+y)|}=e^{C_1}\text{ [exponentiate]}\Rightarrow(x-2)(2+y)=C$$

$$2x+xy-4-2y=C$$

$$2x+xy-2y=C+4\Rightarrow 2x+xy-2y=C\text{ [}C+4\text{ is a constant just as }C\text{ is]}$$

17. $(x-xy^2)\,dx=-(x^2y+y)\,dy$

$$x(1-y^2)dx=-y(x^2+1)dy$$

$$\frac{x\,dx}{x^2+1}=\frac{-y\,dy}{1-y^2}\text{ [separate variables]}$$

$$\frac{1}{2}\int\frac{2x\,dx}{x^2+1}=\frac{1}{2}\int\frac{-2y\,dy}{1-y^2}$$

$$\frac{1}{2}\ln|x^2+1|=\frac{1}{2}\ln|1-y^2|+C_1$$

$$\ln|x^2+1|=\ln|1-y^2|+2C_1\Rightarrow\ln|x^2+1|-\ln|1-y^2|=2C_1$$

$$\ln\left|\frac{x^2+1}{1-y^2}\right|=2C_1$$

$$e^{\ln\left|\frac{x^2+1}{1-y^2}\right|}=e^{2C_1}\text{ [exponentiate]}\Rightarrow\frac{x^2+1}{1-y^2}=C$$

$$x^2+1=C(1-y^2)\Rightarrow x^2+1=C-Cy^2$$

$$x^2+Cy^2=C-1$$

With Exponential Functions

19. $ye^{2x}=(1+e^{2x})y'$

$$ye^{2x}=(1+e^{2x})\frac{dy}{dx}\Rightarrow ye^{2x}=\frac{(1+e^{2x})dy}{dx}\Rightarrow e^{2x}=\frac{(1+e^{2x})dy}{y\,dx}$$

$$\frac{e^{2x}\,dx}{1+e^{2x}} = \frac{dy}{y}\ \text{[separate variables]}$$

$$\frac{1}{2}\int \frac{2e^{2x}\,dx}{1+e^{2x}} = \int \frac{dy}{y}$$

$$\frac{1}{2}\ln\left|1+e^{2x}\right| = \ln y + C_1$$

$$\ln\sqrt{1+e^{2x}} = \ln y + C_1$$

$$\ln\sqrt{1+e^{2x}} - C_1 = \ln y \Rightarrow e^{\ln\sqrt{1+e^{2x}}-C_1} = e^{\ln y}\ \text{[exponentiate]} \Rightarrow \sqrt{1+e^{2x}}\,e^{C} = y$$

$$C\sqrt{1+e^{2x}} = y$$

21. $e^{x-y}\,dx + e^{y-x}\,dy = 0$

$$\left(e^{x+y}\right)e^{x-y}\,dx + \left(e^{x+y}\right)e^{y-x}\,dy = 0$$

$$e^{2x}\,dx + e^{2y}\,dy = 0$$

$$\frac{1}{2}\int 2e^{2x}\,dx + \frac{1}{2}\int 2e^{2y}\,dy = 0$$

$$\frac{1}{2}e^{2x} + \frac{1}{2}e^{2y} + C_1 = 0 \Rightarrow \frac{1}{2}e^{2x} + \frac{1}{2}e^{2y} = C_1$$

$$e^{2x} + e^{2y} = 2C_1 \Rightarrow e^{2x} + e^{2y} = C$$

With Trigonometric Functions

23. $\tan y\,dx + (1+x)\,dy = 0$

$$\tan y\,dx = -(1+x)dy$$

$$\frac{-dx}{1+x} = \frac{dy}{\tan y} \Rightarrow \frac{-dx}{1+x} = \cot y\,dy\ \left[\begin{array}{l}\text{separate variables by dividing by } \tan y \text{ and } (1+x)\\ \dfrac{1}{\tan y} = \cot y\end{array}\right]$$

$$\int \frac{-dx}{1+x} = \int \cot y\,dy$$

$$-\ln(1+x) = \ln(\sin y) + C_1$$

$$\ln(\sin y) + \ln(1+x) = C_1 \Rightarrow \ln\left[(1+x)\sin y\right] = C_1$$

$$e^{\ln\left[(1+x)\sin y\right]} = e^{C_1}\ \text{[exponentiate]} \Rightarrow (1+x)\sin y = C$$

By calculator [the calculator solution will simplify to $(1+x)\sin y$]

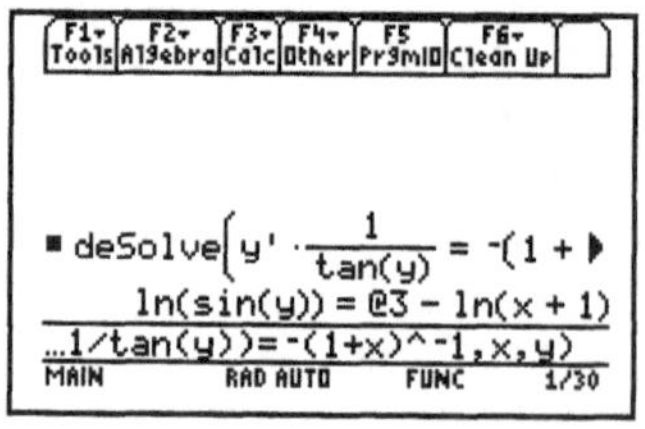

25. $\cos x \sin y\,dy + \sin x \cos y\,dx = 0$

$$\cos x(\sin y\,dy) + \sin x(\cos y\,dx) = 0$$

$$\frac{\cos x \sin y}{\cos x \cos y}dy + \frac{\sin x \cos y}{\cos x \cos y}dx = 0 \Rightarrow \frac{\sin y}{\cos y}dy + \frac{\sin x}{\cos x}dx = 0 \Rightarrow \tan y\,dy + \tan x\,dx = 0$$

$$\int \tan y\,dy + \int \tan x\,dx = 0$$

$$-\ln|\cos y| - \ln|\cos x| + C_1 = 0 \Rightarrow \ln|(\cos y)(\cos x)| = C_1$$

$$e^{\ln|(\cos y)(\cos x)|} = e^{C_1}\ \text{[exponentiate]} \Rightarrow |(\cos y)(\cos x)| = C \Rightarrow \cos x \cos y = C$$

27. $4\sin x \sec y\,dx = \sec x\,dy$

$$\frac{4\sin x\sec y\,dx}{\sec x}=\frac{\sec x\,dy}{\sec x}\Rightarrow\frac{4\sin x\sec y\,dx}{\sec x}=dy$$

$$\frac{4\sin x\,dx}{\sec x}=\frac{dy}{\sec y}\;[\text{divide by }\sec y]$$

$$\int\frac{4\sin x\,dx}{\sec x}=\int\frac{dy}{\sec y}\Rightarrow\int\frac{4\sin x\,dx}{\frac{1}{\cos x}}=\int\cos y\,dy\Rightarrow 4\int\sin x(\cos x)\,dx=\int\cos y\,dy$$

$$\frac{4\sin^2 x}{2}=\sin y+C\Rightarrow 2\sin^2 x=\sin y+C$$

$$2\sin^2 x-\sin y=C$$

Particular Solution

29. $y^2y'=x^2, x=0$ when $y=1$

$$y^2\frac{dy}{dx}=x^2\Rightarrow\frac{y^2\,dy}{dx}=x^2\Rightarrow y^2\,dy=x^2\,dx$$

$$\int y^2\,dy=\int x^2\,dx\Rightarrow\frac{y^3}{3}=\frac{x^3}{3}+C$$

$$\frac{y^3}{3}-\frac{x^3}{3}=C$$

$$\text{At }(0,1)\Rightarrow\frac{1^3}{3}-\frac{0^3}{3}=C\Rightarrow\frac{1}{3}=C$$

$$\text{Thus, }\frac{y^3}{3}=\frac{x^3}{3}+\frac{1}{3}\Rightarrow y^3-x^3=3$$

31. $y'\sin y=\cos x,\ x=\frac{\pi}{2}$ when $y=0$

$$\frac{dy}{dx}\sin y=\cos x\Rightarrow\frac{dy\sin y}{dx}=\cos x\Rightarrow\sin y\,dy=\cos x\,dx$$

$$\int\sin y\,dy=\int\cos x\,dx\Rightarrow-\cos y=\sin x+C$$

$$C=\sin x+\cos y$$

$$\text{At }\left(\frac{\pi}{2},0\right)\Rightarrow C=\sin\left(\frac{\pi}{2}\right)+\cos(0)\Rightarrow C=1+1\Rightarrow C=2$$

Thus, $\sin x+\cos y=2$

Exercise 4 ◊ Exact First-Order DE

Integrable Combinations

1. $y\,dx+x\,dy=7\,dx$

Recognizing the left side as the derivative of xy, integrate both sides

$$\int(y\,dx+x\,dy)=\int 7\,dx$$

$$xy=7x+C$$

$$xy-7x=C$$

3. $x\frac{dy}{dx}=3-y$

$$x\,dy=(3-y)\,dx$$

$$x\,dy=3\,dx-y\,dx$$

$$x\,dy+y\,dx=3\,dx$$

$$\int(x\,dy+y\,dx)=\int 3\,dx$$

$$xy=3x+C$$

$$xy-3x=C$$

5. $2xy'=x-2y$

$$2x\frac{dy}{dx} = x - 2y$$
$$\frac{2x\,dy}{dx} = x - 2y$$
$$2x\,dy = (x - 2y)dx \text{ [multiply both sides by } dx]$$
$$2x\,dy = x\,dx - 2y\,dy$$
$$2x\,dy + 2y\,dy = x\,dx \Rightarrow 2(x\,dy + y\,dy) = x\,dx$$
$$\int 2(x\,dy + y\,dy) = \int x\,dx$$
$$2xy = \frac{x^2}{2} + C_1$$
$$4xy = x^2 + 2C_1 \Rightarrow 4xy - x^2 = C$$

7. $x\,dy = (3x^2 + y)\,dx$
$$x\,dy = 3x^2\,dx + y\,dx$$
$$x\,dy - y\,dx = 3x^2\,dx$$
$$\frac{x\,dy - y\,dx}{x^2} = \frac{3x^2\,dx}{x^2} \Rightarrow \frac{x\,dy - y\,dx}{x^2} = 3\,dx \quad \left[\text{recognizing the left side as the derivative of } \frac{y}{x}\text{, integrate both sides}\right]$$
$$\int \frac{x\,dy - y\,dx}{x^2} = \int 3\,dx$$
$$\frac{y}{x} = 3x + C$$
$$\frac{y}{x} - 3x = C$$

9. $3x^2 + 2y + 2xy' = 0$
$$3x^2 + 2y + 2x\frac{dy}{dx} = 0$$
$$3x^2\,dx + 2y\,dx + 2x\,dy = 0 \text{ [multiply both sides by } dx]$$

Recognizing $2y\,dx + 2x\,dx$ as the derivative of $2xy$, integrate each term
$$\int(2x\,dy + 2y\,dx) = \int -3x^2\,dx$$
$$2xy = -x^3 + C$$
$$x^3 + 2xy = C$$

11. $(2x - y)y' = x - 2y$
$$2x\frac{dy}{dx} - y\frac{dy}{dx} = x - 2y$$
$$2x\,dy - y\,dy = (x - 2y)dx \text{ [multiply through by } dx]$$
$$2x\,dy - y\,dy = x\,dx - 2y\,dx \Rightarrow 2x\,dy + 2y\,dx = x\,dx + y\,dy \Rightarrow 2(x\,dy + y\,dx) = x\,dx + y\,dy$$
$$\int 2(x\,dy + y\,dx) = \int x\,dx + \int y\,dy$$
$$2xy = \frac{x^2}{2} + \frac{y^2}{2} + C_1$$
$$4xy = x^2 + y^2 + 2C_1 \Rightarrow 4xy = x^2 + y^2 + C$$
$$x^2 + y^2 - 4xy = C$$

13. $y\,dx - x\,dy = 2y^2\,dx$
$$\frac{y\,dx - x\,dy}{y^2} = \frac{2y^2\,dx}{y^2} \Rightarrow \frac{y\,dx - x\,dy}{y^2} = 2\,dx$$

Recognizing the left side as the derivative of $\frac{x}{y}$, integrate both sides

$\int \frac{y\,dx - x\,dy}{y^2} = \int 2\,dx$

$\frac{x}{y} = 2x + C$

15. $(4y^3 + x)\frac{dy}{dx} = y$

$4y^3\frac{dy}{dx} + x\frac{dy}{dx} = y$

$4y^3\,dy + x\,dy = y\,dx$ [multiply through by dx]

$-x\,dy + y\,dx = 4y^3\,dy$

$\frac{y\,dx - x\,dy}{y^2} = \frac{4y^3\,dy}{y^2} \Rightarrow \frac{y\,dx - x\,dy}{y^2} = 4y\,dy$ [divide through by y^2]

Recognizing the left side as the derivative of $\frac{x}{y}$, integrate both sides

$\int \frac{y\,dx - x\,dy}{y^2} = \int 4y\,dy$

$\frac{x}{y} = 2y^2 + C_1$

$x = 2y^3 + C_1 y \Rightarrow 2y^3 = x - Cy$

17. $3x - 2y^2 - 4xyy' = 0$

$3x - 2y^2 - 4xy\frac{dy}{dx} = 0$

$3x\,dx - 2y^2\,dx - 4xy\,dy$ [multiply through by dx]

$3x\,dx - (2y^2\,dx + 4xy\,dy) = 0$ [factor out the minus sign in the last two terms]

Recognizing $2y^2\,dx + 4xy\,dy$ as the derivative of $2xy^2$, integrate both sides

$\int 3x\,dx - \int (2y^2\,dx + 4xy\,dy) = 0$

$\frac{3x^2}{2} - 2xy^2 + C_1 = 0 \Rightarrow \frac{3x^2}{2} - 2xy^2 = C_1$

$4xy^2 - 3x^2 = 2C_1 \Rightarrow 4xy^2 - 3x^2 = C$

19. $4x = y + xy'$, $x = 3$ when $y = 1$

$4x = y + x\frac{dy}{dx}$

$4x\,dx = y\,dx + x\,dy$

$\int y\,dx + x\,dy = \int 4x\,dx$

$xy = 2x^2 + C_1$

$C = 2x^2 - xy$

At $(3,1) \Rightarrow C = 2(3^2) - (3)(1) \Rightarrow C = 15$

Thus, $2x^2y - xy = 15$

21. $y = (3y^3 + x)\frac{dy}{dx}$, $x = 1$ when $y = 1$

$y = 3y^3\frac{dy}{dx} + x\frac{dy}{dx}$

$y\,dx = 3y^3\,dy + x\,dy$

$y\,dx - x\,dy = 3y^3\,dy$

$\frac{y\,dx - x\,dy}{y^2} = \frac{3y^3\,dy}{y^2} \Rightarrow \frac{y\,dx - x\,dy}{y^2} = 3y\,dy$

$\int \frac{y\,dx - x\,dy}{y^2} = \int 3y\,dy$

$\frac{x}{y} = \frac{3y^2}{2} + C$

$C = \frac{x}{y} - \frac{3y^2}{2}$

At $(1,1) \Rightarrow C = \frac{1}{1} - \frac{3(1^2)}{2} \Rightarrow C = 1 - \frac{3}{2} \Rightarrow C = -\frac{1}{2}$

Thus, $-\frac{1}{2} = \frac{x}{y} - \frac{3y^2}{2} \Rightarrow 3y^3 - y = 2x$

$-2xy = \frac{4x^3}{3} + C_1$

$C = 2xy + \frac{4x^3}{3}$

At $(5, 2) \Rightarrow C = 2(5)(2) + \frac{4(5^3)}{3} \Rightarrow C = 20 + \frac{500}{3} \Rightarrow C = 186.7$

Thus, $2xy + \frac{4x^3}{3} = 186.7 \Rightarrow 6xy + 4x^3 = 560$

23. $3x - 2y = (2x - 3y)\frac{dy}{dx}$, $x = 2$ when $y = 2$

$3x - 2y = 2x\frac{dy}{dx} - 3y\frac{dy}{dx}$

$3x\,dx - 2y\,dx = 2x\,dy - 3y\,dy$

$3x\,dx + 3y\,dy = 2x\,dy + 2y\,dx$

$3(x\,dx + y\,dy) = 2(x\,dy + y\,dx)$

$\int 2(x\,dy + y\,dx) = \int 3(x\,dx + y\,dy)$

$2xy = \frac{3x^2}{2} + \frac{3y^2}{2} + C_1$

$\frac{3x^2}{2} + \frac{3y^2}{2} - 2xy = C \Rightarrow 3x^2 + 3y^2 - 4xy = C$

At $(2, 2) \Rightarrow 3(2^2) + 3(2^2) - 4(2)(2) = C \Rightarrow 8 = C$

Thus, $3x^2 + 3y^2 - 4xy = 8$

Exercise 5 ◊ First-Order Homogeneous DE

1. $(x - y)\,dx - 2x\,dy = 0$

This is a first-order homogenous DE:

$(tx - ty)\,dx - 2tx\,dy = 0 \Rightarrow t[(x - y)\,dx - 2x\,dy]$

Let $y = vx$, $dy = v\,dx + x\,dv$, and substitute $\Rightarrow (x - vx)\,dx - 2x(v\,dx + x\,dv) = 0$

$x\,dx - vx\,dx - 2xv\,dx - 2x^2\,dv = 0 \Rightarrow x(1 - 3v)\,dx - 2x^2\,dv = 0$

$\frac{dx}{x} - \frac{2\,dv}{1 - 3v} = 0$ [separate variables]

$\int \frac{dx}{x} - \int \frac{2\,dv}{1 - 3v} = 0 \Rightarrow \int \frac{dx}{x} - \frac{2}{3}\int \frac{3\,dv}{1 - 3v} = 0$

$$\ln x + \frac{2}{3}\ln(1-3v) = C_1 \Rightarrow \ln x + \ln(1-3v)^{2/3} = \ln C_1$$

$$\ln x(1-3v)^{2/3} = \ln C_1$$

$$x(1-3v)^{2/3} = C \text{ [take the antilog]}$$

$$x\left(1-3\frac{y}{x}\right)^{2/3} = C\left[\text{substitute } v = \frac{y}{x}\right]$$

$$x^{1/3}(x-3y)^{2/3} = C \Rightarrow x(x-3y)^2 = C \text{ [cube both sides]}$$

3. $(x^2 - xy)y' + y^2 = 0$

Let $y = vx$, $y' = v\,dx + x\,dv$, and substitute $\Rightarrow \left[x^2 - x(vx)\right]\left(v + x\frac{dv}{dx}\right) + v^2x^2 = 0$

$$(x^2 - x^2v)\left(v + x\frac{dv}{dx}\right) + v^2x^2 = 0 \Rightarrow x^2(1-v)\left(v + x\frac{dv}{dx}\right) + v^2x^2 = 0$$

$$v + x\frac{dv}{dx} - v^2 - vx\frac{dv}{dx} + v^2 = 0 \Rightarrow x\frac{dv}{dx}(1-v) = -v$$

$$\frac{v-1}{v}dv = \frac{dx}{x} \Rightarrow \left(1 - \frac{1}{v}\right)dv = \frac{dx}{x} \text{ [separate variables]}$$

$$\int\left(1-\frac{1}{v}\right)dv = \int\frac{dx}{x}$$

$$v - \ln v = \ln x + C$$

$$\frac{y}{x} - \ln\left(\frac{y}{x}\right) = \ln x + C\left[\text{substitute } \frac{y}{x} \text{ for } v\right]$$

$$\frac{y}{x} - (\ln y - \ln x) = \ln x + C \Rightarrow \frac{y}{x} - \ln y + \ln x = \ln x + C$$

$$\ln y - \frac{y}{x} = C$$

$$x\ln y - y = Cx$$

5. $xy^2\,dy - (x^3 + y^3)\,dx = 0$

Let $y = vx$, $dy = x\,dv + v\,dx$, and substitute $\Rightarrow x(v^2x^2)(x\,dv + v\,dx) - (x^3 + v^3x^3)\,dx = 0$

$$x^3v^3(x\,dv + v\,dx) - (x^3 + v^3x^3)\,dx = 0 \Rightarrow x^4v^2\,dv + x^3v^3\,dx - x^3dx - x^3v^3\,dx = 0$$

$$v^2\,dv - \frac{1}{x}dx = 0 \text{ [separate variables]}$$

$$\int v^2\,dv - \int\frac{1}{x}dx = 0$$

$$\frac{v^3}{3} - \ln x + C = 0 \Rightarrow \frac{v^3}{3} - \ln x = C_1$$

$$\frac{\left(\frac{y}{x}\right)^3}{3} - \ln x = C_1 \Rightarrow \frac{y^3}{3x^3} - \ln x = C_1\left[\text{substitute } \frac{y}{x} \text{ for } v\right]$$

$$y^3 = x^3(3\ln x + C)\left[\text{solve for } y^3\right]$$

7. $x - y = 2xy'$, $x = 1$, $y = 1$

Let $y = vx$, $y' = v + xv'$, and substitute $\Rightarrow x - vx = 2x(v + xv')$

$$x - vx = 2xv + 2x^2v'$$

$$x(1-v) = 2x(v + xv') \Rightarrow 1 - v = 2(v + xv') \text{ [divide through by } x\text{]}$$

$$1 - v = 2v + 2xv' \Rightarrow 1 - 3v = 2xv'$$

$$\frac{dx}{2x} = \frac{dv}{1-3v}$$

$$\int \frac{dx}{2x} = \int \frac{dv}{1-3v} \Rightarrow \frac{1}{2}\int \frac{dx}{x} = -3\int \frac{-3dv}{1-3v}$$

$$\frac{1}{2}\ln x = -\left(\frac{1}{3}\right)\ln|1-3v| + C_1$$

$$3\ln x + 2\ln|1-3v| = C$$

$$\ln x^3 + 2\ln\left|1-3\left(\frac{y}{x}\right)\right| = C \left[\text{substitute } \frac{y}{x} \text{ for } v\right]$$

$$\ln\left|x^3\left(\frac{x-3y}{x}\right)^2\right| = C \Rightarrow \ln\left|x(x-3y)^2\right| = C$$

At $(1, 1) \Rightarrow \ln\left|1(1-3(1))^2\right| = C \Rightarrow \ln|4| = C \Rightarrow \ln 4 = C$

Thus, $\ln\left|x(x-3y)^2\right| = \ln 4$

$x(x-3y)^2 = 4$ [take antilog]

9. $(x^3 + y^3)\,dx - xy^2\,dy = 0, x = 1, y = 0$
   Let $y = vx$, $dy = vdx + xdv$ and substitute

$$x^3\,dx + v^3x^3\,dx - x^3v^2v\,dx - x^3v^2x\,dv = 0$$

$$dx + v^3\,dx - v^3\,dx - v^2x\,dv = 0$$

$$\frac{dx}{x} = v^2\,dv$$

$$\int \frac{dx}{x} = \int v^2\,dv \Rightarrow \ln|x| = \frac{v^3}{3} + C$$

$$\ln|x| = \frac{\left(\frac{y}{x}\right)^3}{3} + C \Rightarrow \ln|x| = \frac{y^3}{3x^3} + C \Rightarrow \ln|x| - \frac{y^3}{3x^3} = C$$

At $(1, 0) \Rightarrow \ln|1| - \frac{0^3}{3(1)^3} = C \Rightarrow 0 - 0 = C \Rightarrow C = 0$

Thus, $\ln|x| - \frac{y^3}{3x^3} = 0 \Rightarrow y^3 = 3x^3\ln|x|$

Exercise 6 ◊ First-Order Linear DE

1. $y' + \frac{y}{x} = 4 \Rightarrow y' + \left(\frac{1}{x}\right)y = 4$

   This is a first-order linear differential equation in standard form with $P(x) = \frac{1}{x}$ and $Q(x) = 4$

   Calculating $\int P\,dx \Rightarrow \int \frac{1}{x}dx = \ln x \Rightarrow e^{\int P\,dx} = e^{\ln x} = x$; therefore x is the integrating factor

   Substituting into $ye^{\int P\,dx} = \int Qe^{\int P\,dx}\,dx \Rightarrow yx = \int 4x\,dx$

   $yx = 2x^2 + C$

   $y = \frac{2x^2}{x} + \frac{C}{x} \Rightarrow y = 2x + \frac{C}{x}$

3. $xy' = 4x^3 - y \Rightarrow y' + \left(\frac{1}{x}\right)y = 4x^2$ with $P(x) = \frac{1}{x}$ and $Q(x) = 4x^2$

   Calculating $\int P\,dx \Rightarrow \int \frac{1}{x}dx = \ln x \Rightarrow e^{\int P\,dx} = e^{\ln x} = x$; therefore x is the integrating factor

$yx = \int 4x^2(x)dx \Rightarrow yx = \int 4x^3\,dx$

$yx = x^4 + C \Rightarrow y = x^3 + \dfrac{C}{x}$

5. $y' = x^2 - x^2y \Rightarrow y' + x^2y = x^2 \Rightarrow \dfrac{dy}{dx} + x^2y = x^2$ [standard form of first-order linear differential equation]

$P(x) = x^2$ and $Q(x) = x^2 \Rightarrow \int P\,dx = \int x^2\,dx = \dfrac{x^3}{3} \Rightarrow e^{\int P\,dx} = e^{x^3/3}$, the integrating factor

$ye^{x^3/3} = \int e^{x^3/3}(x^2)dx$

$ye^{x^3/3} = e^{x^3/3} + C$

$y = 1 + \dfrac{C}{e^{x^3/3}} \Rightarrow y = 1 + Ce^{-x^3/3}$

7. $y' = \dfrac{3 - xy}{2x^2} \Rightarrow y' = \dfrac{3}{2x^2} - \dfrac{xy}{2x^2} \Rightarrow y' + \dfrac{1}{2x}y = \dfrac{3x^{-2}}{2}$ with $P(x) = \dfrac{1}{2x}$ and $Q(x) = \dfrac{3x^{-2}}{2}$

$\int P\,dx = \int \dfrac{1}{2x}dx = \dfrac{1}{2}\ln x = \ln\sqrt{x};\ e^{\int P\,dx} = e^{\ln\sqrt{x}} = \sqrt{x}$, the integrating factor

$y\sqrt{x} = \int \dfrac{3x^{-2}}{2}(\sqrt{x})dx$

$y\sqrt{x} = -3x^{-1/2} + C$

$y = -\dfrac{3x^{-1/2}}{\sqrt{x}} + \dfrac{C}{\sqrt{x}} \Rightarrow y = \dfrac{C}{\sqrt{x}} - \dfrac{3}{x}$

9. $xy' = 2y - x \Rightarrow \dfrac{dy}{dx} - \dfrac{2}{x}y = -1$ with $P(x) = -\dfrac{2}{x}$ and $Q(x) = -1$

$\int P\,dx = \int -\dfrac{2}{x}dx = -2\ln x = \ln x^{-2} \Rightarrow e^{\int P\,dx} = e^{\ln x^{-2}} = x^{-2}$; the integrating factor

Substituting into $ye^{\int P\,dx} = \int Q(x)e^{\int P\,dx}\,dx \Rightarrow yx^{-2} = \int -1(x^{-2})dx$

$yx^{-2} = \int(-x^{-2})dx \Rightarrow yx^{-2} = x^{-1} + C$

$y = \dfrac{x^{-1}}{x^{-2}} + \dfrac{C}{x^{-2}} \Rightarrow y = x + Cx^2$

11. $y' = \dfrac{2 - 4x^2y}{x + x^3} \Rightarrow y' = \dfrac{2}{x + x^3} - \dfrac{4x^2y}{x + x^3} \Rightarrow y' + \dfrac{4x^2}{x + x^3}y = \dfrac{2}{x + x^3}$ with $P(x) = \dfrac{4x^2}{x + x^3}$ and $Q(x) = \dfrac{2}{x + x^3}$

$\int P\,dx = \int \dfrac{4x^2}{x + x^3}dx = 2\ln(1 + x^2) = \ln(1 + x^2)^2 \Rightarrow e^{\int P\,dx} = e^{\ln(1+x^2)^2} = (1 + x^2)^2$; the integrating factor

$y(1 + x^2)^2 = \int \dfrac{2}{x + x^3}\left[(1 + x^2)^2\right]dx \Rightarrow y(1 + x^2)^2 = \int \dfrac{2}{x}(1 + x^2)dx \Rightarrow y(1 + x^2) = 2\int\left(\dfrac{1}{x} + x\right)dx$

$y(1 + x^2)^2 = 2\ln x + x^2 + C$

13. $xy' + x^2y + y = 0 \Rightarrow x\dfrac{dy}{dx} + x^2y + y = 0 \Rightarrow \dfrac{dy}{dx} + xy + \dfrac{y}{x} = 0 \Rightarrow \dfrac{dy}{dx} + \left(x + \dfrac{1}{x}\right)y = 0$

With $P(x) = x + \dfrac{1}{x}$ and $Q(x) = 0; \int P\,dx = \int x + \dfrac{1}{x}dx = \dfrac{x^2}{2} + \ln x$

$e^{\int P\,dx} = e^{(x^2/2) + \ln x} = \left(e^{x^2/2}\right)\left(e^{\ln x}\right) = xe^{x^2/2}$, the integrating factor

$y\left(xe^{x^2/2}\right) = \int 0\left(xe^{x^2/2}\right)dx \Rightarrow y\left(xe^{x^2/2}\right) = C \Rightarrow y = \dfrac{C}{\left(xe^{x^2/2}\right)}$

15. $y' + y = e^x$ with $P(x) = 1$ and $Q(x) = e^x$

$\int P\,dx = \int dx = x \Rightarrow e^{\int P\,dx} = e^x$; the integrating factor

$$ye^x = \int e^x\left(e^x\right)dx \Rightarrow ye^x = \frac{1}{2}\int e^{2x}(2)dx \Rightarrow ye^x = \frac{1}{2}e^{2x} + C$$

$$y = \frac{e^{2x}}{2}\left(e^{-x}\right) + Ce^{-x} \Rightarrow y = \frac{e^x}{2} + \frac{C}{e^x}$$

17. $y' = 2y + 4e^{2x} \Rightarrow \dfrac{dy}{dx} - 2y = 4e^{2x}$ with $P(x) = -2$ and $Q(x) = 4e^{2x}$

$\int P\,dx = \int -2\,dx = -2x \Rightarrow e^{\int P\,dx} = e^{-2x}$; the integrating factor

$$ye^{-2x} = \int 4e^{2x}\left(e^{-2x}\right)dx \Rightarrow ye^{-2x} = \int 4\,dx \Rightarrow ye^{-2x} = 4x + C$$

$$y = 4xe^{2x} + Ce^{2x} \Rightarrow y = e^{2x}(4x + C)$$

19. $y' = \dfrac{4\ln x - 2x^2 y}{x^3} \Rightarrow y' + \dfrac{2x^2 y}{x^3} = \dfrac{4\ln x}{x^3} \Rightarrow y' + \left(\dfrac{2}{x}\right)y = \dfrac{4\ln x}{x^3}$ with $P(x) = \dfrac{2}{x}$ and $Q(x) = \dfrac{4\ln x}{x^3}$

$\int P\,dx = \int \frac{2}{x}dx = 2\ln x = \ln x^2 \Rightarrow e^{\int P\,dx} = e^{\ln x^2} = x^2$; the integrating factor

$$yx^2 = \int \frac{4\ln x}{x^3}\left(x^2\right)dx \Rightarrow yx^2 = \int 4\ln x\frac{dx}{x} \Rightarrow yx^2 = 2\ln^2 x + C$$

With Trigonometric Expressions

21. $y' + y = \sin x \Rightarrow \dfrac{dy}{dx} + y = \sin x$ with $P(x) = 1$ and $Q(x) = \sin x$

$\int P\,dx = \int dx = x \Rightarrow e^{\int P\,dx} = e^x$; the integrating factor

$$ye^x = \int \sin x\left(e^x\right)dx \Rightarrow ye^x = \frac{e^x}{2}(\sin x - \cos x) + C$$

$$y = \frac{e^x}{2}(\sin x - \cos x)\left(e^{-x}\right) + Ce^{-x} \Rightarrow y = \frac{1}{2}(\sin x - \cos x) + Ce^{-x}$$

23. $y' = 2\cos x - y \Rightarrow \dfrac{dy}{dx} + y = 2\cos x$ with $P(x) = 1$ and $Q(x) = 2\cos x$

$\int P\,dx = \int dx = x \Rightarrow e^{\int P\,dx} = e^x$; the integrating factor

$$ye^x = \int 2\cos x\left(e^x\right)dx \Rightarrow ye^x = 2\int e^x\cos x\,dx \Rightarrow ye^x = e^x(\cos x + \sin x) + C$$

$$y = e^x(\cos x + \sin x)\left(e^{-x}\right) + Ce^{-x} \Rightarrow y = \cos x + \sin x + Ce^{-x}$$

Bernoulli's Equation

25. $y' + \dfrac{y}{x} = 3x^2y^2 \Rightarrow$ this is a Bernoulli equation with $P(x) = \dfrac{1}{2}$, $Q(x) = 3x^2$ and $n = 2$

$$\left(\frac{1}{y^2}\right)\frac{dy}{dx} + \left(\frac{1}{y^2}\right)\frac{y}{x} = \left(\frac{1}{y^2}\right)3x^2y^2 \Rightarrow \left(\frac{1}{y^2}\right)\frac{dy}{dx} + \frac{y^{-1}}{x} = 3x^2 \left[\text{divide through by } \frac{1}{y^2}\right]$$

Let $z = y^{1-2} = y^{-1}$ and $\dfrac{dz}{dx} = -y^{-2}\dfrac{dy}{dx} \Rightarrow \dfrac{dy}{dx} = -y^2\dfrac{dz}{dx}$

$$-\frac{dz}{dx} = \frac{z}{x} = 3x^2 \Rightarrow \frac{dz}{x} - \frac{z}{x} = -3x^2 \left[\text{this is now a first-order DE with } P(x) = -\frac{1}{x} \text{ and } Q(x) = -3x^2\right]$$

$$\int P\,dx = \int -\frac{1}{x}dx = -\ln x \Rightarrow e^{\int P\,dx} = e^{-\ln x} = x^{-1}$$

Substitute into $ze^{\int P\,dx} = \int Qe^{\int P\,dx}dx \Rightarrow zx^{-1} = \int -3x^2\left(x^{-1}dx\right) \Rightarrow \dfrac{z}{x} = -\dfrac{3x^2}{2} + C$

$z = -3x^3 + Cx$ [solve for z]

Substitute $z = \dfrac{1}{y}$ into $z = -3x^3 + Cx : \dfrac{1}{y} = -3x^3 + Cx$

$$1 = -3x^3y + Cxy \Rightarrow 1 = xy\left(C - \frac{3x^2}{2}\right)$$

27. $y' = y - xy^2(x+2)$

$$\left(\frac{1}{y^2}\right)\frac{dy}{dx} = \left(\frac{1}{y^2}\right)y - \left(\frac{1}{y^2}\right)\left[xy^2(x+2)\right] \Rightarrow \left(\frac{1}{y^2}\right)\frac{dy}{dx} = \frac{1}{y} - x(x+2)\left[\text{divide through by } \frac{1}{y^2}\right]$$

Let $z = \frac{1}{y}$ and $\frac{dz}{dx} = -\frac{1}{y^2}\frac{dy}{dx} \Rightarrow \frac{dy}{dx} = -y^2\frac{dz}{dx}$ and substitute into $\left(\frac{1}{y^2}\right)\frac{dy}{dx} = \frac{1}{y} - x(x+2)$

$$-\frac{1}{z^2}\left(-z^2\frac{dz}{dx}\right) = z - x(x+2) \Rightarrow -\frac{dz}{dx} - z = -x(x+2) \Rightarrow \frac{dz}{dx} + z = x(x+2)$$

This is now a first-order DE with $P(x) = 1$ and $Q(x) = x(x+2)$

$\int P\,dx = \int dx = x \Rightarrow e^{\int P\,dx} = e^x$; the integating factor

$$\frac{1}{y}e^x = e^x x^2 + C \Rightarrow \frac{1}{y} = \frac{e^x x^2 + C}{e^x} \Rightarrow y = \frac{e^x}{e^x x^2 + C}$$

Particular Solution

29. $xy' + y = 4x,\ x = 1$ when $y = 5$

$$xy' + y = 4x \Rightarrow x\left(\frac{dy}{dx}\right) + y = 4x \Rightarrow \frac{dy}{dx} + \frac{y}{x} = 4;\ P(x) = \frac{1}{x},\ Q(x) = 4$$

$\int P\,dx = \int \frac{1}{x}dx = \ln x \Rightarrow e^{\int P\,dx} = e^{\ln x} = x$; the integrating factor

$$yx = \int 4(x)dx \Rightarrow yx = 2x^2 + C \Rightarrow C = yx - 2x^2$$

$$\text{At } (1, 5) \Rightarrow C = (5)(1) - 2(1^2) \Rightarrow C = 3$$

Thus, $yx = 2x^2 + 3 \Rightarrow y = 2x + \frac{3}{x}$

31. $y' + \frac{y}{x} = 5,\ x = 1$ when $y = 2$

$P(x) = \frac{1}{x},\ Q(x) = 5;\ \int P\,dx = \int \frac{1}{x}dx = \ln x \Rightarrow e^{\int P\,dx} = e^{\ln x} = x$; the integrating factor

$$yx = \int 5x\,dx \Rightarrow yx = \frac{5x^2}{2} + C \Rightarrow C = yx - \frac{5x^2}{2}$$

$$\text{At } (1, 2) \Rightarrow C = (2)(1) - \frac{5(1)^2}{2} \Rightarrow C = -\frac{1}{2}$$

Thus, $yx = \frac{5x^2}{2} - \frac{1}{2} \Rightarrow y = \frac{5x}{2} - \frac{1}{2x}$

33. $y' = \tan^2 x + y\cot x,\ x = \frac{\pi}{4}$ when $y = 2$

$$\frac{dy}{dx} - y\cot x = \tan^2 x;\ P(x) = \cot x,\ Q(x) = \tan^2 x$$

$$\int P\,dx = \int \cot x\,dx = -\ln\sin x \Rightarrow e^{\int P\,dx} = e^{-\ln\sin x} = \csc x$$

$$y\csc x = \int \tan^2 x\csc x\,dx \Rightarrow y\csc x = \int \frac{\sin^2 x}{\cos^2 x\sin x}dx \Rightarrow y\csc x = -\int \cos^{-2}x(-\sin x)dx \Rightarrow y\csc x = \frac{1}{\sec x} + C$$

$$y = \frac{1}{\sec x\csc x} + \frac{C}{\csc x} \Rightarrow y = \tan x + C\sin x$$

$$\text{At } \left(\frac{\pi}{4}, 2\right) \Rightarrow 2 = \tan\left(\frac{\pi}{4}\right) + C\sin\left(\frac{\pi}{4}\right) \Rightarrow 2 = 1 + C\left(\frac{\sqrt{2}}{2}\right) \Rightarrow 1 = C\left(\frac{\sqrt{2}}{2}\right) \Rightarrow C = \sqrt{2}$$

Thus, $y = \tan x + \sqrt{2}\sin x$

Exercise 7 ◊ Geometric Applications of First-Order Des
Slopes of Curves

1. Expressing $y' = x + \frac{1}{x} + \frac{y}{x}$ into standard form $\Rightarrow \frac{dy}{dx} - \left(\frac{1}{x}\right)y = x + \frac{1}{x}$; $P(x) = -\frac{1}{x}$, $Q(x) = x + \frac{1}{x}$

$$\int P\,dx = \int -\frac{1}{x}dx = -\ln x = \ln\frac{1}{x} \Rightarrow e^{\int P\,dx} = e^{\ln 1/x} = \frac{1}{x}$$

$$y\left(\frac{1}{x}\right) = \int\left(x + \frac{1}{x}\right)\left(\frac{1}{x}dx\right) \Rightarrow \frac{y}{x} = \int\left(1 + x^{-2}\right)dx \Rightarrow \frac{y}{x} = x - \frac{1}{x} + C$$

$$y = x\left(x - \frac{1}{x} + C\right) \Rightarrow y = x^2 - 1 + Cx;\text{ at } (2, 9) \Rightarrow 9 = 2^2 - 1 + C(2) \Rightarrow 9 = 3 + 2C \Rightarrow 2C = 6 \Rightarrow C = 3$$

Thus, $y = x^2 + 3x - 1$

3. $\frac{dy}{xd} = \frac{x}{y} \Rightarrow y\,dy = x\,dx \Rightarrow \int y\,dy = \int x\,dx \Rightarrow \frac{y^2}{2} = \frac{x^2}{2} + C \Rightarrow y^2 = x^2 + 2C \Rightarrow y^2 = x^2 + C$

At $(3, 4) \Rightarrow 4^2 = 3^2 + C \Rightarrow 16 - 9 = C \Rightarrow C = 7$

Thus, $y^2 = x^2 + 7$

5. $\frac{dy}{dx} = x + y \Rightarrow \frac{dy}{dx} - y = x$; $P(x) = -1$, $Q(x) = x$

$$\int P\,dx = \int -1\,dx = -x \Rightarrow e^{\int P\,dx} = e^{-x}$$

$ye^{-x} = \int xe^{-x}\,dx \Rightarrow$ integrating by parts with $u = x$ $\quad dv = e^{-x}\,dx$

$du = dx \quad v = -e^{-x}$

$$ye^{-x} = -xe^{-x} + \int e^{-x}\,dx \Rightarrow ye^{-x} = -xe^{-x} - e^{-x} + C$$

$$y = -x - 1 + Ce^x$$

At $(2, 3) \Rightarrow 3 = -2 - 1 + Ce^2 \Rightarrow 3 = -3 + Ce^2 \Rightarrow 6 = Ce^2 \Rightarrow C = \frac{6}{e^2}$

Thus, $y = -x - 1 + \left(\frac{6}{e^2}\right)e^x \Rightarrow y = 0.812e^x - x - 1$

Tangents and Normals

7. $AP = -\left(\frac{y}{y'}\right)\sqrt{1 + (y')^2}$ and $BP = x\sqrt{1 + (y')^2}$

$$-\frac{y\sqrt{1+m^2}}{m} = x\sqrt{1+m^2} \Rightarrow x\frac{dy}{dx} + y = 0$$

$$\frac{dy}{y} + \frac{dx}{x} = 0 \Rightarrow \int\left(\frac{dy}{y} + \frac{dx}{x}\right) = 0 \Rightarrow \ln y + \ln x + C_1 = 0 \Rightarrow \ln xy = C_1 \Rightarrow e^{\ln xy} = e^C \Rightarrow xy = C$$

At $(4, 1) \Rightarrow (4)(1) = C \Rightarrow C = 4$

Thus, $xy = 4$

9. $PC = y\sqrt{1 + (y')^2}$; since PC equals the square of the ordinate of $P \Rightarrow PC = y\sqrt{1 + (y')^2} = y^2$

$$y^2 = y\sqrt{1 + (y')^2} \Rightarrow y = \sqrt{1 + (y')^2} \Rightarrow y^2 = 1 + (y')^2 \Rightarrow y^2 - 1 = (y')^2 \Rightarrow y' = \sqrt{y^2 - 1}$$

Separating variables: $\frac{dy}{dx} = \sqrt{y^2 - 1} \Rightarrow dy = dx\sqrt{y^2 - 1} \Rightarrow \frac{dy}{\sqrt{y^2 - 1}} = dx$

$$\int\frac{dy}{\sqrt{y^2 - 1}} = \int dx \Rightarrow \ln\left(y + \sqrt{y^2 - 1}\right) = x + C$$

At $(0, 1) \Rightarrow \ln\left(1 + \sqrt{1^2 - 1}\right) = 0 + C \Rightarrow \ln 1 = C \Rightarrow C = 0$

$$\ln\left(y+\sqrt{y^2-1}\right)=x \Rightarrow e^{\ln\left(y+\sqrt{y^2-1}\right)}=e^x \Rightarrow y+\sqrt{y^2-1}=e^x \Rightarrow e^x-y=\sqrt{y^2-1}$$

$$\left(e^x-y\right)^2=\left(\sqrt{y^2-1}\right)^2 \Rightarrow e^{2x}-2e^xy+y^2=y^2-1 \Rightarrow e^{2x}-2e^xy=-1$$

$$-2e^xy=-1-e^{2x} \Rightarrow 2e^xy=1+e^{2x} \Rightarrow y=\frac{\left(1+e^{2x}\right)}{2e^x} \Rightarrow y=\left(\frac{1}{2}\right)\left(e^x+\frac{1}{e^x}\right) \Rightarrow y=\frac{e^x+e^{-x}}{2}$$

Orthogonal Trajectories

11. $x^2+y^2=r^2 \Rightarrow 2x+2yy'=0 \Rightarrow 2yy'=-2x \Rightarrow y'=-\frac{x}{y}$

$$y'=\frac{y}{x}=\frac{dy}{dx} \Rightarrow y\,dx=x\,dy \Rightarrow \frac{dy}{y}=\frac{dx}{x}$$

$$\int\frac{dy}{y}=\int\frac{dx}{x} \Rightarrow \ln y=\ln x+C$$

$$\ln y-\ln x=C \Rightarrow \ln\left(\frac{y}{x}\right)=C \Rightarrow e^{\ln(y/x)}=e^C \Rightarrow \frac{y}{x}=C \Rightarrow y=Cx$$

$$2y\,dy=-x\,dx \Rightarrow \int 2y\,dy=\int -x\,dx \Rightarrow y^2=-\frac{x^2}{2}+C \Rightarrow y^2+\frac{x^2}{2}=C \Rightarrow x^2+2y^2=2C \Rightarrow x^2+2y^2=C$$

13. $x^2-y^2=Cy$

Differentiating and solving for $\frac{dy}{dx}$ to find the slope:

$$2x-2y\frac{dy}{dx}=C\frac{dy}{dx} \Rightarrow \frac{dy}{dx}=\frac{2x}{2y+C}$$

Solve $x^2-y^2=Cy$ for $C \Rightarrow C=\frac{x^2-y^2}{y}$

Substitute for C in $\frac{dy}{dx}=\frac{2x}{2y+C} \Rightarrow \frac{dy}{dx}=\frac{2x}{2y+\left(\frac{x^2-y^2}{y}\right)} \Rightarrow \frac{dy}{dx}=\frac{2xy}{2y^2+x^2-y^2} \Rightarrow \frac{dy}{dx}=\frac{2xy}{x^2+y^2}$

Take negative reiprocal of this slope to obtain the slope of the orthogonal trajectory:

$$\frac{dy}{dx}=-\frac{x^2+y^2}{2xy} \Rightarrow 2xy\,dy+\left(x^2+y^2\right)dx=0$$

Let $y=vx$ and $dy=v\,dx+x\,dv \Rightarrow 2x(vx)(v\,dx+x\,dv)+\left(x^2+x^2v^2\right)dx=0$

$$2x^2v^2\,dx+2x^3v\,dv+x^2\,dx+x^2v^2\,dx=0$$

$$3x^2v^2\,dx+x^2\,dx+2x^3v\,dv=0$$

$$x^2\left(3v^2+1\right)dx+2x^3v\,dv=0$$

$$\frac{1}{x}dx+\frac{2v\,dv}{3v^2+1}=0$$

$$\int\frac{1}{x}dx+\int\frac{2v\,dv}{3v^2+1}=0 \Rightarrow \int\frac{1}{x}dx+\frac{1}{3}\int\frac{6v\,dv}{3v^2+1}=0$$

$$\ln|x|+\frac{1}{3}\ln\left|3v^2+1\right|=C_1 \Rightarrow \ln\left|x^3\left(3v^2+1\right)\right|=3C_1 \Rightarrow \ln\left|x^3\left(3v^2+1\right)\right|=C$$

$$x^3\left(3v^2+1\right)=k;\text{ substituting } v=\frac{y}{x} \Rightarrow x^3\left(3\left(\frac{y}{x}\right)^2+1\right)=k \Rightarrow 3xy^2+x^3=k$$

Exercise 8 ◊ Exponential Growth and Decay

Exponential Growth

1. $\frac{dy}{dt}=ny$ where n is the constant of proportionality

$$\frac{dy}{dt}=ny \Rightarrow dy=ny\,dt \Rightarrow \frac{dy}{y}=n\,dt \text{ [separate variables]}$$

$$\int\frac{dy}{y}=\int n\,dt \Rightarrow \int\frac{dy}{y}=n\int dt \Rightarrow \ln y=nt+C \Rightarrow e^{\ln y}=e^{nt+C} \Rightarrow y=e^{nt}\left(e^{C}\right) \Rightarrow y=ae^{nt} \text{ where } a=e^{C}$$

3. Using $y=ae^{nt}$ with $a=158$, $n=0.069$ and $t=20$; $y=158e^{(0.069)(20)} \Rightarrow y=628.0344 \Rightarrow y=628$ million barrels/day

Exponential Decay

5. This is exponential decay where $T=T_0e^{-nt}$ where

   T_0 = temperature above room temperature at $t=0$; $T_0=1850$
   T = temperature above room temperature at time
   t = time; $t=2.5(60)=150$ minutes
   n = constant; $n=0.035$

   $$T=1850e^{-(0.035)(150)} \Rightarrow T=9.7079 \Rightarrow T=9.71°\text{F (above room temperature)}$$

Exponential Growth to an Upper Limit

7. Using $T=a\left(1-e^{-nt}\right)$ where T is temperature, $a=1550$, $n=0.065$ and $t=25$

   $$T=1550\left[1-e^{-(0.065)(25)}\right] \Rightarrow T=1550[0.8030883248] \Rightarrow T=1244.7869 \Rightarrow T=1240°\text{F}$$

Motion in a Resisting Medium

9. $\frac{dv}{dt}=13.7-2.83v$ Using Newton's Second Law, $F=ma$

   $F=\frac{W}{g}\left(\frac{dv}{dt}\right)$ where $F=19.4-4v$

   Therefore, $19.4-4v=\frac{45.5\text{lb}}{32.2\text{ft/s}^2}\left(\frac{dv}{dt}\right)$

   $$(19.4-4v)\left(\frac{32.2\text{ ft/s}^2}{45.5\text{lb}}\right)=\frac{dv}{dt} \Rightarrow \frac{624.68-128.8v}{45.5}=\frac{dv}{dt} \Rightarrow 13.7292-2.8307v=\frac{dv}{dt}$$

   $$\frac{dv}{dt}=13.7-2.83v$$

11. $w-kv=\left(\frac{w}{g}\right)\frac{dv}{dt} \Rightarrow gw-kgv=w\frac{dv}{dt} \Rightarrow \frac{dv}{dt}=\frac{gw-kgv}{w} \Rightarrow \frac{dv}{dt}=g-\left(\frac{kg}{w}\right)v$

    But $\frac{dv}{dt}=0$ when $v=155$ ft/sec, thus $0=g-\left(\frac{kg}{w}\right)(155) \Rightarrow g=\frac{kg}{w}(155) \Rightarrow \frac{g}{155}=\frac{kg}{w} \Rightarrow gw=155kg$

    $w=155k \Rightarrow k=\frac{w}{155}$

    Thus, substituting for k into $\frac{dv}{dt}=g-\left(\frac{kg}{w}\right)v \Rightarrow \frac{dv}{dt}=g-\left(\frac{\left(\frac{w}{155}\right)g}{w}\right)v \Rightarrow \frac{dv}{dt}=g-\frac{gv}{155}$

13. Integrate acceleration in Problem 11 twice to find the equation for displacement

    a equals $\frac{dv}{dt}=\frac{g(155-v)}{155} \Rightarrow \frac{dv}{155-v}=\frac{g}{155}dt$

    $$\int\frac{dv}{155-v}=\int\frac{g}{155}dt \Rightarrow \int\frac{dv}{155-v}=\frac{g}{155}\int dt \Rightarrow -\ln(155-v)=\frac{g}{155}t+C$$

    At $t=0$, $v=0 \Rightarrow -\ln(155-0)=\left(\frac{g}{155}\right)(0)+C \Rightarrow C=-\ln 155$

    Thus, $\ln(155-v)-\ln 155=\frac{-g}{155}t \Rightarrow \ln\frac{155-v}{155}=\frac{-g}{155}t \Rightarrow e^{\ln\frac{155-v}{155}}=e^{\frac{-g}{155}t} \Rightarrow \frac{155-v}{155}=e^{-gt/155}$

    $155-v=155e^{-gt/155} \Rightarrow v=155-155e^{-gt/155}$

Substitute $v = \frac{ds}{dt} \Rightarrow \frac{ds}{dt} = 155 - 155e^{-gt/155} \Rightarrow \int ds = \int 155 - 155e^{-gt/155}\, dt$

$\int ds = \int 155\, dt + \frac{155^2}{g}\int e^{-gt/155}\left(\frac{-g}{155}\right) dt \Rightarrow s = 155t + \frac{155^2}{g}e^{-gt/155} + C_1$

At $t = 0$, $s = 0 \Rightarrow 0 = 155(0) + \frac{155^2}{g}e^{-g(0)/155} + C_1 \Rightarrow 0 = \frac{155^2}{g} + C_1 \Rightarrow C_1 = -\frac{155^2}{g}$

Thus, $s = 155t + \frac{155^2}{g}e^{-gt/155} - \frac{155^2}{g}$; thus $s(1) = 155 + \frac{155^2}{32.2}e^{-32.2/155} - \frac{155^2}{32.2} \Rightarrow s(1) = 15.0406 \Rightarrow s(1) = 15.0$ ft

15. $\frac{dy}{g - \frac{g}{15,625}v^2} = dt \Rightarrow \int \frac{dy}{g - \frac{g}{15,625}v^2} = \int dt \Rightarrow -\frac{15,625}{g}\int \frac{dv}{v^2 - 125^2} = \int dt \Rightarrow -\frac{15,625}{g}\left(\frac{1}{250}\right)\ln\left|\frac{v-125}{v+125}\right| = t + C$

But $v = 0$ at $t = 0$; $C = -\frac{125}{2g}\ln\left|-\frac{125}{125}\right| \Rightarrow C = 0$

$t = -\frac{15,625}{250g}\ln\left|\frac{v-125}{v+125}\right| \Rightarrow$ at $v = 60$ ft/s $\Rightarrow t = -\frac{15,625}{250(32.2)}\ln\left|\frac{60-125}{60+125}\right|$

$t = (-1.94)(-1.045968) \Rightarrow t = 2.03021 \Rightarrow t = 2.03$ s

17. $V = 135 - 114e^{-t/4.19} \Rightarrow 70 = 135 - 114e^{-t/4.19} \Rightarrow -114e^{-t/4.19} = -65 \Rightarrow 114e^{-t/4.19} = 65$

$\frac{114e^{-t/4.19}}{114} = \frac{65}{114} \Rightarrow e^{-t/4.19} = \frac{65}{114} \Rightarrow \ln\left(e^{-t/4.19}\right) = \ln\left(\frac{65}{114}\right) \Rightarrow \frac{-t}{4.19} = \ln\left(\frac{65}{114}\right)$

$t = -4.19\ln\left(\frac{65}{114}\right) \Rightarrow t = 2.353988 \Rightarrow t = 2.36$ s

Exercise 9 ◊ Series *RL* and *RC* Circuits

Series *RL* Circuit

1. Using Kirchhoff's voltage law $\Rightarrow Ri + L\frac{di}{dt} = 0 \Rightarrow \frac{di}{dt} + \frac{R}{L}i = 0$

   This is a first-order DE with $\int P(t)\,dt = \int \frac{R}{L}dt = \frac{R}{L}t \Rightarrow e^{\int P\,dt} = e^{Rt/L}$

   Therfore $ie^{Rt/L} = \int ie^{Rt/L}\,dt = \int e^{Rt/L}(0)\,dt = k$

   At $t = 0$, $i = \frac{E}{R} \Rightarrow k = \frac{E}{R}$, thus $ie^{Rt/L} = \frac{E}{R} \Rightarrow i = \frac{E}{R}e^{-Rt/L}$

3. $v = Ee^{-Rt/L} \Rightarrow v = L\frac{di}{dt} = L\left(\frac{E}{R}\right)\left[0 - \left(-\frac{R}{L}\right)e^{-(Rt/L)}\right] \Rightarrow v = Ee^{-(Rt/L)}$

Series *RC* Circuit

5. Summing the voltages around the loop: $E - v_r = v$ or $v_r = E - v$

   Since i is the same in the reisitor and the capacitor: $i = \frac{v_r}{R} = C\frac{dv}{dt}$

   But $v_r = E - v$, thus $\frac{E - v}{R} = C\frac{dv}{dt}$ or $\frac{E}{R} - \frac{v}{R} = C\frac{dv}{dt} \Rightarrow \frac{dv}{dt} + \frac{1}{RC}v = \frac{E}{RC}$

   This is a first-order linear DE with $P(x) = \frac{1}{RC}$, $Q(x) = \frac{E}{RC}$

   $\int P\,dt = \int \frac{1}{RC}dt = \frac{t}{RC} \Rightarrow e^{\int P\,dt} = e^{t/RC}$, integrating factor

   $ve^{t/RC} = \int \frac{E}{RC}e^{t/RC}\,dt \Rightarrow ve^{t/RC} = E\int e^{t/RC}\left(\frac{1}{RC}\right)dt \Rightarrow ve^{t/RC} = Ee^{t/RC} + k$

   At $t = 0$, $v = 0$, thus $(0)e^{0/RC} = Ee^{0/RC} + k \Rightarrow 0 = E + k \Rightarrow k = -E$

   Then $ve^{t/RC} = Ee^{t/RC} - E \Rightarrow v = e^{-t/RC}\left(Ee^{t/RC} - E\right) \Rightarrow v = E\left(1 - e^{-t/RC}\right)$, the voltage across the capacitor

7. $\frac{t}{RC}=\frac{0.002}{538\left(525\times10^{-6}\right)}=0.007081\Rightarrow e^{-(t/RC)}=e^{-0.007081}=0.992944=0.9929$

$$i=\frac{E}{R}e^{-(t/RC)}=\left(\frac{125}{538}\right)(0.9929)=0.23069\text{ A}=231\text{ mA}$$

$$v=E\left(1-e^{-t/RC}\right)=125(1-0.992944)=0.882\text{ V}$$

Circuits in Which *R*, *L*, or *C* Are Not Constant

9. $i=10(0.00155+1)-10(0.00155+1)^{-2}\Rightarrow i=10.0155-9.9690719\Rightarrow i=0.046428\text{ A}\Rightarrow i=46.4\text{ mA}$

11. $i(4.82\text{ ms})=10-\frac{40}{\left[5(0.00482)+2\right]^2}=10-9.763287=0.236712\text{ A}=237\text{ mA}$

13. When $t=1.85$ ms $\Rightarrow i=\frac{100(0.00185)\left[(0.00185)^2+12\right]}{\left[(0.00185)^2+4\right]^2}=\frac{2.22}{16}=0.13875\text{ A}=139\text{ mA}$

Series RL or RC Circuit with Alternating Current

15. $i=I\sin(\omega t-\phi)+\left(\frac{IX_L}{Z}\right)e^{-Rt/L}; R=233\Omega,\ L=5.82\text{ H},\ e=58.0\sin377t,\ \omega=377\text{ rad/s}$

$$X_L=\omega L=377(5.82)=2194.14=2194\Omega$$

$$Z=\sqrt{R^2+x_L^2}=\sqrt{(233)^2+(2194)^2}=2206.337=2206\Omega$$

$$\tan\phi=\frac{X_L}{R}=\frac{2194}{233}=9.4163=9.42;\quad \phi=\tan^{-1}(9.42)=83.9403°=83.9°$$

$$I=\frac{E}{Z}=\frac{58}{2206}=0.02629=0.0263\text{ A};\quad \frac{IX_L}{Z}=\frac{0.0263(2194)}{2206}=0.0261489=0.0261\text{ A}$$

$$\frac{R}{L}=\frac{233}{5.82}=40.03=40$$

Thus substituting into $i=I\sin(\omega t-\phi)+\left(\frac{IX_L}{Z}\right)e^{-Rt/L}\Rightarrow i=26.3\sin(377t-83.9°)+26.1e^{-40t}$ mA

17. Summing the voltage drops around the loop yields $Ri+\frac{1}{C}\int i\,dt=E\sin\omega t$

Taking the derivative; $R\frac{di}{dt}+\frac{i}{C}=\omega E\cos\omega t$ since the derivative of $E\sin\omega t$ is $\omega E\cos\omega t$

Dividing by *R*, the differential equation is then

$\frac{di}{dt}+\frac{1}{RC}i=\frac{\omega E}{R}\cos\omega t$, a first-order linear DE; with the integrating factor $e^{\int P\,dt}=e^{\int 1/RC\,dt}=e^{t/RC}$

The solution is then $ie^{t/RC}=\int\frac{\omega E}{R}e^{t/RC}\cos\omega t\,dt\Rightarrow ie^{t/RC}=\frac{\omega E}{R}\int e^{t/RC}\cos\omega t\,dt$

Using integral 42, with $a=\frac{1}{RC}$ and $b=\omega$,

$$ie^{t/RC}=\frac{\omega E}{R}\left[\frac{e^{t/RC}}{\frac{1}{R^2C^2}+\omega^2}\left(\frac{1}{RC}\cos\omega t+\omega\sin\omega t\right)\right]+C=\frac{\omega Ee^{t/RC}}{\frac{R}{R^2C^2}+R\omega^2}\left(\frac{1}{RC}\cos\omega t+\frac{\omega RC}{RC}\sin\omega t\right)+C$$

Dividing by $e^{t/RC}$, $i=\frac{E}{\left(\frac{1}{\omega RC^2}+R\omega\right)RC}(\cos\omega t+\omega RC\sin\omega t)+\frac{C}{e^{t/RC}}$

$$i=\frac{E}{R^2\omega C+\frac{1}{\omega C}}(\cos\omega t+\omega RC\sin\omega t)+Ce^{-t/RC}=\frac{1}{\omega C}\frac{E}{R^2+\frac{1}{\omega^2C^2}}(\cos\omega t+\omega RC\sin\omega t)+Ce^{-t/RC}$$

$$i = \frac{E}{Z}\left[\frac{X_c}{Z}\cos\omega t + \frac{R}{Z}\sin\omega t\right] + Ce^{-t/RC} \text{ since } X_c = \frac{1}{\omega C} \text{ and } Z^2 = R^2 + \frac{1}{\omega^2 C^2}$$

From the impedance triangle, $\frac{X_c}{Z} = \sin\phi$ and $\frac{R}{Z} = \cos\phi$, so $i = \frac{E}{Z}\left[\cos\omega t\sin\phi + \sin\omega t\cos\phi\right] + Ce^{-t/RC}$

$$i = \frac{E}{Z}\sin(\omega t + \phi) + Ce^{-t/RC}$$

when $t = 0$, $i = \frac{E}{R}\sin\omega t = 0$, thus $0 = \frac{E}{R}\sin\omega t + C \Rightarrow C = -\frac{E}{R}\sin\omega t$

Substituting back, $i = \frac{E}{Z}\left[\underbrace{\sin(\omega t + \phi)}_{\text{steady state current}} - \underbrace{e^{-t/RC}\sin\phi}_{\text{transient current}}\right]$

CHAPTER 30 REVIEW PROBLEMS

1. $xy + y + xy' = e^x \Rightarrow y(x+1) + x\frac{dy}{dx} = e^x \Rightarrow \frac{dy}{dx} + \left(\frac{x+1}{x}\right)y = \frac{e^x}{x}$ [first-order linear DE]

$$\int P\,dx = \int\frac{x+1}{x}dx = \int\left(1 + \frac{1}{x}\right)dx = x + \ln x \Rightarrow e^{\int P\,dx} = e^{x+\ln x} = e^x e^{\ln x} = xe^x$$

$$yxe^x = \int\frac{e^x}{x}\left(xe^x\right)dx \Rightarrow yxe^x = \frac{1}{2}\int e^{2x}\,2\,dx \Rightarrow yxe^x = \frac{1}{2}e^{2x} + C \Rightarrow yxe^x = \frac{e^{2x}}{2} + C$$

$$y = \frac{e^{2x}}{2xe^x} + \frac{C}{xe^x} \Rightarrow y = \frac{e^x}{2x} + \frac{C}{xe^x}$$

3. $y' + y - 2\cos x = 0$ [first-order linear DE] $\Rightarrow y' + y = 2\cos x$

$\int P\,dx = \int dx = x \Rightarrow e^{\int P\,dx} = e^x$, integrating factor

$$ye^x = \int 2\cos x\left(e^x\right)dx \Rightarrow ye^x = 2\int e^x\cos x\,dx \Rightarrow ye^x = 2\left(\frac{e^x}{2}\right)(\cos x + \sin x) + C \Rightarrow ye^x = e^x(\cos x + \sin x) + C$$

$$y = e^{-x}\left[e^x(\cos x + \sin x) + C\right] \Rightarrow y = \cos x + \sin x + Ce^{-x}$$

5. $2y + 3x^2 + 2xy' = 0 \Rightarrow 2y + 3x^2 + 2x\frac{dy}{dx} = 0$

$$2y\,dx + 3x^2\,dx + 2x\,dy = 0 \Rightarrow 2x\,dy = -2y\,dx - 3x^2\,dx \Rightarrow 2x\,dy = -\left(2y + 3x^2\right)dx \Rightarrow \frac{2x\,dy}{dx} = -2y - 3x^2$$

$$\frac{dy}{dx} = \frac{-2y}{2x} - \frac{3x^2}{2x} \Rightarrow \frac{dy}{dx} + \frac{y}{x} = -\frac{3x}{2}$$

Since $P = \frac{1}{x} \Rightarrow \int P\,dx = \int\frac{1}{x}dx = \ln x \Rightarrow e^{\int P\,dx} = e^{\ln x} = x$, the integrating factor

$$yx = \int -\frac{3x}{2}(x\,dx) \Rightarrow yx = -3\int\frac{x^2}{2}dx \Rightarrow yx = -\frac{x^3}{2} + C_1$$

$$y = -\frac{x^2}{2} + \frac{C}{x} \Rightarrow 2xy = -x^3 + 2C \Rightarrow x^3 + 2xy = C$$

7. $y^2 + \left(x^2 - xy\right)y' = 0$ [homogeneous]

Let $y = vx$, then $y' = v + xv'$

$$v^2x^2 + \left(x^2 - x^2v\right)(v + xv') = 0 \Rightarrow v^2x^2 + x^2(1 - v)(v + xv') = 0 \Rightarrow v^2 + (1 - v)(v + xv') = 0$$

$$v^2 + v + xv' - v^2 - xvv' = 0 \Rightarrow v + xv' - xvv' = 0 \Rightarrow v + xv'(1 - v) = 0$$

$$\frac{dx}{x} + \frac{1 - v}{v}dv = 0 \Rightarrow \frac{dx}{x} + \frac{dv}{v} - dv = 0$$

$$\int\left(\frac{dx}{x} + \frac{dv}{v} - dv\right) = 0 \Rightarrow \ln x + \ln v - v + C = 0 \Rightarrow \ln xv = v + C$$

Subsitituting $v=\frac{y}{x}\Rightarrow \ln x\left(\frac{y}{x}\right)=\frac{y}{x}+C\Rightarrow \ln y=\frac{y}{x}+C\Rightarrow x\ln y=y+Cx\Rightarrow x\ln y-y=Cx$

9. $(1-x)\frac{dy}{dx}=y^2$ [variables separable]

$\frac{dy}{dx}=\frac{y^2}{1-x}\Rightarrow\frac{dy}{y^2}=\frac{dx}{1-x}\Rightarrow y^{-2}\,dy=\frac{dx}{1-x}$

$\int y^{-2}\,dy=\int\frac{dx}{1-x}\Rightarrow\int y^{-2}\,dy=-\int\frac{-dx}{1-x}\Rightarrow -y^{-1}=-\ln(1-x)+C_1\Rightarrow\frac{1}{y}=\ln|1-x|+C\Rightarrow 1=y\ln|1-x|+Cy$

$y\ln|1-x|+Cy=1$

11. $y+2xy^2+(y-x)y'=0\Rightarrow y+2xy^2+(y-x)\frac{dy}{dx}=0\Rightarrow y+2xy^2+y\frac{dy}{dx}-x\frac{dy}{dx}$

$y\,dx+2xy^2\,dx+y\,dy-x\,dy=0\Rightarrow y\,dx-x\,dy+2xy^2\,dx+y\,dy=0\Rightarrow\frac{y\,dx-x\,dy}{y^2}+2x\,dx+\frac{dy}{y}=0$ [exact]

$\int\frac{y\,dx-x\,dy}{y^2}+\int 2x\,dx+\int\frac{dy}{y}=0\Rightarrow x^2+\frac{x}{y}+\ln y+C=0\Rightarrow x^2+\frac{x}{y}+\ln y=C$

13. $y'\sin y=\cos x,\qquad x=\frac{\pi}{4}$ when $y=0$

$\frac{dy}{dx}\sin y=\cos x\Rightarrow\sin y\,dy=\cos x\,dx$

$\int\sin y\,dy=\int\cos x\,dx\Rightarrow-\cos y=\sin x+C$

At $\left(\frac{\pi}{4},0\right)\Rightarrow-\cos(0)=\sin\left(\frac{\pi}{4}\right)+C\Rightarrow-1=0.707106+C\Rightarrow C=-1.707$

Therefore $-\cos y=\sin x-1.707\Rightarrow\sin x+\cos y=1.707$

15. $y\,dx=(x-2x^2y)\,dy,\qquad x=2$ when $y=1$

$y\,dx=x\,dy-2x^2y\,dy\Rightarrow y\,dx-x\,dy=-2x^2y\,dy\Rightarrow\frac{x\,dy-y\,dx}{x^2}=2y\,dy$ [exact]

$\int\frac{x\,dy-y\,dx}{x^2}=\int 2y\,dy\Rightarrow\frac{y}{x}=y^2+C$

At $(2,1)\Rightarrow\frac{1}{2}=1^2+C\Rightarrow C=-\frac{1}{2}$

Thus $\frac{y}{x}=y^2-\frac{1}{2}\Rightarrow y=y^2x-\frac{1}{2}x\Rightarrow 2y=2y^2x-x\Rightarrow x+2y=2y^2x$

17. Exponential decay that follows the formula $y=ae^{-nt}$ where $a=1550$ rev/min, $n=0.095$ and $t=6$ sec

$y=1550e^{-(0.095)6}\Rightarrow y=\left(\frac{1550}{60}\right)e^{-0.57}$ [needed to change 1550 rev/min to seconds]

$y=25.83(0.56552548)\Rightarrow y=14.6075\Rightarrow 14.6$ rev/s

19. $x^2=4y\Rightarrow 2x=4\frac{dy}{dx}\Rightarrow\frac{1}{2}x=\frac{dy}{dx}\Rightarrow\frac{dy}{dx}=\frac{x}{2}\Rightarrow$ for the orthogonal trajectory, $\frac{dy}{dx}=-\frac{2}{x}$

$\int\frac{dy}{dx}=\int-\frac{2}{x}dx\Rightarrow\int\frac{dy}{dx}=-2\int\frac{1}{x}dx\Rightarrow y=-2\ln|x|+C$

21. $y'=2+\frac{y}{x}\Rightarrow\frac{dy}{dx}=2+\frac{y}{x}\Rightarrow\frac{dy}{dx}-\frac{y}{x}=2\Rightarrow\frac{dy}{dx}-\left(\frac{1}{x}\right)y=2;\ P(x)=-\frac{1}{x},\ Q(x)=2$

$\int P\,dx=\int-\frac{1}{x}dx=-\ln x=\ln\frac{1}{x}\Rightarrow e^{\int P\,dx}=e^{1/x}=\frac{1}{x}$, integrating factor

$y\left(\frac{1}{x}\right)=\int 2\left(\frac{1}{x}dx\right)\Rightarrow\frac{y}{x}=2\ln x+C\Rightarrow y=2x\ln x+Cx\Rightarrow Cx=y-2x\ln x$

At $(1, 2) \Rightarrow C(1) = 2 - 2(1)\ln 1 \Rightarrow C = 2 - 2(0) \Rightarrow C = 2$
Thus $y = 2x\ln x + 2x$

23. Using the formula for exponential growth, $y = ae^{nt} \Rightarrow y = 500e^{15(0.15)}$
$y = 500(9.487755) \Rightarrow y = 4743.867 = 4740$ g

Chapter 31: Second-Order Differential Equations

Exercise 1 ◊ Second-Order DE

1. $y'' = 5 \Rightarrow \frac{d}{dx}y' = 5 \Rightarrow dy' = 5dx$

   Integrate; $\int \frac{d}{dx}y' = \int 5dx \Rightarrow \frac{dy}{dx} = 5x + C_1 \Rightarrow dy = (5x + C_1)dx$

   Integrate again; $\int dy = \int (5x + C_1)dx \Rightarrow y = \frac{5}{2}x^2 + C_1x + C_2$

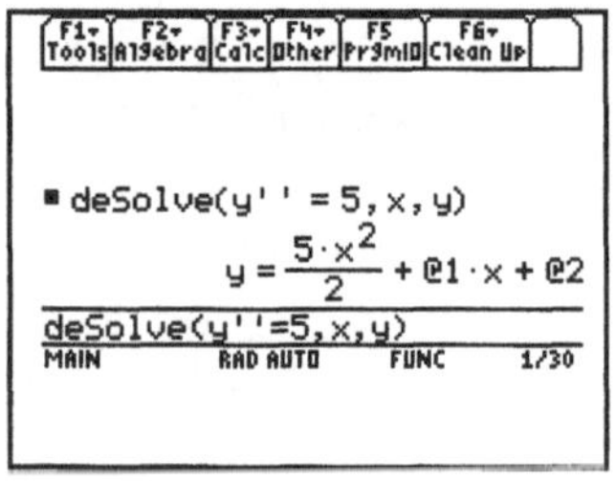

3. $y'' = 3e^x \Rightarrow \frac{d}{dx}y' = 3e^x \Rightarrow dy' = 3e^x\,dx$

   $\int dy' = \int 3e^x\,dx \Rightarrow \frac{dy}{dx} = 3e^x + C_1 \Rightarrow dy = (3e^x + C_1)dx$

   $\int dy = \int (3e^x + C_1)dx \Rightarrow y = 3e^x + C_1x + C_2$

5. $y'' - x^2 = 0$ where $y' = 1$ at the point (0, 0)

   $\frac{d}{dx}y' = x^2 \Rightarrow dy' = x^2\,dx$

   $\int dy' = \int x^2\,dx \Rightarrow \frac{dy}{dx} = \frac{x^3}{3} + C_1 \Rightarrow dy = \left(\frac{x^3}{3} + C_1\right)dx;$ at $y' = 1, x = 0 \Rightarrow 1 = \frac{0^3}{3} + C_1 \Rightarrow C_1 = 1$

   $\int dy = \int\left(\frac{x^3}{3} + 1\right)dx \Rightarrow y = \frac{x^4}{12} + x + C_2;$ at $(0, 0) \Rightarrow 0 = \frac{0^4}{12} + 0 + C_2 \Rightarrow C_2 = 0$

   Thus $y = \frac{x^4}{12} + x$

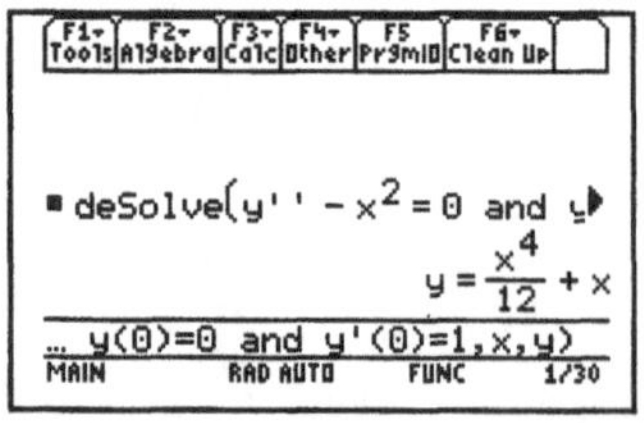

Exercise 2 ◊ Constant Coefficient and Right Side Zero
Second-Order DE, Roots of Auxiliary Equation Real and Unreal

1. $y'' - 6y' + 5y = 0;$ the auxilary equation is $m^2 - 6m + 5 = 0$

   $m^2 - 6m + 5 = 0 \Rightarrow (m-5)(m-1) = 0;\ m = 5,\ m = 1$

   Substituting into $y = C_1e^{m_1x} + C_2e^{m_2x} \Rightarrow y = C_1e^x + C_2e^{5x}$

3. $y'' - 3y' + 2y = 0;$ the auxiliary equation is $m^2 - 3m + 2 = 0$

   $m^2 - 3m + 2 = 0 \Rightarrow (m-2)(m-1) = 0;\ m = 2,\ m = 1$

   Substituting into $y = C_1e^{m_1x} + C_2e^{m_2x} \Rightarrow C_1e^x + C_2e^{2x}$

5. $y'' - y' - 6y = 0;$ auxiliary equation: $m^2 - m - 6 = 0$

$(m-3)(m+2)=0;\ m=3,\ m=-2 \Rightarrow y=C_1e^{3x}+C_2e^{-2x}$

7. $5y''-2y'=0;$ auxiliary equation: $5m^2-2m=0$

$$m(5m-2)=0;\ m=0,\ m=\frac{2}{5} \Rightarrow y=C_1e^{0x}+C_2e^{2x/5} \Rightarrow y=C_1+C_2e^{2x/5}$$

9. $6y''+5y'-6y=0;$ auxiliary equation: $6m^2+5m-6=0$

$$(2m+3)(3m-2)=0;\ m=-\frac{3}{2},\ m=\frac{2}{3} \Rightarrow y=C_1e^{2x/3}+C_2e^{-3x/2}$$

Second-Order DE, Roots of Auxiliary Equation Real and Equal

11. $y''-4y'+4y=0;$ auxiliary equation: $m^2-4m+4=0$
$(m-2)(m-2)=0;\ m=2$ [equal roots]
Substituting into $y=C_1e^{mx}+C_2xe^{mx} \Rightarrow y=C_1e^{2x}+C_2xe^{2x} \Rightarrow y=(C_1+C_2x)e^{2x}$

13. $y''-2y'+y=0;$ auxiliary equation: $m^2-2m+1=0$
$(m-1)(m-1)=0 \Rightarrow m=1$ [equal roots] $\Rightarrow y=C_1e^{x}+C_2xe^{x}$

15. $y''+4y'+4y=0;$ auxiliary equation: $m^2+4m+4=0$
$(m+2)(m+2)=0 \Rightarrow m=-2$ [equal roots] $\Rightarrow y=C_1e^{-2x}+C_2xe^{-2x} \Rightarrow y=(C_1+C_2x)e^{-2x}$

17. $y''+2y+y=0;$ auxiliary equation: $m^2+2m+1=0$
$(m+1)(m+1)=0 \Rightarrow m=-1$ [equal root] $\Rightarrow y=C_1e^{-x}+C_2xe^{-x}$

Second-Order DE, Roots of Auxiliary Equation Not Real

19. $y''+4y'+13y=0;$ auxiliary equation: $m^2+4m+13=0$

$$m=\frac{-4\pm\sqrt{(4)^2-4(1)(13)}}{2(1)} \Rightarrow m=-2\pm 3i$$

Substitute into $y=e^{ax}(C_1\cos bx+C_2\sin bx)$ with $a=-2,\ b=3 \Rightarrow y=e^{-2x}(C_1\cos 3x+C_2\sin 3x)$

21. $y''-6y'+25y=0;$ auxiliary equation: $m^2-6m+25=0$

$$m=\frac{-(-6)\pm\sqrt{(-6)^2-4(1)(25)}}{2(1)} \Rightarrow m=3\pm 4i$$

With $a=3,\ b=4 \Rightarrow y=e^{3x}(C_1\cos 4x+C_2\sin 4x)$

23. $y''+4y=0;$ auxiliary equation: $m^2+4=0$

$$m=\frac{0\pm\sqrt{(0)^2-4(1)(4)}}{2(1)} \Rightarrow m=\pm 2i$$

With $a=0,\ b=2 \Rightarrow y=e^{(0)x}(C_1\cos 2x+C_2\sin 2x) \Rightarrow y=C_1\cos 2x+C_2\sin 2x$

You may also solve for m by $m^2+4=0 \Rightarrow m^2=-4 \Rightarrow m=\pm 2i$

25. $y''-4y'+5y=0;$ auxiliary equation: $m^2-4m+5=0$

$$m=\frac{-(-4)\pm\sqrt{(-4)^2-4(1)(5)}}{2(1)} \Rightarrow m=2\pm i$$

With $a=2,\ b=1 \Rightarrow y=e^{2x}(C_1\cos x+C_2\sin x)$

27. $y''+6y'+9y=0,$ $\quad y=0$ and $y'=3$ when $x=0$
Auxiliary equation: $m^2+6m+9=0 \Rightarrow (m+3)(m+3)=0 \Rightarrow m=-3$ [equal roots]

Thus $y=(C_1+C_2x)e^{-3x}$

At $x=0,\ y=0 \Rightarrow 0=\left[C_1+C_2(0)\right]e^{-3(0)} \Rightarrow C_1=0;$ therefore $y=(0+C_2x)e^{-3x} \Rightarrow y=C_2xe^{-3x}$

Taking the derivative of $y=C_2xe^{-3x} \Rightarrow y'=-3C_2xe^{-3x}+C_2e^{-3x}$

At $y'=3,\ x=0,\ C_1=0 \Rightarrow 3=-3C_2(0)e^{-3(0)}+C_2e^{-3(0)} \Rightarrow 3=0+C_2 \Rightarrow C_2=3$

Thus $y=3xe^{-3x}$

29. $y''-2y'+y=0,\qquad y=5$ and $y'=-9$ when $x=0$

Auxiliary equation: $m^2-2m+1=0 \Rightarrow (m-1)(m-1)=0 \Rightarrow m=1$ [equal roots]

Thus $y=C_1e^x+C_2xe^x$

At $x=0,\ y=5 \Rightarrow 5=C_1e^0+C_2(0)e^0 \Rightarrow C_1+0=5 \Rightarrow C_1=5;$ therefore $y=5e^x+C_2xe^x$

Taking the derivative of $y=5e^x+C_2xe^x \Rightarrow y'=5e^x+C_2xe^x+C_2e^x$

At $y'=-9,\ x=0,\ C_1=5 \Rightarrow -9=5e^0+C_2(0)e^0+C_2e^0 \Rightarrow -9=5+0+C_2 \Rightarrow C_2=-14$

Thus $y=5e^x-14xe^x$

31. $y''-2y'=0,\qquad y=1+e^2$ and $y'=2e^2$ when $x=1$

Auxiliary equation: $m^2-2m=0 \Rightarrow m(m-2)=0 \Rightarrow m=0,\ m=2$

Thus $y=C_1e^0+C_2e^{2x} \Rightarrow y=C_1+C_2e^{2x}$

At $x=1,\ y=1+e^2 \Rightarrow 1+e^2=C_1+C_2e^{2(1)} \Rightarrow 1+e^2=C_1+C_2e^2$

Taking the derivative of $y=C_1+C_2e^{2x} \Rightarrow y'=2C_2e^{2x}$

At $y'=2e^2,\ x=1 \Rightarrow 2e^2=2C_2e^{2(1)} \Rightarrow 2e^2=2C_2e^2 \Rightarrow C_2=1$

Thus $1+e^2=C_1+(1)e^2 \Rightarrow 1+e^2=C_1+e^2 \Rightarrow C_1=1$

Therefore $y=1+(1)e^{2x} \Rightarrow y=1+e^{2x}$

33. $y''-4y'=0,\qquad y=1$ and $y'=-1$ when $x=0$

Auxiliary equation: $m^2-4=0 \Rightarrow (m+2)(m-2)=0 \Rightarrow m=-2,\ m=2$

Thus $y=C_1e^{2x}+C_2e^{-2x}$

At $x=0,\ y=1 \Rightarrow 1=C_1e^{2(0)}+C_2e^{-2(0)} \Rightarrow 1=C_1+C_2$

Taking the derivative of $y=C_1e^{2x}+C_2e^{-2x} \Rightarrow y'=2C_1e^{2x}-2C_2e^{-2x}$

At $x=0,\ y'=-1 \Rightarrow -1=2C_1e^{2(0)}-2C_2e^{-2(0)} \Rightarrow -1=2C_1-2C_2$

Solving $C_1+C_2=1$ and $2C_1-2C_2=-1$ simultaneously

$$\begin{array}{lcl} C_1+C_2=1 & \Rightarrow & 2C_1+2C_2=2 \\ \underline{2C_1-2C_2=-1} & \Rightarrow & \underline{2C_1-2C_2=-1} \\ & & 4C_1 = 1 \Rightarrow C_1=\frac{1}{4} \end{array}$$

Thus $\frac{1}{4}+C_2=1 \Rightarrow C_2=\frac{3}{4}$

Thus $y=\frac{1}{4}e^{2x}+\frac{3}{4}e^{-2x}$

35. $y''+2y'+2y=0,\qquad y=0$ and $y'=1$ when $x=0$

Auxiliary equation: $m^2+2m+2=0 \Rightarrow m=-1\pm i$ [by the quadratic formula]

At $a=-1,\ b=1 \Rightarrow y=e^{-x}(C_1\cos x+C_2\sin x)$

At $x=0,\ y=0 \Rightarrow 0=e^0(C_1\cos(0)+C_2\sin(0)) \Rightarrow 0=C_1(1)+C_2(0) \Rightarrow C_1=0$

Thus $y=e^{-x}((0)\cos x+C_2\sin x) \Rightarrow y=e^{-x}C_2\sin x$

Taking the derivative of $y=e^{-x}C_2\sin x \Rightarrow y'=-C_2(\sin x)e^{-x}+C_2(\cos x)e^{-x}$

At $y' = 1$, $x = 0 \Rightarrow 1 = -C_2(\sin(0))e^0 + C_2e^{-x}(\cos(0))e^0 \Rightarrow 1 = -C_2(0)(1) + C_2(1)(1) \Rightarrow C_2 = 1$

Thus $y = e^{-x}((0)\cos x + (1)\sin x) \Rightarrow y = e^{-x}\sin x$

Third-Order DE

37. $y''' - 2y'' - y' + 2y = 0$

Auxiliary equation: $m^3 - 2m^2 - m + 2 = 0 \Rightarrow (m^3 - 2m^2) - (m-2) = 0 \Rightarrow m^2(m-2) - (m-2) = 0$

$$(m^2 - 1)(m-2) = 0 \Rightarrow (m+1)(m-1)(m-2) = 0 \Rightarrow m = -1,\ m = 1,\ m = 2$$

Therefore $y = C_1e^x + C_2e^{-x} + C_3e^{2x}$

39. $y''' - 6y'' + 11y' - 6y = 0$

Auxiliary equation: $m^3 - 6m^2 + 11m - 6 = 0 \Rightarrow (m-1)(m-2)(m-3) = 0 \Rightarrow m = 1,\ m = 2,\ m = 3$

Therefore $y = C_1e^x + C_2e^{2x} + C_3e^{3x}$

Auxiliary equation: $m^3 + m^2 - 4m - 4 = 0 \Rightarrow (m^3 + m^2) - (4m + 4) = 0 \Rightarrow m^2(m+1) - 4(m+1) = 0$

41. $y''' - 3y'' - y' + 3y = 0$

Auxiliary equation: $m^3 - 3m^2 - m + 3 = 0 \Rightarrow (m^3 - 3m^2) - (m-3) = 0 \Rightarrow m^2(m-3) - (m-3) = 0$

$$(m^2 - 1)(m-3) = 0 \Rightarrow (m+1)(m-1)(m-3) = 0 \Rightarrow m = -1,\ m = 1,\ m = 3$$

Therefore $y = C_1e^x + C_2e^{-x} + C_3e^{3x}$

43. $4y''' - 3y' + y = 0$

Auxiliary equation: $4m^3 - 3m + 1 = 0 \Rightarrow (m+1)(2m-1)(2m-1) = 0 \Rightarrow m = -1,\ m = \frac{1}{2},\ m = \frac{1}{2}$

Therefore $y = C_1e^{x/2} + C_2e^{x/2}x + C_3e^{-x} \Rightarrow y = e^{x/2}(C_1 + C_2x) + C_3e^{-x}$

Exercise 3 ◊ Right Side Not Zero

With Algebraic Expressions

1. $y'' - 4y = 12$

Auxiliary equation: $m^2 - 4 = 0$; roots $m = 2$, $m = -2$

Complementary equation: $y_c = C_1e^{2x} + C_2e^{-2x}$

Terms of $f(x)$ and their derivative (less coefficients): $f(x) = \text{constant},\ f'(x) = 0,\ f''(x) = 0$

Thus $y_p = C,\ y_p' = 0,\ y_p'' = 0$

Substitute into $y'' - 4y = 12$

Equating constant $\Rightarrow 0 - 4C = 12 \Rightarrow C = -3$

Thus $y_p = -3$

Therefore $y = y_c + y_p \Rightarrow y = C_1e^{2x} + C_2e^{-2x} + (-3) \Rightarrow y = C_1e^{2x} + C_2e^{-2x} - 3$

3. $y'' - y' - 2y = 4x$

Auxiliary equation: $m^2 - m - 2 = 0$; roots $m = -1$, $m = 2$

Complementary equation: $y_c = C_1e^{2x} + C_2e^{-x}$

Terms of $f(x)$ and their derivative (less coefficients): $f(x) = x,\ f'(x) = \text{constant},\ f''(x) = 0$

Thus $y_p = Ax + B;\ y_p' = A,\ y_p'' = 0$

Substitute into $y'' - y' - 2y = 4x \Rightarrow 0 - A - 2(Ax + B) = 4x \Rightarrow -2Ax + (-A - 2B) = 4x$

Equating coefficients of $x \Rightarrow -2A = 4 \Rightarrow A = -2$

Equating constant $\Rightarrow -A - 2B = 0 \Rightarrow -(-2) - 2B = 0 \Rightarrow 2B = 2 \Rightarrow B = 1$

Thus $y_p = -2x + 1$ [from $Ax + B \Rightarrow -2x + 1$]

Therefore $y = y_c + y_p \Rightarrow y = C_1e^{2x} + C_2e^{-x} - 2x + 1$

5. $y'' - 4y = x^3 + x$

Auxiliary equation: $m^2 - 4 = 0$; roots $m = 2$, $m = -2$

Complementary equation: $y_c = C_1e^{2x} + C_2e^{-2x}$

Terms of $f(x)$ and their derivative (less coefficients): $f(x) = Ax^3 + Bx,\ f'(x) = Cx^2 + D,$

$f''(x) = Ex$ [duplicate of $f(x)$]. So,

$$y_p = Ax^3 + Cx^2 + Bx + D,\ y_p' = 3Ax^2 + 2Cx + B,\ y_p'' = 6Ax + 2C$$

Substitute into $y'' + 4y = x^3 + x \Rightarrow 6Ax + 2C - 4(Ax^3 + Cx^2 + Bx + D) = x^3 + x$

$-4Ax^3 - 4Cx^2 + (6A - 4B)x + 2C - 4D = x^3 + x$

Equating coefficients of $x \Rightarrow -4A = 1 \Rightarrow A = -\dfrac{1}{4}$

$$-4Cx^2 = 0 \Rightarrow C = 0$$

$$6A - 4B = 1 \Rightarrow -\frac{6}{4} - 4B = 1 \Rightarrow -4B = \frac{5}{2} \Rightarrow B = -\frac{5}{8}$$

$$D = 0$$

$$\text{Thus } y_p = -\frac{1}{4}x^3 - \frac{5}{8}x\left[\text{from } Ax^3 + Cx^2 + Bx + D \Rightarrow -\frac{1}{4}x^3 + 0x^2 - \frac{5}{8}x + 0\right]$$

Therefore $y = y_c + y_p \Rightarrow y = C_1e^{2x} + C_2e^{-2x} - \dfrac{1}{4}x^3 - \dfrac{5}{8}x$

With Exponential Expressions

7. $y'' - y' - 2y = 6e^x$

Auxiliary equation: $m^2 - m - 2 = 0$; roots $m = 2$, $m = -1$

Complementary equation: $y_c = C_1e^{-x} + C_2e^{2x}$

Terms of $f(x)$ and their derivative (less coefficients): $f(x) = e^x,\ f'(x) = e^x,\ f''(x) = e^x$, so

$$y_p = Ae^x,\ y_p' = Ae^x,\ y_p'' = Ae^x$$

Substitute into $y'' - y' - 2y = 6e^x \Rightarrow Ae^x - Ae^x - 2(Ae^x) = 6e^x \Rightarrow Ae^x - Ae^x - 2Ae^x = 6e^x$

Equating coefficients of $x \Rightarrow (1 - 1 - 2)A = 6 \Rightarrow -2A = 6 \Rightarrow A = -3$

$$\text{Thus } y_p = -3e^x \left[\text{from } Ae^x \Rightarrow -3e^x\right]$$

Therefore $y = y_c + y_p \Rightarrow y = C_1e^{2x} + C_2e^{-x} - 3e^x$

9. $y'' - 4y = 4x - 3e^x$

Auxiliary equation: $m^2 - 4 = 0$; roots $m = 2$, $m = -2$

Complementary equation: $y_c = C_1e^{2x} + C_2e^{-2x}$

Terms of $f(x)$ and their derivative (less coefficients) are x, e^x and a constant. So,

$$y_p = Ax + Be^x + C,\ y_p' = A + Be^x,\ y_p'' = Be^x$$

Substitute into $y'' - 4y = 4x - 3e^x \Rightarrow Be^x - 4(Ax + Be^x + C) = 4x - 3e^x \Rightarrow Be^x - 4Ax - 4Be^x - 4C = 4x - 3e^x$

$-4Ax - 3Be^x - 4C = 4x - 3e^x$

Equating coefficients of x and $e^x \Rightarrow -4A = 4 \Rightarrow A = -1;\ -3B = -3 \Rightarrow B = 1$

Equating constant $\Rightarrow -4C = 0 \Rightarrow C = 0$

$$\text{Thus } y_p = -x + e^x + 0 \Rightarrow y_p = -x + e^x$$

Therefore $y = y_c + y_p \Rightarrow y = C_1e^{2x} + C_2e^{-2x} + e^x - x$

11. $y'' + 4y' + 4y = 8e^{2x} + x$

Auxiliary equation: $m^2 + 4m + 4 = 0$; roots $m = -2$, $m = -2$

Complementary equation: $y_c = C_1e^{-2x} + C_2xe^{-2x}$

Terms of $f(x)$ and their derivative (less coefficients) are e^{2x}, x and a constant. So,

$$y_p = Ae^{2x} + Bx + C,\ y'_p = 2Ae^{2x} + B,\ y''_p = 4Ae^{2x}$$

Substitute into $y'' + 4y' + 4y = 8e^{2x} + x \Rightarrow 4Ae^{2x} + 4\left(2Ae^{2x} + B\right) + 4\left(Ae^{2x} + Bx + C\right) = 8e^{2x} + x$

$$4Ae^{2x} + 8Ae^{2x} + 4B + 4Ae^{2x} + 4Bx + 4C = 8e^{2x} + x \Rightarrow 16Ae^{2x} + 4Bx + 4\left(B + C\right) = 8e^{2x} + x$$

Equating coefficients of e^{2x} and $x \Rightarrow 16A = 8 \Rightarrow A = \frac{1}{2};\ 4B = 1 \Rightarrow B = \frac{1}{4}$

Equating constant $\Rightarrow 4\left(B + C\right) = 0 \Rightarrow 4\left(\frac{1}{4}\right) + 4C = 0 \Rightarrow 1 + 4C = 0 \Rightarrow 4C = -1 \Rightarrow C = -\frac{1}{4}$

$$\text{Thus } y_p = \frac{1}{2}e^{2x} + \frac{1}{4}x - \frac{1}{4} \Rightarrow y_p = \frac{e^{2x}}{2} + \frac{x}{4} - \frac{1}{4}$$

Therefore $y = y_c + y_p \Rightarrow y = C_1e^{-2x} + C_2xe^{-2x} + \frac{e^{2x}}{2} + \frac{x}{4} - \frac{1}{4} \Rightarrow y = \left(C_1 + C_2x\right)e^{-2x} + \frac{2e^{2x} + x - 1}{4}$

With Trigonometric Expressions

13. $y'' + 4y = \sin 2x$

Auxiliary equation: $m^2 + 4 = 0$; roots $m = 2i,\ m = -2i$

Complementary equation: $y_c = C_1 \cos 2x + C_2 \sin 2x$

Terms of $f(x)$ and their derivative (less coefficients) are $f(x) = \sin 2x, f'(x) = \cos 2x,\ f''(x) = \sin 2x$

But $\sin 2x$ and $\cos 2x$ are duplicates of the terms in y_c, thus multiply each by x. So,

$$y_p = Ax\sin 2x + Bx\cos 2x$$

$$y'_p = 2Ax\cos 2x + A\sin 2x - 2Bx\sin 2x + B\cos 2x$$

$$y''_p = -4Ax\sin 2x + 4A\cos 2x - 4Bx\cos 2x - 4B\sin 2x$$

Substitute into $y'' + 4y = \sin 2x \Rightarrow -4Ax\sin 2x + 4A\cos 2x - 4Bx\cos 2x - 4B\sin 2x$

$$+4\left(Ax\sin 2x + Bx\cos 2x\right) = \sin 2x \Rightarrow 4A\cos 2x - 4B\sin 2x = \sin 2x$$

Equating coefficients of $\sin 2x \Rightarrow -4B = 1 \Rightarrow B = -\frac{1}{4}$

Equating constant $\Rightarrow\ 4A = 0 \Rightarrow A = 0$

$$\text{Thus } y_p = (0)x\sin 2x - \left(\frac{1}{4}\right)x\cos 2x \Rightarrow y_p = -\frac{x\cos 2x}{4}$$

Therefore $y = y_c + y_p \Rightarrow y = C_1\cos 2x + C_2\sin 2x - \frac{x\cos 2x}{4}$

15. $y'' + 2y' + y = \cos x$

Auxiliary equation: $m^2 + 2m + 1 = 0$; roots $m = -1,\ m = -1$

Complementary equation: $y_c = C_1e^{-x} + C_2xe^{-x} \Rightarrow y_c = \left(C_1 + C_2x\right)e^{-x}$

Terms of $f(x)$ and their derivative (less coefficients) are $\sin x$ and $\cos x$, so

$$y_p = A\cos x + B\sin x$$

$$y'_p = -A\sin x + B\cos x$$

$$y''_p = -A\cos x - B\sin x$$

Substitute into $y'' + 2y' + y = \cos x \Rightarrow -A\cos x - B\sin x + 2\left(-A\sin x + B\cos x\right) + A\cos x + B\sin x = \cos x$

$$-A\cos x - B\sin x - 2A\sin x + 2B\cos x + A\cos x + B\sin x = \cos x$$

Equating coefficients of $\cos x \Rightarrow 2B = 1 \Rightarrow B = \frac{1}{2}$

Equating coefficients of $\sin x \Rightarrow\ -2A = 0 \Rightarrow A = 0$

$$\text{Thus } y_p = (0)\cos x + \left(\frac{1}{2}\right)\sin x \Rightarrow y_p = \left(\frac{1}{2}\right)\sin x$$

Therefore $y = y_c + y_p \Rightarrow y = \left(C_1 + C_2x\right)e^{-x} + \left(\frac{1}{2}\right)\sin x$

17. $y'' + y = 2\cos x - 3\cos 2x$

Auxiliary equation: $m^2 + 1 = 0$; roots $m = i, m = -i$

Complementary equation: $y_c = C_1 \cos x + C_2 \sin x$

Terms of $f(x)$ and their derivative (less coefficients) are $\sin x$, $\cos x$, $\sin 2x$ and $\cos 2x$, so

$$y_p = Ax\cos x + B\cos 2x + Cx\sin x + D\sin 2x$$

$$y_p' = -Ax\sin x + A\cos x - 2B\sin 2x + Cx\cos x + C\sin x + 2D\cos 2x$$

$$y_p'' = -Ax\cos x - 2A\sin x - 4B\cos 2x - Cx\sin x + 2C\cos x - 4D\sin 2x$$

Substitute into $y'' + y = 2\cos x - 3\cos 2x \Rightarrow -Ax\cos x - 2A\sin x - 4B\cos 2x - Cx\sin x + 2C\cos x - 4D\sin 2x +$
$Ax\cos x + B\cos 2x + Cx\sin x + D\sin 2x = 2\cos x - 3\cos 2x$

$-3B\cos 2x - 3D\sin 2x + 2C\cos x - 2A\sin x = 2\cos x - 3\cos 2x$

Equating coefficients of $\cos x \Rightarrow\ 2C = 2 \Rightarrow C = 1$

Equating coefficients of $\cos 2x \Rightarrow\ -3B = -3 \Rightarrow B = 1$

Equating coefficients of $\sin x \Rightarrow -2A = 0 \Rightarrow A = 0$

Equating coefficients of $2\sin x \Rightarrow -3D = 0 \Rightarrow D = 0$

Thus $y_p = (0)x\cos x + (1)\cos 2x + (1)x\sin x + (0)\sin 2x \Rightarrow y_p = \cos 2x + x\sin x$

Therefore $y = y_c + y_p \Rightarrow y = C_1\cos x + C_2\sin x + \cos 2x + x\sin x$

With Exponential and Trigonometric Expressions

19. $y'' + y = e^x \sin x$

Auxiliary equation: $m^2 + 1 = 0$; roots $m = i, m = -i$

Complementary equation: $y_c = C_1\cos x + C_2\sin x$

Terms of $f(x)$ and their derivative (less coefficients) are $e^x\sin x$ and $e^x\cos x$. So,

$$y_p = Ae^x\cos x + Be^x\sin x$$

$$y_p' = -Ae^x\sin x + Ae^x\cos x + Be^x\cos x + Be^x\sin x$$

$$y_p'' = -Ae^x\cos x - Ae^x\sin x - Ae^x\sin x + Ae^x\cos x - Be^x\sin x + Be^x\cos x + Be^x\cos x + Be^x\sin x$$

$$\Rightarrow y_p'' = -2Ae^x\sin x + 2Be^x\cos x$$

Substitute into $y'' + y = e^x\sin x \Rightarrow -2Ae^x\sin x + 2Be^x\cos x + Ae^x\cos x + Be^x\sin x = e^x\sin x$

Equating coefficients of $e^x\sin x \Rightarrow\ -2A + B = 1$

Equating coefficients of $e^x\cos x \Rightarrow\ A + 2B = 0$

Solving simultaneous equations: $A = -\dfrac{2}{5}, B = \dfrac{1}{5}$

Thus $y_p = -\left(\dfrac{2}{5}\right)e^x\cos x + \left(\dfrac{1}{5}\right)e^x\sin x \Rightarrow y_p = \dfrac{e^x}{5}(\sin x - 2\cos x)$

Therefore $y = y_c + y_p \Rightarrow y = C_1\cos x + C_2\sin x + \dfrac{e^x}{5}(\sin x - 2\cos x)$

21. $y'' - 4y' + 5y = e^{2x}\sin x$

Auxiliary equation: $m^2 - 4m + 5 = 0$; roots $m = 2 + i, m = 2 - i$

Complementary equation: $y_c = e^{2x}(C_1\sin x + C_2\cos x)$

Terms of $f(x)$ and their derivative (less coefficients) are $e^{2x}\sin x$ and $e^{2x}\cos x$. To eliminate duplication, multiply each term by x, so

$$y_p = Axe^{2x}\sin x + Bxe^{2x}\cos x$$

$$y_p' = Axe^{2x}\cos x + 2Axe^{2x}\sin x + Ae^{2x}\sin x - Bxe^{2x}\sin x + 2Bxe^{2x}\cos x + Be^{2x}\cos x$$

$$= (A + 2B)xe^{2x}\cos x + (2A - B)xe^{2x}\sin x + Ae^{2x}\sin x + Be^{2x}\cos x$$

$$y_p'' = (A + 2B)\left[-xe^{2x}\sin x + 2xe^{2x}\cos x + e^{2x}\cos x\right] + (2A - B)\left[xe^{2x}\cos x + 2xe^{2x}\sin x + e^{2x}\sin x\right]$$
$$+ Ae^{2x}\cos x + 2Ae^{2x}\sin x - Be^{2x}\sin x + 2Be^{2x}\cos x$$

Substitute into $y''-4y'+5y=e^{2x}\sin x \Rightarrow (A+2B)\left[-xe^{2x}\sin x+2xe^{2x}\cos x+e^{2x}\cos x\right]$

$+(2A-B)\left[xe^{2x}\cos x+2xe^{2x}\sin x+e^{2x}\sin x\right]+Ae^{2x}\cos x+2Ae^{2x}\sin x-Be^{2x}\sin x+2Be^{2x}\cos x$

$-4\left[(A+2B)xe^{2x}\cos x+(2A-B)xe^{2x}\sin x+Ae^{2x}\sin x+Be^{2x}\cos x\right]+5\left[Axe^{2x}\sin x+Bxe^{2x}\cos x\right]=e^{2x}\sin x$

$=e^{2x}\{(A+2B)[-x\sin x+2x\cos x+\cos x]+(2A-B)[x\cos x+2x\sin x+\sin x]+A\cos x+2A\sin x$

$-B\sin x+2B\cos x-4[(A+2B)x\cos x+(2A-B)x\sin x+A\sin x+B\cos x]+5[Ax\sin x+Bx\cos x]\}$

$=e^{2x}\sin x$

Equating coefficients of $\sin x \Rightarrow$ $2A-B+2A-B-4A=1 \Rightarrow -2B=1 \Rightarrow B=-\frac{1}{2}$

Equating coefficients of $\cos x \Rightarrow$ $A+2B+A+2B-4B=0 \Rightarrow 2A=0 \Rightarrow A=0$

Thus $y_p=(0)xe^{2x}\sin x-\left(\frac{1}{2}\right)xe^{2x}\cos x \Rightarrow y_p=\left(-\frac{x}{2}\right)e^{2x}\cos x$

Therefore $y=y_c+y_p \Rightarrow y=e^{2x}(C_1\sin x+C_2\cos x)-\left(\frac{x}{2}\right)e^{2x}\cos x \Rightarrow y=e^{2x}\left[C_1\sin x+C_2\cos x-\left(\frac{x}{2}\right)\cos x\right]$

Particular Solutions

23. $y''-4y'=8,\qquad y=y'$ when $x=0$

Auxiliary equation: $m^2-4m=0$; roots $m=0$, $m=4$

Complementary equation: $y_c=C_1+C_2e^{4x}$

Terms of $f(x)$ and their derivative (less coefficients) are constants, which duplicates the constant in y_c. To eliminate duplicates, multiply each term by x. So,

$y_p=Cx+D,\ y_p'=C,\ y_p''=0$

Substitute into $y''-4y'=8 \Rightarrow 0-4(C)=8$

Equating constant $\Rightarrow -4C=8 \Rightarrow C=-2$

Thus $y_p=-2x+0 \Rightarrow y_p=-2x$

Therefore $y=y_c+y_p \Rightarrow y=C_1+C_2e^{4x}-2x$

$y'=4C_2e^{4x}-2$

At $x=0,\ y=0 \Rightarrow 0=C_1+C_2e^{4(0)}-2(0) \Rightarrow 0=C_1+C_2$

At $x=0,\ y=0,\ y'=0 \Rightarrow 4C_2e^{4(0)}-2 \Rightarrow 0=4C_2-2 \Rightarrow 2=4C_2 \Rightarrow C_2=\frac{1}{2};\quad 0=C_1+C_2 \Rightarrow 0=C_1+\frac{1}{2} \Rightarrow C_1=-\frac{1}{2}$

Thus $y=-\frac{1}{2}+\frac{1}{2}e^{4x}-2x \Rightarrow y=\frac{e^{4x}}{2}-2x-\frac{1}{2}$

25. $y''+4y=2,\qquad y=0$ when $x=0$ and $y=\frac{1}{2}$ when $x=\frac{\pi}{4}$

Auxiliary equation: $m^2+4=0$; roots $m=2i$, $m=-2i$

Complementary equation: $y_c=C_1\cos 2x+C_2\sin 2x$

Terms of $f(x)$ and their derivative (less coefficients) are constants. So,

$y_p=C,\ y_p'=0,\ y_p''=0$

Substitute into $y''+4y=2 \Rightarrow 0+4(0)+4C=2$

Equating constant $\Rightarrow 4C=2 \Rightarrow C=\frac{1}{2}$

Thus $y_p=\frac{1}{2}$

Therefore $y=y_c+y_p \Rightarrow y=C_1\cos 2x+C_2\sin 2x+\frac{1}{2}$

At $x=0,\ y=0 \Rightarrow 0=C_1\cos 2(0)+C_2\sin 2(0)+\frac{1}{2} \Rightarrow 0=C_1(1)+C_2(0)+\frac{1}{2} \Rightarrow C_1=-\frac{1}{2}$

At $x=\frac{\pi}{4},\ y=\frac{1}{2}\Rightarrow\frac{1}{2}=-\frac{1}{2}\cos 2\left(\frac{\pi}{4}\right)+C_2\sin 2\left(\frac{\pi}{4}\right)+\frac{1}{2}\Rightarrow\frac{1}{2}=-\frac{1}{2}(0)+C_2(1)+\frac{1}{2}\Rightarrow C_2=0$

Thus $y=\left(-\frac{1}{2}\right)\cos 2x+(0)\sin 2x+\frac{1}{2}\Rightarrow y=\frac{1-\cos 2x}{2}$

25. $y''-2y'+y=2e^x,\qquad y'=2e$ at $(1,0)$

Auxiliary equation: $m^2-2m+1=0$; roots $m=1,\ m=1$

Complementary equation: $y_c=(C_1+C_2x)e^x$

Terms of $f(x)$ and their derivative (less coefficients) are e^x, which duplicates a term in y_c. But xe^x is a duplicate also. To eliminate duplicates, multiply each term by x. So,

$$y_p=Ax^2e^x,\ y_p'=Ax^2e^x+2Axe^x,\ y_p''=Ax^2e^x+4Axe^x+2Ae^x$$

Substitute into $y''-2y'+y=2e^x\Rightarrow Ax^2e^x+4Axe^x+2Ae^x-2\left(Ax^2e^x+2Axe^x\right)+Ax^2e^x=2e^x$

$$\Rightarrow 2Ae^x=2e^x$$

Equating e^x term $\Rightarrow 2A=2\Rightarrow A=1$

Thus $y_p=x^2e^x$

Therefore $y=y_c+y_p\Rightarrow y=(C_1+C_2x)e^x+x^2e^x$

$$y'=(C_1+C_2x)e^x+C_2e^x+x^2e^x+2xe^x$$

At $x=1,\ y=0\Rightarrow 0=\left(C_1+C_2(1)\right)e^1+(1)^2e^1\Rightarrow 0=C_1+C_2+1=0\Rightarrow C_1+C_2=-1$

At $x=1,\ y=0,\ y'=2e\Rightarrow 2e=\left(C_1+C_2(1)\right)e^1+C_2e^1+(1)^2e^1+2(1)e^1\Rightarrow 2e=(C_1+C_2)e+C_2e+e+2e$

$$\Rightarrow 2=C_1+C_2+C_2+1+2\Rightarrow C_1+2C_2=-1$$

Solving simultaneous equations: $C_1=-1,\ C_2=0$

Thus $y=\left(-1+(0)x\right)e^x+x^2e^x\Rightarrow y=e^x\left(x^2-1\right)$

27. $y''+y=-2\sin x,\qquad y=0$ at $x=0$ and $\frac{\pi}{2}$

Auxiliary equation: $m^2+1=0$; roots $m=i,\ m=-i$

Complementary equation: $y_c=C_1\cos x+C_2\sin x$

Terms of $f(x)$ and their derivative (less coefficients) are $\sin x$ and $\cos x$, which duplicate the terms in y_c. To eliminate the duplicates, multiply each term by x. Thus,

$$y_p=Ax\sin x+Bx\cos x$$
$$y_p'=Ax\cos x+A\sin x+B\cos x-Bx\cos x$$
$$y_p''=2A\cos x-2B\sin x-Ax\sin x-Bx\cos x$$

Substitute into $y''+y=-2\sin x\Rightarrow 2A\cos x-2B\sin x-Ax\sin x-Bx\cos x+Ax\sin x+Bx\cos x=-2\sin x$

$$\Rightarrow 2A\cos x-2B\sin x=-2\sin x$$

Equating $\cos x$ term $\Rightarrow 2A=0\Rightarrow A=0$

Equating $\sin x$ term $\Rightarrow -2B=-2\Rightarrow B=1$

Thus $y_p=(0)x\sin x+(1)x\cos x\Rightarrow y_p=x\cos x$

Therefore $y=y_c+y_p\Rightarrow y=C_1\cos x+C_2\sin x+x\cos x$

At $x=0,\ y=0\Rightarrow 0=C_1\cos(0)+C_2\sin(0)+(0)\cos(0)\Rightarrow 0=C_1$

At $x=\frac{\pi}{2},\ y=0\Rightarrow C_1\cos\left(\frac{\pi}{2}\right)+C_2\sin\left(\frac{\pi}{2}\right)+\left(\frac{\pi}{2}\right)\cos\left(\frac{\pi}{2}\right)\Rightarrow 0=C_1(0)+C_2(1)+\left(\frac{\pi}{2}\right)(0)C_2=0\Rightarrow C_2=0$

Thus $y=(0)\cos x+(0)\sin x+x\cos x\Rightarrow y=x\cos x$

Exercise 4 ◊ Mechanical Vibrations

Simple Harmonic Motion

1. (a) Since period is the reciprocal of frequency, find the frequency first

$$f = \frac{\omega_n}{2\pi}, \text{ where } \omega_n = 182 \left[\text{from } x = x_0 \cos\omega_n t\right] \Rightarrow \frac{182}{2\pi} = 28.966\,\text{Hz}$$

$$\text{period} = \frac{1}{f} \Rightarrow \frac{1}{28.966} = 0.034522\,\text{s} = 34.5\text{ ms}$$

(b) amplitude = 3.75 $\left[\text{from } x = x_0 \cos\omega_n t \text{ where } x_0 \text{ is the amplitude}\right]$

3. Using $x = x_0 \cos\omega_n t$ with $x = 0 \Rightarrow x_0 \sin\omega_n t$

   At $t = 0 \Rightarrow \omega_n = 15$ $(\text{at } t = 0) \Rightarrow x = \sin 15t$

5. $x = x_0 \cos\omega_n t$ where $x_0 = 1$ ["pulled down an additional 1.00 in."] and $\omega_n = \sqrt{\frac{kg}{W}}$

   $$k = 0.25\left[\text{from } k = \frac{\text{force}}{\text{distance}} \Rightarrow \frac{2}{8}\right],\ g = 386\text{ in./s}^2,\ W = 2 \Rightarrow \omega_n = \sqrt{\frac{0.25(386)}{2}} = 6.9462 = 6.95$$

   Therefore $x = \cos 6.95t$

7. Using $x = x_0 \cos\omega_n t$ with $x_0 = 1.50$

   $$\omega_n = \sqrt{\frac{kg}{W}} \Rightarrow \sqrt{\frac{1.5}{3}} = \sqrt{2} \Rightarrow x = 1.50\cos\sqrt{2}t$$

Damped Vibrations

9. $x = 2.50e^{-3t}\cos 55t$

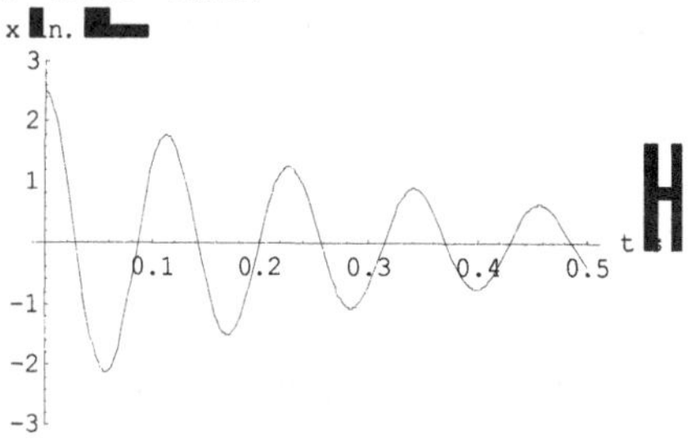

11. $x = 2.50e^{-3t}\cos 55t$ when $t = 0.30$ seconds $\Rightarrow x = 2.50e^{-3(0.30)}\cos 55(0.30) = -0.71393 \Rightarrow -0.714$ in.

13. Auxiliary equation: $m^2 + 5m + 4 = 0$; roots $m = -1$, $m = -4$

    Complementary equation: $x = C_1e^{-t} + C_2e^{-4t}$

    $$x' \Rightarrow v = -C_1e^{-t} - 4C_2e^{-4t}$$

    Substituting $x = 1.50$ and $v = 15.3$:

    $$C_1 + C_2 = 1.50$$
    $$-C_1 - 4C_2 = 15.3$$

    Solving simultaneous equations: $C_1 = 7.10$, $C_2 = -5.60$

    Therefore $x = 7.10e^{-t} - 5.60e^{-4t}$

15. (a) Since $a > \omega_n$; overdamped

    (b) at $a = 24.1$ lbs

Exercise 5 ◊ *RLC* Circuits

Series *LC* Circuits with dc Source

1. Since $R = 0$, obtain i by using the formula $i = \frac{E}{\omega_n L}\sin\omega_n t = 100$

   $$\omega_n = \frac{1}{\sqrt{LC}} = \frac{1}{\sqrt{1(1)(10^{-6})}} = 10^3 = 1000$$

Therefore $i = \frac{100}{1000(1)} \sin 1000t = 0.1 \sin 1000t$ A

3. Summing the voltage drop around the loop,

$Ri + L\frac{di}{dt} + \frac{q}{C} = E$; but $R = E = 0$; therefore $L\frac{di}{dt} + \frac{q}{C} = 0$

But $i = \frac{dq}{dt}$, so $\frac{di}{dt} = q''$, thus $Lq'' + \frac{q}{C} = 0$

Auxiliary equation: $Lm^2 + \frac{1}{C} = 0$; roots $m = \pm\sqrt{-\frac{1}{LC}}$; $m = \omega_n i$, $m = -\omega_n i$

Complementary equation (the charge): $q = C_1 \sin \omega_n t + C_2 \cos \omega_n t$

Since $q(0) = 255 \times 10^{-6}$ C, $C_2 = 255 \times 10^{-6}$ C

Taking the derivative, $i = \frac{dq}{dt} = \omega_n C_1 \cos \omega_n t - \omega_n C_2 \sin \omega_n t$

But $i(0) = 0$, thus $C_1 = 0$

Thus $q = 255 \cos 1000t\ \mu$C

Series *RLC* Circuits with dc Source

5. (a) First calculate $a = \frac{R}{2L}$ and $\omega_n = \frac{1}{\sqrt{LC}}$ to determine if the circuit is underdamped or overdamped

$a = \frac{1.55}{2(0.125)} = 6.2$; $\omega_n = \frac{1}{\sqrt{(0.125)(250) \times 10^{-6}}} = 178.88 = 179 \Rightarrow \omega_n > a$, thus the circuit is underdamped

(b) $i = \frac{E}{\omega_d L} e^{-at} \sin \omega_d t$ where $\omega_d = \sqrt{\omega_n^2 - a^2} \Rightarrow \sqrt{(179)^2 - (6.2)^2} = 178.89 = 179$

Thus $i = \frac{100}{179(0.125)} e^{-6.2t} \sin 179t \Rightarrow i = 4.47 e^{-6.2t} \sin 179t$

7. (a) First calculate $a = \frac{R}{2L}$ and $\omega_n = \frac{1}{\sqrt{LC}}$ to determine if the circuit is underdamped or overdamped

$a = \frac{1.75}{2(1.50)} = 0.583$; $\omega_n = \frac{1}{\sqrt{(1.5)(4.25)}} = 0.396 \Rightarrow a > \omega_n$, thus the circuit is overdampled

(b) $i = \frac{E}{2i\omega_d L}\left(e^{(-a+\omega_d i)t} - e^{(-a-\omega_d i)t}\right)$ where $\omega_d = \sqrt{\omega_n^2 - a^2} \Rightarrow \sqrt{(0.396)^2 - (0.583)^2} = 0.4278703i = 0.428i$

Thus $\omega_d = 0.428i$

$-a - \omega_d i = -0.583 - 0.428 = -1.01$

$-a + \omega_d i = -0.583 + 0.428 = -0.155$

$\frac{E}{2\omega_d L} = \frac{100}{2(-0.428)(1.50)} = -77.8$

Therefore $i = -77.8e^{-1.01t} + 77.8e^{-0.155t}$ A

Series *RLC* Circuits with ac Source

9. $X = \omega L - \frac{1}{\omega C} = 175(0.125) - \frac{10^6}{175(225)} = -3.52$

$Z^2 = R^2 + X^2 = (10.5)^2 + (-3.52)^2 = 122.6404 \Rightarrow Z = \sqrt{122.6404} = 11.07\ \Omega$

$\phi = \arctan\frac{X}{R} = \arctan\left(\frac{-3.52}{10.5}\right) = -18.5°$

$\frac{E}{Z} = \frac{100}{11.07} = 9.03$ A

$i = \frac{E}{Z}\sin(\omega t - \phi) \Rightarrow i = 9.03 \sin[175t - (-18.5°)] = 9.03 \sin(175t + 18.5°)$

11. $X_L = \omega L = 377(0.35) = 131.95 = 132\ \Omega$

$$X_c = \frac{1}{\omega C} = \frac{10^6}{377(20)} = 132.9\ \Omega$$

$X = X_L - X_c = 132 - 132.9 = -0.9\ \Omega$

$Z^2 = R^2 + X^2 = (1550)^2 + (-0.9)^2 = 2{,}402{,}500.81 \Rightarrow Z = \sqrt{2{,}402{,}500.81} = 1550\ \Omega$

$$\phi = \arctan\frac{X}{R} = \arctan\left(\frac{-0.9}{1550}\right) = -0.033$$

$$I_{\max} = \frac{E}{Z} = \frac{250}{1550} = 0.16129 = 0.161\text{ A}$$

$i = I_{\max}\sin(\omega t - \phi) \Rightarrow i = 0.161\sin(377t\text{ A} + 0.033) = 161\sin(377t\text{ mA} + 0.033)$

13. Resonant frequency $\Rightarrow$ first find $\omega = \dfrac{1}{\sqrt{LC}} = \dfrac{1}{\sqrt{0.175(1.5\times10^{-3})}} = 61.721 = 61.7$ rads/s

Resonant frequency $= \dfrac{\omega}{2\pi} \Rightarrow \dfrac{61.7}{2\pi} = 9.82$ Hz

15. $\omega = \dfrac{1}{\sqrt{LC}}$; solving for $C \Rightarrow C = \dfrac{1}{\omega^2 L} = \dfrac{1}{(6\times10^5)^2(5.6\times10^{-6})} = 4.960317\times10^{-7} = 0.496\ \mu$F

17. At resonance the current will be a maximum $\Rightarrow i = \dfrac{E}{Z}\sin(\omega t - \phi) = \dfrac{E}{Z} = \dfrac{E}{R} = \dfrac{100}{10} = 10$ A

CHAPTER 31 REVIEW PROBLEMS

1. $y'' - 2y = 2x^3 + 3x$

Auxiliary equation: $m^2 - 2 = 0$; roots: $m = \sqrt{2},\ m = -\sqrt{2}$

Complementary equation: $y_c = C_1e^{\sqrt{2}x} + C_2e^{-\sqrt{2}x}$

Terms of $f(x)$ and its derivatives (less coeficients) are $f(x) = Ax^3 + Bx;\ f'(x) = Cx^2 + D;$

$f''(x) = Ex$ [duplicate of $f(x)$], so

$$y_p = Ax^3 + Cx^2 + Bx + D,\ y_p' = 3Ax^2 + 2Cx + B,\ y_p'' = 6Ax + 2C$$

Substituting into $y'' - 2y = 2x^3 + 3x \Rightarrow 6Ax + 2C - 2(Ax^3 + Cx^2 + Bx + D) = 2x^3 + 3x$

$$\Rightarrow 6Ax + 2C - 2Ax^3 - 2Cx^2 - 2Bx - 2D = 2x^3 + 3x$$

Equating coeficients $\Rightarrow A = -1,\ B = -\dfrac{9}{2},\ C = 0,\ D = 0$

$$y_p = -x^3 - \frac{9}{2}x$$

Thus $y = y_c + y_p \Rightarrow C_1e^{\sqrt{2}x} + C_2e^{-\sqrt{2}x} - x^3 - \dfrac{9}{2}$

3. $y''' - 2y' = 0$

Auxiliary equation: $m^3 - 2m = 0$; roots: $m = \pm\sqrt{2},\ m = 0$

Thus $y_c = C_1e^{\sqrt{2}x} + C_2e^{-\sqrt{2}x} + C_3e^0 \Rightarrow C_1e^{\sqrt{2}x} + C_2e^{-\sqrt{2}x} + C_3$

5. $y'' + 2y' - 3y = 0$

Auxiliary equation: $m^2 + 2m - 3 = 0$; roots: $m = 1,\ m = -3$

Thus $y_c = C_1e^x + C_2e^{-3x}$

7. $y'' + 6y' + 9y = 0$ $\qquad$ $y = 0$ and $y' = 2$ when $x = 0$

Auxiliary equation: $m^2 + 6m + 9 = 0$; roots: $m = -3,\ m = -3$

Complementary equation: $y_c = C_1e^{-3x} + C_2xe^{-3x}$ and $y' = -3C_1e^{-3x} - 3C_2xe^{-3x} + C_2e^{-3x}$

At $x=0$, $y=0 \Rightarrow 0=C_1e^{-3(0)}+C_2e(0)^{-3(0)} \Rightarrow C_1=0$

At $x=0$, $y=0$, $y'=2 \Rightarrow 2=-3C_1e^{-3(0)}-3C_2(0)e^{-3(0)}+C_2e^{-3(0)} \Rightarrow -3C_1+C_2=2$

Solving simultaneous equation: $C_1=0$, $C_2=2$

Thus $y=2xe^{-3x}$

9. $y''-2y=3x^3e^x$

Auxiliary equation: $m^2-2=0$; roots $m=\sqrt{2}$, $m=-\sqrt{2}$

Complementary equation: $y_c=C_1e^{\sqrt{2}x}+C_2e^{-\sqrt{2}x}$

Terms of $f(x)$ and their derivative (less coefficients) are x^3e^x, x^2e^x, xe^x and e^x

$$y_p=Ax^3e^x+Bx^2e^x+Cxe^x+De^x$$

$$y_p'=Ax^3e^x+3Ax^2e^x+Bx^2e^x+2Bxe^x+Cxe^x+Ce^x+De^x$$

$$=Ax^3e^x+(3A+B)x^2e^x+(2B+C)xe^x+(C+D)e^x$$

$$y_p''=Ax^3e^x+3Ax^2e^x+(3A+B)x^2e^x+2(3A+B)xe^x+(2B+C)xe^x+(2B+C)e^x+(C+D)e^x$$

$$=Ax^3e^x+(6A+B)x^2e^x+(6A+4B+C)xe^x+(2B+2C+D)e^x$$

Substitute into $y''-2y=3x^3e^x \Rightarrow Ax^3e^x+(6A+B)x^2e^x+(6A+4B+C)xe^x+(2B+2C+D)e^x$

$$-2\left(Ax^3e^x+Bx^2e^x+Cxe^x+De^x\right)=3x^3e^x$$

$$\Rightarrow -Ax^3e^x+(6A-B)x^2e^x+(6A+4B-C)xe^x+(2B+2C-D)e^x=3x^3e^x$$

Equating coefficients $\Rightarrow A=-3$, $B=-18$, $C=-90$, $D=-216$

$$\text{Thus } y_p=-3x^3e^x-18x^2e^x-90xe^x-216e^x \Rightarrow -3e^x\left(x^3+6x^2+30x+72\right)$$

Therefore $y=y_c+y_p \Rightarrow y=C_1e^{\sqrt{2}x}+C_2e^{-\sqrt{2}x}-3e^x\left(x^3+6x^2+30x+72\right)$

11. $y''+y'=5\sin 3x$

Auxiliary equation: $m^2+m=0$; roots $m=0$, $m=-1$

Complementary equation: $y_c=C_1+C_2e^{-x}$

Terms of $f(x)$ and their derivative (less coefficients) are $f(x)=\sin 3x$, $f'(x)=\cos 3x$, $f''(x)=-\sin 3x$

$$y_p=A\sin 3x+B\cos 3x$$

$$y_p'=3A\cos 3x-3B\sin 3x$$

$$y_p''=-9A\sin 3x-9B\cos 3x$$

Substitute into $y''+y'=5\sin 3x \Rightarrow -9A\sin 3x-9B\cos 3x+3A\cos 3x-3B\sin 3x=5\sin 3x$

$$\Rightarrow (-9A-3B)\sin x+(-9B+3A)\cos 3x=5\sin 3x$$

Equating coefficients $\Rightarrow -9A-3B=5$, $3A-9B=0$

Solving simultaneous equations: $A=-\frac{1}{2}$, $B=-\frac{1}{6}$

$$\text{Thus } y_p=-\frac{1}{2}\sin 3x-\frac{1}{6}\cos 3x$$

Therefore $y=y_c+y_p \Rightarrow y=C_1+C_2e^{-x}-\frac{1}{2}\sin 3x-\frac{1}{6}\cos 3x$

13. $y''+9y=0$

Auxiliary equation: $m^2+9=0$; roots: $m=3i$, $m=-3i$

Thus $y_c=C_1\cos 3x+C_2\sin 3x$

15. $y''-2y'+y=0$ $\quad$ $y=0$ and $y'=2$ when $x=0$

Auxiliary equation: $m^2-2m+1=0$; roots: $m=1$, $m=1$

Complementary equation: $y_c=C_1e^x+C_2xe^x$ and $y'=C_1e^x+C_2xe^x+C_2e^x$

At $x = 0,\ y = 0 \Rightarrow = C_1 e^{(0)} + C_2(0)e^{(0)} \Rightarrow C_1 = 0$

At $x = 0,\ y = 0,\ y' = 2 \Rightarrow 2 = C_1 e^{(0)} + C_2(0)e^{(0)} + C_2 e^0 \Rightarrow C_1 + C_2 = 2$

Solving simultaneous equation: $C_1 = 0,\ C_2 = 2$

Thus $y = 2xe^x$

17. $y'' - 4y' + 4y = 0$

Auxiliary equation: $m^2 - 4m + 4 = 0$; roots: $m = 2,\ m = 2$

Thus $y = C_1 e^{2x} + C_2 xe^{2x}$

19. $y'' + 4y' + 4y = 0$

Auxiliary equation: $m^2 + 4m + 4 = 0$; roots: $m = -2,\ m = -2$

Thus $y = C_1 e^{-2x} + C_2 xe^{-2x}$

21. $y'' - 2y' + y = 0$

Auxiliary equation: $m^2 - 2m + 1 = 0$; roots: $m = 1,\ m = 1$

Thus $y = C_1 e^x + C_2 xe^x$

23. $x = x_0 \cos \omega_n t$ where $x_0 = 1.5$ ["pulled down an additional 1.50 in."] and $\omega_n = \sqrt{\dfrac{kg}{W}}$

$k = 0.714\left[\text{from } k = \dfrac{\text{force}}{\text{distance}} \Rightarrow \dfrac{5}{7}\right],\ g = 386 \text{ in/s}^2,\ W = 5 \Rightarrow \omega_n = \sqrt{\dfrac{0.714(386)}{5}} = 7.4243 = 7.42$

Therefore $x = 1.5\cos 7.42t$

25. $\omega_n = \dfrac{1}{\sqrt{LC}} = \dfrac{1}{\sqrt{2.20(1.75)(10^{-6})}} = 509.64 = 510$ rads/s

$i = \left(\dfrac{E}{\omega_n L}\right)\sin \omega_n t = \dfrac{110}{510(2.20)}\sin 510t \Rightarrow i = 0.0981\sin 510t \text{ A} \Rightarrow i = 98.1\sin 510t \text{ mA}$

27. $a = \dfrac{R}{2L} = \dfrac{3.75}{2(0.1)} = 18.75;\quad \omega_n = \dfrac{1}{\sqrt{LC}} = \dfrac{1}{\sqrt{0.1(150)(10^{-6})}} = 258$

$i = \dfrac{E}{\omega_d L}e^{-at}\sin \omega_d t$ where $\omega_d = \sqrt{\omega_n^2 - a^2} = \sqrt{(258)^2 - (18.75)^2} = 257.3$

$i = \dfrac{120}{257.3(0.1)}e^{-18.75t}\sin 257.3t \Rightarrow i = 4.66e^{-18.75t}\sin 257.3t \text{ A}$